PLANT CELL WALLS

PLANT CELL WALLS

Edited by

N.C. CARPITA

Dept. Botany & Plant Pathology Purdue University, West Lafayette, IN, USA

M. CAMPBELL

Dept. Plant Sciences University of Oxford, Oxford, UK

and

M. TIERNEY

Dept. Botany University of Vermont, Burlington, VT, USA

Reprinted from *Plant Molecular Biology*, Volume 47 (I, II), 2001

KLUWER ACADEMIC PUBLISHERS

DORDRECHT / BOSTON / LONDON

A C.I.P. Catalogue record for this book is available from the Library of Congress

Library of Congress Catalogue-in-Publication Data

Plant cell walls / edited by Nicholas C. Carpita, Malcolm Campbell and Mary Tierney
    p. cm.
  Includes bibliographical references (p.).
  ISBN 0-7923-7179-8 (hb : alk. paper)
  1. Plant cell walls. I. Carpita, Nicholas C. II. Campbell, Malcolm. III. Tierney, Mary.

  QK725.P558 2001
  571.6'82–dc21

                                                        2001046208

Published by Kluwer Academic Publishers,
P.O. Box 17, 3300 AA Dordrecht, The Netherlands

Sold and distributed in North, Central and South America
by Kluwer Academic Publishers,
101 Philip Drive, Norwell, MA 02061, U.S.A.

In all other countries, sold and distributed
by Kluwer Academic Publishers,
P.O. Box 322, 3300 AH Dordrecht, The Netherlands

Printed on acid-free paper

Printed in the Netherlands

CONTENTS

Cover illustration

A functionally important aspect of the *in muro* modification of the pectic matrix is the regulation of the degree and pattern of methyl esterification of the homogalactouronan (HG) backbone. The image shows a junction between three tobacco stem cortical cells that have been immunolabelled with the monoclonal antibodies LM7 (red) and PAM1 (green) and stained with the cellulose-binding reagent Calcofluor (blue). PAM1 and LM7 are methylester pattern-specific antibodies and bind to unesterified and partially methyl-esterified HG respectively. In this issue, both antibodies bind to a region of cell wall that lines intercellular spaces, but the discrete locations of LM7 and PAM1 labelling indicates that the distribution pattern of methylesters along the HG backbone is differentially regulated within cell wall microdomains. (Courtesy of Willats *et al.*, Centre of Plant Sciences, Leeds, UK)

Plant Molecular Biology **47**: 1–5, 2001.
© 2001 *Kluwer Academic Publishers. Printed in the Netherlands.*

Overview

Molecular biology of the plant cell wall: searching for the genes that define structure, architecture and dynamics

Nick Carpita[1,*], Mary Tierney[2] and Malcolm Campbell[3]
[1]*Department of Botany and Plant Pathology, Purdue University, West Lafayette, IN 47907-1155, USA (*author for correspondence; e-mail carpita@btny.purdue.edu);* [2]*Department of Botany, University of Vermont, Marsh Life Building, Burlington, VT 05405, USA;* [3]*Department of Plant Sciences, University of Oxford, South Parks Road, Oxford OX1 3RB, UK*

Introduction

The plant cell wall is a highly organized composite that may contain many different polysaccharides, proteins, and aromatic substances. These complex matrices define the features of individual cells within the plant body. Ultimately, the plant wall functions as the determinant of plant morphology. The importance of the plant cell wall is revealed in the shear number of genes that are likely to be involved in cell wall biogenesis, assembly, and modification. For example, over 17% of the 25 498 *Arabidopsis* genes have signal peptides, and over 400 proteins have been identified that reside in the wall (Arabidopsis Genome Initiative, 2000). If just one-half of the proteins with signal peptides function in the biosynthesis, assembly, and modification of the walls, then well over 2000 genes are likely to participate in wall biogenesis during plant development. This number is considerably larger if all the cytosolic proteins that function in substrate generation are included. Beyond this, some integral membrane-associated proteins, such as cellulose synthase, obviously function in cell wall biogenesis but do not contain signal peptides. Thus, it is likely that some 15% of the *Arabidopsis* genome is dedicated to cell wall biogenesis and modification. Of these, only small subsets have been characterized.

Recently, forward and reverse genetic approaches have provided insight into the genes relevant to cell wall metabolism. Forward genetic approaches have historically been hampered by technical problems associated with characterization of polymer synthesis *in vitro* and of higher-order architectural assembly and rearrangement during growth. On the other hand, reverse genetic and molecular biological approaches, based on discovery of homologous genes from bacteria, fungal, and animal systems, have augmented the collection of recognized wall-relevant genes considerably, but the functions of many of these genes still remain elusive.

The major steps in wall biogenesis and modification can be divided into six specific stages: (1) the synthesis of monomer building blocks, such as nucleotide sugars and monolignols, (2) the biosynthesis of oligomers and polysaccharides at the plasma membrane and ER-Golgi apparatus, (3) the targeting and secretion of Golgi-derived materials, (4) the assembly and architectural patterning of polymers, (5) dynamic rearrangement during cell growth and differentiation, and (6) wall disassembly and catabolism of the spent polymers. For some of these stages, such as the generation of known substrates, complete knowledge of the biochemical pathways has led to discovery of many of the genes encoding the enzymes involved in the catalysis. For other stages, such as wall assembly, the kinds of proteins that might participate remain purely speculative. To put into perspective the challenges of gene discovery and determination of function, we have assembled articles by leading cell-wall researchers that illustrate the most recent advances in this field and the long road of discovery that lies ahead.

Visualizing gene expression

After a century's work by carbohydrate chemists and biochemists, we now have a fairly complete catalog of the major polysaccharides of the walls of higher plants. The vast majority of these studies are a result of bulk chemical analysis and do not give many ideas for the dynamic changes that occur in walls of different tissues, different cells of the tissue, and even within domains of a single cell wall. *In situ* hybridization studies have been influential in beginning to unravel cell specificity of wall-relevant gene expression, and antibodies directed against wall-relevant enzymes and specific epitopes of their substrate have afforded us a glimpse of the sub-domains of a single cell wall. Willats *et al.* (this issue) provide a comprehensive summary of the complexities of pectin fine structure and how the use of monoclonal antibodies against pectin epitopes has revolutionized our knowledge of their cell and wall domain specificity and their dynamics during growth and development. In particular, antibodies directed against two neutral sugar side-groups, arabinans and galactans, have revealed a remarkable sub-domain distribution that will now allow more refined determinations of structural-functional and dynamic relationships of these transient components during cell growth and development.

Defining pathways to synthesis

The selection of *Arabidopsis* mutants in which a cell-wall sugar is over- or under-represented led to the discovery of two genes involved in nucleotide-sugar interconversion pathways. Reiter and Vanzin (this issue) describe the molecular genetics of these important pathways for *de novo* substrate production and the salvage of certain sugars after they are excised from polysaccharides. The 4,6-dehydratase involved in the synthesis of GDP-L-fucose, and the C-4 epimerase that interconverts UDP-Xyl and UDP-Ara represent just two of a minimum of 11 enzymes that function in the *de novo* pathways of nucleotide sugar synthesis from GDP- or UDP-Glc. Comparative genomics have given important clues on identification of the remainder, and this article identified several candidate genes that assure that the genes that encode the entire nucleotide-interconversion pathways will be deduced very quickly. Another group of C-1 kinases, NDP-pyrophosphorylases, and other carbohydrate-generating enzymes are involved in the salvage of sugars back into the nucleotide-sugar pool. A gene encoding only one of these enzymes, an arabinokinase, has been identified.

Haigler *et al.* (this issue) delve further into the pathways of carbon into the nucleotide-sugar path-

ways and ultimately to cellulose via a UDP-glucose shuttle. Several years ago, they discovered that sucrose synthase, or 'SuSy', is associated with the plasma membrane, and they presented evidence that sucrose may provide glucose directly to cellulose synthase. In their article, they present several other biochemical and cellular mechanisms that might directly impact cellulose synthesis from the cytosolic side of the plasma membrane.

Genomic approaches to define wall-relevant genes

Genomic approaches have provided a global view of gene expression related to primary and secondary cell wall synthesis. Henrissat *et al.* (this issue) provide a robust census of *Arabidopsis* glycosidases and glycosyltransferases derived from knowledge of the entire *Arabidopsis* genome sequence. One surprise of this census is that *Arabidopsis* encodes many more of these enzymes than does *Saccharomyces cerevisiae*, *Drosophila melanogaster* or *Caenorhabditis elegans*. Over 600 genes are involved in polysaccharide synthesis and turnover, and almost one-quarter of them (140) are involved in the turnover of pectins. One of the major surprises that resulted from the sequencing of the *Arabidopsis* genome was the percentage of gene families with more than 5 members (Arabidopsis Genome Initiative, 2000), and this also holds true for hydrolase and glycosyltransferase gene families. In addition, 80 different hydrolase and 45 different glycosyltransferase gene families were identified, and a great many of them define families found only in plants. Of these, well over half are thought to function within the secretory pathway or the cell wall. Representative enzymes from many of these families have been crystallized, and their 3-dimensional structure has been determined, so the families are beginning to be defined on the basis of their amino acid sequences as well as the structure of their active sites. Hrmova and Fincher (this issue) focus on a specific subset of the hydrolase families genes from barley and other cereals for which the 3-dimensional structures are known. Through activity studies, they begin to define the substrate specificities of individual family members of β-glucan exo- and endohydrolases. Through expression studies, the function of some of these hydrolases involved in the turnover of storage polymers and in cell growth may be inferred.

Perrin *et al.* (this issue) outline two principal empirical routes to identify synthases and glycosyl transferases and to characterize their functions. Classical means to purify and identify these enzymes relied on biochemical schemes that were difficult at best and, in many instances, impossible to accomplish. They demonstrate how bioinformatics and functional genomics can provide a powerful means to identify and evaluate candidate genes through database searches and 'expression profiling' by microarray analyses. In this article, they nicely review the recent advances using genetic, reverse-genetic, biochemical, and heterologous expression methods that can be employed to determine the function of these families of genes.

Cellulose synthase is arguably the most important enzyme involved in plant cell wall biosynthesis. Richmond and Somerville (this isssue) discuss the enormity of the cellulose synthase superfamily of *Arabidopsis* and how a powerful multidisciplinary approach can be used to determine gene function within this large superfamily. They show how cellulose synthase-related functions might be deciphered using a systematic analysis of individual cellulose synthase family members. The systematic analysis melds a number of approaches, including bioinformatics, classical and reverse genetics coupled with chemical analysis of mutants, and gene expression analysis using microarrays and promoter::reporter fusions.

The genes that are at the core of cell wall biogenesis are those that encode polysaccharide synthases and glycosyl transferases. Synthases are defined as processive glycosyltransferases that iterate linkage of mono- or disaccharide units into the backbone polymer, whereas glycosyltransferases decorate the backbone with addition of specific sugars. An enormous task lies ahead to define the function of all the candidate genes that comprise this stage of wall biogenesis. Twelve genes define the 'true' cellulose synthase (*CesA*) gene family, and 6 additional families encode 30 more cellulose synthase-like (*Csl*) genes. Of the 12 *CesA*s, only three have actually been confirmed biochemically, defined through selection of mutants lacking their function and rescue of wild-type cellulose synthesis by complementation. Vergara and Carpita (this issue) provide a phylogenetic comparison of the *CesA* genes from two grass species, rice and maize, with those of *Arabidopsis* and two additional dicotyledonous species. From analysis of the amino acid sequences of what was originally thought to be a hypervariable region, they discovered that this region was not really variable but contained family-specific combinations of motifs that probably function in catalysis or processivity. Their work raises the possibility

that not all *CesA* genes encode synthases of cellulose and underscores the need to define the function of each synthase gene by refined biochemical techniques.

The genomics of cell specialization

The secondary cell walls provide excellent examples of how cell wall modification confers specific properties upon a cell to allow it to fulfill specialized functions. Secondary cell walls are frequently a feature of cells that provide support for the plant body, and cells involved in the transport of water and solutes from the roots to the aerial tissues. Secondary cell walls allow these cells to resist the forces of gravity and/or the tensional forces associated with the transpirational pull on a column of water. Turner *et al.* (this issue) summarize how a clever mutant screen was used to define genes specifically involved in cellulose synthesis and lignification during secondary cell wall formation. The mutant screen was based on the fact that the inability to produce secondary cell wall components in cells that would normally have a secondary cell wall, like xylem cells, would cause these cells to collapse. The mutants uncovered by this screen continue to reveal much about the function of cellulose synthase family members, and mechanisms involved in the control of lignin biosynthesis.

As wood is essentially a collection of secondary cell walls, many cell-wall-relevant genes have also emerged from genomics research associated with wood formation. For example, Mellerowicz *et al.* (this isssue) have examined gene expression associated with xylem development in poplar. Using an approach which fuses expressed sequence tag (EST) analysis, genetic modification and microarray analysis of gene expression patterns, Mellerowicz *et al.* are developing a 'genetic roadmap' to secondary cell wall formation. Whetten *et al.* (this issue) also combined EST analysis with microarray assessment of transcript accumulation to develop an understanding of wood formation in loblolly pine. The findings of these two groups point to the power of using trees to develop a comprehensive picture of secondary cell wall formation in the context of xylem development. Future studies which make interspecific comparisons of this process (between pine and poplar, for example) should create a picture of the general mechanisms underpinning secondary cell wall formation.

One of the few model systems to study the precise development of a single cell type *in vitro* is that of the transdifferentiation of *Zinnia* mesophyll cells into tracheary elements. Milioni *et al.* (this issue) optimized the time-course of *trans*-differentiation to 48 h to permit selection of time-points for comparative gene expression studies. They exploited a powerful AFLP-cDNA approach to document the dynamic expression of over 600 genes involved in signaling, wall-polymer synthesis and degradation, lignification, and programmed cell death in this system. The *Zinnia* system is a powerful tool with which to uncover candidate genes involved in cell wall formation *in planta*.

How are polymers secreted and assembled into a cell-specific architecture?

Gaspar *et al.* (this issue) present an in-depth analysis of the gene families that comprise the arabinogalactan-proteins (AG-Ps). The precise function of these proteoglycans is still unknown, but they are associated with several developmental events, such as differentiation, cell-cell recognition, embryogenesis, and programmed cell death. They discovered several years ago that some of the AG-Ps contain glycosylphosphatidylinositol anchoring domains, the so-called 'GPI anchors'. Because this structural feature is associated with signaling in animal cells, its presence indicates a potentially new function for AG-Ps. Their review presents models of proteoglycan function in animals and yeast that may shed light on special functions of AG-Ps in plants.

While specific knowledge of the proteins involved in assembly and rearrangement during growth are some of the least understood, the xyloglucan endotransglycosylases and α- and β-expansins, have greatly modified long-held views about how plant growth regulators controlled growth-related wall expansion. Darley *et al.*'s article (this issue) summarizes how these two enzymes might function coordinately during wall expansion and addresses an equally important question that heretofore has rarely been broached: how does the growth stop?

The cell wall is more than an extensible box

The six stages of wall development might reasonably be used to classify the fundamental structural elements of the wall, but they are far from a comprehensive set of genes whose products function in the plant's 'extracellular matrix'. Because plant cells are fixed

spatially with respect to their neighbors, plant development relies on discrete and coordinate changes in the cell wall to direct the final shape of each cell that, ultimately, defines the morphology of the entire plant body. All living cells contain also cell wall molecules that affect patterns of development, mark a cell's position within the plant, or participate in cell-cell and wall-nucleus communication. Another surprise that emerged from the analysis of the *Arabidopsis* genome is the relative richness of the protein kinase gene families, a great many of which reside in the plasma membrane with external facing receptor domains (Arabidopsis Genome Initiative, 2000). Anderson *et al.* (this issue) explored a unique group of plasma membrane-associated protein kinases called WAKs. At least some of the WAKs appear to be directly associated with pectins and glycine-rich proteins within the wall. Through this interaction, WAKs may function in a range of cellular processes, from cell growth and cell anchoring to resistance against pathogens. They undoubtedly represent the tip of the iceberg with respect to understanding how and what messages plant cells communicate.

Remodeling the wall for plant improvement

A major practical goal of plant cell wall research is to generate plants with genetically defined variation in composition and architecture to permit assessment of modifications on wall properties and plant development. As the range of products produced by transgenic plants continues to broaden, plant cell walls have now become key targets for plant improvement. Examples include the modification of pectin-cross-linking or cell-cell adhesion to increase shelf-life of fruits and vegetables, the enhancement of dietary fiber contents of cereals, the improvement of yield and quality of fibers, and the relative allocation of carbon to wall biomass for use as biofuels. The reviews that comprise this special issue highlight a few of the advances in the identification of the relevant genes and gene products that are being or could be manipulated to alter cell wall structures in our crop plants and trees. Brummell and Harpster (this issue) review many of the potential enzymes and proteins that are potential determinants of wall softening and swelling that accompany some types of fruit ripening, such as tomato. They explain how antisense inhibition and over-expression has been used to dissect the temporal requirements involved in wall depolymerization during fruit ripening.

With the advent of biotechnology, agricultural researchers are investigating particular enzymes involved in cell wall metabolism in the hope of producing crops with desired characteristics by enhancing commercially valuable traits, such as fiber production in flax, cotton, ramie and sisal, or abolishing costly ones, such as lignification in some plant tissues. For example, the pulp and paper industry and the livestock industry each would benefit by selective reduction the lignin content in their respective sources of raw material. Reducing lignin content would reduce organochlorine wastes and cut costs tremendously for the paper industry, which currently uses chemical extractions to purify cellulose from wood. Halpin *et al.* (this issue) describe their novel approaches to determine which biosynthetic steps are both necessary and sufficient to alter lignin content and composition for desired end uses. In one approach, they have developed a novel system which uses the self-cleaving 2A peptide from the hoof-and-mouth virus to simultaneously express two individual proteins under the control of one promoter. While this system is quite useful for the simultaneous analysis of multiple biosynthetic steps in the lignin biosynthetic pathway, it is also likely to be broadly applicable to the analysis of many proteins, beyond those involved in cell wall biogenesis. This is an excellent example of how science designed to cope with the problems associated with cell wall analysis it likely to be of benefit to all plant scientists.

Cell-wall functional genomics in the coming decade

The completion of the *Arabidopsis* genome sequence culminates the first century of genetics research since the rediscovery of Mendel's experiments. Now that we have a complete inventory of the genes sufficient to make a higher plant, what will be the next step for cell-wall biologists? We estimate that about 15% of the genome is connected in some way with the biogenesis, rearrangement, and turnover of a cell wall. About 45% of the genome encodes proteins for which no known function can be deduced. The remaining 55% for which database comparisons provide putative function, but only about 1000 genes have been assigned a function by direct experimental evidence (Somerville and Dangl, 2000). This year, an ambitious new initiative was launched with a goal to know the function of every *Arabidopsis* gene by the year 2010 (Chory *et al.*, 2000).

The research of the 2010 Program is expected to radiate from the perspective of the gene but, as cell-wall biologists are acutely aware, the cell wall is, literally and figuratively, the farthest cellular structure from the gene. The reason for this is that the wall is an amalgam of a great number of molecules that are synthesized by an as yet unknown cellular machinery, encoded by genes that are far from being fully characterized. Hence, the scientific problems to be solved extend beyond expression arrays and three-dimensional protein structures. To achieve the goal of understanding the function of every wall-relevant gene will require new biochemical and cellular methodologies to parallel and even exceed the advances in gene and protein technologies that are embodied by the 2010 Program.

The *Arabidopsis* genome has become a springboard for comparative genetics with the genomes of many other plant species, including our important crop plants. Although *Arabidopsis* has proven itself to be a superior model plant for genetic studies, many other species are far more suitable for cellular and biochemical studies that will unveil gene function. The articles constituting this issue not only illustrate the enormous progress that has been made in identifying the wealth of wall-related genes but they also point to future directions and how far we have to go.

References

Arabidopsis Genome Initiative. 2000. Analysis of the genome sequence of the flowering plant *Arabidopsis thaliana*. Nature 408: 796–815.

Chory, J., Ecker, J.R., Briggs, S., Caboche, M., Coruzzi, G., *et al.* 2000. Functional genomics and the virtual plant. A blueprint for understanding how plants are built and how to improve them. Plant Physiol. 123: 423–425.

Somerville, C. and Dangl, J. 2000. Plant biology in 2010. Science 290: 2077–2078.

SECTION 1

CYTOLOGY AND METABOLISM

Plant Molecular Biology **47**: 9–27, 2001.
© 2001 *Kluwer Academic Publishers. Printed in the Netherlands.*

9

Pectin: cell biology and prospects for functional analysis

William G.T. Willats[1], Lesley McCartney[1], William Mackie[2] and J. Paul Knox[1,*]
[1]*Centre for Plant Sciences and* [2]*School of Biochemistry & Molecular Biology, University of Leeds, Leeds LS2 9JT, UK* (*author for correspondence; e-mail j.p.knox@leeds.ac.uk)*

Key words: cell wall, homogalacturonan, pectin, pectinases, polysaccharide, rhamnogalacturonan

Abstract

Pectin is a major component of primary cell walls of all land plants and encompasses a range of galacturonic acid-rich polysaccharides. Three major pectic polysaccharides (homogalacturonan, rhamnogalacturonan-I and rhamnogalacturonan-II) are thought to occur in all primary cell walls. This review surveys what is known about the structure and function of these pectin domains. The high degree of structural complexity and heterogeneity of the pectic matrix is produced both during biosynthesis in the endomembrane system and as a result of the action of an array of wall-based pectin-modifying enzymes. Recent developments in analytical techniques and in the generation of anti-pectin probes have begun to place the structural complexity of pectin in cell biological and developmental contexts. The *in muro* de-methyl-esterification of homogalacturonan by pectin methyl esterases is emerging as a key process for the local modulation of matrix properties. Rhamnogalacturonan-I comprises a highly diverse population of spatially and developmentally regulated polymers, whereas rhamnogalacturonan-II appears to be a highly conserved and stable pectic domain. Current knowledge of biosynthetic enzymes, plant and microbial pectinases and the interactions of pectin with other cell wall components and the impact of molecular genetic approaches are reviewed in terms of the functional analysis of pectic polysaccharides in plant growth and development.

Abbreviations: HGA, homogalacturonan; PME, pectin methyl esterase; PG, polygalacturonase; RG-I, rhamnogalacturonan-I; RG-II, rhamnogalacturonan-II; XGA, xylogalacturonan.

Introduction to the primary cell wall matrix

Plant cell walls are major contributors to both the form and properties of plant structures. Most current discussions suggest that the primary cell walls that surround all plant cells are fibrous composites in which cellulose microfibrils are tethered together by cross-linking glycans (Carpita and Gibeaut, 1993; Cosgrove, 1999, 2000). In dicotyledons these links are predominantly xyloglucan polysaccharides whereas glucuronoarabinoxylans dominate in the Poaceae and other monocots (Carpita and Gibeaut, 1993; Carpita, 1996). In general terms, it is useful to think of the load-bearing cellulose-cross-linking glycan network as being embedded in a more soluble matrix of polysaccharides, glycoproteins, proteoglycans, low-molecular-weight compounds and ions. Pectin is the most abundant class of macromolecule within this matrix and, in addition, it is also abundant in the middle lamellae between primary cell walls where it functions in regulating intercellular adhesion. Pectin is one of the major components of primary cell walls and it is generally thought to account for about one third of all primary cell wall macromolecules, although lower levels occur in primary cell walls of certain families belonging to the Poales (Mohnen, 1999; Smith and Harris, 1999). Pectin is greatly reduced or absent in non-extendable secondary cell walls and is the only major class of plant polysaccharide to be largely restricted to primary cell walls. The pectic matrix provides an environment for the deposition, slippage and extension of the cellulosic-glycan network, is involved in the control of

10

cell wall porosity and is the major adhesive material between cells.

The aim of this review is to cover our current knowledge of the structure-function relationships of the diverse and complex polysaccharides that comprise pectin and, in particular, to survey the prospects for the analysis of their functions in the context of plant growth and development.

Pectic polysaccharide composition and primary structure

The most useful starting biochemical definition of pectin is that it is a group of polysaccharides that are rich in galacturonic acid (GalA). GalA occurs in two major structural features that form the backbone of three polysaccharide domains that are thought to be found in all pectin species: homogalacturonan (HGA), rhamnogalacturonan-I (RG-I) and rhamnogalacturonan-II (RG-II) (O'Neill *et al.*, 1990; Albersheim *et al.*, 1996, Mohnen, 1999). It is thought that these three polysaccharide domains can be covalently linked to form a pectic network throughout the primary cell wall matrix and middle lamellae. This network has considerable potential for modulation of its structure by the action of cell wall-based enzymes.

HGA is a linear homopolymer of $(1\rightarrow4)$-α-linked-D-galacturonic acid and is thought to contain some 100–200 GalA residues (Thibault *et al.*, 1993; Zhan *et al.*, 1998). HGA is an abundant and widespread domain of pectin and appears to be synthesized in the Golgi apparatus and deposited in the cell wall in a form that has 70–80% of GalA residues methyl-esterified at the C-6 carboxyl (O'Neill *et al.*, 1990; Mohnen, 1999). The removal of methyl ester groups within the cell wall matrix results in HGA capable of being cross-linked by calcium and the formation of supramolecular assemblies and gels (see below). Other biosynthetic modifications and substitutions of HGA also occur but are not as widespread as methyl esterification. GalA residues in HGA can be *O*-acetylated and this is predominantly at C-3, although C-2 substitution can also occur. Acetylated HGA appears to be particularly abundant in sugar beet roots and potato tubers (Ishii, 1997; Mohnen, 1999; Pauly and Scheller, 2000). GalA of HGA may be substituted at C-3 with residues of xylose producing a domain known as xylogalacturonan (XGA) that appears to be quite widespread and has been isolated from pea seed coats, apple pectin, watermelon fruit and carrot

cells (Schols *et al.*, 1995; Kikuchi *et al.*, 1996; Yu and Mort, 1996; Renard *et al.*, 1997; Le Goff *et al.*, 2000). Substitution of GalA with apiose at C-2 or C-3 results in apiogalacturonan which has been found in the duckweeds *Lemna minor* and *Spirodela polyrrhiza* (Hart and Kindel, 1970; Longland *et al.*, 1989). These biosynthetic modifications, or combinations of modifications, are likely to modify the functional properties of the HGA domain, but in precisely what ways is not known.

In addition to HGA, an acidic pectic domain consisting of as many as 100 repeats of the disaccharide $(1\rightarrow2)$-α-L-rhamnose-$(1\rightarrow4)$-α-D-galacturonic acid has been isolated from a wide range of plants and is known as RG-I (Albersheim *et al.*, 1996; O'Neill *et al.*, 1990). RG-I is abundant and heterogeneous and generally thought to be glycosidically attached to HGA domains. This may indicate a biosynthetic switch from one type of galacturonan backbone to the other. In most cases, 20–80% of rhamnose residues in RG-I are substituted at C-4 with side chains in which neutral residues predominate and these can vary in size from a single glycosyl residue to 50 or more, resulting in a large and highly variable family of polysaccharides (Albersheim *et al.*, 1996). Common structural features of the side chains include polymeric $(1\rightarrow4)$-β-linked D-galactosyl and $(1\rightarrow5)$-α-linked L-arabinosyl residues (Mohnen, 1999). Pectic $(1\rightarrow4)$-β-linked D-galactan with non-reducing terminal-arabinose (*t*-Ara) substituted at the O-3 of some of the Gal units is known as type I arabinogalactan (Carpita and Gibeaut, 1993). A range of other linkages involving these and other sugars, including uronic acids, can also be present (Lerouge *et al.*, 1993; An *et al.*, 1994; Ros *et al.*, 1996; Mohnen, 1999). Arabinans can become branched by links through O-2 and O-3. Arabinogalactans of type II with $(1\rightarrow3)$-β- and $(1\rightarrow6)$-β-linked-D-galactosyl residues also occur on pectic backbones (Guillon and Thibault, 1989; Renard *et al.*, 1991). Type II arabinogalactans also occur in a complex family of proteoglycans known as arabinogalactan-proteins (Nothnagel, 1997). The highly branched nature of RG-I has led to it being known as the hairy region of pectin, in contrast to HGA domains which are known as the smooth region (Schols and Voragen, 1996). Whether GalA residues within RG-I can also be methyl-esterified as in the HGA domain is unknown. A small number of GalA residues in the RG-I backbone of sugar beet pectins are substituted with single glucuronic acid residues (Renard *et al.*, 1999).

Despite its name, RG-II is not structurally related to RG-I but is a branched pectic domain containing an HGA backbone. RG-II is a highly conserved and widespread domain isolated from cell walls by *endo*polygalacturonase cleavage indicating covalent attachment to HGA. RG-II has a backbone of around 9 GalA residues that are $(1\rightarrow4)$-α-linked and is substituted by 4 heteropolymeric side chains of known and consistent lengths (O'Neill *et al.*, 1996; Vidal *et al.*, 2000). The side chains contain eleven different sugars including apiose, aceric acid and 2-keto-3-deoxy-D-*manno*-octulosonic acid (KDO) (O'Neill *et al.*, 1996; Vidal *et al.*, 2000). A significant feature of RG-II is that it can dimerize by means of borate ester links through apiosyl residues (Kobayashi *et al.*, 1996; O'Neill *et al.*, 1996; Ishii *et al.*, 1999). RG-II appears to be the only major pectic domain that does not have significant structural diversity or modulation of its fine structure.

Figure 1 presents a highly schematic diagram of the structure of HGA, RG-I and RG-II and lists the major potential variations within their fine structure. An unknown aspect of pectic network structure is the distribution of HGA, RG-I and RG-II within pectin chains. Both RG-I and RG-II have been isolated by *endo*polygalacturonase treatment of pectins suggesting that they are both attached to HGA, but whether RG-I occurs on the same chain as RG-II and whether the domains are attached to HGA at reducing or non-reducing ends is unknown (Albersheim *et al.*, 1996).

Conformational aspects of pectic polysaccharides

From a conformational aspect, HGA is the best understood pectic domain. Experimental studies (X-ray fibre diffraction, circular dichroism, NMR spectroscopy) have amply demonstrated that in the solid/gel state, HGA is extended but flexible and may occur as 2_1 and/or 3_1 helices depending on factors such as the degree of hydration and the nature of the counterion (Walkinshaw and Arnott, 1981; Morris *et al.*, 1982; Powell *et al.*, 1982; Alagna *et al.* 1986; Jarvis and Apperley, 1995). Molecular modelling studies are consistent with this but, interestingly, indicate that the conformational flexibility may be even greater than supposed. Thus, in addition to the 2_1 and 3_1 helical conformations, a left-handed threefold (3_2) and a right-handed fourfold (4_1) conformation are also possibilities (Pérez *et al.*, 2000). Both the experimental and theoretical approaches have been used to explore

the association of HGA with calcium ions since this is responsible for the characteristic properties of pectins. Early conclusions were that, by analogy with the polyguluronate component of alginate (Atkins *et al.*, 1973), calcium coordination resulted in association of 2_1 chains in dimers (the egg-box model) which formed the junction zones in the gel network. This model requires the chains to be in exact register (as in polyguluronate) but a recent modelling study suggests that this is not the case and that for effective coordination of calcium to be achieved in 2_1 dimers, the chains must be staggered (S. Pérez and I. Braccini, personal communication). Interestingly, an X-ray diffraction study of a complex of a single crystal of a pectate lyase and a pentameric HGA fragment showed that, in the binding site, the oligosaccharide was a mixture of the 2_1 and 3_1 conformations (Scavetta *et al.*, 1999). Clearly, the biosynthetic modifications to the core HGA structure and subsequent modifications may influence conformation of HGA chains *in planta*.

The RG-I component arises from the incorporation of rhamnose residues into a galacturonan backbone. Although the insertion of a single rhamnose residue causes a sterical deviation ('kink') compared to the relatively extended HGA backbone, modelling studies suggest that the insertion of more than two or three rhamnose residues results in the overall chain having a fully extended conformation similar to HGA (Engelsen *et al.*, 1996). The predicted low-energy conformation for the strictly alternating sequence of RG-I is one of three-fold helices (Engelsen *et al.*, 1996). As HGA chains are generally uninterrupted by rhamnose residues (Zhan *et al.*, 1998), the importance of the rhamnose residues appears to rest with their function as sites for further glycosylation (cf. amino acids such as asparagine and serine in glycoproteins). Insight into the stereostructural features of RG-I has been gained by modelling studies and these have provided some interesting considerations and possibilities. For example, it is possible to consider the construction of RG-I as the assembly of a range of structural motifs and it has become evident that the relatively complex side chains can be accommodated into the RG-I backbone without serious steric restrictions no matter what the size (Engelsen *et al.*, 1996; Pérez *et al.*, 2000). These insights, along with NMR studies that indicate that components of RG-I side chains are highly mobile elements in cell walls (Foster *et al.*, 1996; Renard and Jarvis, 1999b), suggest that even relatively long RG-I side chains can be accommodated with efficient packing and may be capable of wrapping around the

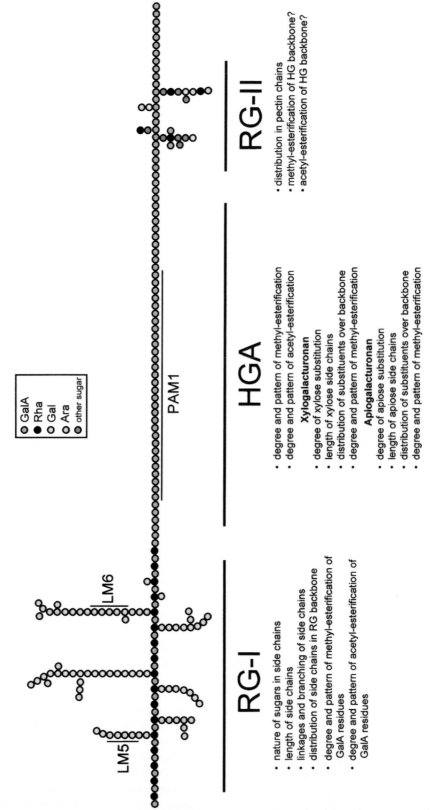

Figure 1. Simplified schematic diagram to indicate some of the features of the three major polysaccharide domains of pectin: homogalacturonan (HGA), rhamnogalacturonan-I (RG-I) and rhamnogalacturonan-II (RG-II). Both RG-I and RG-II are thought to be covalently attached to HGA domains but at present there is no evidence for a direct linkage between RG-I and RG-II. Listed beneath each schematic representation are some of the important ways in which the fine structure of the domains can vary. Oligosaccharide epitopes recognised by anti-HGA monoclonal antibody PAM1 and anti-RG-I monoclonal antibodies LM5 and LM6 are indicated.

RG-I backbone. The range of galactan linkages that are known to occur in RG-I side chains results in various distinct helical structures but the functional significance of any distinct conformations is unknown (Engelsen *et al.*, 1996; Pérez *et al.*, 2000). On account of the size (ca. 5 kDa) and its highly conserved nature, RG-II is a good target for the analysis of secondary structure. However, the complexity of its monosaccharide composition makes the elucidation of its conformational parameters a formidable task. Recent studies have addressed this problem by NMR spectroscopy (du Penhoat *et al.*, 1999) and molecular modelling (Pérez *et al.*, 2000), although a complete 3-dimensional model has not yet been proposed.

It is not claimed that the proposed structures indicated by modelling studies are in any sense conformationally fixed, but they represent a realistic starting point for considering aspects of the structure and function of the complex pectic components. For example, if it turns out that, in some instances at least, monoclonal antibodies such as the anti-HGA PAM1 (discussed below) recognize conformational features of pectic polysaccharides, then this raises the exciting prospect that future studies may be able to correlate biological changes with conformational features *in planta*.

Pectin is multifunctional

It is certain that pectin is multifunctional. An important goal is to ascribe physical properties to specific structural domains within the pectic network and to relate these to wider cell biological properties in growth and development. This task is difficult. In addition to the properties of oligosaccharide domains per se, additional factors such as the position and linkage between domains and interaction with other macromolecular systems and small entities such as calcium, borate, polyamines and phenolic compounds are all likely to contribute to pectin-based matrix properties.

Pectin is almost certain to have distinct roles during cell wall deposition and assembly and also subsequently during cell expansion. Modulation of pectin structure within the cell walls may therefore reflect progressive changes in roles during cell development. The primary cell wall matrix plays a crucial role during expansive cell growth in the deposition of cell wall components and in allowing the load-bearing fibrils to slide apart (Cosgrove, 2000). Recent work has shown that pectin is crucial for the formation of synthetic composites that reflect plant cell walls and that pectins can influence the deposition of cellulose microfibrils and that this in turn has impact upon mechanical properties (Chanliaud and Gidley, 1999). It is also of interest that, once formed, the pectin can be removed from the synthetic composites and specific mechanical properties are maintained (Chanliaud and Gidley, 1999).

The pectic network influences the pH and ionic status of the matrix and, through its capacity to form gels, is also intimately involved in the generation of mechanical and porosity properties of cell walls. These properties in turn are likely to influence cell wall links and hydration status and influence the movement and access of cell wall-modifying proteins through the matrix. No mechanisms are known for the regulated and specific movement of macromolecules through the primary cell wall matrix, but such mechanisms would seem to be required to account for accumulation of proteins in culture media and at regions of cell walls surrounding intercellular spaces. The pectic network in the regions of the middle lamella has roles in intercellular adhesion (Knox, 1992) and bonds linking middle lamella pectin to pectin within primary cell walls will also be important here. The controlled disassembly of these links will be required for the formation of intercellular space in parenchymatous tissues. Furthermore, HGA is a source of oligosaccharide signals that act in the co-ordination of development and defence (Dumville and Fry, 2000).

The pectic network is clearly a target for specific developmental modifications such as cell wall swelling and softening during fruit ripening (Fischer and Bennett, 1991) and cell separation during leaf and fruit abscission, pod dehiscence and root cap cell differentiation (Wen *et al.*, 1999; Roberts *et al.*, 2000).

Cell biology of pectin: where are pectic domains within the cell wall matrix and what do they do?

Two important questions concerning the polysaccharide domains that comprise pectin, and their modified derivatives, are: how widespread across tissues, organs and species are pectic domains, and what is their location within cell walls? HGA, RG-II and RG-I are proposed to be components of all primary cell walls, and considerable evidence supports this (Albersheim *et al.*, 1996). However, the occurrence of variations on these domains and the distribution of modified forms is less clear. For example, highly acetylated HGA is known to be abundant in sugar beet roots and potato tubers and XGA has been isolated from apple fruit and pea testa. Do these apparently restricted occurrences merely reflect a taxonomic diversity of pectic polysaccharide structure or the modulation of structure correlating with specific functions within these particular organs and species, or both of these possibilities? Furthermore, does acetylated HGA, for example, occur in low abundance at specific locations in a wide range of plant organs but is its presence lost upon homogenization of organs for compositional analysis? One way to address these questions is by the development and use of technologies for the analysis of pectic structure *in planta*. One of the most powerful ways to study pectin in its physiological and developmental contexts is by the use of anti-pectin antibody probes, and antibody-based analyses of HGA, RG-II and RG-I have now been carried out.

Homogalacturonan

The best understood structure-function relationship of a pectic domain is HGA and its involvement in calcium-mediated gel formation. The formation of gels is probably important for maintaining the integrity of cell walls, both within and between cell layers and across the middle lamella although, in many cases, factors in addition to calcium cross-links are responsible for maintaining the integrity of the pectic network (Jarvis, 1984, 1992).

The capacity of HGA to participate in gel formation and to contribute to cell wall stiffening is regulated *in muro* largely by the action of pectin methyl esterases (PMEs), although we know so little about HGA biosynthesis that differential methyl esterification during biosynthesis may also lead to varied methyl-esterification states of HGA *in muro*. PMEs remove methyl-ester groups from HGA resulting in stretches of acidic residues that can associate with other HGA chains by calcium cross-links. The de-esterification of HGA appears to be a complex regulated process that does not occur uniformly throughout tissues or cell walls (Liberman *et al.*, 1999). Some of the first indications of the variation of methyl esterification of HGA within cell walls was obtained using the first generation of anti-pectin monoclonal antibodies (Table 1). The anti-HGA probe 2F4 recognizes a calcium-cross-linked dimer of HGA, while JIM5 and JIM7 bind to a range of methyl-esterification states of HGA. All three antibodies have been used widely in immunolocalization studies and have been

discussed extensively elsewhere (Knox, 1997; Jauneau *et al.*, 1998; Willats *et al.*, 2000a). These probes are HGA-specific, but they have not been able to provide precise information on the extent or pattern of the de-esterification of HGA chains within cell walls.

It is becoming apparent that PME isoforms can have differing action patterns on HGA and that action patterns can also be influenced by the extent of methyl esterification of the substrate and pH (Catoire *et al.*, 1998; Limberg *et al.*, 2000a; Denès *et al.*, 2000). Broadly, PMEs are thought to have two types of action patterns. The first is a block-wise action (single chain mechanism) in which the enzyme removes methyl groups from contiguous GalA residues on a HGA backbone resulting in relatively long stretches or blocks of de-esterified residues. Blocks in the region of 14 contiguous residues are thought to be needed for most effective calcium cross-linking of HGA chains (Jarvis, 1984). A block-wise action is thought to be the predominant mechanism of plant PMEs. Large blocks of non-esterified HGA have been located in cell walls with a phage display monoclonal antibody, PAM1, that requires about 30 contiguous de-esterified residues for binding (Willats *et al.*, 1999a). This is much larger epitope than is recognized by most antibodies and PAM1 may recognize a length-dependent conformational epitope of HGA, a pheomenon that has been observed in other polysaccharide systems (Brisson *et al.*, 1997). Initial studies with PAM1 have indicated that non-esterified blocks of HGA occur in distinct regions of cell walls, often in cell junctions and this probe will be very useful to look at PME action in cell development. PMEs are also produced by some plant pathogens but the action pattern of these microbial PMEs are thought to be predominantly non-block-wise (multiple-chain mechanism) resulting in the removal in single, or a limited number of, methyl ester groups at a time (Catoire *et al.*, 1998; Limberg *et al.*, 2000a; Denès *et al.*, 2000).

Distinct isoforms of mung bean PMEs have differences in activities relating to pH and the extent of methyl esterification of the substrate (Catoire *et al.*, 1998), but the significance of the range of PMEs occurring in multigene families is far from clear (Micheli *et al.*, 1998). One possibility is that PMEs are differentially regulated throughout development and that they may have different action patterns at different location within cell walls, resulting in differences in cell wall stiffening or other properties. We have recently accumulated evidence that supports this idea in that we have identified an HGA epitope (bound by

Table 1. Antibodies specific to oligosaccharide domains in pectic polysaccharides. All antibodies are hybridoma monoclonal antibodies unless indicated to be phage display monoclonal antibodies (PD) or antisera (AS).

Pectic domain	Antibody	Epitope	References
HGA	JIM5	unknown/relatively low methyl-esterified epitope	VandenBosch *et al.*, 1989; Knox *et al.*, 1990; Willats *et al.*, 2000
HGA	JIM7	unknown/relatively high methyl-esterified epitope	Knox *et al.*, 1990; Willats *et al.*, 2000
HGA	2F4	calcium-mediated dimer of non-esterified chains	Liners *et al.*, 1992
HGA	PAM1 (PD)	ca 30 non-esterified GalA residues (block-wise)	Willats *et al.*, 1999a
HGA	LM7	non-block-wise pattern of de-esterification	Willats *et al.*, 2001, *See Fig 2*
XGA	LM8	undefined	*unpub. See Fig. 3*
RG-II	(PD)	undefined/incl. part of HGA backbone	Williams *et al.*, 1996
RG-II	(AS)	undefined/incl. part of HGA backbone	Matoh *et al.* 1998
RG-I	CCRC-M2	undefined	Puhlmann *et al.*, 1994
RG-I	LM5	four $(1\rightarrow4)$-β-D-galactosyl residues	Jones *et al.*, 1997
RG-I	LM6	five $(1\rightarrow5)$-α-L-arabinosyl residues	Willats *et al.*, 1998

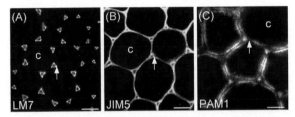

Figure 2. Immunofluorescent labelling of transverse sections of pea stem parenchyma with anti-HGA monoclonal antibodies LM7, JIM5 and PAM1 indicates that the degree and pattern of methyl esterification of HGA is spatially-regulated in cell walls in relation to intercellular spaces. A. LM7 binds specifically to HGA with a non-block-wise distribution of methyl esters and its binding is restricted to regions of cell walls lining intercellular spaces, being particularly abundant at the junction between adhered and separated primary cell walls (Willats *et al.*, 2001). B. JIM5 binds to a range of partially methyl-esterified epitopes of HGA (Knox *et al.*, 1990; Willats *et al.*, 2000) and binds throughout the cell walls. C. PAM1 is specific to long non-esterified blocks of HGA (Willats *et al.*, 1999) and its binding is restricted to the lining of intercellular spaces and regions of the cell wall close to the plasma membrane, other than those near intercellular spaces. Arrows indicate a corner of an intercellular space and c indicates the interior of a cell. Scale bar = 100 μm.

monoclonal antibody LM7) that is produced *in vitro* by the non-block-wise action of an *Aspergillus* PME and by alkaline de-esterification, but not by block-wise action of a plant PME. However, this epitope occurs *in planta* and, most significantly, this epitope occurs specifically at corners of intercellular spaces and at the junction between cells in a wide range of species, suggesting that some process *in vivo* produces this epitope (Willats *et al.*, 2001). Figure 2 shows the pattern of the LM7 epitope in parenchyma cell walls in a section of pea stem. This is the first HGA epitope to be described that has a precise and consistent location in a range of organs and species and therefore probably reflects a specific function. This observation suggests that some plant PMEs may have a non-block-wise action pattern. Furthermore, *in vitro* analysis of the F31 model pectin (see Willats *et al.*, 1999, 2000a) containing high levels of the LM7 epitope (but low levels of the PAM1 epitope), indicates that it can form a calcium-mediated gel and that such a gel has elasticity, compressive and porosity properties different from one formed from pectin with abundant large PAM1-reactive blocks (Willats *et al.*, 2001). Overall, these observations indicate that PMEs may have the potential to modify or fine-tune matrix properties at a local level in specific regions of the cell wall in response to functional requirements. Continued refinement of the techniques for the analysis of the fine structure of HGA and patterns of HGA methyl esterification will be important in elucidating this (van Alebeek *et al.*, 2000; Daas *et al.*, 2000; Limberg *et al.*, 2000b). A report of a protein inhibitor of PME from kiwi fruit (Camardella *et al.*, 2000) is another factor that may influence networks that regulate HGA structure in these systems. The recent identification of a PME as a receptor for a virus movement protein (Dorokhov *et al.*, 1999; Chen *et al.*, 2000) indicates that proteins that participate in networks that regulate HGA structure may also be multifunctional. Highly methyl-esterified HGA chains can associate *in vitro* at low pH in the presence of sucrose by means of H-bonding and hydrophobic interactions to form acid gels. Although conditions in the cell wall matrix are not suitable for the formation of such gels, intermediate types of gel have been proposed (Jarvis, 1984) and it is possible that chain association of highly methyl-esterified regions of HGA may contribute to pectic gel structure *in planta*.

PME activity is a key control point for both the assembly and disassembly of the pectic network. As discussed above, the action of PMEs can promote the formation of supramolecular pectin gels. However, the degree and pattern of methyl esterification is also important in regulating the cleavage of HGA by pectinolytic enzymes. In addition to PME action regulating the occurrence of de-esterified HGA regions that can be cross-linked by calcium, the occurrence of such sites can be regulated by enzymes that can cleave the HGA backbone. These enzymes have the capacity to degrade potential sites of HGA chain interaction and also have the capacity to generate oligosaccharide fragments that have physiological functions (Dumville and Fry, 2000). Polygalacturonases (PGs) are the most abundant and extensively studied of the HGA degrading enzymes, and typically exist in multi-gene families and may have both *endo* and *exo* activities (Hadfield and Bennett, 1998; Torki *et al.*, 2000). Although pollen-specific and pod dehiscence zone-specific forms have been characterized (Roberts *et al.*, 2000) the involvement of PGs in organ development, cell expansion and intercellular space formation are far from clear. Several cDNAs with homology to microbial pectate lyases have been reported from plant systems and a pectate lyase capable of cleaving non-esterified regions of HGA has been characterised from *Zinnia elegans* (Domingo *et al.*, 1998).

Methyl esterification is just one biochemical modification that can influence HGA structure. Acetylation and xylose substitution of HGA are thought to occur during biosynthesis and before deposition in the cell wall. Knowing where these structurally modified forms occur within organs or within cell walls may provide insight into their functions. In certain cases, up to 90% of GalA residues may be *O*-acetylated, and this is known to hinder enzymatic breakdown of HGA and also alter solubility and gelation properties (Rombouts and Thibault, 1986; Renard and Jarvis, 1999a). The precise function of HGA (or RG-I) acetylation in the context of cell or tissue properties is not known and

no specific probes have yet been developed to localize acetylated HGA within cell walls. However, 2F4 has been used in combination with PME and alkaline de-esterification to indicate that acetylated HGA occurs throughout cell walls of suspension-cultured sugar beet cells and that higher levels are associated with decreased intercellular adhesion of these cells (Liners *et al.*, 1994).

The substitution of HGA with xylose is also known to inhibit gel formation and in addition it is likely to alter its ability to act as a substrate for pectin-modifying enzymes, such as *endo*polygalacturonase (Yu and Mort, 1996), but this is not necessarily the case for *exo*polygalacturonases (Kester *et al.*, 1999). Some insight into the occurrence of XGA has recently been obtained using immunochemical techniques. A XGA-rich preparation from pea testa (Renard *et al.*, 1997; Le Goff *et al.*, 2000) has been used to generate monoclonal antibodies. Monoclonal antibody LM8 binds to pea testa XGA and the antibody indicates that the LM8 epitope does not occur throughout the testa but is restricted to an inner layer of crushed parenchyma cells next to the cotyledon as shown in Figure 3. The LM8-reactive cells have the distinctive feature of being loosely attached to the inner testa surface and subsequently becoming completely unattached during seed development. These thin-walled parenchyma cells are known to act as transfer cells and are involved in the exchange of sucrose across the extracellular space between the testa and cotyledons (Tegeder *et al.*, 1999). Examination of range of systems and species indicates that the LM8 XGA epitope is widespread in that it occurs in a range of species, including maize, but in all cases is restricted to loosely attached cells such as root cap cells and certain cells in suspension cultures. Compositional analysis of pectins in carrot cell cultures have indicated that levels of XGA are correlated with the size of cell clusters (Kikuchi *et al.*, 1996). The restriction of the XGA epitopes to different cell types sharing a common morphological feature indicates a common functional role in these cells. The presence of XGA in these cells is likely to change matrix properties but how this relates to cell separation is not yet clear. The observations with LM8 indicate that pectic domains that are thought to be restricted to certain species or organs may be more widespread than currently appreciated. Furthermore, they have very precise locations (and functions) and occur at a level of abundance that may not be readily detected during compositional analyses of whole organs.

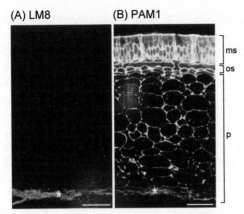

(A) LM8   **(B) PAM1**

Figure 3. Immunofluorescent micrographs of radial sections of resin-embedded pea testa labelled with anti-xylogalacturonan (XGA) antibody LM8 and anti-HGA PAM1. A. A sample of XGA isolated from pea testa (Le Goff *et al.*, 2000) was used to generate the LM8 monoclonal antibody. LM8 binds specifically to the crushed parenchyma (*) cells that line the entire inner surface of the testa and which become loosened from the rest of testa parenchyma during development. LM8 does not bind to any other cell type in the testa. B. Treatment of an equivalent section with alkali to remove methyl ester groups from HGA and probing with PAM1 indicates the presence of HGA in all tissues of the testa including macrosclereid (ms), osteosclereid (os) and parenchyma (p) cell walls. Scale bars = 100 μm.

Rhamnogalacturonan-II

RG-II has a short HGA backbone with complex, but structurally conserved, side chains (O'Neill *et al.*, 1996; Vidal *et al.*, 2000). Antibody-based studies have indicated that RG-II occurs widely in all primary cell walls, that it may be absent from the middle lamella region, and that there is a base-labile modification of RG-II within the cell wall (Williams *et al.*, 1996; Matoh *et al.*, 1998). RG-II appears to be attached to HGA as it is released by PG action. RG-II exists as a dimer cross-linked by a borate ester between two apiosyl residues (Kobayashi *et al.*, 1996; O'Neill *et al.*, 1996). The presence of boron in RG-II dimers may account for the requirement of boron for plant growth and development and boron-mediated cross-linking of RG-II may be involved in the regulation of the porosity of the pectic matrix. Boron depletion of suspension-cultured *Chenopodium album* cells led to larger weakly attached cells with increased cell wall porosity (Fleischer *et al.*, 1998, 1999). RG-II dimers formed immediately upon addition of boron to these cells with a concurrent reduction in cell wall porosity (Fleischer *et al.*, 1999). Some indications were obtained that calcium cross-linking was also important for RG-II cross-links and that they may participate in

maintaining the structural integrity of the cell walls (Fleischer *et al.*, 1999; Kobayashi *et al.*, 1999). Boron depletion of bean root nodules led to loss of covalently bound hydroxyproline-rich proteins which may indicate some connection of RG-II to proteins that may contribute to wall strength and integrity (Bonilla *et al.*, 1997). The widespread occurrence of RG-II, its structural conservation and the fact that it is resistant to degradation by all the known pectin-degrading enzymes (Vidal *et al.*, 1999) may indicate that it has a distinct structural role and may provide stable links at key regions of HGA chains within the pectic network.

Rhamnogalacturonan-I

RG-I, in contrast to RG-II, is highly variable both in its fine structure and in its occurrence within cell walls. Although RG-I components are now implicated in a range of processes, there is currently virtually no knowledge of how its structure relates to function. The current view of RG-I domains as dynamic and variable has largely been brought about by the generation and use of monoclonal antibodies to epitopes occurring in the side chains of RG-I. The monoclonal antibody LM5 binds to four residues of $(1{\rightarrow}4)$-β-D-galactan (Jones *et al.*, 1997) and the monoclonal antibody LM6 binds to five residues of $(1{\rightarrow}5)$-α-L-arabinan (Willats *et al.*, 1998). These probes are to defined neutral oligosaccharide structures occurring in pectin and were made with synthetic neoglycoprotein immunogens. Strategies useful in the development of anti-pectin probes are discussed elsewhere (Knox, 1997; Jauneau *et al.*, 1998; Willats *et al.*, 2000b).

The two antibodies, LM5 and LM6, have provided two important insights into the nature of RG-I. Firstly, galactan and arabinan epitopes can occur in different regions of organs indicating that there is no single RG-I structure that occurs throughout the cell walls of an organ or plant (Willats *et al.*, 1998, 1999b; Bush and McCann, 1999; Ermel *et al.*, 2000). We can postulate that galactan-rich and arabinan-rich forms of RG-I occur at distinct locations within developing organs. Intriguingly, the arabinan epitope is most abundant in meristematic cells of the carrot root apex and the galactan epitope appears at certain points during cell differentiation in this system (Willats *et al.*, 1999b). It is not clear at present how widespread this differential occurrence of these two epitopes is, although a similar pattern of galactan occurrence has been observed at the flax root apex (Vicré *et al.*, 1998). Moreover, during secondary growth of aspen the galactan epitope is present in meristematic cells of the cambial zone and is lost during phloem differentiation (Ermel et al., 2000). The significance of these developmental regulations of RG-I side chains is far from clear but may relate to mechanical properties. During pea cotyledon development the galactan epitope is also developmentally regulated (and in this non-iterative developmental system this can be observed temporally) and it is deposited in the cell wall relatively late in seed development (McCartney et al., 2000). In this case, the appearance of galactan has been correlated with an increase in firmness of cotyledons as assessed by compressive tests, indicating that galactan may have a role in modulating the mechanical properties of parenchymatous tissues (McCartney et al., 2000).

The second significant feature of RG-I-type polymers revealed by LM5 and LM6 is that, when galactan and arabinan do occur in the same cell, they can have distinct, spatially restricted locations within the cell wall. In the systems so far examined in which the galactan epitope appears during cell differentiation, the galactan is observed to occur in a narrow region of cell wall next to the plasma membrane (Jones et al., 1997; Vicré et al., 1998; Bush and McCann, 1999; McCartney et al., 2000). In these cases, the galactan epitope occurs during cell differentiation and is not present at earlier proliferative stages. Exceptions to this include the tomato fruit pericarp where galactan appears to be present from early stages of development and to occur throughout the cell wall (Jones et al., 1997; Orfila and Knox, 2000). In the meristematic cells of the aspen cambial zone the galactan epitope is present prior to differentiation but appears in some cell walls in regions close to the plasma membrane (Ermel et al., 2000).

A consistent feature of galactan occurrence is that it is specifically absent from cell walls in regions of pit fields containing plasmodesmata (Bush and McCann, 1999; Orfila and Knox, 2000). HGA domains are spatially arrayed at the inner face of tomato pericarp cell walls in relation to pit fields and are particularly abundant at pit fields themselves (Casero and Knox, 1995). The galactan epitope shows some distinct patterning at equivalent surfaces of tomato cell walls, but is absent from pit fields themselves, while the arabinan epitope is abundant in regions surrounding pit fields (Orfila and Knox, 2000). These patterns of structurally distinct RG-I domains may reflect specific mechanical properties required of the cell wall matrix surrounding plasmodesmata.

LM5 and LM6 bind to just two of the structural features of RG-I side chains that are known to be complex, heteropolymeric branched structures. Other structural features of RG-I are likely to be developmentally or spatially regulated within the cell wall matrix. An undefined RG-I-specific epitope (Puhlmann et al., 1994) is developmentally regulated at the Arabidopsis root apex (Freshour et al., 1996). RG-I side chains may also contain uronic acids (An et al., 1994). Type II arabinogalactan side chains that occur on RG-I also occur on cell surface arabinogalactan-protein proteoglycans. In this class of macromolecules, uronic acids, probably in terminal positions, are associated with cell positioning and possible fate determination although the biosynthetic or functional relationships concerning these structures in RG-I domains and in AGPs are unknown (Nothnagel, 1997; Knox, 1997; McCabe et al., 1997, Steele-King et al., 2000).

Interactions and links within the pectic network

The connections and interactions of domains within the pectic network are likely to be highly complex and understanding the precise nature of these interactions is a major challenge. RG-I and RG-II are thought to be covalently attached to HGA, and possibly each other, by glycosidic linkages as indicated schematically in Figure 1, but relative positions are unknown. The calcium cross-linking of HGA domains and borate ester dimerization of RG-II domains may contribute to the integrity of the pectic network surrounding cellulose microfibrils. Calcium-HGA cross-links also appear to be important for intercellular adhesion in some cases, although other pectic links are also involved, as removal of calcium leads to cell separation in only a few cases. In the Chenopodiaceae, ferulic acid attached to arabinosyl and galactosyl residues may be involved in cross-linking pectin (Wallace and Fry, 1994).

Observations made with antibodies indicate that the proportions of HGA, RG-II and RG-I vary within a particular region of the pectic network. An additional level of complexity of the pectic network is that differing populations of pectic domains are linked into the cell wall matrix by a range of bonds. This is most clearly indicated by the sequential solubilization of cell wall components with water, calcium chelators, mild alkali that will cleave ester links and strong alkali that is thought to remove non-cellulosic glycan chains, leaving a cellulose residue (Redgwell and Selvendran, 1986). Populations of pectic polysaccharides with varying compositions are solubilized to different degrees by all these extractants. There is evidence for the existence of covalent links between RG-I and extensin protein in cotton (Qi *et al.*, 1995) and for a significant proportion of xyloglucan to be covalently attached to the pectic network through arabinan/galactan of RG-I (Thompson and Fry, 2000). Evidence from NMR studies suggests that galactan, characteristic of RG-I, is tightly associated with cellulose residues after extensive solubilizaton by alkali (Foster *et al.*, 1996). The capacity for proteins to bind to pectic domains and influence their properties and functions has not been studied extensively, but may be an important aspect of the functionality of the pectic matrix. A protein-pectin interaction functioning during pollen tube adhesion in the lily style has recently been described (Mollet *et al.*, 2000).

In addition to possible interaction and covalent attachment to other macromolecules in the cell wall, pectic polysaccharides have considerable potential for interaction with ions and low-molecular-weight compounds. Acidic domains can bind aluminium and other cations in addition to calcium and boron and this may be important in factors such as the resistance to aluminium toxicity (Schmohl and Horst, 2000). Cation binding may influence conformational states and thus the properties of the pectic network (Gillet *et al.*, 1998) and can influence the osmotic and hydration swelling properties of cell walls that may impact upon turgor relations of cells (Tibbits *et al.*, 1998). Polyamines can compete for calcium-binding sites on HGA domains and thus influence not only the supramolecular assembly of pectins, but also the generation and action of oligogalacturonides and PME action with consequences upon physiology and growth (Charnay *et al.*, 1992; Messiaen *et al.*, 1997; Messiaen and Van Cutsem, 1999).

Molecular genetics and the functional analysis of pectin

Molecular genetics provides powerful approaches for the functional analyses of biological systems. The complexity, heterogeneity and multifunctionality of pectin and the indirect link between genes and complex carbohydrate structure makes using these approaches a daunting challenge. However, there is considerable potential for the application of molecular genetic approaches to the functional analysis of pectin including manipulation of its structure *in planta*, iso-

lation of mutants and elucidation of biosynthetic pathways.

At least 46 glycosyltransferases are required for the synthesis of the pectic polysaccharide structures that can be found in HGA, RG-I and RG-II, based on the assumption of one linkage for one enzyme (Mohnen, 1999). None of these are isolated, as yet, at the gene level, but HGA galacturonosyltransferase (Doong and Mohnen, 1998), HGA methyltransferase (Goubet and Mohnen, 1999; Ishikawa *et al.*, 2000), RG-I acetyltransferase (Pauly and Scheller, 2000) and $(1\rightarrow4)$-β-galactosyltransferase (Geshi *et al.*, 2000) activities have been identified. (See Table 2 for an overview of proteins involved in pectin synthesis, modification and degradation.) At present, classical enzyme isolation procedures are required as it is unlikely in the short term that specific activities can be determined from sequences (Mohnen, 1999). As seen in Table 2, knowledge is significantly more advanced about pectinases (proteins that are able to modify or degrade pectins). The structure-function relationships are particularly well advanced for the wide range of pectinases that have been isolated from saprophytic and plant pathogenic fungi and bacteria. These, alongside plant pectinases, are extremely powerful tools for the eludication of the fine structure of pectin domains (Schols and Voragen, 1996; Limberg *et al.*, 2000a).

A series of over-expression or antisense inhibition of expression studies of pectin-modifying enzymes have been carried out. The solubilization and swelling of cell walls during tomato fruit ripening has been a major focus for the analysis of pectin-degrading enzymes, including PG and PME, although it has not been easy to elucidate precise relationships between pectic enzyme activities and cell wall structure and properties (Smith *et al.*, 1990; Carrington *et al.*, 1993; Tieman and Handa, 1994). Galactosidases capable of degrading RG-I galactan are known to occur in plant tissues. At least seven β-galactosidase genes are expressed during tomato fruit development but the specificity of these towards pectin or other molecules is unknown and their functions in relation to cell wall softening during fruit ripening far from clear (Smith and Gross, 2000).

The role of modifying or degradative enzymes in wider aspects of plant growth and development are now being explored and two studies have indicated that PMEs may have diverse and complex roles during growth and development. Inhibition of the expression of a PME during pea root development led to reduced root elongation, altered root cell morphology and re-

Table 2. Pectic enzymes. Current status of knowledge of major enzyme activities involved in pectic polysaccharide domain synthesis, modification and degradation. This tabulation is not exhaustive in terms of either enzyme activities that have been detected or in the citation of the appropriate literature but aims to provide source references and a guide to major enzyme activities. A full discussion of pectin biosynthesis can be found elsewhere (Mohnen, 1999). A, activity has been identified; Sq, protein/gene sequences are known; St, some information on protein structure has been determined.

Enzyme	A	Sq	St	Recent source references
Biosynthetic				
Galacturonosyltransferase (HGA)	×			Doong and Mohnen, 1998; Scheller *et al.*, 1999; Mohnen 1999
Pectin methyltransferase (HGA)	×			Goubet and Mohnen, 1999; Mohnen, 1999; Ishikawa *et al.*, 2000
O-Acetyl transferase (RG-I)	×			Pauly and Scheller, 2000
(1→4)-β-D-galactosyltransferase	×			Geshi *et al.*, 2000
Modifying/degradative – Plant				
Polygalacturonase	×	×		Hadfield and Bennett, 1998; Torki *et al.*, 2000
Pectate lyase	×	×		Domingo *et al.*, 1998
Pectin methyl esterase	×	×		Catoire *et al.*, 1998; Micheli *et al.*, 1998
Acetyl esterase (HGA)	×			Williamson, 1991, Bordenave *et al.*, 1995, Christensen *et al.*, 1996
(1→4)-β-D-galactanase	×	×		Buckeridge and Reid, 1994, Carey *et al.*, 1995; Smith and Gross 2000
Modifying/degradative – Microbial				
Polygalacturonase	×	×	×	van Santen *et al.*, 1999
Pectate lyase	×	×	×	Herron *et al.*, 2000
Pectin lyase	×	×	×	Vitali *et al.*, 1998
Pectin methyl esterase	×	×		Kester *et al.*, 2000
Xylogalacturonase	×	×		van der Vlugt-Bergmans *et al.*, 2000
Acetyl esterase (HGA)	×	×		Schevchik *et al.*, 1997
RG acetyl esterase (RG)	×	×	×	Mølgaard et al., 2000
RG hydrolase (RG)	×	×	×	Peterson 1997; Mutter *et al.*, 1998a,c
RG lyase	×	×		Mutter *et al.*, 1998b, c
(1→4)-β-D-galactanase	×	×	×	Bonnin *et al.*, 1997; Ryttersgaard *et al.*, 1999
(1→5)-α-L-arabinanase	×	×	×	Pitson *et al.*, 1997; Scott *et al.*, 1999

duced root cap cell separation (Wen *et al.*, 1999). Over-expression of a *Petunia* PME during the growth of potato plants led to reduced tuber yield and the surprising observation of a reduction of PME activity and enhanced elongation in shoot apical tissues in comparison with wild type (Pilling *et al.*, 2000). A fungal endo-(1→4)-β-D-galactanase has been expressed in potato tubers resulting in loss of over 70% of galactosyl residues from RG-I fractions, but no altered phenotype of tubers was detected, although the galactanase did result in increased pectin solubility (Sørensen *et al.*, 2000). This is a promising approach and it will be of interest to see the effects of the expression of fungal arabinanases on plant development, in light of the abundant occurrence of pectic arabinan

in cell walls of meristematic cells in some systems (Willats *et al.*, 1999b).

Another approach to relate pectic structure to function is the isolation of mutated plants that show altered pectin structure or identification of mutated plants with altered morphology that are subsequently shown to have altered pectin or that can provide insight into pectin function. This approach is in its infancy in relation to pectin and cell walls and no examples of the former have yet been reported (Fagard *et al.*, 2000). Determining subtle alterations to pectin structure has been difficult, as discussed above, until the advent of novel and defined probes. The *quartet* mutation in *Arabidopsis* produces pollen in tetrads as microspores fail to separate because, although callose is degraded as normal, pectin is not and its persistence maintains intercellular adhesion (Rhee and Somerville, 1998). Pectin with a low level of methyl-esterification has been implicated in the abnormal tissue fusions resulting from the *adherent1* mutation in maize (Sinha and Lynch, 1998) and the transgenic expression of a fungal cutinase in *Arabidopsis* has also provided evidence for a function for pectin in organ fusions (Sieber *et al.*, 2000). The *sku5* mutation in *Arabidopsis* results in an exaggerated root growth pattern when grown on agar and preliminary evidence indicates that disruption of a pectin esterase-like gene is the cause, indicating a subtle role for HGA in root growth (Rutherford and Masson, 1996; Sedbrook *et al.*, 1998). The impact of genetic factors on complex systems may be hard to understand if components of the system cannot be seen clearly in context. The anti-$(1\rightarrow5)$-α-L-arabinan probe LM6 has recently indicated that in *Cnr*, a ripening mutant of tomato (Thompson *et al.*, 1999), which has reduced intercellular adhesion and firmer, non-swollen cell walls in the pericarp, the deposition of the arabinan component in the cell wall is severely disrupted (Orfila *et al.*, 2001). Whether the lack of appropriately deposited arabinan impacts more directly upon cell wall swelling or intercellular adhesion is unknown at this stage.

Genetic approaches are likely to elucidate the complex networks and interplay of factors that influence pectic domain structure and properties. For example, PME action is required before PGs or pectate lyases can act on methyl-esterified HGA chains, while the reduction of galactan content in potato tubers has consequences for other aspects of pectin structure (Sørensen *et al.*, 2000). The reduction of cellulose levels by chemicals and virus-induced silencing approaches results in elevated levels of pectin (Shedletzky *et al.*, 1992; Burton *et al.*, 2000) suggesting the existence of feedback loops that interconnect the cellular machinery controlling cellulose and pectin biosynthesis.

Pectins, pectinases and plant pathogenesis

In addition to its diverse roles in cell physiology and growth, pectin is in the front line of any attack upon plant cells by pathogens and subject to challenge by an array of pectinases with diverse specificities (Table 2). The literature on pectin in relation to plant pathogen interactions is considerable (Collmer and Keen, 1986; Barras *et al.*, 1994; Prade *et al.*, 1999; Esquerré-Tugayé *et al.*, 2000). The pectic network is

a major target for degradation by bacterial and fungal enzymes and pectin-degrading enzymes can be factors that determine the virulence of a pathogen (Rogers *et al.*, 2000). The extent and patterns of the substitution of pectic domains may have defensive functions in altering their ability to be degraded. Moreover, the oligosaccharides released by degradation of pectins can function as signalling molecules during defence responses. Oligogalacturonides, derived from HGA, act as endogenous elicitors and can induce the expression of a range of genes including those encoding proteinase inhibitors, pathogenesis-related proteins and enzymes leading to phytoalexins. Little is known of the transduction pathways and the most effective size of oligogalacturonides may vary between signalling systems (Simpson *et al.*, 1998). The biology of these systems is complex and in certain cases oligogalacturonides can act as endogenous suppressors of plant resistance responses (Moerschbacher *et al.*, 1999). In addition, structurally complex uronides, other than oligoglacturonides, may also be important in signalling events (Boudart *et al.*, 1998).

PG-inhibiting proteins (PGIPs) are known in a wide range of species and exist in multigene families. PGIPs can interact with *endo*PGs from phytopathogenic fungi and inhibit their activities *in vitro* with distinct specificities (Cook *et al.*, 1999; Leckie *et al.*, 1999). The interplay of proteins that can inhibit PGs may also be of interest in relation to signalling in response to pathogens (Marty *et al.*, 1997; Boudart *et al.*, 1998). In apple, gene expression can be induced by wounding and fungal infection (Yao *et al.*, 1999) and fungal PGs and PG-PGIP complexes may be involved in early stages of recognition and molecular dialogue between plants and pathogens (Esquerré-Tugayé *et al.*, 2000). The interplay of these and other factors is likely to influence the nature and the potential activity of any oligosaccharide products.

Oligogalacturonides are also involved in the complex pathways that signal wound induction of gene expression in tomato and *Arabidopsis* (Simpson *et al.*, 1998; Rojo *et al.*, 1999) and a wound-induced PG has been identified in tomato that may participate in defences against herbivores (Bergey *et al.*, 1999).

Conclusion

The term 'pectin' covers a group of acidic heteropolysaccharides with distinct structural domains that are subject to both biosynthetic and cell wall-based modifications. These polysaccharides appear to be linked together to form a pectic network throughout cell walls and possibly throughout a plant. The pectic network is clearly involved in a range of functions relating to physiology, growth, development and defence. The extensive structural and conformational variety and the dynamic nature of the pectic network presumably reflects the range of properties it provides to the cell wall matrix, in terms of mechanics, ionic and hydration conditions, signals, potential for molecular interactions and capacity to be degraded by plant and microbial pectinases. Overall, the modulation of pectic structure can be viewed as the fine-tuning of conditions and capacities within the cell wall matrix providing both mechanical properties and an operating environment for the activities of cell wall-modifying enzymes and other factors. Although the pectic network may be tightly structurally integrated throughout a plant, it may be important or useful in some cases to be somewhat wary of pectin as a catch-all name, because of the implied existence of a single structural form. Instead it may be more useful to consider populations of pectic domains and aim to ascribe particular functions to the major structural domains or their modified derivatives. Pectin occurs in the cell walls of all land plants and an extended examination of its complexity in the context of taxonomic differences in pectic occurrence and structure (Jarvis *et al.*, 1988; Potgieter and van Wyk, 1992; Smith and Harris, 1999) is also likely to be valuable. The increased awareness of the complexity and dynamic nature of the pectic network has been due largely to the development of appropriate tools to determine its structural complexity and to dissect this complexity at the cell biological level. Although recent work has placed some structural features in cell biological context, there is much still to be done in this area, particularly concerning the occurrence and function of micro-domains of the primary cell wall matrix.

Acknowledgements

We thank the UK Biotechnology and Biological Sciences Research Council and the EU Framework V Quality of Life Programme for financial support. We are grateful to our numerous colleagues for provision of pectin samples and enlightening discussions.

References

Alagna, L., Prosperi, T., Tomlinson, A.A.G. and Rizzo, R. 1986. Extended X-ray absoprtion fine structure investigation of solid and gel forms of calcium poly(α-D-galacturonate). J. Phys. Chem. 90: 6853–6857.

Albersheim, P., Darvill, A.G., O'Neill, M.A., Schols, H.A. and Voragen, A.G.J. 1996. An hypothesis: the same six polysaccharides are components of the primary cell walls of all higher plants. In: J. Visser and A.G.J. Voragen (Eds) Pectins and Pectinases, Elsevier Science, Amsterdam, pp. 47–55.

An, J., O'Neill, M.A., Albersheim, P. and Darvill, A.G. 1994. Isolation and structural characterization of β-D-glucosyluronic acid and 4-O-methyl β-D-glucosyluronic acid-containing oligosaccharides from the cell wall pectic polysaccharide, rhamnogalacturonan I. Carbohydrate Res. 252: 235–243.

Atkins, E.D.T., Nieduszynski, I.A., Mackie, W., Parker, K.D. and Smolko E.E. 1973. Structural components of alginic acid. II. The crystalline structure of poly-α-L-guluronic acid. Results of X-ray diffraction and polarized infrared studies. Biopolymers 12: 1879–1887.

Barras, F., Van Gijsegem, F. and Chatterjee, A.K. 1994. Extracellular enzymes and pathogenesis of soft-rot *Erwinia*. Annu. Rev. Phytopath. 32: 201–234.

Bergey, D.R., Orozco-Cardenas, M., de Moura, D.S. and Ryan, C.A. 1999. A wound- and systemin-inducible polygalacturonase in tomato leaves. Proc. Natl. Acad. Sci. USA 96: 1756–1760.

Bonilla, I., Mergold-Villasenor, C., Campos, M.E., Sanchez, N., Pérez, H., Lopez, L., *et al.* 1997. The aberrant cell walls of boron-deficient bean root nodules have no covalently bound hydroxyproline/proline-rich proteins. Plant Physiol. 115: 1329–1340.

Bonnin E., Vigouroux J. and Thibault, J.F. 1997. Kinetic parameters of hydrolysis and transglycosylation catalyzed by an exo-β-(1,4)-galactanase. Enzyme Microbial Tech. 20: 516–522.

Bordenave, M., Goldberg, R., Huet, J.C. and Pernollet, J.C. 1995. A novel protein from mung bean hypocotyl cell walls with acetyl esterase activity. Phytochemistry 38: 315–319.

Boudart, G., Lafitte, C., Barthe, J.P., Frassez, D. and Esquerré-Tugayé, M.-T. 1998. Differential elicitation of defense responses by pectic fragments in bean seedlings. Planta 206: 86–94.

Brisson, J.R., Uhrinova, S., Woods, R.J., van der Zwan, M., Jarrell, H.C., Paoletti, L.C., Kasper, D.L. and Jennings, H.J. 1997. NMR and molecular dynamics studies of the conformational epitope of the type III group B *Streptococcus* capsular polysaccharide and derivatives. Biochemistry 36: 3278–3292.

Buckeridge, M.S. and Reid, J.S.G. 1994. Purification and properties of a novel β-galactosidase or exo-(1→4)-β-D-galactanase from the cotyledons of germinated *Lupinus angustifolius* L. seeds. Planta 192: 502–511.

Burton, R.A., Gibeaut, D.M., Bacic, A., Findlay, K., Roberts, K., Hamilton, A., Baulcombe, D.C. and Fincher, G.B. 2000. Virus-induced silencing of a plant cellulose synthase gene. Plant Cell 12: 691–705.

Bush, M.S. and McCann, M.C. 1999. Pectic epitopes are differentially distributed in the cell walls of potato (*Solanum tuberosum*) tubers. Physiol. Plant. 107: 201–213.

Camardella, L., Carratore, V., Ciardiello, M.A., Servillo, L., Balestrieri, C. and Giovane, A. 2000. Kiwi protein inhibitor of pectin methylesterase. Amino-acid sequence and structural importance of two disulfide bridges. Eur. J. Biochem. 267: 4561–4565.

Carey, A.T., Holt, K., Picard, S., Wilde, R., Tucker, G.A., Bird, C.R., Schuch, W. and Seymour, G.B. 1995. Tomato exo-(1→4)-

β-D-galactanase: isolation, changes during ripening in normal and mutant tomato fruit, and characterization of a related cDNA clone. Plant Physiol. 108: 1099–1107.

Carpita, N.C. 1996. Structure and biogenesis of the cell walls of grasses. Annu. Rev. Plant Physiol. Plant Mol. Biol. 47: 445–476.

Carpita, N.C. and Gibeaut, D.M. 1993. Structural models of primary cell walls in flowering plants: consistency of molecular structure with the physical properties of the walls during growth. Plant J. 3: 1–30.

Carrington, C.M.S., Greve, L.C. and Labavitch, J.M. 1993. Cell-wall metabolism in ripening fruit. 6. Effect of the antisense polygalacturonase gene on cell-wall changes accompanying ripening in transgenic tomatoes. Plant Physiol. 103: 429–434.

Casero, P.J. and Knox, J.P. 1995. The monoclonal antibody JIM5 indicates patterns of pectin deposition in relation to pit fields at the plasma-membrane-face of tomato pericarp cell walls. Protoplasma 188: 133–137.

Catoire, L., Pierron, M., Morvan, C., Hervé du Penhoat, C. and Goldberg, R. 1998. Investigation of the action patterns of pectinmethylesterase isoforms through kinetic analyses and NMR spectroscopy. Implications in cell wall expansion. J. Biol. Chem. 273: 33150–33156.

Chanliaud, E. and Gidley, M.J. 1999. In vitro synthesis and properties of pectin/Acetobacter xylinus cellulose composites. Plant J. 20: 25–35.

Charnay, D., Nari, J. and Noat, G. 1992. Regulation of plant cell wall pectin methyl esterase by polyamines: interaction with the effect of metal ions. Eur. J. Biochem. 205: 711–714.

Chen, M.H., Sheng, J.S., Hind, G., Handa, A.K. and Citovsky, V. 2000. Interaction between the tobacco mosaic virus movement protein and host cell pectin methylesterases is required for viral cell-to-cell movement. EMBO J. 19: 913–920.

Christensen, T.M.I.E., Nielsen, J.E. and Mikkelsen, J.D. 1996. Isolation, characterization and immunolocalization of orange fruit acetyl esterase. In: J. Visser and A.G.J. Voragen (Eds.) Pectins and Pectinases, Elsevier Science, Amsterdam, pp. 723–730.

Collmer, A. and Keen, N.T. 1986. The role of pectic enzymes in plant pathogenesis. Annu. Rev. Phytopath. 24: 383–409.

Cook, B.J., Clay, R.P., Bergmann, C.W., Albersheim, P. and Darvill, A.G. 1999. Fungal polygalacturonases exhibit different substrate degradation patterns and differ in their susceptibilities to polygalacturonase inhibiting proteins. Mol. Plant-Microbe Interact. 12: 703–711.

Cosgrove, D.J. 1999. Enzymes and other agents that enhance cell wall extensibility. Annu. Rev. Plant Physiol. Plant Mol. Biol. 50: 391–417.

Cosgrove, D.J. 2000. Expansive growth of plant cell walls. Plant Physiol. Biochem. 38: 109–124.

Daas, P.J.H., Voragen, A.G.J. and Schols, H.A. 2000. Characterization of non-esterified galacturonic acid sequences in pectin with endopolygalacturonase. Carbohydrate Res. 326: 120–129.

Denès, J.-M., Baron, A., Renard, C.M.G.C., Péan, C. and Drilleau, J.-F. 2000. Different action patterns for apple pectin methylesterase at pH 7.0 and 4.5. Carbohydrate Res. 327: 385–393.

Domingo, C., Roberts, K., Stacey, N.J., Connerton, I., Ruiz-Teran, F. and McCann, M.C. 1998. A pectate lyase from Zinnia elegans is auxin inducible. Plant J. 13: 17–28.

Doong, R.L. and Mohnen, D. 1998. Solubilization and characterization of a galacturonosyltransferase that synthesizes the pectic polysaccharide homogalacturonan. Plant J. 13: 363–374.

Dorokhov, Y.L., Mäkinen, K., Frolova, O.Y., Merits, A., Saarinen, J., Kalkkinen, N., et al. 1999. A novel function for a unbiquitous plant enzyme pectin methylesterase: the host cell receptor for the tobacco mosaic virus movement protein. FEBS Lett. 461: 223–228.

Dumville, J.C. and Fry, S.C. 2000. Uronic acid-containing oligosaccharins: their biosynthesis, degradation and signalling roles in non-diseased plant tissues. Plant Physiol. Biochem. 38: 125–140.

du Penhoat, C.H., Gey, C., Pellerin, P. and Pérez, S. 1999. An NMR solution study of the mega-oligosaccharide, rhamnogalacturonan II. J. Biomol. NMR 14: 253–271.

Engelsen, S.B., Cros, S., Mackie, W. and Pérez, S. 1996. A molecular builder for carbohydrates: applications to polysaccharides and complex carbohydrates. Biopolymers 39: 417–433.

Ermel, F.F., Follet-Gueye, M.L., Cibert, C., Vian, B., Morvan, C., Catesson, A.N. and Goldberg, R. 2000. Differential localization of arabinan and galactan side chains of rhamnogalacturonan 1 in cambial derivatives. Planta 210: 732–740.

Esquerré-Tugayé, M.-T., Boudart, G. and Dumas, B. 2000. Cell wall degrading enzymes, inhibitory proteins, and oligosaccharides participate in the molecular dialogue between plants and pathogens. Plant Physiol. Biochem. 38: 157–163.

Fagard, M., Höfte, H. and Vernhettes, S. 2000. Cell wall mutants. Plant Physiol. Biochem. 38: 15–25.

Fischer, R.L. and Bennett, A.B. 1991. Role of cell wall hydrolases in fruit ripening. Annu. Rev. Plant Physiol. Plant Mol. Biol. 42: 675–703.

Fleischer, A., O'Neill, M.A. and Ehwald, R. 1999. The pore size of non-graminaceous plant cell walls is rapidly decreased by borate ester cross-linking of the pectic polysaccharide rhamnogalacturonan II. Plant Physiol. 121: 829–838.

Fleischer, A., Titel, C. and Ehwald, R. 1998. The boron requirement and cell wall properties of growing and stationary suspension-cultured Chenopodium album L. cells. Plant Physiol. 117: 1401–1410.

Foster, T.J., Ablett, S., McCann, M.C. and Gidley, M.J. 1996. Mobility resolved C-13 NMR spectroscopy of primary plant cell walls. Biopolymers 39: 51–66.

Freshour, G., Clay, R.P., Fuller, M.S., Albersheim, P., Darvill, A.G. and Hahn, M.G. 1996. Developmental and tissue-specific structural alterations of the cell wall polysaccharides of Arabidopsis thaliana roots. Plant Physiol. 110: 1413–1429.

Geshi, N., Jorgensen, B., Scheller, H.V. and Ulvskov, P. 2000. In vitro biosynthesis of 1,4,-β-galactan attached to rhamnogalacturonan I. Planta 210: 622–629.

Gibeaut, D.M. 2000. Nucleotide sugars and glycosyltransferases for synthesis of cell wall matrix polysaccharides. Plant Physiol. Biochem. 38: 69–80.

Gillet, C., Voué, M. and Cambier, P. 1998. Site-specific counter-ion binding and pectic chains conformational transitions in the Nitella cell wall. J. Exp. Bot. 49: 797–805.

Goubet, F. and Mohnen, D. 1999. Solubilization and partial characterization of homogalacturonan-methyltransferase from microsomal membranes of suspension-cultured tobacco cells. Plant Physiol. 121: 281–290.

Guillon, F. and Thibault, J.-F. 1989. Methylation analysis and mild acid hydrolysis of the 'hairy' fragments of sugar beet pectins. Carbohydrate Res. 190: 85–96.

Hadfield, K.A. and Bennett, A.B. 1998. Polygalacturonases: many genes in search of a function. Plant Physiol. 117: 337–343.

Hart, D.A. and Kindel, P.K. 1970. Isolation and partial characterization of apiogalacturonans from the cell wall of Lemna minor. Biochem. J. 116: 569–579.

Herron, S.R., Benen, J.A.E., Scavetta, R.D., Visser, J. and Jurnak, F. 2000. Structure and function of pectic enzymes: virulence factors of plant pathogens. Proc. Natl. Acad. Sci. USA 97: 8762–8769.

Ishii, T. 1997. O-Acetylated oligosaccharides from pectins of potato tuber cell walls. Plant Physiol. 113: 1265–1272.

Ishii, T., Matsunaga, T., Pellerin, P., O'Neill, M.A., Darvill, A. and Albersheim, P. 1999. The plant cell wall polysaccharide rhamnogalacturonan II self-assembles into a covalently cross-linked dimer. J. Biol. Chem. 274: 13098–13104.

Ishikawa, M., Kuroyama, Y., Takeuchi, Y. and Tsumuraya, Y. 2000. Characterization of pectin methyltransferase from soybean hypocotyls. Planta 210: 782–791.

Jarvis, M.C. 1984. Structure and properties of pectin gels in plant cell walls. Plant Cell Envir. 7: 153–164.

Jarvis, M.C. 1992. Control of thickness of collenchyma cell walls by pectins. Planta 187: 218–220.

Jarvis, M.C. and Apperley, D. 1995. Chain conformation in concentrated pectin gels: evidence from ^{13}C NMR. Carbohydrate Res. 275: 131–145.

Jarvis, M.C., Forsyth, W. and Duncan, H.J. 1988. A survey of the pectic content of nonlignified monocot cell walls. Plant Physiol. 88: 309–314.

Jauneau, A., Roy, S., Reis, D. and Vian, B. 1998. Probes and microscopical methods for the localization of pectins in plant cells. Int. J. Plant Sci. 159: 1–13.

Jones, L., Seymour, G.B. and Knox, J.P. 1997. Localization of pectic galactan in tomato cell walls using a monoclonal antibody specific to $(1{\rightarrow}4)$-β-D-galactan. Plant Physiol. 113: 1405–1412.

Kester, H.C.M., Benen, J.A.E. and Visser, J. 1999. The exopolygalacturonase from Aspergillus tubingensis is also active on xylogalacturonan. Biotechnol. Appl. Biochem. 30: 53–57.

Kester, H.C.M., Benen, J.A.E., Visser, J., Warren, M.E., Orlando, R., Bergmann, C., et al. 2000. Tandem mass spectrometric analysis of Aspergillus niger pectin methylesterase: mode of action on fully methyl-esterified oligogalacturonates. Biochem. J. 346: 469–474.

Kikuchi, A., Edashige, Y., Ishii, T. and Satoh, S. 1996. A xylogalacturonan whose level is dependent on the size of cell clusters is present in the pectin from cultured carrot cells. Planta 200: 369–372.

Knox, J.P. 1992. Cell adhesion, cell separation and plant morphogenesis. Plant J. 2: 137–141.

Knox, J.P. 1997. The use of antibodies to study the architecture and developmental regulation of plant cell walls. Int. Rev. Cytol. 171: 79–120.

Knox, J.P., Linstead, P.J., King, J., Cooper, C. and Roberts, K. 1990. Pectin esterification is spatially regulated both within cell walls and between developing tissues of root apices. Planta 181: 512–521.

Kobayashi, M., Matoh, T. and Azuma, J. 1996. Two chains of rhamnogalacturonan II are cross-linked by borate-diol ester bonds in higher plant cell walls. Plant Physiol. 110: 1017–1020.

Kobayashi, M., Nakagawa, H., Asaka, T. and Matoh, T. 1999. Borate-rhamnogalacturonan II bonding re-inforced by Ca^{2+} retains pectic polysaccharides in higher plant cell walls. Plant Physiol. 119: 199–203.

Leckie, F., Mattei, B., Capodicasa, C., Hemmings, A., Nuss, L., Aracri, B., De Lorenzo, G. and Cervone, F. 1999. The specificity of polygalacturonase-inhibiting protein (PGIP): a single amino acid substitution in the solvent-exposed β-strand/β-turn region of the leucine-rich repeats (LRRs) confers a new recognition capability. EMBO J. 18: 2352–2363.

Le Goff, A., Renard, C.M.G.C., Bonnin, E. and Thibault, J.-F. 2001. Extraction, purification and chemical characterization of xylogalacturonans from pea hulls. Carbohydrate Polymers 45: 325–334.

Lerouge, P., O'Neill, M.A., Darvill, A.G. and Albersheim, P. 1993. Structural characterization of endo-glycanase-generated oligoglycosyl sie chains of rhamnogalacturonan I. Carbohydrate Res. 243: 359–371.

Liberman, M., Mutaftschiev, S., Jauneau, A., Vian, B., Catesson, A.M. and Goldberg, R. 1999. Mung bean hypocotyl homogalacturonan: localization, organization and origin. Ann. Bot. 84: 225–233.

Limberg, G., Körner, R., Bucholt, H.C., Christensen, T.M.I.E., Roepstorff, P. and Mikkelsen, J.D. 2000a. Analysis of different de-esterification mechanisms for pectin by enzymatic fingerprinting using endopectin lyase and endopolygalacturonase II from A. niger. Carbohydrate Res. 327: 293–307.

Limberg, G., Körner, R., Bucholt, H.C., Christensen, T.M.I.E., Roepstorff, P. and Mikkelsen, J.D. 2000b. Quantification of the amount of galacturonic acid residues in block sequences in pectin homogalacturonan by enzymatic fingerprinting with exo- and endo-polygalacturonase II from Aspergillus niger. Carbohydrate Res. 327: 321–332.

Liners, F., Gaspar, T. and van Cutsem, P. 1994. Acetyl- and methylesterification of pectins of friable and compact sugar-beet calli: consequences for intercellular adhesion. Planta 192: 545–556.

Liners, F., Thibault, J.-F. and Van Cutsem, P. 1992. Influence of the degree of polymerization of oligogalacturonates and of esterification pattern on pectin on their recognition by monoclonal antibodies. Plant Physiol. 99: 1099–1104.

Longland, J.M., Fry, S.C. and Trewavas, A.J. 1989. Developmental control of apiogalacturonan biosynthesis and UDP-apiose production in a duckweed. Plant Physiol. 90: 972–976.

McCabe, P.F., Valentine, T.A., Forsberg, L.S. and Pennell, R.I., 1997. Soluble signals from cells identified at the cell wall establish a developmental pathway in carrot. Plant Cell 9: 2225–2241.

McCartney, L., Ormerod, A.P., Gidley, M.J. and Knox, J.P. 2000. Temporal and spatial regulation of pectic $(1{\rightarrow}4)$-β-D-galactan in cell walls of developing pea cotyledons: implications for mechanical properties. Plant J. 22: 105–113.

Marty, P., Jouan, B., Bertheau, Y., Vian, B. and Goldberg, R. 1997. Charge density in stem cell walls of Solanum tuberosum genotypes and susceptibility to blackleg. Phytochemistry 44: 1435–1441.

Matoh, T., Takasaki, M., Takabe, K. and Kobayashi, M. 1998. Immunocytochemistry of rhamnogalacturonan II in cell walls of higher plants. Plant Cell Physiol. 39: 483–491.

Messiaen, J. and Van Cutsem, P. 1999. Polyamines and pectins. II. Modulation of pectic-signal transduction. Planta 208: 247–256.

Messiaen, J., Cambier, P. and Van Cutsem, P. 1997. Polyamines and pectins. I. Ion exchange and selectivity. Plant Physiol. 113: 387–395.

Micheli, F., Holliger, C., Goldberg, R. and Richard, L. 1998. Characterization of the pectin methylesterase-like gene AtPME3: a new member of a gene family comprising at least 12 genes in Arabidopsis thaliana. Gene 220: 13–20.

Moerschbacher, B.M., Mierau, M., Graessner, B., Noll, U. and Mort, A.J. 1999. Small oligomers of galacturonic acid are endogenous suppressors of disease resistance reactions in wheat leaves. J. Exp. Bot. 50: 605–612.

Mohnen, D. 1999. Biosynthesis of pectins and galactomannans. In: D. Barton, K. Nakanishi and O. Meth-Cohn (Eds.) Comprehensive Natural Products Chemistry, vol. 3, Elsevier Science, Amsterdam, pp. 497–527.

Mølgaard, A., Kauppinen, S. and Larsen, S. 2000. Rhamnogalacturonan acetylesterase elucidates the structure and function of a new family of hydrolases. Struct. Fold. Design 8: 373–383.

Mollet, J.-C., Park, S.-Y., Nothnagel, E.A. and Lord, E.M. 2000. A lily stylar pectin is necessary for pollen tube adhesion to an in vitro stylar matrix. Plant Cell 12: 1737–1749.

Morris, E.R., Powell, D.A., Gidley, M.J. and Rees, D.A. 1982. Conformation and interactions of pectins I. Polymorphism between gel and solid states of calcium polygalacturonate. J. Mol. Biol. 155: 507–516.

Mutter, M., Beldman, G., Pitson, S.M., Schols, H.A. and Voragen, A.G.J. 1998a. Rhamnogalacturonan α-D-galactopyranosyluronohydrolase: an enzyme that specifically removes the terminal nonreducing galacturonosyl residue in rhamnogalacturonan regions of pectin. Plant Physiol. 117: 153–163.

Mutter, M., Colquhoun, I.J., Beldman, G., Schols, H,A., Bakx, E.J. and Voragen, A.G.J. 1998b. Characterization of recombinant rhamnogalacturonan α-L-rhamnopyranosyl-(1,4)-α-D-galactopyranosyluronide lyase from Aspergillus aculeatus: an enzyme that fragments rhamnogalacturonan I regions of pectin. Plant Physiol. 117: 141–152.

Mutter, M., Renard, C.M.G.C., Beldman, G., Schols, H.A. and Voragen, A.G.J. 1998c. Mode of action of RG-hydrolase and RG-lyase toward rhamnogalacturonan oligomers. Characterization of degradation products using RG-rhamnohydrolase and RG-galacturonohydrolase. Carbohydrate Res. 311: 155–164.

Nothnagel, E. 1997. Proteoglycans and related components in plant cells. Int. Rev. Cytol. 174: 195–291.

O'Neill, M.A., Albersheim, P. and Darvill, A. 1990. The pectic polysaccharides of primary cell walls. In: P.M. Dey (Ed.) Methods in Plant Biochemistry, vol. 2, Academic Press, London, pp. 415–441.

O'Neill, M.A., Warrenfeltz, D., Kates, K., Pellerin, P., Doci, T., Darvill, A.G. and Albersheim, P. 1996. Rhamnogalacturonan-II, a pectic polysaccharide in the walls of growing plant cell, forms a dimer that is covalently-linked by a borate ester. J. Biol. Chem. 271: 22923–22930.

Orfila, C. and Knox, J.P. 2000. Spatial regulation of pectic polysaccharides in relation to pit fields in cell walls of tomato fruit pericarp. Plant Physiol. 122: 775–781.

Orfila, C., Seymour, G.B., Willats, W.G.T., Huxham, I.M., Jarvis, M.C., Dover, C.J., Thompson, A.J. and Knox, J.P. 2001. Altered middle lamella homogalacturonan and disrupted deposition of (1→5)-α-L-arabinan in the pericarp of Cnr, a ripening mutant of tomato. Plant Physiol. 126: 210–221.

Pauly, M. and Scheller, H.V. 2000. O-Acetylation of plant cell wall polysaccharides: identification and partial characterization of a rhamnogalacturonan O-acetyl-transferase from potato suspension-cultured cells. Planta 210: 659–667.

Petersen, T.N., Kauppinen, S. and Larsen, S. 1997. The crystal structure of rhamnogalacturonase A from Aspergillus aculeatus: a right-handed parallel β helix. Structure 5: 533–544.

Pérez, S., Mazeau, K. and du Penhoat, C.H. 2000. The three-dimensional structures of the pectic polysaccharides. Plant Physiol. Biochem. 38: 37–55.

Pilling, J., Willmitzer, L. and Fisahn, J. 2000. Expression of a Petunia inflata pectin methyl esterase in Solanum tuberosum L. enhances stem elongation and modifies cation distribution. Planta 210: 391–399.

Pitson, S.M., Voragen, A.G.J., Vincken, J.P. and Beldman, G. 1997. Action patterns and mapping of the substrate-binding regions of endo-(1→5)-α-L-arabinanases from Aspergillus niger and Aspergillus aculeatus. Carbohydrate Res. 303: 207–218.

Potgieter, M.J. and van Wyk, A.E. 1992. Intercellular pectic protuberances in plants: their structure and taxonomic significance. Bot. Bull. Acad. Sin. 33: 295–316.

Powell, D.A., Morris, E.R., Gidley, M.J. and Rees, D.A. 1982. Conformation and interactions of pectins II. Influence of residue sequence on chain association in calcium pectate gels. J. Mol. Biol. 155: 517–531.

Prade, R.A., Zhan, D.F., Ayoubi, P. and Mort, A.J. 1999. Pectins, pectinases and plant-microbe interactions. Biotech. Genet. Eng. Rev. 16: 361–391.

Puhlmann, J., Bucheli, E., Swain, M.J., Dunning, N., Albersheim, P., Darvill. A.G. and Hahn, M.G. 1994. Generation of monoclonal antibodies against plant cell wall polysaccharides. I. Characterization of a monoclonal antibody to a terminal α-(1→2)-linked fucosyl-containing epitope. Plant Physiol. 104: 699–710.

Qi, X.Y., Behrens, B.X., West, P.R. and Mort, A.J. 1995. Solubilization and partial characterization of extensin fragments from cell-walls of cotton suspension cultures: evidence for a covalent cross-link between extensin and pectin. Plant Physiol. 108: 1691–1701.

Redgwell, R.J. and Selvendran, R.R. 1986. Structural features of the cell wall polysaccharides of onion Allium cepa. Carbohydrate Res. 157: 183–199.

Renard, C.M.G.C., Crépeau, M.-J. and Thibault, J.-F. 1999. Glucuronic acid is directly linked to galacturonic acid in the rhamnogalacturonan backbone of beet pectins. Eur. J. Biochem. 266: 566–574.

Renard, C.C. and Jarvis, M.C. 1999a. Acetylation and methylation of homogalacturonans 1. Optimisation of the reaction and characterization of the products. Carbohydrate Polymers 39: 201–207.

Renard, C.M.G.C. and Jarvis, M.C. 1999b. A cross-polarization, magic-angle-spinning, ^{13}C-NMR study of polysaccharides in sugar beet cell walls. Plant Physiol 119: 1315–1322.

Renard, C.M.G.C., Voragen, A.G.J., Thibault, J.-F. and Pilnik, W. 1991. Studies on apple protopectin V: structural studies on enzymatically extracted pectins. Carbohydr Polym 16: 137–154.

Renard, C.M.G.C., Weightman, R.M. and Thibault, J.F. 1997. The xylose-rich pectins from pea hulls. Int. J. Biol. Macromol. 21: 155–162.

Rhee, S. and Somerville, C. 1998. Tetrad pollen formation in quartet mutants of Arabidopsis thaliana is associated with persistence of pectic polysaccharides of the pollen mother cell wall. Plant J. 15: 79–88.

Roberts, J.A., Whitelaw, C.A., Gonzalez-Carranza, Z.H. and McManus, M.T. 2000. Cell separation processes in plants: models, mechanisms and manipulation. Ann. Bot. 86: 223–235.

Rogers, L.M., Kim, Y.-K., Guo, W., González-Candelas, L., Li, D. and Kolattukudy, P.E. 2000. Requirement for either a host- or pectin-induced pectate lyase for infection of Pisum sativum by Nectria hematococca. Proc. Natl. Acad. Sci. USA 97: 9813–9818.

Rojo, E., León, J. and Sánchez-Serrano, J.J. 1999. Cross-talk between wound signalling pathways determines local versus systemic gene expression in Arabidopsis thaliana. Plant J. 20: 135–142.

Rombouts, F.M. and Thibault, J.F. 1986. Enzymatic and chemical degradation of the fine structure of pectins from sugar beet pulp. Carbohydrate Res. 256: 83–95.

Ros, J.M., Schols, H.A. and Voragen, A.G.J. 1996. Extraction, characterization, and enzymatic degradation of lemon peel pectins. Carbohydrate Res. 282: 271–284.

Rutherford, R. and Masson, P. 1996. Arabidopsis thaliana sku mutant seedlings show exaggerated surface dependent alteration in root growth vector. Plant Physiol. 111: 987–998.

Ryttersgaard, C., Poulsen, J.C.N., Christgau, S., Sandal, T., Dalboge, H. and Larsen, S. 1999. Crystallization and preliminary X-ray studies of β-l,4-galactanase from *Aspergillus aculeatus*. Acta Crystallogr. D55: 929–930.

Scavetta, R.D., Herron, S.R., Hotchkiss, A.T., Kita, N., Keen, N.T., Benen, J.A.E., *et al.* 1999. Structure of a plant cell wall fragment complexed to pectate lyase C. Plant Cell 11: 1081–1092.

Scheller, H.V., Doong, R.L., Ridley, B.L. and Mohnen, D. 1999. Pectin biosynthesis: a solubilized α-1,4-galacturonosyltransferase from tobacco catalyzes the transfer of galacturonic acid from UDP-galacturonic acid onto the non-reducing end of homogalacturonan. Planta 207: 512–517.

Schols, H.A., Bakx, E.J., Schipper, D. and Voragen, A.G.J. 1995. A xylogalacturonan subunit present in the modified hairy regions of apple pectin. Carbohydrate Res. 279: 265–279.

Schols, H.A. and Voragen, A.G.J. 1996. Complex pectins: structure elucidation using enzymes. In: J. Visser and A.G.J. Voragen (Eds.) Pectins and Pectinases, Elsevier Science, Amsterdam, pp. 3–19.

Schmohl, N. and Horst, W.J. 2000. Cell wall pectin content modulates aluminium sensitivity of *Zea mays* (L.) cells grown in suspension culture. Plant Cell Envir. 23: 735–742.

Scott, M., Pickersgill, R.W., Hazlewood, G.P., Gilbert, H.J. and Harris, G.W. 1999. Crystallization and preliminary X-ray analysis of arabinanase A from *Pseudomonas fluorescens* subspecies *cellulosa*. Acta Crystallogr. D55: 544–546.

Sedbrook, J., Hung, K., Carroll, K. and Masson, P. 1998. *sku5*, a mutation in a pectin esterase-like gene, confers an exaggerated right slanting phenotype on agar surfaces. 9th International Conference on Arabidopsis Research, University of Wisconsin-Madison, USA, abstract p109.

Shedletzky, E., Shmuel M., Trainin. T., Kalman, S. and Delmer, D. 1992. Cell wall structure in cells adapted to growth on the cellulose-synthesis inhibitor 2,6-dichlorobenzonitrile. Plant Physiol. 100: 120–130.

Shevchik, V.E. and HugouvieuxCottePattat, N. 1997. Identification of a bacterial pectin acetyl esterase in *Erwinia chrysanthemi* 3937. Mol. Microbiol. 24: 1285–1301.

Sieber, P., Schorderet, M., Ryser, U., Buchala, A., Kolattukudy, P., Métraux, J.-P. and Nawrath, C. 2000. Transgenic arabidopsis plants expressing a fungal cutinase show alterations in the structure and properties of the cuticle and postgenital organ fusions. Plant Cell 12: 721–737.

Simpson, S.D., Ashford, D.A., Harvey, D.J. and Bowles, D.J. 1998. Short chain oligogalacturonides induce ethylene production and expression of the gene encoding aminocyclopropane 1-carboxylic acid oxidase in tomato plants. Glycobiology 8: 579–583.

Sinha, N. and Lynch, M. 1998. Fused organs in the *adherent1* mutation in maize show altered epidermal walls with no perturbations in tissue identities. Planta 206: 184–195.

Smith, B.G. and Harris, P.J. 1999. The polysaccharide composition of Poales cell walls: Poaceae cell walls are not unique. Biochem. System. Ecol. 27: 33–53.

Smith, C.J.S, Watson, C.F., Morris, P.C., Bird, C.R., Seymour, G.B., Gray, J.E., *et al.* 1990. Inheritance and effect on ripening of antisense polygalacturonase genes in transgenic tomatoes. Plant Mol. Biol. 14: 369–379.

Smith, D.L. and Gross, K.C. 2000. A family of at least seven β-galactosidase genes is expressed during tomato fruit development. Plant Physiol. 123: 1173–1183.

Sørensen, S.O., Pauly, M., Bush. M., Skjøt, M., McCann, M.C., Borkhardt, B. and Ulvskov, P. 2000. Pectin engineering: modification of potato pectin by *in vivo* expression of an endo-1,4-β-D-galactanase. Proc. Natl. Acad. Sci. USA 97: 7639–7644.

Steele-King, C.G., Willats, W.G.T. and Knox, J.P. 2000. Arabinogalactan-proteins and cell development in roots and somatic embryos. In: E.A. Nothnagel, A. Bacic and A.E. Clarke (Eds.) Cell and Developmental Biology of Arabinogalactan Proteins, Kluwer Academic Publishers/Plenum, pp. 95–108.

Tegeder, M., Wang, X.-D., Frommer, W.B., Offler, C.E. and Patrick, J.W. 1999. Sucrose transport into developing seeds of *Pisum sativum* L. Plant J. 18: 151–161.

Thibault, J.F., Renard, C.M.G.C., Axelos, M.A.V., Roger, P. and Crepeau, M.J. 1993. Studies of the length of homogalacturonic regions in pectins by acid-hydrolysis. Carbohydrate Res. 238: 271–286.

Thompson, A.J., Tor, M., Barry, C.S., Vrebalov, J., Orfila, C., Jarvis, M.C., *et al.* 1999. Molecular and genetic characterization of a novel pleiotropic tomato-ripening mutant. Plant Physiol. 120: 383–389.

Thompson, J.E. and Fry, S.C. 2000. Evidence for covalent linkage between xyloglucan and acidic pectins in suspension-cultured rose cells. Planta 211: 275–286.

Tibbits, C.W., MacDougall, A.J. and Ring, S.G. 1998. Calcium binding and swelling behaviour of a high methoxyl pectin gel. Carbohydrate Res. 310: 101–107.

Tieman, D.M. and Handa, A.K. 1994. Reduction in pectin methylesterase activity modifies tissue integrity and cation levels in ripening tomato (*Lycopersicon esculentum* Mill.) fruits. Plant Physiol. 106: 429–436.

Torki, M., Mandaron, P., Mache, R. and Falconet, D. 2000. Characterization of a ubiquitous expressed gene family encoding polygalacturonase in *Arabidopsis thaliana*. Gene 242: 427–436.

van Alebeek, G.J.,W.M., Zabotina, O., Beldman, G., Schols, H.A. and Voragen, A.G.J. 2000. Esterification and glycosidation of oligogalacturonides: examination of the reaction products using MALDI-TOF MS and HPAEC. Carbohydrate Polymers 43: 39–46.

VandenBosch, K.A., Bradley, D.J., Knox, J.P., Perotto, S., Butcher, G.W. and Brewin, N.J. 1989. Common components of the infection thread matrix and the intercellular space identified by immunocytochemical analysis of pea nodules and uninfected roots. EMBO J. 8: 335–342.

van der Vlugt-Bergmans, C.J.B., Meeuwsen, P.J.A., Voragen, A.G.J. and van Ooyen, A.J.J. 2000. Endo-xylogalacturonan hydrolase, a novel pectinolytic enzyme. Appl. Envir. Microbiol. 66: 36–41.

van Santen, Y., Benen, J.A.E., Schröter, K.-H., Kalk, K.H., Armand, S., Visser, J. and Dijkstra, B.W. 1999. 1.68-Å crystal structure of endopolygalacturonase II from *Aspergillus niger* and identification of active site residues by site-directed mutagenesis. J. Biol. Chem. 274: 30474–30480.

Vicré, M., Jauneau, A., Knox, J. P. and Driouich A. 1998. Immunolocalization of β(1→4)- and β(1→6)-D-galactan epitopes in the cell wall and Golgi stacks of developing flax root tissues. Protoplasma 203: 26–34.

Vidal, S., Salmon, J.M., Williams, P. and Pellerin, P. 1999. *Penicillium daleae*, a soil fungus able to degrade rhamnogalacturonan II, a complex pectic polysaccharide. Enzyme Microbiol. Technol. 24: 283–290.

Vidal, S., Doco, T., Williams, P., Pellerin, P., York, W.S., O'Neill, M.A., *et al.* 2000. Structural characterization of the pectic polysaccharide rhamnogalacturonan II: evidence for the backbone location of the aceric acid-containing oligoglycosyl side chain. Carbohydrate Res. 326: 277–294.

Vitali, J., Schick, B., Kester, H.C.M., Visser, J. and Jurnak, F. 1998. The three-dimensional Structure of *Aspergillus niger* pectin lyase B at 1.7 Å resolution. Plant Physiol. 116: 69–80.

Wallace, G. and Fry, S.C 1994. Phenolic components of the plant cell wall. Int. Rev. Cytol. 151: 229–267.

Walkinshaw, M.D. and Arnott, S. 1981. Conformations and interactions of pectins. II. Models of junction zones in pectinic acid and calcium pectate gels. J. Mol. Biol. 53: 1075–1085.

Wen, F., Zhu, Y. and Hawes, M.C. 1999. Effect of pectin methylesterase gene expression on pea root development. Plant Cell 11: 1129–1140.

Willats, W.G.T., Marcus, S.E. and Knox J.P. 1998. Generation of a monoclonal antibody specific to (1→5)-α-L-arabinan. Carbohydrate Res. 308: 149–152.

Willats, W.G.T., Gilmartin P.M., Mikkelsen, J.D. and Knox, J.P. 1999a. Cell wall antibodies without immunization: Generation and use of de-esterified homogalacturonan block-specific antibodies from a naive phage display library. Plant J. 18: 57–65.

Willats, W.G.T., Steele-King, C.G., Marcus, S.E. and Knox, J.P. 1999b. Side chains of pectic polysaccharides are regulated in relation to cell proliferation and cell differentiation. Plant J. 20: 619–628.

Willats, W.G.T., Limberg, G., Bucholt, H.C., van Alebeek, G.-J., Benen, J., Christensen, T. M.I.E., *et al.* 2000a. Analysis of pectic epitopes recognised by hybridoma and phage display monoclonal antibodies using defined oligosaccharides, polysaccharides and enzymatic degradation. Carbohydrate Res. 327: 309–320.

Willats, W.G.T., Steele-King, C.G., McCartney, L., Orfila, C., Marcus S.E. and Knox, J.P. 2000b. Making and using antibody probes to study plant cell walls. Plant Physiol. Biochem. 38: 27–36.

Willats, W.G.T., Orfila, C., Limberg, G., Buchholt, H.C., van Alebeek, G.-J.W.M., Voragen, A.G.J., *et al.* 2001. Modulation of the degree and pattern of methyl-esterification of pectic homogalacturonan in plant cell walls: implications for pectin methyl esterase action, matrix properties and cell adhesion. J. Biol. Chem. 276: 19404–19413.

Williams, M.N.V., Freshour, G., Darvill, A.G., Albersheim, P. and Hahn, M.G. 1996. An antibody Fab selected from a recombinant phage display library detects deesterified pectic polysaccharide rhamnogalacturonan II in plant cells. Plant Cell 8: 673–685.

Williamson, G. 1991. Purification and characterization of pectin acetylesterase from orange peel. Phytochemistry 30: 445–449.

Yao, C., Conway, W.S., Ren, R., Smith, D., Ross, G.S. and Sams, C.E. 1999. Gene encoding polygalacturonase inhibitor in apple fruit is developmentally regulated and activated by wounding and fungal infection. Plant Mol. Biol. 39: 1231–1241.

Yu, L. and Mort, A.J. 1996. Partial characterization of xylogalacturonans from cell walls of ripe watermelon fruit: inhibition of endopolygalacturonase activity by xylosylation. In: J Visser and A.G.J. Voragen (Eds.) Pectins and Pectinases, Elsevier Science, Amsterdam, pp. 79–88.

Zhan, D., Janssen, P. and Mort, A.J. 1998. Scarcity or complete lack of single rhamnose residues interspersed within the homogalacturonan regions of citrus pectin. Carbohydrate Res. 308: 373–380.

Plant Molecular Biology **47**: 29–51, 2001.
© 2001 *Kluwer Academic Publishers. Printed in the Netherlands.*

Carbon partitioning to cellulose synthesis

Candace H. Haigler[1,*], Milka Ivanova-Datcheva[2,3], Patrick S. Hogan[2], Vadim V. Salnikov[1], Sangjoon Hwang[1], Kirt Martin[1,4] and Deborah P. Delmer[2]
[1]*Department of Biological Sciences, Texas Tech University, Box 43131, Lubbock, TX 79409-3131, USA (*author for correspondence; e-mail candace.haigler@ttu.edu);* [2]*Section of Plant Biology, University of California, Davis, CA 95616, USA; current addresses:* [3]*Novartis Seeds, Inc., 7240 Holsclaw Road, Gilroy, CA 95020, USA;* [4]*Department of Natural Science, Lubbock Christian University, Lubbock, TX 79407, USA*

Key words: calcium, carbon partitioning, cellulose, cotton fiber, sucrose synthase, phosphorylation

Abstract

This article discusses the importance and implications of regulating carbon partitioning to cellulose synthesis, the characteristics of cells that serve as major sinks for cellulose deposition, and enzymes that participate in the conversion of supplied carbon to cellulose. Cotton fibers, which deposit almost pure cellulose into their secondary cell walls, are referred to as a primary model system. For sucrose synthase, we discuss its proposed role in channeling UDP-Glc to cellulose synthase during secondary wall deposition, its gene family, its manipulation in transgenic plants, and mechanisms that may regulate its association with sites of polysaccharide synthesis. For cellulose synthase, we discuss the organization of the gene family and how protein diversity could relate to control of carbon partitioning to cellulose synthesis. Other enzymes emphasized include UDP-Glc pyrophosphorylase and sucrose phosphate synthase. New data are included on phosphorylation of cotton fiber sucrose synthase, possible regulation by Ca^{2+} of sucrose synthase localization, electron microscopic immunolocalization of sucrose synthase in cotton fibers, and phylogenetic relationships between cellulose synthase proteins, including three new ones identified in differentiating tracheary elements of *Zinnia elegans*. We develop a model for metabolism related to cellulose synthesis that implicates the changing intracellular localization of sucrose synthase as a molecular switch between survival metabolism and growth and/or differentiation processes involving cellulose synthesis.

Abbreviations: CesA, cellulose synthase; *Csl*, cellulose-like synthase (genes); DCB, dichlobenil; DPA, days after anthesis; SPS, sucrose phosphate synthase; SuSy, sucrose synthase; P-SuSy, particulate SuSy; S-SuSy, soluble SuSy

Significance of carbon partitioning to cellulose synthesis

Despite its natural and economic importance, we are only at the threshold of understanding how plants allocate carbon to synthesis of cellulose, or $(1\rightarrow4)\beta$-D-glucan. Cellulose synthesis is a strong, essentially irreversible, carbon sink in plants. Cellulose accounts for 28–30% of dry matter in typical forage grasses (Theander, 1993) and 42–45% of wood (Smook, 1992). The abundance of cellulose is explained by its role as the foundational polymer in plant cell walls. Cell walls

(1) form the plant body, (2) constrain the direction of plant morphogenesis, and (3) confer specialized functions such as water conduction and support (in xylem) and control of transpiration (in guard cells). Consequently, cellulose is the most abundant renewable biomass synthesized on earth, with about 10^{11} tons being synthesized and destroyed each year (Preston, 1974). Throughout history, man has used cellulose extensively in the form of fuel, timber, fiber, forage, and chemical cellulose. Except for burning and building with wood, most large-scale industrial applications place the most value on the cellulose in cellulosic

plant products. Therefore, understanding how carbon partitioning to cellulose synthesis is controlled is a critical question in plant biology, biotechnology, and sustainable agriculture.

This short review will discuss recent research that implicates sucrose synthase (SuSy) as having a major role in channeling substrate UDP-Glc to cellulose synthase, especially during high-rate secondary wall cellulose synthesis. We will develop a general metabolic model for carbon partitioning to cellulose synthesis and discuss the possible roles of SuSy, cellulose synthase, and other enzymes including sucrose phosphate synthase and UDP-Glc pyrophosphorylase. Cotton fibers will be emphasized as a model because their storage metabolism is almost exclusively directed toward cellulose synthesis during secondary wall deposition (see below). Tracheary elements, which represent one of the major cell types constituting wood, will be discussed as another model. In this case, extensive cellulose synthesis occurs along with substantial xylan and lignin synthesis (Haigler, 1985; Smook, 1992). We will also discuss other cases in which genes implicated in control of carbon partitioning to cellulose synthesis have been well described or manipulated in transgenic plants. The emphasis of this chapter will be on genetic regulation placed within its essential developmental and biochemical context. Other recent reviews provide more details about biochemistry and cellular energetics related to carbon partitioning to cellulose synthesis and to other aspects of cellulose synthesis (Ross *et al.*, 1991; Delmer and Amor, 1995; Volman *et al.*, 1995; Blanton and Haigler, 1996; Brown *et al.*, 1996; Kawagoe and Delmer, 1997; Delmer, 1999a, b). Gene families that encode cellulose synthases (*CesA*s) and other cellulose-like synthases (*Csl*s) are covered to a certain extent here, as well as the article by Richmond and Somerville (2001, in this issue).

Carbon partitioning to cellulose increases during cotton fiber development

The development of cotton fibers of *Gossypium hirsutum* and *Gossypium barbadense* has been well reviewed (Basra and Malik, 1984; Ryser, 1985, 1999), and only essential points are mentioned here. Cotton fibers are extraordinary, elongated, trichomes of the cotton seed epidermis; each cell becomes over 2.5 cm long in about 21 days. Toward the end of elongation, secondary wall deposition via enhanced cellulose synthesis begins, and it continues until the fiber dies about

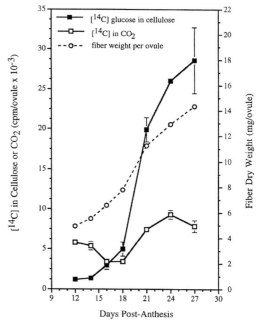

Figure 1. Change in rate of fiber cellulose synthesis, rate of CO_2 evolution (in ovule plus fiber), and fiber weight over time in cultured cotton ovules (*Gossypium hirsutum* cv. Acala SJ-1). The transition between primary and secondary wall synthesis began about 14 DPA, and secondary wall synthesis had not ended when the experiment was terminated on 27 DPA. Methods were as described previously (Roberts *et al.*, 1992). Briefly, ovules with attached fiber were incubated for 4 h in ^{14}C-glucose, CO_2 was trapped in 1 M KOH, and cellulose in fibers stripped from ovules was determined after acetic/nitric digestion (Updegraff, 1969). Fibers were stripped from ovules and freeze-dried before weighing. Data points represent means and standard deviations of 3 replicates containing 6 ovules each. (Graph modified from Martin, 1999.)

40 days after anthesis (DPA). (The precise timing of these developmental transitions and stages depends on genotype and environment.) The thin primary wall is typical of dicots, containing about 20–25% cellulose and non-cellulosic polymers such as xyloglucan and uronic acid-rich polymers (Meinert and Delmer, 1977).

At about 16 DPA in *Gossypium hirsutum*, transient deposition of $(1 \rightarrow 3)\beta$-D-glucan (callose) occurs, and the rate of cellulose synthesis increases more than 100-fold as the wall begins to thicken. At about 24 DPA, elongation ceases and the rate of deposition of virtually pure cellulose into the secondary cell wall increases further, continuing at its maximum rate for at least 10 more days. At this stage, cotton fibers do not store starch or synthesize matrix glycans or lignin; about 80% of the imported carbon is directed to cellulose. This ratio for carbon partitioning during secondary wall deposition can be calculated from the *in planta* data of Mutsaers (1976) and observed directly by comparing incorporation of radiolabel from exogenous ^{14}C-glucose into cellulose and CO_2 in cultured cotton ovules plus fiber (Figure 1; modified from Martin, 1999). Fiber differentiation on cultured ovules includes the same developmental stages as exist *in planta*, and developmental shifts occur in the rate of cellulose synthesis within fibers alone (Carpita and Delmer, 1981; Figure 1) and in respiration of the ovule/fiber system (Figure 1). The scale of increased cellulose synthesis in cultured fibers is reduced because many fibers fail to turn on high-rate secondary wall synthesis, although some cultured fiber walls do become quite thick (Haigler and coworkers, unpublished). At the end of secondary wall deposition, the fibers may undergo programmed cell death (not yet proven) after which they dehydrate. The mature fiber, which is mostly secondary wall, contains at least 90% crystalline cellulose by weight.

Although fibers may import translocated sucrose symplastically through plasmodesmata at the fiber foot (Ryser, 1992, Ruan *et al.*, 1997), cellulose biosynthesis may start from sucrose or glucose and fructose. Invertases are present in cotton fibers in addition to SuSy (Buchala, 1987; Basra *et al.*, 1990; Wäfler and Meier, 1994), and they cannot be excluded as an additional means of sucrose degradation. UDP-glucose (UDP-Glc) was implicated as the probable immediate substrate for cotton fiber secondary wall cellulose synthesis (Franz, 1969; Carpita and Delmer, 1981) and proven to be so for bacterial cellulose synthesis (Ross *et al.*, 1991). However, secondary wall cellulose synthesis may not rely directly on the free pool of cytoplasmic UDP-Glc, but instead use UDP-Glc directly channeled from particulate SuSy in the cell cortex acting degradatively. In this model, developed from work on cotton fibers during secondary wall synthesis (Amor *et al.*, 1995), sucrose is the substrate that is initially required for high-rate secondary-wall cellulose synthesis.

The role of sucrose synthase in intracellular carbon partitioning to cellulose

Overview

The enzyme sucrose synthase (SuSy; EC 2.4.1.13; sucrose + UDP ↔ UDP-Glc + fructose) plays a major role in the degradation of sucrose in plant sink tissues. Although the reaction is freely reversible, under most conditions SuSy catalyzes sucrose cleavage. An advantage of sucrose degradation by SuSy compared to invertase is that the energy of the glycosidic bond is conserved in UDP-Glc. In many non-photosynthetic tissues, soluble SuSy (S-SuSy) exists at high levels in the cytoplasm, where its products may be used in general metabolism and for synthesis of storage polymers such as starch (ap Rees, 1984; Copeland, 1990; Quick and Schaffer, 1996). A recent comprehensive review (Winter and Huber, 2000) includes many aspects of SuSy activity that will not be dealt with here. We will focus on SuSy-mediated carbon metabolism as it relates to intracellular carbon partitioning to cellulose and related polymers.

Years ago degradative SuSy activity was associated with cell wall synthesis. Based on feeding of ^{14}C-sucrose to etiolated pea epicotyls, a coupled reaction between SuSy and glucan synthases was proposed (Rollit and Maclachlan, 1974). This approach was extended to build a model in which soluble SuSy in pea roots (stele, cortex, and apex) was proposed to supply UDP-Glc for polysaccharide biosynthesis, whereas invertases supplied hexoses for respiration (Dick and ap Rees, 1976). Consistent with this model, glucan synthesis in detergent-permeabilized cotton fibers (*Gossypium arboreum*) showed a 140-fold preference for sucrose compared to glucose or UDP-Glc. There was little competition between sucrose and UDP-Glc, which served primarily as a substrate for callose, or $(1{\rightarrow}3)\beta$-D-glucan, synthesis (Pillonel *et al.*, 1980). Sucrolysis via SuSy was also shown to feed glycolysis in several heterotropic systems (Huber and Akazawa, 1986; Xu *et al.*, 1989), and SuSy activity was proposed to mark sink strength in developing bean seeds, potato tubers, and growing roots (Sung *et al.*, 1989). Other evidence for a role for SuSy in providing UDP-Glc to synthesis of cellulose and/or other cell wall polymers was found in maize endosperm (Chourey *et al.*, 1991), tomato roots which store little starch (Wang *et al.*, 1993), and cotton fibers during initiation (Nolte *et al.*, 1995). Even though it was suggested that SuSy might be membrane-associated while acting degradatively in heterotrophic bean cells (Delmer and Albersheim, 1970), soluble SuSy was the only form studied for many years.

SuSy associated with the plasma membrane and/or cortical cytoskeletal elements (P-SuSy) has a special role in contributing UDP-Glc to secondary wall cellulose synthesis and to callose synthesis

We made the surprising finding that a substantial proportion of the total SuSy protein is associated with particulate fractions of developing cotton fibers – in cell wall pellets that sediment at low centrifugal forces ($2000 \times g$) and in high-speed ($100\,000 \times g$) pellets that contain membranes and cytoskeletal elements (Amor *et al.*, 1995). This special, particulate form of SuSy is referred to hereafter as P-SuSy, and soluble SuSy is designated S-SuSy. SuSy is not an integral membrane protein, but its association with the cotton microsomal fraction is quite strong; it could not be removed by high-salt or mild detergent treatments. Semi-permeabilized fibers also used sucrose more efficiently than UDP-Glc as a substrate for cellulose synthesis. Immunofluorescence showed that SuSy could exist in the cortex/cell wall area of secondary wall-stage cotton fibers in patterns consistent with both oriented cellulose microfibril synthesis and predicted punctate sites of callose synthesis (Amor *et al.*, 1995). Therefore, we proposed that P-SuSy channels UDP-Glc to cellulose or callose synthesis at the plasma membrane.

Using cryogenic electron microscopic methods that should preserve *in vivo* protein localization (Nicolas and Bassot, 1993), we obtained further support for this model. These results show that SuSy is abundant all along the cotton fiber surface and just above the cortical microtubules in close proximity to the predicted site of cellulose synthesis in the plasma membrane (Figure 2a). However, SuSy in this location could also be related to the callose synthesis that occurs in cotton fibers (Maltby *et al.*, 1979) and that can also proceed using sucrose as substrate in *in vitro* assays (Pillonel *et al.*, 1980; Amor *et al.*, 1995). The deposition of callose fibrils has been shown to occur on plasma membrane sheets with attached microtubules isolated from tobacco BY-2 protoplasts (Hirai *et al.*, 1998). SuSy labeling with two polyclonal antisera, one made against cotton SuSy (Figure 2a) and one made against bean SuSy (data not shown), was also observed in the cotton fiber wall, most densely in a zone just outside the membrane. The same pattern was observed in true cross-sections of cotton fibers and by using affinity-purified antiserum to cotton SuSy, although labeling density in the latter case was much reduced in all cellular locations (data not shown). Labeling of the fiber wall was not observed using the pre-immune serum for the cotton SuSy antibody or other rabbit polyclonal antibodies (data not shown), so non-specific binding is not a likely explanation. Since SuSy lacks an identifiable signal sequence, localization in the cell wall is not expected, and further work will be required to establish any significance for this observation. We are testing the hypothesis that stress on cotton fibers as they are unavoidably handled before freezing could cause movement of SuSy into the cell wall, possibly attached to the ends of terminated microfibrils.

Coupling between P-SuSy and glucan synthases has the advantages of (1) promoting synthesis of cellulose from sucrose with no additional energy input, (2) avoiding competition for use of UDP-Glc by other pathways, and (3) allowing immediate recycling of UDP, a compound that inhibits the reactions catalyzed by cellulose synthase (Ross *et al.*, 1991) and callose synthase (Morrow and Lucas, 1986). The dependence of cellulose synthesis on sucrose may relate to why the ratio of cellulose to callose synthesized *in vitro* by cotton fiber membrane preparations was enhanced when a combination of MOPS buffer and sucrose was used during membrane isolation and assay (Kudlicka *et al.*, 1995, 1996). Cellulose synthesis was also enhanced when the bacterium *Acetobacter xylinus* was transformed with mung bean SuSy that had been modified to have less regulation via phosphorylation, making the bacterium able to use sucrose for cellulose synthesis (Nakai *et al.*, 1999). Although it is unlikely that the plant SuSy associated directly with the bacterial cellulose synthase, this research supports the idea that the ability to recycle UDP rapidly can enhance cellulose synthesis.

Data from several plant cell systems now support a role for P-SuSy in secondary wall cellulose synthesis. Developing, thick-walled, transfer cells in cotton seeds have high levels of SuSy, although its precise intracellular location is unknown (Ruan *et al.*, 1997). Immunofluorescence of tracheary elements sometimes showed SuSy over developing secondary wall thickenings, which are patterned sites of high-rate cellulose synthesis. However, in other tracheary elements or even in different regions of the same cell immunofluorescence indicating SuSy was diffuse or punctate over the cell surface (Figure 3a). Use of superior cryogenic electron microscopic methods (Nicolas and Bassot, 1993) showed SuSy consistently enriched near the plasma membrane underlying the tracheary element thickenings, whereas it labeled less frequently

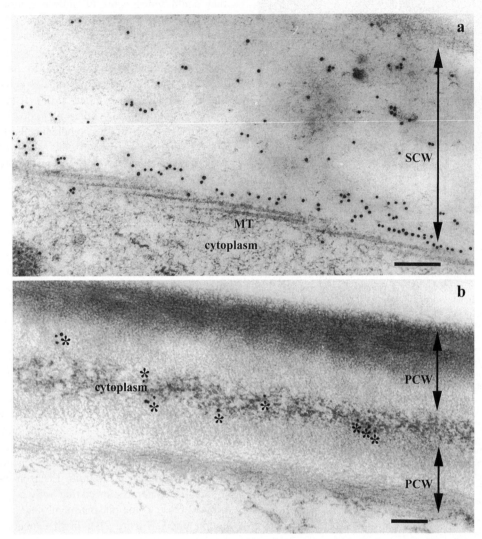

Figure 2. Thin sections of plant-grown cotton fibers labeled with anti-serum to SuSy and 20 nm colloidal gold. A. Secondary wall stage fiber at 30 DPA. SuSy exists between the cortical microtubules (MT) and the thick, secondary cell wall (SCW), which has a width approximately as indicated by the double-headed arrow. SuSy does label in a zone near the plasma membrane, which is the predicted location of cellulose synthases. Labeling was also sometimes observed deeper within the cell wall. B. Primary wall stage fiber at 10 DPA. Here the section is near the top surface of the tubular fiber; the plane of section has just entered the cytoplasm and thin primary wall (PCW) is shown on both sides in a tangential section as marked by double-headed arrows. Sparse colloidal gold/SuSy is highlighted by asterisks. Cryogenic methods were used (Haigler, Grimson and coworkers, in preparation), which preserve cellular structure accurately and greatly hinder molecular movement (Nicolas and Bassot, 1993). Briefly, fibers still attached to a seed fragment were plunged into re-solidifying propane (near $-168\,°C$) cooled by liquid nitrogen, transferred to acetone ($-80\,°C$ for 4 days, then $-40\,°C$ overnight), infiltrated at $-40\,°C$ over 2 weeks by drop-wise addition of resin (a 1:1 mix of Lowicryl K4M and HM20 acrylic resins, Electron Microscopy Sciences, Ft. Washington, PA) followed by 3 changes of 100% resin, flat embedded between two slides (Reymond and Pickett-Heaps, 1983), and polymerized by 360 nm UV irradiation for 2 days at $-20\,°C$ then for 1 day at $4\,°C$. Areas of fiber for sectioning were chosen in a light microscope, cut out, and mounted on a blank resin block. Immunolabeling was performed by standard methods (Brewin *et al.*, 1986) using anti-SuSy (Amor *et al.*, 1995) at 1:1000 and secondary antibody/colloidal gold at 1:100. Similar results on secondary wall stage fibers (with less dense labeling, but including labeling in the cell wall) were obtained with a second polyclonal serum raised against bean SuSy and with anti-serum to cotton SuSy that was affinity-purified with recombinant SuSy bound to a column (data not shown). Substitution of the SuSy pre-immune serum at 1:1000 for anti-SuSy yielded unlabeled sections or sections with very sparse labeling that appeared non-specific (data not shown). Bar is 0.2 μm.

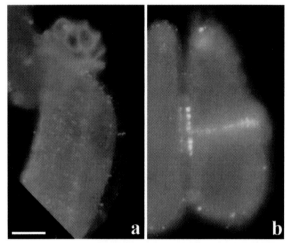

Figure 3. Whole mounts of tracheary elements differentiating in culture from isolated mesophyll cells of *Zinnia elegans* labeled with anti-serum to SuSy and FITC. A. At the top of the cell, SuSy underlies the secondary wall thickenings, but, at the bottom of the cell, SuSy has a punctate pattern. This experimentally induced variability was common in immunofluorescence preparations, and cryogenic immunoelectron microscopy was used to show that SuSy was consistently enriched over the secondary wall thickenings *in vivo* (Salnikov *et al.*, 2001). B. This cell divided twice before differentiation commenced so that 3 cells exist (left, and right top and bottom). SuSy is localized over the mature cell plates, seemingly corresponding to patches of callose-containing plasmodesmata. Fixatives used with equivalent results were: (a) Histochoice, an acidic, buffered solution containing glyoxal, a derivative of formaldehyde, and no chelators to affect calcium concentration (Amresco, Solon, OH), and (b) 3.7% formaldehyde, 1 mM $MgSO_4$, and 5 mM EGTA in 50 mM PIPES, pH 6.9 (closely related to a previously described buffer optimized for microtubules; Doonan and Clayton, 1986). Immunolabeling was performed by standard methods (Doonan and Clayton, 1986) using anti-cotton SuSy (Amor *et al.*, 1995; proved to recognize one band in western blots of *Zinnia* protein; Salnikov *et al.*, in press) at 1:200 and secondary antibody/FITC at 1:200. Substitution of the SuSy pre-immune serum for anti-SuSy resulted in only dull, generalized fluorescence (data not shown). Bar is 30 μm.

and densely between thickenings (Salnikov *et al.*, in press). Since plasma membrane rosettes, which have been labeled with antibodies to cellulose synthase (Kimura *et al.*, 1999), are enriched beneath these same secondary wall thickenings (Haigler and Brown, 1986), we can state that SuSy is preferentially localized near patterned aggregates of cellulose synthase. The changeable immunofluorescence results probably reflect artifacts of specimen preparation and demonstrate the labile nature of *in vivo* SuSy localization (see further discussion below). Double labeling also showed that SuSy was in the same plane as cortical actin in tracheary elements (Salnikov *et al.*, in press).

Evidence is also accumulating that P-SuSy is associated with callose synthesis, in addition to the sometimes punctate SuSy localization in cotton fibers that could reflect sites of callose synthesis (Amor *et al.*, 1995). It was proposed that SuSy localized in companion cells of maize and citrus could provide UDP-Glc for rapid synthesis of callose in sieve elements (Nolte and Koch, 1993). SuSy has also been localized in callose-secreting tapetal cells of maize anthers (Chourey and Miller, 1995). It has been co-localized with callose synthase in callose-containing, forming cell plates of dividing plant cells (Hong *et al.*, 2001). Cultured mesophyll cells of *Zinnia elegans* also show punctate SuSy on mature cell plates, seemingly corresponding to patches of callose-containing plasmodesmata (Figure 3B).

Does P-SuSy have the same role during primary wall cellulose synthesis?

Available data support the hypothesis that a smaller amount of P-SuSy supports primary wall cellulose synthesis, but this relationship may be more facultative than during secondary wall synthesis. The ratio of S-SuSy to P-SuSy has not been quantified in young cotton fibers undergoing elongation via primary wall synthesis, but high levels of total SuSy do exist from the very beginning of fiber initiation (Nolte *et al.*, 1995; Ruan *et al.*, 1997). Contrary to the wild type, a fiberless mutant of cotton does not show enriched SuSy in seed epidermal cells (Ruan and Chourey, 1998), although the specific mutation may exist upstream of the expression of genes encoding SuSy in the fiber initiation program. Electron microscopic immunolocalization showed that SuSy is sparsely distributed close to the plasma membrane of elongating cotton fibers (Figure 2B); in this location it could be associated with primary wall cellulose synthases that are typically more widely spaced in the plasma membrane than secondary wall cellulose synthases (Haigler, 1985).

P-SuSy was also detected in other cell types containing cells engaged in primary wall synthesis including cotton roots, pea and bean stems, and cultured tobacco cells (Amor *et al.*, 1995). In a careful developmental and experimental analysis, the amounts of SuSy mRNA and SuSy activity were highest in various tissues of growing carrot leaves, stems, and tap roots that would have been synthesizing primary and secondary walls (Sturm *et al.*, 1995; 1999). In protoplasts of growing carrot suspension cultures, 25–50% of total SuSy was associated with the membrane fraction (Sturm *et al.*, 1999). On an overall basis, maize endosperms had 14% P-SuSy during cell expansion and only 3% P-SuSy during rapid starch synthesis (Carlson and Chourey, 1996). Maize pulvini (a special region of the internode involved in gravitropism) that were gravi-stimulated so that cell expansion via primary wall synthesis was induced showed a 10-fold increase in P-SuSy (Winter et al., 1997).

Transformed plants with changed SuSy expression provide a powerful means of testing the role of SuSy in primary wall synthesis. The expression of a reporter gene linked to the promoter of a maize gene for SuSy (Sh1) increased over 400-fold when maize protoplasts began to regenerate primary walls, and this increase could be blocked by the cellulose-synthesis inhibitor dichlobenil (DCB) (Mass et al., 1990). However, DCB did not inhibit expression of the gene for the main form of carrot SuSy (Susy*Dc1) (Sturm et al., 1999), indicating species-specific SuSy regulation. The sh1 mutation of maize causes degeneration of some, but not all, expanding primary cell walls in developing endosperm. The cells that degenerate have particularly thin, fragile, primary walls, and the synthesis of these is hypothesized to depend on plasma-membrane-associated Sh1 (Chen and Chourey, 1989; Carlson and Chourey, 1996). In dwarfed carrot plants transformed with an antisense gene construct for SuSy, tap roots having 18–23% of normal SuSy activity contained only about 63% of the normal crystalline cellulose (per gram fresh weight). They also contained expanded parenchyma cells, probably due to insufficient reinforcement of their primary walls by cellulose microfibrils (Tang and Sturm, 1999). Tomato fruits with only 1% residual soluble SuSy activity after antisense suppression showed no change in cellulose or cross-linking glycans, but the methanolysis procedure used would not be the most definitive for this evaluation (Chengappa et al., 1999). Tomato fruits with almost no residual, soluble SuSy activity do finally grow to normal size (D'Aoust et al., 1999), but it was not determined whether their primary wall composition might have become different.

We recently determined that potato tubers with SuSy expression repressed by antisense techniques showed reduced cellulose content (Figure 4). The reduction was substantial in the line T-112 that had only 4% of the SuSy activity and 66% of the starch content found in wild-type tubers (Zrenner et al., 1995). The transgenic tuber cell walls also showed increased uronic acids, but no increases in other non-cellulosic sugars were observed, suggesting compensation for reduced primary wall cellulose by synthesis of additional pectin. Similar compensation has been observed in primary walls of DCB-adapted plant cell cultures with reduced cellulose (Shedletzky et al., 1992) and in tobacco leaves where expression of a cellulose synthase gene was lowered by virus-mediated gene silencing (Burton et al., 2000). These compensatory changes suggest that, although the wall polysaccharides are extracellular, there still may be complex feedback events determining how carbon is partitioned between various wall polymers. In this regard, we recently observed a substantial increase in the level of cellulose synthase (CesA) protein in cotton fibers treated with a novel herbicide, CGA 325'615, which inhibits synthesis of crystalline cellulose (Peng et al., 2001).

In addition to a possible role in facilitating primary wall cellulose synthesis, it has also been proposed that SuSy can associate with Golgi membranes to facilitate mixed-linkage $(1\rightarrow3),(1\rightarrow4)\beta$-D-glucan synthesis in maize (Buckeridge et al., 1999). Since these authors did not detect SuSy in the Golgi fraction of soybeans, they hypothesized that the β-glucan synthase could have a cellulose synthase ancestor so that it uniquely retained the ability to associate with SuSy. We also were unable to localize SuSy in the Golgi apparatus of tracheary elements of the dicot Zinnia elegans (Salnikov et al., in press), which, in contrast to maize, was expected to be synthesizing a type of xylan that would not require UDP-Glc as substrate.

How might the SuSy gene family relate to control of carbon partitioning to polysaccharide synthesis?

Most plants contain at least two, and perhaps three, genes encoding isoforms of SuSy (reviewed in Winter and Huber, 2000). There is ongoing work in several laboratories on mechanisms that control the intracellular location of SuSy. First, one might consider that S- and P-SuSy could be encoded by different genes subject to different modes of regulation. Although the data are limited, it would appear that no single gene is responsible for encoding P-SuSy. At least two isozymes from distinct genes encoding SuSy were associated with the plasma membrane of maize (Carlson and Chourey, 1996). When two genes encoding SuSy are expressed in the same cell, the proteins form homo- or heterotetramers (Chen and Chourey, 1989; Koch et al., 1992; Guerin and Carbonero, 1997), suggesting that the isozymes are interchangeable in at least some cellular roles. However, because both classes of maize genes (Sh1 and Sus1) are conserved among grasses

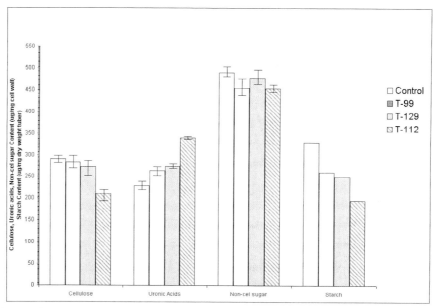

Figure 4. Cell wall composition and starch content of potato tubers expressing anti-sense SuSy constructs. Antisense constructs, growth and harvesting of tubers were as described in Zrenner *et al.* (1995). Tubers were freeze-dried, ground to a powder, and the resulting wall pellet was extracted with DMSO to remove starch, 2× in methanol/chloroform/H_2O (14:4:3), 1× in 10% methanol, and 2× for 1 h at 70 °C in 90% methanol, followed by extensive H_2O washing. Cellulose is that fraction of carbohydrate that remained insoluble after acetic-nitric digestion (Updegraff, 1969). Non-cellulosic neutral sugars were determined by GLC of alditol acetates (Blakeney *et al.*, 1983), and uronic acids were determined by the method of Blumenkrantz and Asboe Hansen (1973). Results are the average of 5 independent replicates. The levels of starch indicated are taken from the data of Zrenner *et al.* (1995) for these transgenic and control lines.

and show different rates of divergence in their exons compared to their introns, it has been argued that both have critical roles at least under some physiological conditions (Shaw *et al.*, 1994).

Dicots have two additional classes of genes encoding SuSy that are distinct from the *Sh1* and *Sus1* classes of maize and other monocots (Shaw *et al.*, 1994; Sebkova *et al.*, 1995; Fu and Park, 1995; Sturm *et al.*, 1999). However, no previous phylogenetic analyses could be interpreted with specific knowledge of genes encoding SuSy that were expressed in secondary-wall-synthesizing cells, especially because vascular expression was attributed to a role in phloem unloading. In a cotton fiber cDNA library, we detected at least 3 distinct genes encoding SuSy, but the deduced amino acid sequences are all 93% identical to each other and tryptic peptide profiles on HPLC for S- and P-SuSy are virtually indistinguishable (Delmer and coworkers, unpublished). Since cotton is an allotetraploid (Wendel *et al.*, 1999), at least two of these genes encoding SuSy could represent alleles from the two distinct genomes. Figure 5 shows an unrooted cladogram derived from comparing the deduced amino acid sequence of one of the representative cotton genes encoding SuSy with other SuSy proteins. The SuSy proteins from graminaceous monocots fall into a distinct clade, while cotton SuSy shows most similarity with one *Arabidopsis* SuSy and with two other dicot proteins that are expressed during nodulation of soybean or actinorhizal associations in alder. While these comparisons clearly indicate some divergence between genes encoding SuSy from dicots and grasses, they offer no insight into possible distinct roles for individual genes coding for S-Susy vs. P-SuSy. Furthermore, the N-terminal sequences in the predicted region of serine phosphorylation (see below) are conserved in nearly all SuSy proteins, again offering no clue to possible distinct motifs for differential phosphorylation (inset in Figure 5). In addition, deduced SuSy proteins do not show obvious sites for prenylation or myristoylation that might promote membrane association. One or two transmembrane helices might exist in maize SuSy sequences, but this alone could not explain why both isoforms exist in soluble or membrane-associated forms (Carlson and Chourey, 1996). In summary, the data at present do not support the idea that distinct genes encode P- and S-SuSy.

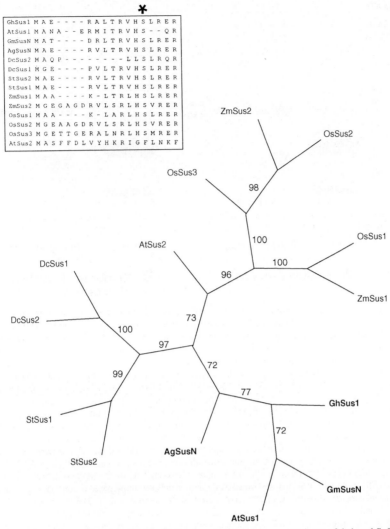

Figure 5. An unrooted cladogram constructed as described by Holland *et al.* (2000) from comparisons of deduced SuSy amino acid sequences. Since the 3 distinct SuSy genes expressed in cotton fibers are extremely similar, only one is included in this analysis. Gh, *Gossypium hirsutum* (accession number U73588); At, *Arabidopsis thaliana* (P49040 and Q00917); Gm, *Glycine max* (P49034); Dc, *Daucus carota* (P49035 and P049039); St, *Solanum tuberosum* (P10691 and P94039); Zm, *Zea mays* (P04712 and P49036); Os, *Oryza sativa* (P30298, P31924 and Q43009). The inset shows a pile-up of the SuSy N-terminal amino acids with predicted site of phosphorylation marked by an asterisk (*).

What other mechanisms may regulate intracellular location of SuSy?

Phosphorylation

Winter and Huber (2000) discuss extensively the possibility that SuSy localization is controlled by its state of phosphorylation. It is possible that SuSy phosphorylation, which depends on a conserved N-terminal domain as documented in many species and tissues, is an important part of SuSy regulation in sink tissues (Zhang *et al.*, 1999; Loog *et al.*, 2000). Maize S-SuSy was phosphorylated *in vivo* and *in vitro* on Ser-

15 (of the protein encoded by *ZmSus2*) by a Ca^{2+}- and phospholipid-dependent protein kinase (Shaw *et al.*, 1994; Koch *et al.*, 1995; Huber *et al.*, 1996; Lindblom *et al.*, 1997; Loog *et al.*, 2000). Consequently, the K_m for UDP and sucrose, but not UDP-Glc or fructose, was lowered, thus favoring the sucrose cleavage reaction (Huber *et al.*, 1996). However, later research showed that the magnitude of the effect was so low that it might be physiologically insignificant (Winter *et al.*, 1997). Soybean nodule S-SuSy was phosphorylated *in vitro* by a Ca^{2+}-dependent protein kinase (CDPK), but phosphorylation of the recombinant or native en-

38

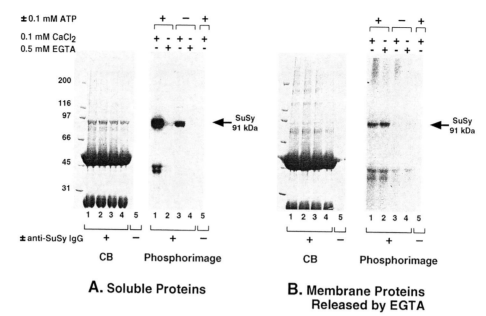

± 0.1 mM ATP
0.1 mM CaCl$_2$
0.5 mM EGTA

200
116
97
66
45
31

← SuSy 91 kDa

← SuSy 91 kDa

1 2 3 4 5   1 2 3 4 5   1 2 3 4 5   1 2 3 4 5

± anti-SuSy IgG   +   −   +   −   +   −   +   −

CB   Phosphorimage   CB   Phosphorimage

A. Soluble Proteins   **B. Membrane Proteins Released by EGTA**

Figure 6. In vitro phosphorylation of SuSy by cotton fiber extracts. Extracts were prepared from fibers harvested from greenhouse-grown cotton plants (*Gossypium hirsutum* cv. Coker 130) at 21 DPA. Locules were frozen in liquid N$_2$ and stored at −80 °C until fibers were detached from seeds and ground to a fine powder under liquid N$_2$. The frozen powder was extracted at 4 °C with extraction buffer (EB; 2 ml per gram fiber fresh weight) that contained 50 mM Hepes/KOH pH 7.3, 0.5 M sucrose, 1 mM DTT, 0.03% Brij 58, 0.5% insoluble PVP, a protease inhibitor cocktail (PI) (Complete, Mini, EDTA-free; Boehringer Mannheim), and the phosphatase inhibitors (PhI) NaF (10 mM) and Na$_3$VO$_4$ (1 mM) plus Microcystin-London Resin White at 0.5 μM (Calbiochem, La Jolla, CA). The soluble fraction (A) and EGTA washes (B) were dialyzed overnight against 50 mM MOPS buffer pH 7.5, 5 mM MgCl$_2$, 0.1 mM CaCl$_2$, 2 mM DTT containing PI and PhI cocktails. Dialysis was performed in a ratio of 1:500 with 3 buffer changes followed by concentration of the extracts on Centricon 50. Extracts (0.5–1 mg total protein per reaction) were incubated with 1–3 μCi [γ-^{32}P]ATP (Amersham Biotech, 6000 Ci/mmol) per reaction in the presence or absence of unlabeled 0.1 mM ATP (as indicated), 0.1 mM CaCl$_2$ and 0.5 μM Microcystin London Resin White. Various concentrations of CaCl$_2$ or EGTA up to 2 mM were applied to the reaction mixture to explore the effect of Ca^{2+} on S-Susy and P-Susy phosphorylation. After 10 min incubation at 25 °C, the reaction was terminated by adding 10 mM EDTA, and the phosphorylated SuSy band was identified after immunoprecipitation, SDS-PAGE and electroblotting and autoradiography of the blotted membranes. SuSy was immunoprecipitated in IP buffer containing 150 mM NaCl, 50 mM Tris pH 7.8, 1 mM EGTA, 1% Triton X-100, 0.5% sodium deoxycholate, 0.05% SDS (Todorov *et al.*, 1995) supplemented with PI and PhI inhibitors as described. Protein A Sepharose (Amersham Pharmacia Biotech; prepared according to the manufacturer's instructions; 50 μl/sample) was incubated with 20 μl antibody in 200 μl IP buffer (4 °C, several hours). The slurry was washed 2× in IP buffer, and incubated overnight at 4 °C with protein extracts. The immune complex was collected by brief centrifugation and washed 3× in PBS, 3× in IP buffer containing 0.5 M NaCl and then 3× with PBS. The washed beads were extracted in 'Novex' sample buffer and resolved in Nu PAGE 4 to 12% gradient gels as described above. CB, Coomassie Blue-stained gels.

zyme did not have a major effect on enzyme kinetics or sucrose degradation (Zhang and Chollet, 1997). In contrast, phosphorylation is clearly important in regulating the kinetic properties of mung bean SuSy (Nakai *et al.*, 1999). Other studies showed that maize S-SuSy was more heavily phosphorylated *in vivo* and that phosphorylated SuSy was less hydrophobic and less likely to associate with membranes (Winter *et al.*, 1997, Zhang *et al.*, 1999).

It is reasonable to assume that a change in the phosphorylation state that affects SuSy conformation and substrate binding may also either promote or prevent membrane association, and we tested this further in cotton fiber extracts. Here, S-SuSy was phosphorylated *in vitro* in a Ca^{2+}-dependent manner by an endogenous kinase present in the soluble fraction (Figure 6A). However, phosphorylation of P-SuSy by an endogenous kinase that elutes along with P-SuSy upon EGTA treatment of the particulate fraction occurred in a Ca^{2+}-independent manner (Figure 6B). Therefore, P-SuSy and S-SuSy might be subject to different control. Unlike the results from maize S-SuSy (Winter *et al.*, 1997), we found no difference in the level of phosphorylation of S-Susy vs. P-SuSy (standardized to amount of SuSy protein) isolated from cotton fibers labeled *in vivo* with ^{32}P-orthophosphate (Figure 7). Further complicating interpretation of these results, profiles from 2-D TLC separations of tryptic peptides showed different patterns when SuSy was phosphorylated *in vitro* vs. *in vivo*. In both cases, S-

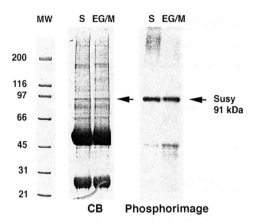

Figure 7. In vivo phosphorylation of SuSy. Ovules with their associated fibers were cultured at 30 °C for 14–18 days (Haigler *et al.*, 1991), then transferred to phosphate-free medium for 24 h, and finally incubated 24 h in medium supplemented with 0.5 mCi/ml [³²P]orthophosphate (carrier- and acid-free, 10 mCi/ml, 8000 Ci/mmol; Amersham Biotech). The ovules were washed briefly in cold phosphate containing PhI cocktail, and the fibers were detached immediately on ice in EB containing the same PI and PhI cocktails plus 0.5 μM Microcystin LRW and homogenized and extracted as described in the legend to Figure 6. S-SuSy (S) and P-SuSy released by EGTA washing of membranes (EG/M) were subjected to immunoprecipitation and SDS-Nu PAGE electrophoresis as described in the legend to Figure 6. Incorporation of ³²P was detected by autoradiography on Bio-Max MR X-ray films (Kodak). Gels were exposed 1–3 days at −80 °C. To verify the SuSy band, half of the immunoprecipitated proteins of each sample were resolved on the same gel and electroblotted onto nitrocellulose membrane and probed as described above using anti-Susy antibody (not shown). CB, Coomassie Blue-stained gels.

and P-SuSy were phosphorylated exclusively on Ser residues such that SuSy from both fractions contained the same major phosphopeptide (presumably that for Ser-11). However, a second phosphorylated peptide was detected in S- and P-SuSy phosphorylated *in vitro* (not shown). Sites in addition to this conserved Ser can also be phosphorylated and may influence at least the kinetic properties of SuSy (Zhang *et al.*, 1999; Anguenot *et al.*, 1999). These results emphasize that conclusions drawn from studies of *in vitro* phosphorylation may be difficult to extend to the *in vivo* situation. Therefore, at present, the role of phosphorylation of SuSy in regulation of intracellular localization remains inconclusive.

Intracellular calcium levels

Our evidence from cotton fibers suggests that Ca^{2+} levels affect the solubility of SuSy in a complex manner. After extraction of cotton fibers in buffer containing 200 nM free Ca^{2+}, both S- and P-SuSy were detected. In most experiments, all or at least a sub-

stantial portion of the P-SuSy could be eluted from the particulate fraction by EGTA treatment (Figure 8A). After extraction of fiber in buffer containing EGTA, very little P-SuSy was detected, presumably because it eluted from the particulate fraction during extraction. Furthermore, this 'eluted P-SuSy' was selectively precipitated by adding Ca^{2+} to the supernatant, leaving only the original S-SuSy in solution (Figure 8B). We do not yet understand why 'eluted P-SuSy' behaves differently in solutions containing Ca^{2+}. However, it is interesting that the EGTA extracts of fiber membranes or the 'eluted P-SuSy' precipitated by Ca^{2+} both contain similar sets of other proteins, including Ca^{2+}-independent kinase, tubulin, actin, and annexins (D. Delmer and M. Datcheva, unpublished). Thus, the level of Ca^{2+} may well influence the particulate nature of SuSy by modulating its association with membranes, with glucan synthases in membranes, with F-actin, and/or with other proteins that may occur in particulate fractions.

These results may relate to a rise in intracellular calcium that could occur at the transition to secondary wall synthesis in cotton fibers and tracheary elements and facilitate formation of a P-SuSy protein complex. Secondary wall deposition in tracheary elements requires uptake of extracellular calcium through calcium channels (Roberts and Haigler, 1990), and the levels of calmodulin and calmodulin-binding proteins increase prior to secondary wall deposition (Kobayashi and Fukuda, 1994). Indirect evidence suggests that similar mechanisms may operate in cotton fibers. Both induction of Ca^{2+}-dependent callose synthesis and an oxidative burst (inferred from onset of H_2O_2 production) occur at the primary to secondary wall transition in cotton fibers (Maltby *et al.*, 1979; Potikha *et al.*, 1999). In addition, expression of a Rac gene is induced (Delmer *et al.*, 1995), and, in animals, Rac activates the NADPH oxidase involved in such an oxidative burst. Since we showed that cotton annexin may interact with callose synthase (and actin in some systems) and may be phosphorylated (Andrawis *et al.*, 1993), one possibility is that annexin acts as a bridge between SuSy and actin and/or glucan synthases in the plasma membrane. Such an association could be regulated by Ca^{2+}, either directly or through changes in protein phosphorylation state.

Interaction with the cytoskeleton

Maize SuSy was detected in the detergent-insoluble fraction of microsomal membranes, actin (but not tubulin) co-immunoprecipitated with S-SuSy, and

40

A. Extraction buffer contains 200nM free Ca^{2+}

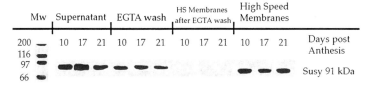

B. Extraction buffer contains 5mM EGTA

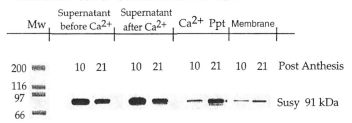

Figure 8. The effect of Ca^{2+} on distribution of S-Susy and P-SuSy in cotton fiber extracts. Cotton locules were harvested from greenhouse plants at 10, 17, or 21 DPA and handled and processed as described in the legend to Figure 6 with the following modifications. For the experiment in A, the extraction buffer also included 0.5 mM EGTA and 0.4 mM $CaCl_2$ (ca. 200 nM free Ca^{2+}; calculated as described in Hayashi *et al.*, 1987). For B, this was replaced by 5 mM EGTA. The fiber homogenate was filtered through several layers of MiraCloth (Calbiochem, La Jolla, CA) and centrifuged at $10\,000 \times g$ for 15 min. The pellet was extracted a second time with EB, and the combined extracts were centrifuged at $100\,000 \times g$ for 1 h. This pellet consisting of membranes and cytoskeletal elements was resuspended in 50 mM BTP-MES pH 7.0 and further extracted ($2\times$, 15 min each) with 10 mM EGTA in 50 mM BTP-MES buffer pH 7.0, 1 mM DTT, and the PI cocktail. EGTA washes were separated from particulate matter by centrifugation at $400\,000 \times g$ for 15 min. Soluble proteins and EGTA washes were concentrated on Centricon 50 (Amicon) prior to electrophoresis. The Ca^{2+} precipitate (Ca^{2+} Ppt) was obtained by adding 15 mM $CaCl_2$ to supernatants, stirring (15 min, 4 °C), and centrifuging ($25\,000 \times g$, 15 min). For western blotting, proteins were separated with 4–12% gradient Nu PAGE Bis-Tris gels (Novex) with SDS-MOPS running buffer under reducing conditions, and electroblotting to membranes was carried out in Nu PAGE transfer buffer. The blots were probed with a 1:10 000 dilution of polyclonal antibody raised against isolated SuSy from *Vicia faba*. Goat anti-rabbit horseradish peroxidase-conjugated second antibody (Sigma) was used as second antibody, and positive bands were detected by enhanced chemiluminescence.

SuSy bound to F-actin polymerized *in vitro* in a 1:5 ratio (possibly indicating that 1 SuSy tetramer binds at intervals of 20 actin monomers). The binding of SuSy to G- and F-actin was apparently not related to SuSy phosphorylation state, but was stimulated by high levels of sucrose (Winter *et al.*, 1998; Winter and Huber, 2000). A peptide in maize SuSy with homology to known actin-binding peptides has also been identified (Winter and Huber, 2000). We have already mentioned two lines of evidence suggesting that SuSy may interact with the cytoskeleton in cotton fibers: (1) immunofluorescence showing SuSy coaligned with helical cellulose microfibrils (Amor *et al.*, 1995), and (2) co-elution of P-SuSy with actin and tubulin in the presence of EGTA. Helical microfibrils in cotton fiber secondary walls are paralleled by the cortical microtubule network, and actin microfilaments exist in the vicinity of the microtubules in an incompletely understood relationship (Seagull, 1993). Increasing tubulin at the onset of secondary wall deposition (Kloth, 1989; Dixon *et al.*, 1994) correlates with the increase in P-

SuSy at the same stage. In addition, a form of tubulin that behaves as an integral membrane protein has been identified in plants (Laporte *et al.*, 1993). This type of tubulin may interact with cellulose synthase such that a solubilized complex isolated with anti-tubulin antibodies can synthesize cellulose using either sucrose or UDP-Glc as substrate (Prof. Mizuno, Tokyo University, personal communication).

Consistent with these observations, SuSy is preferentially localized to patterned sites of secondary wall cellulose synthesis in differentiating tracheary elements where it exists in the same plane with actin between the cortical microtubules and the plasma membrane. However, cortical actin also exists between thickenings, so binding to actin alone cannot explain the preferential localization of SuSy over the thickenings (Salnikov *et al.*, in press). It may well be that other proteins that co-precipitate with P-SuSy (see discussion above) have a role in establishing its patterned localization in tracheary elements. One cannot yet rule out the possibility that SuSy could bind directly to cellulose synthases, which also increase dramatically at the initiation of secondary wall formation (Pear *et al.*, 1996). However, this possibility may be harder to rationalize with the idea that cellulose synthases move rapidly through the plasma membrane as they spin out cellulose microfibrils (reviewed in Blanton and Haigler, 1996).

Can SuSy gene expression regulate the rate and extent of cellulose synthesis?

All available data suggest that SuSy is a critical partner in high-rate secondary wall cellulose synthesis. A second question is to what degree, if any, the rate and extent of secondary wall cellulose synthesis are regulated at the level of expression of genes encoding SuSy. In several heterotrophic systems, SuSy activity has been correlated with mRNA and/or protein levels (Nguyen-Quoc *et al.*, 1990; Crespi *et al.*, 1991; Sebkova *et al.*, 1995). However, SuSy activity often seems to be greatly excessive compared to the actual metabolic requirements for starch synthesis. Each successive mutation of maize *Sh1* or *Sus1* reduced endosperm starch content by only an additional 22–25% (Chourey *et al.*, 1998). In commercial tomato fruits, SuSy enzyme activity assayed *in vitro* was at least 10 times greater than maximum growth rate, including respiration (Sun *et al.*, 1992). Tomato fruits with only 1–2% residual soluble SuSy activity after antisense suppression showed no change in starch or sugar accumulation (Chengappa *et al.*, 1999; D'Aoust *et al.*, 1999). In developing potato tubers, no change in starch content was observed until SuSy was suppressed to below 30% of control levels (Zrenner *et al.*, 1995).

Similarly, SuSy activity may have only weak control over the synthesis of cellulose, at least during primary wall synthesis. Although perhaps partially explained by the discovery of a third maize gene encoding SuSy (Carlson *et al.*, 2000), a double *sh1 sus1-1* maize mutant with only 0.51% of normal SuSy activity in the endosperm grows normally (Chourey *et al.*, 1998). As discussed previously, we showed that potato tubers with only 4% residual soluble SuSy activity had about 70% of the wild-type level of (mostly primary wall) cellulose. Similarly, in developing carrot tap roots, transgenic plants with repressed SuSy activity (0.7–29% of wild type) showed stronger reduction in starch content (about 38% of wild type) than in crystalline cellulose content (about 63% of wild type). Notably, only 0.7% of normal, soluble SuSy activity supported about the same levels of starch

and cellulose content as 29% residual activity (Tang and Sturm, 1999). Such observations can be explained by metabolic plasticity, enzymes in excess of requirements, or low flux control coefficient for a particular enzyme (Chengappa *et al.*, 1999). However, it will also be necessary to manipulate expression and activity of SuSy in cells or tissues engaged primarily in secondary wall cellulose synthesis to obtain more direct evidence about this case. Evidence has been presented that hypoxic wheat roots have increased SuSy activity (as determined by histochemistry) and increased (150–200%) cellulose content, including in endodermal cells with secondary cell walls (Albrecht *et al.*, 2000).

Structure of cellulose synthases that may relate to carbon partitioning to cellulose synthesis

Studies on cellulose synthesis entered a new era with the identification of two cotton genes (now called *GhCesA-1* and *GhCesA-2*) that are homologues of the *CesA* genes encoding the catalytic subunit of bacterial cellulose synthases (Pear *et al.*, 1996; Haigler and Blanton, 1996; reviewed in Delmer, 1999a). These two genes, while clearly distinct, are highly homologous. They also show high homology to *CesA* genes now identified in many plants including an additional one, *GhCesA-3*, that may be expressed preferentially during cotton fiber primary wall synthesis. (The *CesA* gene family has been discussed recently in Holland *et al.*, 2000 and Richmond and Somerville, this issue; for a complete listing, see www.cellwall.stanford.edu/cellwall/.) Northern analysis of total RNA isolated from roots, leaves, and flowers revealed that *GhCesA-1* and *GhCesA-2* had only low expression in non-fiber tissues. Both genes were highly expressed from the onset of cotton fiber secondary wall synthesis, with expression continuing at high levels throughout this phase. The discovery that *Arabidopsis* mutants impaired in cellulose deposition map to *CesA* loci provided key genetic evidence for a role for these genes in cellulose synthesis (Arioli *et al.*, 1998; Taylor *et al.*, 1999).

With the complete sequencing of the *Arabidopsis* genome (Arabidopsis Genome Initiative, 2000) and extensive ESTs available for maize, it is now clear that both of these species contain at least 10 distinct *CesA* genes. This will probably hold true for most other plants (Holland *et al.*, 2000; Figure 9). There may be several reasons for the existence of so

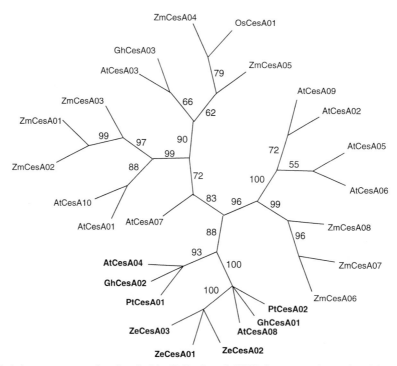

Figure 9. An unrooted cladogram constructed as described by Holland *et al.* (2000) from comparisons of partial deduced CesA amino acid sequences. The partial sequences used spanned the region just downstream of the second conserved D residue to the C-terminal end of the proteins. For GhCesA-1 and AtCesA-2, these regions begin with amino acid residues 469 or 572, respectively. Accession numbers for all sequences are listed in Holland *et al.* (2000), except for the following additional sequences: GhCesA-3 (AAD39534); ZeCesA-1, -2, and -3 (AF323039, AF323040, AF323041).

many *CesA* genes in one species. Emerging data indicate that at least some of them have tissue-specific or developmental-stage-specific expression, and the clustering of deduced proteins in phylogenetic analyses has some predictive value for cell-specific expression (Holland *et al.*, 2000; Richmond and Somerville, 2000). One entire clade (labeled in bold in Figure 9) has now been analyzed in this way, and all members including GhCesA-1 and GhCesA-2, two CesA proteins encoded by genes identified from developing xylem cDNA libraries of poplar, and two *Arabidopsis* CesA proteins are uniquely expressed in cells undergoing secondary wall cellulose synthesis. However, it should be noted in Figure 9 that two other genes associated with secondary wall synthesis in vascular tissues are not included in this clade, *AtCesA-7* (Turner and Somerville, 1997; Taylor *et al.*, 1999) and *ZmCesA-8* (Holland *et al.*, 2000). Partial analysis of deduced proteins in another clade indicates expression predominantly in cells undergoing primary wall synthesis. More recently, we used PCR with primers based on conserved sequences in other *CesA* genes to identify three distinct *CesA* mRNAs that were expressed during rapid, secondary wall cellulose synthesis as tracheary elements differentiated in culture from isolated mesophyll cells of *Zinnia elegans*. The three genes also had different 3′-untranslated regions (S. Hwang and C. Haigler, unpublished), and their deduced protein sequences, when added to the phylogenetic analysis (Figure 9), fall within the same clade as most of the other dicot CesA proteins involved in secondary wall synthesis. This observation further confirms the usefulness of this type of analysis.

It is clear that more than one *CesA* gene can be expressed in the same cell type at the same time in development. One reason for this might be to provide redundancy for protection against mutation in such important genes or to provide a mechanism to enhance rates of cellulose synthesis when, for example, *GhCesA-1* and *-2* are expressed at the same time in the cotton fiber. Another emerging possibility is that two similar, but non-identical, pairs of CesA proteins must function together in order to allow rosette assembly or to create two non-identical catalytic sites that function to catalyze polymerization of alternating residues in the glucan chain. This latter possibility represents a variation on other two-site models for cellulose synthase where the two sites have been proposed to reside within the same catalytic subunit (Vergara and Carpita, 2001, in this issue).

With respect to carbon partitioning, it remains to be determined whether SuSy functions to donate carbon to all forms of CesA. It may be that some CesA proteins have certain motifs that favor direct interaction (or at least involvement) of SuSy as donor for UDP-Glc. If SuSy localizes to the plasma membrane via interaction with a complex containing cortical actin and not directly with CesA, then it may be that any CesA can function coordinately with SuSy. However, as indicated earlier, SuSy seems to assume a more important role for secondary cell wall synthesis, and future work will be needed to see if a *CesA* gene involved in primary wall synthesis can functionally complement a secondary wall *CesA* gene and function in coordination with SuSy.

Lack of homology between cellulose and callose synthases, both of which appear to interact with SuSy, may suggest that factors other than a SuSy-binding motif in the glucan synthases control the interaction with SuSy. A family of at least 10 genes that share distinct homology with yeast $(1\rightarrow3)\beta$-D-glucan synthases (*FKS1*-type genes; see Delmer, 1999) have been identified from *Arabidopsis* genomic sequencing. Other such genes have also been identified through their expression in tobacco pollen tubes (Doblin *et al.*, 2001), cotton fibers (Cui *et al.*, 2001), or during cell plate formation (Hong *et al.*, 2001). The protein encoded by the latter gene was shown to localize to the cell plate during callose synthesis, along with SuSy and several other proteins. Future work will be necessary to determine whether there is any specific sequence within the multiple plant glucan synthase that is important for coordinate function with SuSy.

Other enzymes that may relate to control of carbon partitioning to cellulose

The changing rates of cellulose synthesis at different stages of cotton fiber development allow hypotheses to be formulated about other enzymes that may modulate carbon partitioning to cellulose synthesis. In addition to SuSy and cellulose synthase, enzyme activities that increase with the onset of secondary wall deposition include UDP-Glc pyrophosphorylase (EC 2.7.7.9) (Wäfler and Meier, 1994) and sucrose-phosphate synthase (SPS; EC 2.4.1.14) (Tummala, 1996). Enzyme activities that do not increase consistently with the transition from primary to secondary wall synthesis include invertases (β-fructofuranosidase; EC 3.2.1.26), glucose-6-phosphate 1-dehydrogenase (EC 1.1.1.49), phosphofructokinases including ATP-dependent 6-phosphofructokinase (EC 2.7.1.11) and pyrophosphate:fructose-6-phosphate 1-phosphotransferase (EC 2.7.1.90) (Wäfler and Meier, 1994; Basra *et al.*, 1990). However, total invertase activity, including cell-wall-bound and vacuolar acid invertase and cytoplasmic alkaline invertase, was by far the highest activity measured throughout fiber development among a group of enzymes including UDP-Glc pyrophosphorylase and the phosphofructokinases (but excluding SuSy, SPS, and cellulose synthase) (Wäfler and Meier, 1994). High invertase and/or SuSy activity is consistent with the sucrose pool in cotton fibers depositing secondary wall being quite small compared to the glucose and fructose pools (Carpita and Delmer, 1981; Jacquet, 1982; Basra *et al.*, 1990; Martin, 1999). In other cell types, invertases do affect the size of intracellular sucrose pools, but they appear to be most involved in regulating plant processes such as phloem unloading, control of differentiation, and provision of glycolytic substrates, not in determining sink strength (Sturm *et al.*, 1995; Sturm and Tang, 1999).

UDP-Glc pyrophosphorylase could be involved in cellulose synthesis if it consumed UTP and released PPi while converting glucose-1-phosphate to UDP-Glc. Although the free pool of UDP-Glc is not the immediate substrate for secondary wall cellulose synthesis facilitated by SuSy, free UDP-Glc and fructose-6-phosphate can be used to synthesize more sucrose to support cellulose synthesis (see below). The size of the UDP-Glc pool increases at the secondary wall stage of fiber development (Franz, 1969; Carpita and Delmer, 1981; Martin, 1999). Operation of UDP-Glc pyrophosphorylase in the direction of UDP-Glc synthesis is consistent with the freely reversible nature of this enzyme and with its role in activating glucose residues for glycogen synthesis in animal tissues. However, based on biochemical work, it was hypothesized that UDP-Glc pyrophosphorylase might not be under extensive metabolic control or regulate the UDP-Glc pool primarily through enzyme abundance (Turnquist and Hanson, 1973; Quick and Schaeffer, 1996). The 1.8-fold increase in UDP-Glc pyrophosphorylase activity during secondary wall deposition in cotton fibers (Wäfler and Meier, 1994) has been judged by others to be physiologically insignificant (Quick and Schaffer, 1996). In contrast, the activity of UDP-Glc pyrophosphorylase in the UDP-Glc synthetic direction was hypothesized to be highly related to the abundance of its substrates, UTP and glucose-1-phosphate (Turnquist and Hanson, 1973). This possibility is consistent with the large increase in the glucose-6-phosphate pool, which is readily isomerized to glucose-1-phosphate by phosphoglucomutase (EC 5.4.2.2), in secondary wall stage cotton fibers (Martin, 1999). Supporting the idea that the amount of UDP-Glc pyrophosphorylase is not a strong point of metabolic control, transgenic potato plants with only 4–5% of normal enzyme activity in their tubers due to expression of an antisense gene under control of the 35S CaMV promoter showed no overall phenotypic differences and no change in the growth and development of tubers (Zrenner, 1993). However, these results must be interpreted cautiously in terms of cellulose synthesis because possible changes in cell wall composition were not determined, the extent of gene down-regulation in other parts of the potato plants was not established, and tubers do not contain a strong cellulose sink, being mostly composed of parenchyma cells with primary walls. Further work to manipulate the level of UDP-Glc pyrophosphorylase in secondary-wall-synthesizing cells is needed.

Sucrose phosphate synthase, which synthesizes sucrose phosphate from fructose-6-phosphate and UDP-Glc, when acting coordinately with sucrose-phosphatase (EC 3.1.3.24), is a likely candidate to regulate synthesis of sucrose within cellulose sink cells. Biochemical experiments and analysis of transgenic plants indicate that SPS is the key regulator of the rate and extent of sucrose synthesis in photosynthetic cells. The extensive regulatory mechanisms known for SPS (e.g. glucose-6-phosphate activation and Pi inhibition) could also allow fine control of flux through sucrose in heterotrophic cells (Huber and Huber, 1996; Winter and Huber, 2000). Cotton fibers do synthesize sucrose; much of the glucose supplied to cultured ovules is subsequently converted back to sucrose within the attached fibers (Carpita and Delmer, 1981; Martin, 1999). Sucrose synthesis within fibers may be necessary if invertases cleave part of the translocated sucrose and/or recycle fructose released by the degradative action of SuSy. The latter possibility also advantageously removes cytoplasmic fructose, which is an end-product inhibitor of SuSy (Doehlert, 1987). We have shown that SPS activity increases 2- to 4-fold between primary and secondary wall synthesis in cotton fibers grown *in planta* and in culture (Tummala, 1996) and during deposition of secondary walls in tra-

cheary elements differentiating in culture (Babb and Haigler, 2000). Radiolabel from pulsed ^{14}C-glucose does not accumulate substantially in fructose within cultured cotton fibers (Martin, 1999), suggesting rapid phosphorylation and cycling of the metabolically active fructose pool. These biochemical data, together with increased fiber wall thickness in transgenic cotton plants constitutively over-expressing spinach SPS (Haigler *et al.*, 2000a, b, c), suggest that SPS activity can affect the rate and extent of secondary wall cellulose synthesis. However, further work is needed to separate the relative roles of the source and the sink in causing this effect.

A general model for regulation of intracellular carbon partitioning to cellulose synthesis

We predict that there is tight metabolic control of the timing, rate, and extent of cellulose synthesis due to unusual features of the process. The large amount of cellulose synthesized by plants is a metabolic 'dead end'; it is not degraded *in planta* for the purpose of recycling carbon to alternative metabolic uses. Furthermore, the synthesis of cellulose at any particular moment by an established plant is optional. Additional cellulose synthesis will promote growth and development of specialized cell types, but it is not required for immediate survival that depends on generating ATP and maintaining core metabolism. Therefore, when fixed carbon is limited under stress, evolutionarily successful plants must partition carbon preferentially toward survival and not cellulose synthesis. We envision that plants synthesize cellulose when conditions are more nearly optimal, especially the large amounts of cellulose required for secondary walls.

The relative amount and specific location of P-SuSy could be critical for partitioning carbon to different β-glucan polymers, particularly cellulose and callose. In contrast, S-SuSy may partition carbon to other demands such as respiration, building blocks for growth, and/or deposition of storage materials such as starch. The diagram in Figure 10 shows a model for secondary wall stage cotton fibers in which most SuSy exists as P-SuSy, partitioning at least 80% of incoming carbon into secondary wall cellulose synthesis. In this model, the association of P-SuSy with cellulose synthase could be a molecular on/off switch for cellulose synthesis, with higher P-SuSy/S-SuSy ratios occurring during high-rate secondary wall cellulose synthesis. The latter possibility is supported by our biochemical

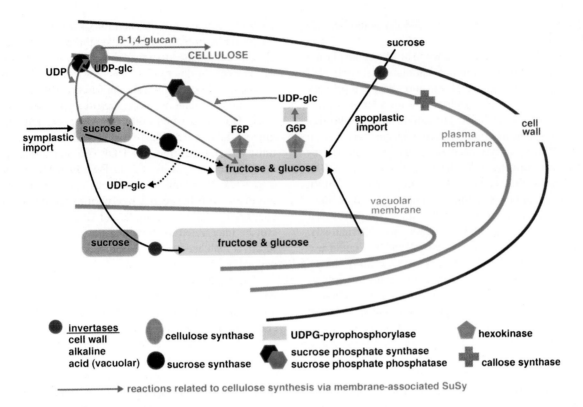

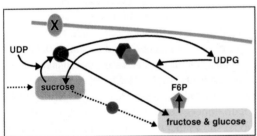

Figure 10. Diagram of a cotton fiber during active secondary wall synthesis. The predominant form of SuSy is P-SuSy (upper left) shown channeling UDP-Glc to the cellulose synthase. All enzymes shown have been shown to exist in cotton fibers. The red arrows indicate reactions that would be involved in cellulose synthesis, including production of cytoplasmic UDP-Glc so that fructose released by SuSy can be used by SPS to synthesize more sucrose. The inset shows a hypothesis about how these metabolic loops might change under stress if P-SuSy became S-SuSy. High-rate cellulose synthesis would be shut down, and S-SuSy could participate with SPS in a cycle of sucrose synthesis and degradation (black arrows) that could supply fructose-6-phosphate (F6P) for glycolysis and UDP-Glc for 'survival' metabolism. When conditions improved, cellulose synthesis and growth could quickly resume if S-SuSy converted to P-SuSy.

and immunolocalization results in primary and secondary wall stage cotton fibers. By cryogenic methods that should have preserved *in vivo* locations of soluble proteins (Nicolas and Bassot, 1993), we did not detect substantial SuSy in the cytoplasm of differentiating cotton fibers (Figure 2) or tracheary elements (Salnikov *et al.*, in press). Even though dilution in the large cytoplasmic volume could partially explain such an observation, in the model we have indicated by a dashed line the possibility that S-SuSy is rare *in vivo*

when conditions are optimal for secondary wall cellulose deposition. During secondary wall synthesis, a small amount of S-SuSy and/or the invertases would cleave sucrose to provide hexoses for maintenance metabolic processes. Similarly, fructose derived from S-SuSy could be phosphorylated to support general metabolism. The diagram does not depict differences in cellulose synthase proteins that are discussed above, which could also help to regulate the conditions under which and extent to which S-SuSy could become P-SuSy.

A callose synthase is shown to be idle in the plasma membrane, but this could be simultaneously activated during the developmental transition between primary and secondary wall synthesis by association with a portion of the P-SuSy. Developmentally controlled callose synthesis could require a specific association with P-SuSy as has been suggested for callose synthase in forming cell plates (Hong *et al.*, 2001). Callose synthesis could also quickly replace cellulose synthesis after wounding or under stress if P-SuSy preferentially associated with callose synthases that were latent in the plasma membrane. Alternatively, wound- or stress-induced callose synthesis could use the free pool of UDP-Glc; in this case, P-SuSy changing to S-SuSy could increase the concentration of cytoplasmic UDP-Glc. In *in vitro* assays, cellulose and callose synthesis are promoted by low and high concentrations of exogenous UDP-Glc, respectively. A labile association of P-SuSy with cellulose synthase could also explain why it has been difficult to synthesize cellulose rather than callose *in vitro* (Delmer 1995, 1999a). Although a distinct family of callose synthase genes has been identified as already discussed, we still cannot exclude the possibility that a cellulose synthase might begin to synthesize callose under wounding conditions, including such factors as high calcium and high UDP-Glc (Delmer and Amor, 1995). This possibility is supported by research on *in vitro* cellulose synthesis by membrane preparations of *Dictyostelium discoideum*; this slime mold is not known to synthesize callose *in vivo*, but about 10% callose is made *in vitro* (Blanton and Northcote, 1990).

The model also takes into account the existence of invertases, UDP-Glc pyrophosphorylase, SPS, and pyrophosphate:fructose-6-phosphate 1-phosphotransferase in cotton fibers. When UDP-Glc pyrophosphorylase functions in the direction of UDP-Glc synthesis as shown, PPi is produced. This could be consumed by conversion of fructose-6-phosphate to fructose 1,6-diphosphate via the pyrophosphate:fructose-6-phosphate 1-phosphotransferase. The activity of this enzyme is over 4-fold higher than the ATP-dependent phosphofructokinase throughout cotton fiber secondary wall deposition (Wäfler and Meier, 1994). The phosphorylation by fructokinase of the fructose released by the degradative action of SuSy consumes one UTP or ATP, but this is a minor energetic cost compared to the potential of one fructose-6-phosphate molecule to yield 36 ATP mole-

cules through glycolysis. Therefore, the channeling of UDP-Glc by SuSy to cellulose synthase can, with the aid of fructokinase, also feed the energy requirements of the fiber. In this model, the free pool of UDP-Glc supports cellulose synthesis only indirectly in so far as it can contribute to more sucrose synthesis in the fiber. However, when cultured ovules are fed ^{14}C-glucose, all of the radioactivity in cellulose and callose must pass through the free UDP-Glc and intracellular sucrose pool on its way to P-SuSy if this model is correct. Therefore, our previous research that did not distinguish between a free pool of UDP-Glc and channeled UDP-Glc (Carpita and Delmer, 1981) can still be interpreted to show that the rate of UDP-Glc turnover matched well the combined rates of cellulose and callose synthesis.

The model also shows synthesis (via SPS and sucrose phosphate phosphatase) and degradation (via P- and S-SuSy and invertases) of sucrose within the same cell, a phenomenon that has been described from other cell types as a 'futile cycle' with energetic cost. However, the cycle is not 'futile' in that it may provide potential for very sensitive regulation of the direction of flux through the sucrose pool according to changing needs for sucrose use (reviewed in Huber and Huber, 1996). The 'futile cycle' was previously described in cells or tissues that would be most enriched in primary wall synthesis and/or starch storage or mobilization, including heterotrophic suspension culture cells (Dancer *et al.*, 1990; Wendler *et al.*, 1990), bean cotyledons (Geigenberger and Stitt, 1991), and potato tubers (Geigenberger and Stitt, 1993). Our model shows that concurrent sucrose degradation (via P-SuSy) and synthesis (via SPS) could be a manifestation of enhancing cellulose synthesis via recycling of the fructose released when P-SuSy channels UDP-Glc to cellulose synthase. The enzymes and metabolites involved are the same as those described for the 'futile cycle', but here P-SuSy pulls use of sucrose toward cellulose. We predict that simultaneous sucrose synthesis and degradation would occur intensively during high-rate cellulose synthesis for secondary wall deposition and at a lower level during primary wall synthesis.

Furthermore, we hypothesize that under stress causing reduced carbohydrate supply, high-rate cellulose synthesis might be stopped as P-SuSy becomes S-SuSy (boxed inset of Figure 10). Here, a simple 'futile cycle' based on S-SuSy, SPS, and fructokinase could provide fructose-6-phosphate for glycolysis and UDP-Glc for survival metabolic processes. This metabolic holding pattern could quickly change to support cellulose synthesis if conditions once more become optimal and S-SuSy reverted to P-SuSy associated with cellulose synthase. By using the same basic metabolic loops but regulating the localization of SuSy, the cell could respond quickly and adaptively to synthesize cellulose only when conditions can support growth in addition to survival. Because of the importance ascribed to allocating carbon to cellulose synthesis, the regulatory mechanisms causing the switch between S-SuSy and P-SuSy could well be quite precise, but variable between different cell types and species. One would expect variability as this regulation has become most advantageously integrated with different signal transduction cascades and different cell biological systems operating at different times in development and in response to different environmental signals. Similar variability has been discussed in the context of control of metabolic flux toward starch synthesis (Smith, 1999). It may well be that continuing research on control of carbon partitioning to cellulose synthesis will reveal a spectrum of regulatory systems based on changing combinations of several regulatory elements.

This review has focused primarily on post-transcriptional/post-translational control of carbon partitioning. However, it is also clear that sugars are important regulators of gene transcription (Koch *et al.*, 1996; Koch, 1996; Sheen *et al.*, 1999). Clearly, some of the genes encoding SuSy and invertase in growing plant organs are regulated by sugar supply (Koch *et al.*, 1992, 1996). Thus, a complete understanding of the complexities of regulation of carbon partitioning will undoubtedly require a deeper understanding of the interplay between regulation of substrate pool sizes, post-translational regulation of activity and localization of important enzymes, and regulation of gene expression. This increased knowledge should allow design of strategies to change the rate and extent of cellulose synthesis for crop improvement. Factors such as feed-back inhibition of elevated substrate pools and product degradation may make it difficult to up-regulate accumulation of products made and stored in the cytoplasm or cytoplasmic organelles (Kinney, 1998). In contrast, we are optimistic that cellulose synthesis and storage can be manipulated because cellulose is removed from its newly polymerized form by crystallization, and the stable, crystalline, microfibrils exists in the cell wall across the plasma membrane from major metabolic pools supporting their synthesis. Therefore, exploring diverse strategies to advantageously manipulate carbon partitioning to cellulose

synthesis provides an intriguing challenge for future research.

Acknowledgements

We gratefully acknowledge the generous gift of rabbit polyclonal antibody prepared against bean SuSy from Heather Ross, Scottish Crop Research Institute, Dundee, UK, that was used for western blotting, immunoprecipitation, and parallel immunolocalization studies. We also thank Uwe Sonnewald for providing the potato tubers used in these studies and Mark Grimson for development of cryogenic methods for cotton fiber fixation. This work was partially supported by NSF grant DBI 9872627 to D.P.D. and C.H.H., by grant DE-FH-03-963ER20238 to D.P.D. from the U.S. Department of Energy, and by grants from Cotton Incorporated, Raleigh, NC and the Texas Advanced Research and Technology Programs to C.H.H.

References

Albrecht, G.O., Klotke, J. and Sophia, B. 2000. The increase in sucrose synthase activity correlates with a higher content of cellulose in wheat roots suffering from oxygen deficiency. Abstract 1020. In: Proceedings of Plant Biology 2000, 15–19 July, San Diego, CA. American Society of Plant Physiologists, Rockville, MD, [http://www.aspp.org/annual_meeting/pb-2000/2000.htm].

Amor, Y., Haigler, C.H., Wainscott, M., Johnson, S. and Delmer, D.P. 1995. A membrane-associated form of sucrose synthase and its potential role synthesis of cellulose and callose in plants. Proc. Natl. Acad. Sci. USA 92: 9353–9357.

Andrawis, A., Solomon, M. and Delmer, D.P. 1993. Cotton fiber annexins: a potential role in the regulation of callose synthase. Plant J. 3: 763–772.

Anguenot, R., Yelle, S. and Nguyen-Quoc, B. 1999. Purification of tomato sucrose synthase phosphorylated isoforms by Fe(III)-immobilized metal affinity chromatography. Arch. Biochem. Biophys. 365: 163–169.

ap Rees, T. 1984. Sucrose metabolism. In: D.H. Lewis (Ed.) Storage Carbohydrates in Vascular Plants: Distribution, Physiology, and Metabolism, Cambridge University Press, Cambridge, UK, pp. 53–73.

Arabidopsis Genome Initiative. 2001. Analysis of the genome sequence of the flowering plant *Arabidopsis thaliana*. Nature 408: 796–815.

Arioli, T., Peng, L., Betzner, A.S., Burn, J., Wittke, W., Herth, W., Camilleri, C., Höfte, H., Plazinske, R., Birch, R., Cork, A., Glover, J., Redmond, J. and Williamson, R.E. 1998. Molecular analysis of cellulose biosynthesis in *Arabidopsis*. Science 279: 717–720.

Babb, V.M. and Haigler, C.H. 2000. Exploration of a role for sucrose phosphate synthase in cellulose synthesis during secondary cell wall deposition. Abstract 319. In: Proceedings of Plant Biology 2000, 15–19 July, San Diego, CA. American Society of Plant Physiologists, Rockville, MD [http://www.aspp.org/annual_meeting/pb-2000/2000.htm].

48

Basra, A.S. and Malik, C.P. 1984. Development of the cotton fiber. Int. Rev. Cytol. 89: 65–113.

Basra, A.S., Sarlach, R.S., Nayyar, H. and Malik, C.P. 1990. Sucrose hydrolysis in relation to development of cotton (*Gossypium* spp.) fibres. Indian J. Exp. Biol. 28: 985–988.

Blakeney, A.B., Harris, P.J., Henry, R.J. and Stone, B.A. 1983. Simple and rapid preparation of alditol acetates for monosaccharide analysis. Carbohydrate Res. 113: 291–299.

Blanton, R.L. and Haigler, C.H. 1996. Cellulose biosynthesis. In: M. Smallwood, J.P. Knox and D.J. Bowles (Eds.) Membranes: Specialized Functions in Plants, BIOS Scientific Publishers, Oxford, UK, pp. 57–75.

Blanton, R.L. and Northcote, D.H. 1990. A 1,4-*β*-sc d-glucan synthase system from *Dictyostelium discoideum*. Planta 180: 324–332.

Blumenkrantz. B. and Asboe-Hansen, G. 1973. New method for quantitative determination of uronic acids. Anal. Biochem. 54: 484–489.

Brewin, N.J., Butcher, G.W., Galfre, G., Larkins, A.P., Wells, B., Wood, E.A. and Robertson, J.G. 1986. Immunochemical analysis of the legume root nodule. In: T.L. Wang (Ed.) Immunology in Plant Science, Cambridge University Press, Cambridge, UK, pp. 155–170.

Brown, R.M. Jr., Saxena, I.M. and Kudlicka, K. 1996. Cellulose biosynthesis in higher plants. Trends Plant Sci. 1: 149–156.

Buchala, A.J. 1987. Acid *β*-fructofuranoside fructohydrolase (invertase) in developing cotton (*Gossypium arboreum* L.) fibres and its relationship to *β*-glucan synthesis from sucrose fed to the fibre apoplast. J. Plant Physiol. 127: 219–230.

Buckeridge, M.S., Vergara, C.E. and Carpita, N.C. 1999. The mechanism of synthesis of a mixed-linkage $(1 \rightarrow 3)$, $(1 \rightarrow 4)\beta$-D-glucan in maize. Evidence for multiple sites of glucosyl transfer in the synthase complex. Plant Physiol 120: 1105–1116.

Burton, R.A., Gibeaut, D.M., Bacic, A., Findlay, K., Roberts, K., Hamilton, A., Baulcombe, D.C. and Fincher, G.B. 2000. Virus-induced silencing of a plant cellulose synthase gene. Plant Cell: 12: 691–706.

Carlson, S. J. and Chourey, P. 1996. Evidence for plasma membrane-associated forms of sucrose synthase in maize. Mol. Gen. Genet. 252: 303–310.

Carlson, S.J., Chourey, P.S. and Helentjaris, T. 2000. Evidence for a third SuSy gene in maize from analysis of sucrose synthase mutants. Abstract 161. In: Proceedings of Plant Biology 2000, 15–19 July, San Diego, CA. American Society of Plant Physiologists, Rockville, MD [http://www.aspp.org/annual_ meeting/pb-2000/2000.htm].

Carpita, N.C. and Delmer, D.P. 1981. Concentration and metabolic turnover of UDP-glucose in developing cotton fibers. J. Biol. Chem. 256: 308–315.

Chen, Y.-C. and Chourey, P.S. 1989. Spatial and temporal expression of the two sucrose synthase genes in maize: immunohistological evidence. Theor. Appl. Genet. 78: 553–559.

Chengappa, S., Guilleroux, M., Phillips, W. and Shields, R. 1999. Transgenic tomato plants with decreased sucrose synthase are unaltered in starch and sugar accumulation in the fruit. Plant Mol. Biol. 40: 213–221.

Chourey P.S., Chen Y.-C. and Miller, M.E. 1991. Early cell degeneration in developing endosperm is unique to the *Shrunken* mutation in maize. Maydica 36: 141–146

Chourey, P.S. and Miller, M.E. 1995. On the role of sucrose synthase in cellulose and callose biosynthesis in plants. In: H.G. Pontis, G.L. Salerno and E.J. Echeverria (Eds.) Sucrose Metabolism, Biochemistry, Physiology, and Molecular Biology, American Society of Plant Physiologists, Rockville, MN, pp. 80–87.

Chourey, P.S., Taliercio, E.W., Carlson, S.J. and Ruan, Y.-L. 1998. Genetic evidence that the two isozymes of sucrose synthase present in developing maize endosperm are critical, one for cell wall integrity and the other for starch biosynthesis. Mol. Gen. Genet. 259: 88–96.

Copeland, L. 1990. Enzymes of sucrose metabolism. Meth. Plant Biochem. 3: 74–85.

Crespi, M.D., Zabaleta, E.J., Pontis, H.G. and Salerno, G.L. 1991. Sucrose synthase expression during cold acclimation in wheat. Plant Physiol. 96: 887–891.

Cui, X., Shin, H., Song, C.C, Laosinchai, W., Amano, Y. and Brown, R.M. Jr. 2001. A putative plant homolog of the yeast *β*-1,3-glucan synthase subunit FKS1 from cotton (*Gossypium hirsutum* L.) fibers. Planta, in press

Dancer, J.E., Hazfield, W.D. and Stitt, M. 1990. Cytosolic cycles regulate the accumulation of sucrose in heterotrophic cell-suspension cultures of *Chenopodium rubrum*. Planta 182: 223–231.

D'Aoust, M.-A., Yelle, S. and Nguyen-Quoc, B. 1999. Antisense inhibition of tomato fruit sucrose synthase decreases fruit setting and the sucrose unloading capacity of young fruit. Plant Cell 11: 2407–2418.

Delmer, D.P. and Albersheim, P. 1970. The biosynthesis of sucrose and nucleoside diphosphate glucoses in *Phaseolus aureus*. Plant Physiol. 45: 782–786.

Delmer, D.P., Pear, J.R., Andrawis, A. and Stalker, D.M. 1995. Genes encoding small GTP-binding proteins analogous to mammalian rac are preferentially expressed in developing cotton fibers. Mol. Gen. Genet. 248: 43–51.

Delmer, D.P. and Amor, Y. 1995. Cellulose biosynthesis. Plant Cell 7: 987–1000.

Delmer, D.P. 1999a. Cellulose biosynthesis: Exciting times for a difficult field of study. Annu. Rev. Plant Physiol. Mol. Biol. 50: 245–276.

Delmer, D.P. 1999b. Cellulose biosynthesis in developing cotton fibers. In: A.S. Basra (Ed.) Cotton Fibers: Developmental Biology, Quality Improvement, and Textile Processing, Haworth Press, New York, pp. 85–112.

Dick, P.S. and ap Rees, T. 1976. Sucrose metabolism by roots of *Pisum sativum*. Phytochemistry 15: 255–259.

Dixon, D.C., Seagull, R.W. and Triplett, B.A. 1994. Changes in the accumulation of *α*- and *β*-tubulin isotypes during cotton fiber development. Plant Physiol. 105: 1347–1353.

Doblin, M.S., De Melis, L., Newbigin, E., Bacic, A. and Read, S.M. 2001. Pollen tubes of *Nicotiana alata* express two genes from different *β*-glucan synthase families. Plant Physiol, in press.

Doehlert, D.C. 1987. Substrate inhibition of maize endosperm sucrose synthase by fructose and its interaction with glucose inhibition. Plant Sci. 52: 153–157.

Doonan, J. and Clayton, L. 1986. Immunofluorescent studies on the plant cytoskeleton. In: T.L. Wang (Ed.) Immunology in Plant Science, Cambridge University Press, Cambridge, UK, pp. 111–136.

Franz, G. 1969. Soluble nucleotides in growing cotton hair. Phytochemistry 8: 737–741.

Fu, H. and Park, W.D. 1995. Sink- and vascular-associated sucrose synthase functions are encoded by different gene classes in potato. Plant Cell 7: 1369–1385.

Geigenberger, P. and Stitt, M. 1991. A 'futile' cycle of sucrose synthesis and degradation is involved in regulating partitioning between sucrose, starch and respiration in cotyledons of germinating *Ricinus communis* L. seedlings when phloem transport is inhibited. Planta 185: 81–90.

Geigenberger, P. and Stitt, M. 1993. Sucrose synthesis catalyzes a readily reversible reaction *in vivo* in developing potato tubers and other plant tissues. Planta 189: 329–339.

Guerin, J. and Carbonero, P. 1997. The spatial distribution of sucrose synthase isozymes in barley. Plant Physiol. 114: 55–62.

Haigler, C.H. 1985. The functions and biogenesis of native cellulose. In: T.P. Nevell and S.H. Zeronian (Eds.) Cellulose Chemistry and its Applications, Ellis Horwood, Chichester, UK, pp. 30–83.

Haigler, C.H. and Blanton, R.L. 1996. New hope for old dreams: evidence that plant cellulose synthase genes have finally been identified. Proc. Natl. Acad. Sci. USA 93: 12082–12085.

Haigler, C.H. and Brown, R.M. Jr. 1986. Transport of rosettes from the Golgi apparatus to the plasma membrane in isolated mesophyll cells of *Zinnia elegans* during differentiation to tracheary elements in suspension culture. Protoplasma 134: 111–120.

Haigler, C.H., Rao, N.R., Roberts, E.M., Huang, J.Y., Upchurch, D.P. and Trolinder, N.L. 1991. Cultured cotton ovules as models for cotton fiber development under low temperatures. Plant Physiol. 95: 88–96.

Haigler, C.H., Hequet, E.F., Krieg, D.R., Strauss, R.E., Wyatt, B.G., Cai, W., Jaradat, T, Srinivas, N.G., Wu, C., Jividen, G.J. and Holaday, A.S. 2000a. Transgenic cotton with improved fiber micronaire, strength, length, and increased fiber weight. In: C.P. Dugger and D.A. Richter (Eds.) Proceedings of the 2000 Beltwide Cotton Conference, 4–8 January 2000, San Antonio, TX. National Cotton Council, Memphis, p. 483.

Haigler, C.H., Cai, W, Martin, L.K., Tummala, J, Anconetani, R, Gannaway, J.G., Jividen, G.J. and Holaday, A.S. 2000b. Mechanisms by which fiber quality and fiber and seed weight can be improved in transgenic cotton growing under cool night temperatures. In: C.P. Dugger and D.A. Richter (Eds.) Proceedings of the 2000 Beltwide Cotton Conference, 4–8 January 2000, San Antonio, TX. National Cotton Council, Memphis, p. 483.

Haigler, C.H., Holaday, A.S., Wu, C., Wyatt, B.G., Jividen, G.J., Gannaway, J.G., Cai, W.X., Hequet, E.F., Jaradat, T.T., Krieg, D.R., Martin, L. K., Strauss, R.E., Nagarur, S. and Tummala, J. 2000c. Transgenic cotton over-expressing sucrose phosphate synthase produces higher quality fibers with increased cellulose content and has enhanced seedcotton yield. Abstract 477. In: Proceedings of Plant Biology 2000, 15–19 July, San Diego, CA. American Society of Plant Physiologists, Rockville, MD [http://www.aspp.org/annual_ meeting/pb-2000/2000.htm].

Hayashi, T., Read, S.M., Bussell, J., Thelen, M., Lin, F.-C., Brown, R.M. Jr. and Delmer, D.P. 1987. UDP-glucose: $(1\rightarrow3)\beta$-glucan synthases from mung bean and cotton. Plant Physiol. 83: 1054–1062.

Hirai, N., Sonobe, S. and Hayashi, T. 1998. *In situ* synthesis of β-glucan microfibrils on tobacco plasma membrane sheets. Proc. Natl. Acad. Sci. USA 95: 15102–15106.

Holland, N., Holland, D., Helentjaris, T., Dhugga, K.S., Xoconostle-Cazares, B. and Delmer, D.P. 2000. A comparative analysis of the plant cellulose synthase (*CesA*) gene family. Plant Physiol. 123: 1313–1324.

Hong, Z., Delauney, A.J. and Verma, D.P.S. 2001. A cell plate-specific callose synthase and its interaction with phragmoplastin. Plant Cell, in press.

Huber, S.C. and Akazawa, T. 1986. A novel sucrose synthase pathway for sucrose degradation in cultured sycamore cells. Plant Physiol. 81: 1008–1013.

Huber, S.C. and Huber, J.L. 1996. Role and regulation of sucrose-phosphate synthase in higher plants. Annu. Rev. Plant Physiol. Plant Mol. Biol. 47: 431–444.

Huber, S.C., Huber, J.L. Gage, D.A., McMichael, R.W., Chourey, P.S, Hannah L.C. and Koch, K. 1996. Phosphorylation of serine-15 of maize leaf sucrose synthase. Plant Physiol. 112: 793–802.

Jaquet, J.P., Buchala, A.J. and Meier, H. 1982. Changes in the nonstructural carbohydrate content of cotton (*Gossypium* spp.) fibres at different stages of development. Planta 156: 481–486.

Kawagoe, Y. and Delmer, D.P. 1997. Pathways and genes involved in cellulose biosynthesis. In: J.K. Setlow (Ed) Genetic Engineering, vol. 19, Plenum Press, New York, pp. 63–87.

Kimura, S., Laosinchai, W., Itoh, T., Cui, X., Linder, R. and Brown, R.M. Jr. 1999. Immunogold labeling of rosette terminal cellulose-synthesizing complexes in the vascular plant *Vigna angularis*. Plant Cell 11: 2075–2085.

Kinney, A.J. 1998. Manipulating flux through plant metabolic pathways. Curr. Opin. Plant Biol. 1: 173–177.

Kloth, R.H. 1989. Changes in the level of tubulin subunits during development of cotton (*Gossypium hirsutum*) fiber. Physiol. Plant. 76: 37–41.

Kobayashi, H. and Fukuda, H. 1994. Involvement of calmodulin and calmodulin-binding proteins in the differentiation of tracheary elements in *Zinnia* cells. Planta 194: 388–394.

Koch, K.E. 1996. Carbohydrate-modulated gene expression in plants. Annu. Rev. Plant Physiol. Plant Mol. Biol. 47: 509–540.

Koch, K.E., Nolte, K.D., Duke, E.R., McCarty, D.R. and Avigne, W.T. 1992. Sugar levels modulate differential expression of maize sucrose synthase genes. Plant Cell 4: 59–69.

Koch, K.E., Xu, J., Duke, E.R., McCarty, D.R., Yuan, C-X., Tan, B.-C. and Avigne, W.T. 1995. Sucrose provides a long distance signal for coarse control of genes affecting its metabolism. In: H.G. Pontis, G. Salerno and E. Echeverria (Eds.) Sucrose Metabolism, Biochemistry, Physiology, and Molecular Biology, American Society of Plant Physiologists, Rockville, MD, pp. 266–277.

Koch, K.E., Wu, Y. and Xu, J. 1996. Sugar and metabolic regulation of genes for sucrose metabolism: potential influence of maize sucrose synthase and soluble invertase responses on carbon partitioning and sugar sensing. J. Exp. Bot. 47: 1179–1185.

Kudlicka, K., Brown, R.M. Jr., Li, L., Lee, J.H., Shen, H. and Kuga, S. 1995. β-glucan synthesis in the cotton fiber. IV. *In vitro* assembly of the cellulose I allomorph. Plant Physiol. 107: 111–123.

Kudlicka, K., Lee, J.H. and Brown, R.M. Jr. 1996. A comparative analysis of in vitro cellulose synthesis from cell-free extracts of mung bean (*Vigna radiata*, Fabaceae) and cotton (*Gossypium hirsutum*, Malvaceae). Am. J. Bot. 83: 274–284.

Laporte, K., Rossignol, M. and Traas, J.A. 1993. Interaction of tubulin with the plasma membrane: tubulin is present in purified plasmalemma and behaves as an integral membrane protein. Planta 191: 413–416.

Lindblom, S., Ek, P., Muszynska, G., Ek, B., Szczegielniak, J. and Engstrom, L. 1997. Phosphorylation of sucrose synthase from maize seedlings. Acta Biochim. Pol. 44: 809–817

Loog, M., Toomik, R., Sak, K., Muszynska, G., Jarv, J. and Ek, P. 2000. Peptide phorphorylation by a calcium-dependent protein kinase from maize seedlings. Eur. J. Biochem. 267: 337–343.

Maas, C., Schaal, S. and Werr, W. 1990. A feedback control element near the transcription start site of the maize *Shrunken* gene determines promoter activity. EMBO J. 9: 3447–3452.

Maltby, D., Carpita, N.C., Montezinos, D., Kulow, C. and Delmer, D.P. 1979. β-1,3-glucan in developing cotton fibers. Plant Physiol. 63: 1158–1164.

Martin, L.K. 1999. Cool-temperature-induced changes in metabolism related to cellulose synthesis in cotton fibers. Ph.D. dissertation, Texas Tech University, Lubbock, TX.

50

Meinert, M.C. and Delmer, D.P. 1977. Changes in biochemical composition of the cell wall of the cotton fiber during development. Plant Physiol. 59: 1088–1097.

Morrow, D.L. and Lucas, W.J. 1986. (1,3)-β-glucan synthase from sugar beet. I. Isolation and solubilization. Plant Physiol. 81: 171–176

Mutsaers, H.J.W. 1976. Growth and assimilate conversion of cotton bolls (*Gossypium hirsutum* L.) 1. Growth of fruits and substrate demand. Ann. Bot. 40: 301–315.

Nakai, T., Tonouchi, N., Konishi, T., Kojima, Y., Tsuchida, T., Yoshinaga, F., Sakae F. and Hayashi, T. 1999. Enhancement of cellulose production by expression of sucrose synthase in *Acetobacter xylinum*. Proc. Natl. Acad. Sci. USA 96: 14–18.

Nguyen-Quoc, B., Krivitzky, M., Huber, S.C. and Lecharny, A. 1990. Sucrose synthase in developing maize leaves: regulation of activity by protein level during the import to export transition. Plant Physiol. 94: 516–523.

Nicolas, T.N. and Bassot, J.M. 1993. Freeze substitution after fast-freeze fixation in preparation for immunocytochemistry. Microscopy Res. Tech. 24: 474–487.

Nolte K.D. and Koch, K.E. 1993. Companion-cell specific localization of sucrose synthase in zones of phloem loading and unloading. Plant Physiol. 101: 899–905.

Nolte, K.D., Hendrix, D.L., Radin, J.W. and Koch, K.E. 1995. Sucrose synthase localization during initiation of seed development and trichome differentiation in cotton ovules. Plant Physiol. 109: 1285–1293.

Pear, J., Kawagoe, Y., Schreckengost, W., Delmer, D.P. and Stalker, D. 1996. Higher plants contain homologs of the CelA genes that encode the catalytic subunit of the bacterial cellulose synthases. Proc. Natl. Acad. Sci. USA 93: 12637–12642.

Peng, L., Xiang, F., Roberts, E., Kawagoe, Y., Greve, L.C., Kreuz, K., Delmer, D.P. 2001. The experimental herbicide cgA 325'615 inhibits synthesis of crystalline cellulose and causes accumulation of non-crystalline β-1,4-glucan associated with CesA protein. Plant Physiol, in press.

Pillonel, C., Buchala, A.J. and Meier, H. 1980. Glucan synthesis by intact cotton fibres fed with different precursors at the stages of primary and secondary wall formation. Planta 149: 306–312.

Potikha, T.S., Collins, C.C., Johnson, D.I., Delmer, D.P. and Levine, A. 1999. The involvement of hydrogen peroxide in the differentiation of secondary walls in cotton fibers. Plant Physiol. 119: 849–858.

Preston, R.D. 1974. The Physical Biology of Plant Cell Walls. Chapman and Hall, London.

Quick, W.P. and Schaffer, A.A. 1996. Sucrose metabolism in sources and sinks. In: E. Zamski and A.A. Schaffer (Eds.) Photoassimilate Distribution in Plants and Crops: Source-Sink Relationships, (AUTHOR: PLEASE PROVIDE PUBLISHER DATA), pp. 115–156.

Reymond, O.L. and Pickett-Heaps, J.D. 1983. A routine flat embedding method for electron microscopy of microorganisms allowing selection and precisely orientated sectioning of single cells by light microscopy. J. Microsc. 130: 79–84.

Richmond, T.A. and Somerville, C.R. 2001. Integrative approaches to determining *Csl* function. Plant Mol. Biol., this issue.

Roberts, A.W. and Haigler, C.H. 1990. Tracheary-element differentiation in suspension cultures of *Zinnia* requires uptake of extracellular Ca^{2+}. Experiments with calcium-channel blockers and calmodulin inhibitors. Planta 180: 502–509.

Roberts, E.M., Nunna, R.R., Huang, J.Y., Trolinder, N.L. and Haigler, C.H. 1992. Effects of cycling temperatures on fiber metabolism in cultured cotton ovules. Plant Physiol. 100: 979–986.

Rollit, J. and Maclachlan, G.A. 1974. Synthesis of wall glucan from sucrose by enzyme preparations from *Pisum sativum*. Phytochemistry 13: 367–374.

Ross, P., Mayer, R. and Benziman, M. 1991. Cellulose biosynthesis and function in bacteria. Microbiol. Rev. 55: 35–58.

Ruan, Y.-L., Chourey, P.S., Delmer, D.P. and Perez-Grau, L. 1997. The differential expression of sucrose synthase in relation to diverse patterns of carbon partitioning in developing cotton seed. Plant Physiol. 115: 375–385.

Ruan, Y.-L. and Chourey, P.S. 1998. A fiberless seed mutation in cotton is associated with lack of fiber cell initiation in ovule epidermis and alterations in sucrose synthase expression and carbon partitioning in developing seeds. Plant Physiol. 118: 399–406.

Ryser, U. 1985. Cell wall biosynthesis in differentiating cotton fibres. Eur. J. Cell Biol. 39: 236–256.

Ryser, U. 1992. Ultrastructure of the epidermis of developing cotton (*Gossypium*) seeds: suberin, pits, plasmodesmata, and their implications for assimilate transport into cotton fibers. Am. J. Bot. 79: 14–22.

Ryser, U. 1999. Cotton fiber initiation and histodifferentiation. In: A.S. Basra (Ed.) Cotton Fibers: Developmental Biology, Quality Improvement, and Textile Processing, Haworth Press, New York, pp. 1–46.

Salnikov, V.V., Grimson, M.J., Delmer, D.P. and Haigler, C.H. 2001. Sucrose synthase localizes to cellulose synthesis sites in tracheary elements. Phytochemistry, in press.

Seagull, R.W. 1993. Cytoskeletal involvement in cotton fiber growth and development. Micron 24: 643–660.

Sebkova, V., Unger, C., Hardegger, M. and Sturm, A. 1995. Biochemical, physiological, and molecular characterization of sucrose synthase from *Daucus carota*. Plant Physiol. 108: 75–83.

Shaw, J.R., Ferl, R.J., Baier, J., St. Clair, D., Carson, C., McCarty, D.R. and Hannah, L.C. 1994. Structural features of the maize sus1 gene and protein. Plant Physiol. 106: 1659–1665.

Shedletzky, E., Shmuel, M., Trainin, T., Kalman, S. and Delmer, D.P. 1992. Cell wall structure in cells adapted to growth on the cellulose-synthesis inhibitor 2,6-dichloro-benzonitrile (DCB): a comparison between two dicotylenonous plants and a graminaceous monocot. Plant Physiol. 100: 120–130.

Sheen, J., Shou, L. and Jang, J.C. 1999. Sugars as signaling molecules. Curr. Opin. Plant Biol. 2: 410-418.

Smith, A.M. 1999. Regulation of starch synthesis in storage organs. In: N.J. Kruger *et al*. (Eds.) Regulation of Primary Metabolic Pathways in Plants, Kluwer Academic Publishers, Dordrecht, Netherlands, pp. 173–193.

Smook, G.A. 1992. Handbook for Pulp and Paper Technologists. Angus Wilde Publications, Vancouver.

Sturm, A., Sebkova, V., Lorenz, K., Hardegger, M., Lienhard, S. and Unger, C. 1995. Development- and organ-specific expression of the genes for sucrose synthase and three isozymes of acid β-fructofuranoside in carrot. Planta 195: 601–610.

Sturm, A. and Tang, G.-Q. 1999. The sucrose-cleaving enzymes of plants are crucial for development, growth, and carbon partitioning. Trends Plant Sci. 4: 401–407.

Sturm, A., Lienhard, S., Schatt, S. and Hardegger, M. 1999. Tissue-specific expression of two genes for sucrose synthase in carrot (*Daucus carota* L.). Plant Mol. Biol. 39: 349–360.

Sun, J., Loboda, T., Sung, S.-J. and Black, C.C. Jr. 1992. Sucrose synthase in wild tomato, *Lycopersicon chmielewskii*, and tomato fruit sink strength. Plant Physiol. 98: 1163–1169.

Sung, S.-J., Xu, D.-P. and Black, C.C. 1989. Identification of actively-filling sucrose sinks. Plant Physiol. 89: 1117–1121.

Tang, G.Q. and Sturm, A. 1999. Antisense repression of sucrose synthase in carrot (*Daucus carota* L.) affects growth rather than sucrose partitioning. Plant Mol. Biol. 41: 465–479.

Taylor, N.G., Scheible, W.R., Cutler, S., Somerville, C.R., and Turner, S.R. 1999. The irregular xylem locus of *Arabidopsis* encodes a cellulose synthase required for secondary wall synthesis. Plant Cell 11: 769–780.

Theander, O. and Westerlund, E. 1993. Quantitative analysis of cell wall components. In: H.G. Jung, D.R. Buxton, R.D. Hatfield and J. Ralph (Eds.) Forage Cell Wall Structure and Digestibility, American Society of Agronomy/CSSA/SSSA, Madison, WI, pp. 83–104.

Todorov, I.T., Attaran, A. and Kearsey, S.E. 1995. BM28, a human member of the MCM2-3-5 family, is displaced from chromatin during DNA replication. J. Cell Biol. 129: 1433–1445.

Tummala, J. 1996. Response of sucrose phosphate synthase activity to cool temperatures in cotton. M.S. thesis, Texas Tech University, Lubbock, TX.

Turner, S. and Somerville, C.R. 1997. Collapsed xylem phenotype of *Arabidopsis* identifies mutants deficient in cellulose deposition in the secondary cell wall. Plant Cell 9: 689–701.

Turnquist, R.L. and Hansen, R.G. 1973. Uridine diphosphphoryl glucose pyrophosphorylase. In: P.D. Boyer (Ed.) The Enzymes, vol. 8, part A, 3rd ed., Academic Press, New York, pp. 51–71.

Updegraff, D.M. 1969. Semi-micro determination of cellulose in biological materials. Anal. Biochem. 32: 420–424.

Vergara, C.E. and Carpita, N.C. 2001. Mixed-linkage β-glucan synthase and the *CesA* gene family in cereals. Plant Mol. Biol., this issue.

Volman, G., Ohana, P. and Benziman, M. 1995. Biochemistry and molecular biology of cellulose synthesis. Carbohydrates Europe 12: 20–27.

Wäfler, U. and Meier, H. 1994. Enzyme activities in developing cotton fibers. Plant Physiol. Biochem. 32: 697–702.

Wang, F., Sanz, A., Brenner, M.L. and Smith, A. 1993. Sucrose synthase, starch accumulation, and tomato fruit sink strength. Plant Physiol. 101: 321–327.

Wendel, J.F., Small, R.L., Cronn, R.C. and Brubaker, C.L. 1999. Genes, jeans, and genomes: reconstructing the history of cotton.

In: L.W.D. van Raamsdonk and J.C.M. den Nijs (Eds.) Plant Evolution in Man-Made Habitats, Proceedings of the 7th Symposium IOPB (Amsterdam, 1998), Hugo de Vries Laboratory, Amsterdam, pp. 133–159.

Wendler, R., Veith, R., Dancer, J., Stitt, M. and Komor, E. 1990. Sucrose storage in cell suspension cultures of *Saccharum* sp. (sugarcane) is regulated by a cycle of synthesis and degradation. Planta 183: 31–39.

Winter, H. and Huber, S.C. 2000. Regulation of sucrose metabolism in higher plants: Localization and regulation of activity of key enzymes. Crit. Rev. Plant Sci. 19: 31–67.

Winter, H., Huber, J.L. and Huber, S.C. 1997. Membrane association of sucrose synthase: changes during graviresponse and possible control by protein phosphorylation. FEBS Lett. 420: 151–155.

Winter, H., Huber, J.L. and Huber, S.C. 1998. Identification of sucrose synthase as an actin-binding protein. FEBS Lett. 430: 205–208.

Xu, D.-P., Sung, S.-J., Loboda, T., Kormanik, P.P. and Black, C.C. 1989. Characterization of sucrolysis via the uridine diphosphate and pyrophosphate-dependent sucrose synthase pathway. Plant Physiol. 90: 635–642.

Zhang, X.-Q. and Chollet, R. 1997. Seryl-phosphorylation of soybean nodule sucrose synthase (nodulin-100) by a Ca^{2+}-dependent protein kinase. FEBS Lett. 410: 126–130.

Zhang, X.Q., Lund, A.A., Sarath, G., Cerny, R.L., Roberts, D.M. and Chollet, R. 1999. Soybean nodule sucrose synthase (nodulin-100): further analysis of its phosphorylation using recombinant and authentic root-nodule enzymes. Arch. Biochem. Biophys. 371: 70–82.

Zrenner, R., Willmitzer, L. and Sonnewald, U. 1993. Analysis of the expression of potato uridinediphosphate-glucose phyrophosphorylase and its inhibition by antisense RNA. Planta 190: 247–252.

Zrenner, R., Salanoubat, M., Willmitzer, L. and Sonnewald, U. 1995. Evidence of the crucial role of sucrose synthase for sink strength using transgenic potato plants (*Solanum tuberosum* L.) Plant J. 7: 97–107.

SECTION 2

GENE AND PROTEIN STRUCTURE

Plant Molecular Biology **47:** 55–72, 2001.
© 2001 *Kluwer Academic Publishers. Printed in the Netherlands.*

A census of carbohydrate-active enzymes in the genome of *Arabidopsis thaliana*

Bernard Henrissat[1,*], Pedro M. Coutinho[2] and Gideon J. Davies[3]

[1]*Architecture et Fonction des Macromolécules Biologiques, UMR 6098, CNRS and Universités d'Aix-Marseille I and II, 31 Chemin Joseph Aiguier, 13402 Marseille cedex 20, France (*author for correspondence; e-mail bernie@afmb.cnrs-mrs.fr); [2]Centre for Biological and Chemical Engineering, Instituto Superior Técnico, Av. Rovisco Pais, 1049-001 Lisboa, Portugal; [3]Structural Biology Laboratory, Department of Chemistry, University of York, Heslington, York YO10 5DD, UK*

Key words: Arabidopsis, genome, glycoside hydrolases, glycosyltransferases

Abstract

The synthesis, modification, and breakdown of carbohydrates is one of the most fundamentally important reactions in nature. The structural and functional diversity of glycosides is mirrored by a vast array of enzymes involved in their synthesis (glycosyltransferases), modification (carbohydrate esterases) and breakdown (glycoside hydrolases and polysaccharide lyases). The importance of these processes is reflected in the dedication of 1–2% of an organism's genes to glycoside hydrolases and glycosyltransferases alone. In plants, these processes are of particular importance for cell-wall synthesis and expansion, starch metabolism, defence against pathogens, symbiosis and signalling. Here we present an analysis of over 730 open reading frames representing the two main classes of carbohydrate-active enzymes, glycoside hydrolases and glycosyltransferases, in the genome of *Arabidopsis thaliana*. The vast importance of these enzymes in cell-wall formation and degradation is revealed along with the unexpected dominance of pectin degradation in *Arabidopsis*, with at least 170 open-reading frames dedicated solely to this task.

Introduction

In plants, glycosides are found as polysaccharides (such as cellulose, cross-linking glycans and starch), disaccharides (sucrose and trehalose, for example), glycolipids, glycoproteins and as conjugates to a variety of non-carbohydrate moieties such as steroids and lignin precursors. In all living organisms, the synthesis of the glycosides is performed by glycosyltransferases which utilize activated sugar donors (typically by nucleotide diphospho sugars, nucleotide monophospho sugars, sugar phosphates or lipid diphospho sugars), or by transglycosylation reactions. The breakdown of glycosidic linkages is performed by glycoside hydrolases and by polysaccharides lyases. In plants, the biosynthetic and hydrolytic enzymes are especially important for cell-wall biosynthesis and modification during growth. Other roles for oligo- and

polysaccharides in plants include energy/carbon storage, signalling and defence. With the recent genomic sequence data (Arabidopsis Genome Initiative, 2000), *Arabidopsis thaliana* becomes the first model plant for the study of higher-organism carbohydrate metabolism. The possibility for functional analysis of gene knockouts together with the availability of numerous mutant phenotypes and the potential for identification of gene activity by tissue make this a particularly alluring system. Here we present an analysis of the carbohydrate-active enzyme content of *Arabidopsis* with an emphasis on the primary synthetic and degradative *apparti*, the glycosyltransferases and glycoside hydrolases.

The sequence-based families of carbohydrate-active enzymes (CAZymes)

Glycoside hydrolases and glycosyltransferases have previously been classified into a number of families based on derived amino-acid sequence similarities (Henrissat, 1991; Henrissat and Bairoch, 1993; Davies and Henrissat, 1995; Henrissat and Bairoch, 1996; Campbell *et al.*, 1997; Henrissat and Davies, 1997; Coutinho and Henrissat, 1999a). As from June 2001, 85 families of glycoside hydrolases and 52 families of glycosyltransferases have been described. We expect the number of glycoside hydrolase families to grow only very slowly as most of these enzymes have probably been already sampled in different organisms. On the other hand, the number of glycosyltransferase families is likely to be an underestimate as these enzymes are particularly difficult to screen for and to characterize. The families of glycoside hydrolases and glycosyltransferases, as well as families of polysaccharide lyases and carbohydrate esterases can be conveniently accessed through the carbohydrate-active enzyme server CAZy at URL: http://afmb.cnrs-mrs.fr/~pedro/CAZY/db.html (Coutinho and Henrissat, 1999a). For each family, this continuously updated resource provides a list of the current members with links to nucleotide, protein and structure data banks.

The sequence-based families proved extremely useful for pre-genomic applications ranging from mechanistic enzymology to structural characterization (Henrissat and Davies, 1997). Within a given sequence-derived family, three-dimensional (3D) structure is conserved. As a consequence, the bioinformatic resource together with known 3D structural data may inform strategies for the analysis of related proteins. Such functional analysis will always remain necessary since the determinants of substrate specificity reside in the fine details of the structure (Davies and Henrissat, 1995).

Whether synthetic or hydrolytic, enzymatic glycosyl transfer occurs at the sugar anomeric centre and may proceed with either retention or inversion of anomeric configuration (Sinnott, 1990; McCarter and Withers, 1994; Zechel and Withers, 2000). Since the mechanism is governed by the spatial orientation of catalytic residues within their 3D template, and since structure itself is dictated by sequence, the stereochemical outcome of the reaction is conserved within any given family of glycoside hydrolases. This observation has suffered no exception since it was first proposed by Withers and colleagues (Gebler *et al.*,

1992). Similarly, the families of glycosyltransferases contain enzymes that operate with the same stereochemistry, regardless of the particular sugar added. In consequence, once established for a member of a family, the mechanism can be safely extended to all other members of the family. Conservation of the catalytic residues within a family is strict; once identified (in position and function) for one member of a family, they can easily be inferred for all members of the family. When a catalytic residue is absent, this indicates either a modification of the molecular mechanism (for instance, myrosinase; Burmeister *et al.*, 2000) or a loss of catalytic function (for example, narbonin, concanavalin B and a putative plant oligosaccharide receptor; Henrissat and Davies, 2000). The power of the sequence family classification, as opposed to *a posteriori* 'superfamily' descriptions, is this predictive power (Henrissat and Davies, 1997).

Enzyme specificity in CAZyme families

One of the exciting features of the family classification is that many of the sequence-based families are polyspecific, that is they contain enzymes of different substrate or product specificity indicative of an evolutionary divergence to acquire new functions. Family 13 of the glycoside hydrolases, for example, contains no less than 20 different enzyme activities. Polyspecificity is not restricted to glycoside hydrolases and can be found in glycosyltransferase families as well. Conversely, many similar enzymatic specificities are found in different families. For example, cellulases are found in glycoside hydrolase families GH5–GH9, GH12, GH26, GH44, GH45, GH48 and GH61 which show a diverse array of unrelated 3D structures (Davies, 1998). In genomic annotations, the polyspecificity of the CAZyme families can lead to both over and under prediction such as 'putative $(1\rightarrow3)\alpha$-GalNAc transferase' (when it is known that only 3 residues are sufficient to alter specificity to $(1\rightarrow4)\alpha$-Gal transferase) or 'putative sugar hydrolase' when over 80 different families are known for these enzymes. CAZyme family classification (e.g. 'glycosyltransferase family X member') provides an excellent, intermediate and conservative annotation system that avoids these problems. Given that both fold and catalytic mechanism are conserved within a family, this can already help in searching the exact substrate and product specificity. The specificity of the characterized members of the

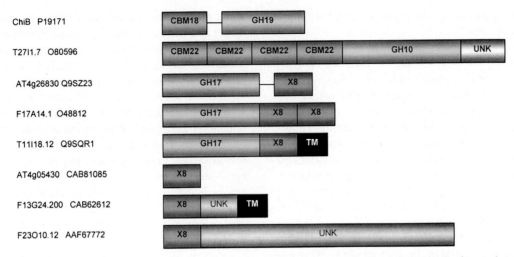

Figure 1. Examples of a few modular ORFs/proteins in *Arabidopsis*. The yellow boxes represent catalytic domains from various glycoside hydrolase families; the blue boxes represent carbohydrate-binding modules from various families (for instance CBM22 represents family 22 of the carbohydrate-binding modules; Charnock *et al.*, 2000); the boxes labelled UNK represent regions of unknown function. The pink boxes represent a module of unknown function forming a family called 'X8' (Henrissat and Davies, 2000). Black boxes represent membrane spanning regions. The name and the accession number (SwissProt or Genpept) of each ORF is given.

same family also constitutes a starting framework for biochemical analysis.

Modularity of CAZymes

CAZymes are frequently modular proteins. A single polypeptide chain may contain one or more catalytic domains as well as one or several non-catalytic modules (Figure 1). The non-catalytic modules are frequently involved in protein-carbohydrate and protein-protein interactions. There are many modules, however, for which no function has yet been assigned. Such modularity, whilst providing a fascinating insight into enzyme function and evolution can severely affect genome annotation. In several cases, a 'putative' assignment to a catalytic activity is based upon sequence similarity with only a non-catalytic module. When compiling a census of the CAZymes of a given organism, the modularity of all the ORFs must first be established and each module treated separately. Sequence classification should not, therefore, be uncoupled from modular annotation (Henrissat and Davies, 2000).

In the present analysis, protein sequences derived from gene and genomic studies of *Arabidopsis* have been gradually retrieved upon public release using the Batch Entrez facility (URL: http://www.ncbi.nlm.nih.gov/Entrez/batch.html). To fully take into account the possible modularity of the proteins, the sequences were typically subjected to gapped BLAST analysis (Altschul *et al.*, 1997) against a library of modules derived from previously classified and/or annotated protein sequences belonging to all CAZyme families. These modules correspond to individually identified catalytic and ancillary modules found in CAZymes, whose sequence limits are derived from (1) available 3D structures, (2) results from deletion studies found in the literature, and (3) definition of conserved regions from sequence analysis by PSI-BLAST (Altschul *et al.*, 1997) and hydrophobic cluster analysis (Callebaut *et al.*, 1997). These annotated protein sequences correspond to each of the more than 8000 protein entries found in the CAZy database as from December 2000.

3-D Structures of CAZymes

The past ten years have seen an explosion of 3D structure for glycoside hydrolases and their constituent modules. More than 1100 structures from CAZymes and of their constituent modules are already present in the CAZy Web site! This reflects advances in cryo-crystallographic techniques, the development of new synchrotron sources and the availability of powerful NMR magnets. Of the 85 glycoside hydrolase families 3D structural representatives are known for around 40. We are just beginning to witness the emergence of 3D structures for glycosyltransferases: 5 structures have been published and we can expect at least 5–10 more in the short term.

Although numerous 3D structures are available and homologous structures are available for many of the families of CAZymes present in *Arabidopsis*, actual 3D structures of CAZymes of plant origins are few and far between. This may reflect the difficulties in isolating pure enzymes from naturally occurring isozymes, the problems with over-expression of plant proteins in heterologous systems and extensive post-translational modifications such as glycosylation. There are no plant glycosyltransferase structures nor a single 3D structure of any CAZyme from *Arabidopsis*. Indeed, plant enzyme structures are available only for glycoside hydrolase families GH1, GH3, GH13, GH14, GH17 and GH18. In other families, 3D structures available from other organisms nevertheless provide valuable structural and mechanistic insights into the homologous plant enzymes.

A fascinating feature of CAZyme structures is their remarkable diversity. In contrast to analogous hydrolytic enzymes, such as lipases and proteases, glycoside hydrolases display a plethora of different 3D folds (Davies and Henrissat, 1995; Henrissat and Davies, 1997). Whether glycosyltransferases will display such a topological variability is unlikely since the constraints of nucleotide binding may limit structural variance. Indeed, the five glycosyltransferase structures reported (Vrielink *et al.*, 1994; Charnock and Davies, 1999; Gastinel *et al.*, 1999; Ha *et al.*, 2000; Pedersen *et al.*, 2000; Ünligil *et al.*, 2000), covering glycosyltransferase families GT2, GT7, GT13, GT28, GT43 and the still unclassified bacteriophage T4 DNA β-glycosyltransferase, display just two different global folds.

Catalytic mechanisms of CAZymes

A brief description of the reactions catalyzed by four classes of CAZyme is given below (Figure 2). Polysaccharide lyases and carbohydrate esterases are also described because of the importance of pectin metabolism to *Arabidopsis* (see below).

Glycoside hydrolases are, without doubt, the best characterized of all CAZymes. They perform acid-base catalysed hydrolysis of glycosidic bonds which can lead to inversion or retention of the anomeric configuration, depending on the enzyme family. Numerous reviews of their structure, function and cat-

alytic mechanisms are available (Sinnott, 1990; McCarter and Withers, 1994; Davies and Henrissat, 1995; Zechel and Withers, 2000). Inversion reactions utilize a classical single displacement mechanism, whereas retention is performed via a double displacement involving the formation and subsequent hydrolysis of a covalent intermediate. An important feature of the retention mechanism, of particular relevance to plant biochemistry, is that the reaction intermediate, a covalent glycosyl-enzyme species, may be intercepted by non-water nucleophiles in a process known as *transglycosylation*. In plants, this reaction can play an important role as exemplified by the xyloglucan endotransferases, which are directly related to glycoside hydrolase family GH16.

Glycosyltransferases catalyse the synthesis of the vast majority of glycosidic bonds in nature. These enzymes utilize activated sugar donors in order to drive glycosidic bond formation. Typically the activating group may be a phosphate, a nucleotide or a lipid phosphate. Both retaining and inverting glycosyltransferases are known. These enzymes are relatively poorly characterized, but reaction mechanisms, similar to those occurring in glycoside hydrolases may be envisaged. Inversion presumably involves a direct single displacement at the anomeric centre of the activated sugar donor. The retention mechanism is less clear, but a double-displacement reaction, through a covalent intermediate is most likely.

Polysaccharide lyases are another class of glycoside-bond-cleaving enzymes, which differ from glycoside hydrolases in that they utilize a fundamentally different reaction mechanism. The majority of polysaccharide lyases abstract an 'activated' proton from a pyranoside-ring C5 atom, which is itself adjacent to a C6 carboxylate moiety. A β-elimination reaction ensues, releasing a 4,5-unsaturated sugar. Such a mechanism demands a C6 carboxylate substitution, such as is found on glucuronic and galacturonic acids, is of particular importance in plant biochemistry in the degradation of pectates, for example. Currently polysaccharide lyases form 10 different sequence-based families.

Carbohydrate esterases catalyse the de-*O* or de-*N*-acetylation of substituted saccharides and they form in 12 families at present. Since an ester consists of an acid plus an alcohol, we may consider two classes: those in which the sugar plays the role of the 'acid',

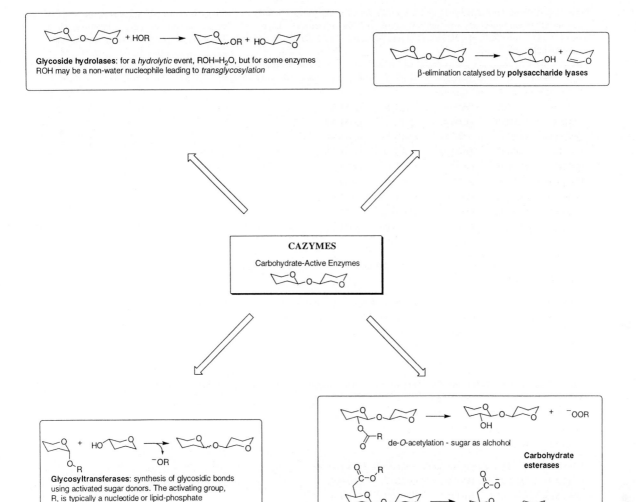

Figure 2. Catalytic mechanisms of CAZymes.

such as pectin methyl esters and those in which the sugar behaves as the alcohol, such as in acetylated xylan. A number of possible reaction mechanisms may be involved: the most common is a Ser-His-Asp catalytic triad-catalysed deacetylation analogous to the action of classical lipase and serine proteases but other mechanisms such as a Zn^{2+}-catalysed deacetylation may also be considered for some families.

CAZymes and the genomic era: a census of *Arabidopsis* CAZymes

In the genomic age, the families described above can aid both ORF annotation and functional prediction. In addition, a listing of all potential carbohydrate-active enzymes of a genome can serve to trace the higher carbohydrate metabolism of an organism. Once the various ORFs of a genome are assigned to these families, it also becomes possible to compare the CAZyme content of different genomes (Coutinho and Henrissat, 1999b). The ca. 130 million base-pair *Arabidopsis* genome, now virtually fully sequenced, contains about 25 500 genes (Kaiser, 2000; Arabidopsis Genome Initiative, 2000). Over 730 of these have been found to belong to glycoside hydrolase or glycosyltransferase families. Tables 1 and 2 list the glycoside hydrolase and glycosyltransferase contents of the *Arabidopsis* genome, respectively, arranged in families; Table 3 list

Table 1. Listing of the *Arabidopsis thaliana* ORFs found in the various families of glycoside hydrolases (accession numbers of the deduced protein sequences are from SwissProt/TrEMBL or GenBank databases). The mechanism (retaining or inverting where known) and the availability of a 3D structural representative in each family are reported. The main enzymatic activities found in each family are indicated.

GH1 (retaining: 3D structure available: β-glucosidases; myrosinases; β-mannosidases: β-galactosidases: 6-P-β-glucosidases: 6-P-β-galactosidases): 49 sequences

Q9FEZ0	O23656	O80689	Q9LQS3	Q9SE50
Q9C8K1	O24433	O80690	Q9LU02	Q9SLA0
Q9C8J9	O48779	O80749	Q9LV33	Q9SR37
Q9C8Y9	O49117	O80750	Q9LV34	Q9SRX8
Q9FIU7	O64879	O82772	Q9LZJ0	Q9STP3
Q9FMD8	064880	P37702	Q9LZJ1	Q9STP4
Q9FH03	O64881	Q42585	Q9M1C9	Q9SVS1
Q9FLU9	O64882	Q42595	Q9M1D0	Q9SX92
Q9FLU8	O64883	Q9LIF9	Q9M1D1	Q9ZUI3
Q9FIW4	O65458	Q9LKR7	Q9SDL5	

GH2 (retaining; 3D structure available; β-galactosidases; β-mannosidases; β-glucuronidase); 2 sequences

O04029	Q9M1I3

GH3 (retaining; 3D structure available; β-glucosidases; β-xylosidases; (1→3)β-D-glucosidases; exo-(1→3), (1→4)β-D-glucanases; cellodextrinases; β-hexosaminidases); 12 sequences

Q9FWY2	Q9LJN4	Q9LZJ4	Q9SD72
Q9FGY1	Q9LXA8	Q9SD68	Q9SD73
Q9FLG1	Q9LXD6	Q9SD69	Q9SGZ5

GH5 (retaining; 3D-structure available; cellulases; β-mannanases; exo-(1→3)β-D-glucanases; endo(1→6)β-D-glucanases; xylanases); 13 sequences

Q9FZ29	Q9LFE7	Q9L W44	Q9SAE6	Q9SKU9
Q9FJZ3	Q9LTM8	Q9LZV3	Q9SG94	
Q9LF52	Q9LTN0	Q9M0H6	Q9SG95	

GH9 (inverting; 3D structure available; cellulases); 27 sequences

Q9FX19	O23697	O64891	Q38817	Q9SF61
Q9C7W3	O48766	O64949	Q38890	Q9STW8
O04478	O49296	O80497	Q9LR07	Q9SUS0
O23134	O64889	O81416	Q9LZG2	Q9SVJ2
O23696	O64890	O82513	Q9M995	Q9SVJ3
Q9SVJ4	Q9SZ90			

GH10 (retaining; 3D structure available; xylanases); 12 sequences

O80596	O81753	O82111	Q9SYE3
O81751	O81754	Q9SM08	Q9SZP3
O81752	O81897	Q9SVF5	Q9ZVK8

GH13 (retaining; 3D structure available; starch/glycogen degradation); 9 sequences

O04196	Q42531	Q9M0S5	Q9SRI7	Q9ZVT2
O23647	Q9LTP8	Q9SGS0	Q9SW26	

GH14 (inverting; 3D structure available; starch/glycogen degradation); 8 sequences

Q9FM68	O23553	O80831	Q9LIR6
Q9FH80	O65258	P25853	Q9ZV58

GH16 (retaining; 3D structure available; (1→3),(1→4)β-D-glucanases; (1→3)β-D-glucanases; xyloglucan endotransferases); 34 sequences

Q9FKL9	P24806	Q39099	Q9SEB0	Q9XIJ7
Q9FKL8	P93046	Q39148	Q9SJL9	Q9XIW1

Table 1. Continued.

Q9F131	Q38857	Q9LJR7	Q9SMP1	Q9ZR10
O04906	Q38908	Q9LK45	Q9SV60	Q9ZSU4
O49412	Q38909	Q9LPY1	Q9SV61	Q9ZV40
O49542	Q38910	Q9M0D1	Q9SVV2	Q9ZVK1
O80803	Q38911	Q9M0D2	Q9T067	

GH17 (retaining; 3D structure available; (1→3)β-D-glucanases); 47 sequences

Q9C8R3	O23562	Q06915	Q9M2T6	Q9SRT4
Q9C7U5	O48727	Q9LJD6	Q9M357	Q9SZ23
Q9FHX5	O48812	Q9LK11	Q9MAQ2	Q9XIR7
Q9FLP4	O49352	Q9LK41	Q9SFW1	Q9ZQG9
Q9FK49	O49353	Q9LP27	Q9SHZ2	Q9ZRI3
Q9FJU9	O49737	Q9LV98	Q9SIX5	Q9ZU91
Q9FGH4	O64577	Q9M046	Q9SNC1	Q9ZUP5
Q9FMZ2	O65399	Q9M088	Q9SQL9	
Q9FGT5	O65675	Q9M0E7	Q9SQM0	
O23473	P33157	Q9M2M0	Q9SQR1	

GH18 (retaining; 3D structure available; chitinases); 10 sequences

O81853	O81855	O81857	O81861	O81863
O81854	O81856	O81858	O81862	P19172

GH19 (inverting; 3D structure available; chitinases); 15 sequences

Q9FZ25	O22842	O24603	P19171	Q9MA41
Q9FXB8	O23248	O24654	Q42042	Q9SN84
O22841	O24598	024658	Q9LSP9	Q9ZSI6

GH20 (retaining; 3D structure available; β-hexosaminidase); 3 sequences

O04477	Q9M3C5	Q9SYK0

GH27 (retaining; α-galactosidases; α-N-acetylgalactosaminidases); 4 sequences

Q9FT98	Q9FT97	Q9LIN8	Q9LYL2

GH28 (inverting; 3D structure available; pectin degradation); 66 sequences

Q9FXC1	O48577	Q9LNG3	Q9SFB7	Q9SUV3
Q9FWX5	O48729	Q9LQD0	Q9SFC9	Q9SUV4
Q9C8C3	O49319	Q9LQD1	Q9SFD0	Q9SVN6
Q9C8C4	O49721	Q9LRY8	Q9SFD1	Q9SY17
Q9C8C2	O65401	Q9LUB8	Q9SFD2	Q9SYK6
Q9FLE3	O65905	Q9L W07	Q9SGY5	Q9SYK7
Q9FF19	O80559	Q9LYJ5	Q9SHH4	Q9SZC1
O04474	O81746	Q9M1L0	Q9SJN7	Q9ZQF5
O22699	O81798	Q9M1R3	Q9SLM8	Q9ZQG6
O22817	O82218	Q9M1Y9	Q9SMT3	Q9ZUE7
O22818	P49062	Q9M318	Q9SRC2	
O22935	P49063	Q9M3B5	Q9SSC2	
O23147	Q38958	Q9M9F6	QSSC5	
O48576	Q9LMA9	Q9S761	Q9SUP5	

GH31 (retaining; α-glucosidases; α-galactosidases; α-xylosidases; α-glucan lyases); 5 sequences

Q9FN05	O22444	Q9LUG2	Q9LZT7	Q9S7Y7

GH32 (retaining; invertases; β-fructosidases; transfructosidases; inulinases); 11 sequences

Q9C721	Q42567	Q9LIB9	Q9SJN3

Table 1. Continued.

Q38801	Q43348	Q9LYI1	Q9SXD2
Q39041	Q43866	Q9SI83	

GH35 (retaining; β-galactosidases); 20 sequences

O23243	O49609	Q9SCU8	Q9SCV1	Q9SCV3
O48836	Q42150	Q9SCU9	Q9SCV2	Q9SCV4
Q9SCV5	Q9SCV7	Q9SCW0	Q9SRH2	Q9SZ15
Q9SCV6	Q9SCV8	Q9SCW1	Q9SSM8	Q9SZN8

GH36 (retaining; α-galactosidases; stachyose synthases; raffinose synthases); 5 sequences

Q9SYJ4	O04607	Q9LFZ7	Q9FND9	Q9SCM1

GH37 (inverting; trehalases); 3 sequences

Q9SU50	O22986	Q9SEH9

GH38 (retaining; α-mannosidases); 4 sequences

Q9FKW9	Q9FFX7	P94078	Q9LFR0

GH43 (inverting; β-xylosidases); 2 sequences

Q9ZRR0	Q9M2X0

GH47 (inverting; α-mannosidases); 5 sequences

Q9C512	Q9C8R9	Q9LJB6	AAK43943	Q9SXC9

GH51 (retaining; α-L-arabinofuranosidases); 2 sequences

Q9SG80	Q9XH04

GH63 (α-glucosidases); 2 sequences

O48699	Q9SR56

GH77 (retaining; starch degradation); 2 sequences

O22198	Q9LV91

GH79 (retaining; endo-β-glucuronidases); 3 sequences

Q9FLK8	Q9FZP1	Q9SDA1

GH81 ((1→3)β-D-glucanases); 2 sequences

Q9LFT3	Q9LPQ0

GH85 (β-N-acetylglucosaminidases); 2 sequences

Q9SRL4	Q9FLA9

other families of enzymes involved in pectin degradation in *Arabidopsis*.

At the present degree of completion, the genome of *Arabidopsis* contains 379 glycoside hydrolases from 29 families; 15 of these 29 families have a 3D structural representative (Figure 3A). The genome contains 356 glycosyltransferases from 27 families; two of these families have a 3D structural representative (Figure 3B).

With a total of over 730 glycoside hydrolases and glycosyltransferases, *Arabidopsis* contains a lot more than any other eukaryotes whose genomes are completely sequenced. Indeed, *Saccharomyces cerevisiae* contains less than 100 of these enzymes, *Drosophila melanogaster* about 200 and *Caenorhabditis elegans* about 250 (Coutinho and Henrissat, unpublished). The large repertoire of glycoside hydrolases and glycosyltransferases of *Arabidopsis* is also apparent when expressed as a percentage of the total coding regions: about 2.8% for *Arabidopsis*, 1.4% for *C. elegans*, 1.4% for *D. melanogaster* and 1.4% for *S. cerevisiae*. Simple primordial organisms, such as the Archae, have only a few glycosyltransferases and glycoside hydrolases (Coutinho and Henrissat, 1999b), whereas in higher, pluricellular organisms large numbers of CAZymes appear as a result of tissue differentiation and the need of finer control mechanisms.

Table 2. Listing of the *Arabidopsis thaliana* ORFs found in the various families of glycosyltransferases (accession numbers of the deduced protein sequences are from SwissProt/TrEMBL or GenBank databases). The mechanism (retaining or inverting where known) and the availability of a 3D structural representative in each family are reported. The main enzymatic activities found in each family are indicated. Asterisks indicate those sequences with two catalytic modules from the same family.

GT1 (inverting; β-Glu transferases, β-Xyl transferases, Rha transferases, β-GluA transferases, β-Gal transferases; REM: transferases adding a single sugar); 116 sequences

Q9FE68	O23406	Q9LML7	Q9M0P3	Q9T080
Q9LML6	O23649	Q9LNE6	Q9M916	Q9T081
Q9FKD1	O48676	Q9LNI1	Q9M9E7	Q9XIG1
Q9FN28	O48715	Q9LNI4	Q9S7R8	Q9XIQ4
Q9FN26	O49492	Q9LPS8	Q9S9P6	Q9XIQ5
Q9F199	O64732	Q9LR44	Q9SCP5	Q9ZQ54
Q9F198	O64733	Q9LS16	Q9SCP6	Q9ZQ94
Q9F197	O81010	Q9LS21	Q9SGA8	Q9ZQ95
Q9F196	O81498	Q9LSM0	Q9SJL0	Q9ZQ96
O04622	O82381	Q9LSY4	Q9SK82	Q9ZQ97
O04930	O82382	Q9LSY5	Q9SKC1	Q9ZQ98
O22182	O82283	Q9LSY6	Q9SKC5	Q9ZQ99
O22183	O82385	Q9LSY8	Q9SNB0	Q9ZQG3
O22186	Q9LEQ4	Q9LSY9	Q9SNB1	Q9ZQG4
O22820	Q9LFJ8	Q9LTA3	Q9SNB2	Q9ZU71
O22822	Q9LFJ9	Q9LTH2	Q9SNB3	Q9ZU72
O23205	Q9LFK0	Q9LTH3	Q9SSJ5	Q9ZUV0
O23270	Q9LHJ2	Q9LVF0	Q9STE3	Q9ZVX4
O23380	Q9LJA6	Q9LVR1	Q9STE4	Q9ZVY5
O23381	Q9LK73	Q9LVW3	Q9STE6	Q9ZWJ3
O23382	Q9LME8	Q9LXV0	Q9SY84	
O23400	Q9LME9	Q9LZD8	Q9SYC4	
O23401	Q9LMF0	Q9M051	Q9SYK8	
O23402	Q9LMF1	Q9M052	Q9SYK9	

GT2 (inverting; 3D structure available; polymerizing enzymes: cellulose synthases, (1→3)β-D-glucan synthases, fungal chitin synthases; enzymes adding a single sugar: β-glycosyltransferases); 47 sequences

Q9FVR3	Q9FN17	O23383	O48947	O80890
Q9FNC3	O22988	O23386	O48948	O80891
Q9FIB9	O22989	O23502	O49323	O80898
Q9FGF9	O22990	O48946	O65338	O80899
Q9LF09	Q9LQC9	Q9SJ22	Q9SZL9	Q9ZQB9
Q9LFL0	Q9LR87	Q9SJA2	Q9T0B2	Q9ZQN8
Q9LG28	Q9LY45	Q9SKJ5	Q9T0B3	Q9ZU5
Q9LJP4	Q9LZR3	Q9SN37	Q9T0B4	
Q9LM93	Q9M9M4	Q9SRT3	Q9T0L2	
Q9LMW5	Q9SB75	Q9SRW9	Q9XHP6	

GT4 (retaining; α-glycosyltransferases; sucrose synthases; sucrose-phosphate synthases); 22 sequences

Q9FXG9	O49464	Q9LSB5	Q9S7D1	Q9XEE9
Q9FWT0	O65264	Q9LXL5	Q9SN30	Q9ZV98
Q9FJ20	P49040	Q9M111	Q9SQZ3	
Q9FGU6	Q00917	Q9M1U9	Q9SSL3	
Q9FY54	Q9LFB4	Q9MAU0	Q9SSP6	

GT5 (retaining; starch/glycogen synthases); 5 sequences

O49727	Q9MAC8	Q9MAQ0	Q9SAA5	Q9SEI7

Table 2. Continued.

GT8 (retaining; α-Glu transferases, α-Gal transferases, glycogenin); 37 sequences, 38 modules

Q9FZ37	O04031	O80649*	Q9LXS3	Q9SVF8
Q9FXB2	O04253	O80766	Q9LZJ9	Q9SZB0
Q9FX71	O04536	Q9LE59	Q9M8J2	Q9ZPZ1
Q9FWY9	O22693	Q9LF35	Q9M9Y5	Q9ZQP4
Q9FWA4	O22893	Q9LHD2	Q9MAB8	Q9ZVI7
Q9FH36	O23503	Q9LN68	Q9S7G2	
Q9FFA1	O48684	Q9LSB1	Q9SKT6	
Q9FIK3	O80518	Q9LSG3	Q9STQ9	

GT10 ((1→3)α-L-fucosyltransferase); 3 sequences

Q9FX97	Q9LJK1	Q9C8W3

GT13 (inverting; glycoprotein (1→2)β-D-GlcNAc transferases); 2 sequences

Q9SZM4	Q9XGM8

GT14 (inverting; glycoprotein (1→6)β-D-GlcNAc transferases); 11 sequences

Q9FLD7	Q9LFQ0	Q9LR71	Q9SZS3
O80927	Q9LIQ3	Q9M8V9	Q9ZQZ7
Q9LE60	Q9LNN5	Q9SS69	

GT17 (inverting; glycoprotein (1→4)β-D-GlcNAc transferases); 6 sequences

Q9LT58	Q9LY81	Q9SAD5	Q9SFP8	Q9SKF0
Q9SS93				

GT19 (inverting; bacterial lipid A dissacharide synthase); 1 sequence

Q9SJB5

GT20 (retaining; trehalose-phosphate synthases); 12 sequences

Q9FZ57	O80738	Q9LRA7	Q9SYM4
O23617	P93653	Q9SFR0	Q9T079
O64608	Q9LM10	Q9SHG0	Q9ZV48

GT22 (mannosyltransferases); 2 sequences

Q9FZ49	Q9LEQ5

GT24 (glucosyltransferases); 1 sequence

Q9M981

GT28 (inverting; 3D structure available; monogalactosyldiacylglycerol synthases, β-GlcNAc transferases); 4 sequences

Q9FZL5	O81770	O82730	Q9SFN2

GT29 (inverting; sialyltransferases); 2 sequences

Q9FRR9	Q9M301

GT30 (bacterial KDO-transferases); 1 sequence

Q9LZR2

GT31 (inverting; β-Gal transferases, β-GalNAc transferases); 33 sequences

Q9C809	O23050	Q9LKA9	Q9MAP8	Q9SSI9
Q9C881	O23072	Q9LM60	Q9SAA4	Q9SUA8
Q9FN55	O23378	Q9LMM4	Q9SH19	Q9SUS6
Q9FKM1	O49381	Q9LQQ7	Q9SIR3	Q9SZ33
Q9FK08	O80941	Q9LQX1	Q9SQL7	Q9ZV71
O04561	O80987	Q9LV16	Q9SQU7	
O23042	O81745	Q9MAH2	Q9SSG4	

Table 2. Continued.

GT32 *(retaining; α-Man transferases, α-GlcNAc transferases); 5 sequences, 6 modules*

O80440*	Q9LFB1	Q9M8H2	Q9S790	Q9SNA8

GT33 *(inverting; β-Man transferases); 1 sequence*
Q9FX74

GT34 *(retaining; α-Gal transferases); 8 sequences*

Q9LF80	Q9LZJ3	Q9SG68	Q9SZG1
O81007	Q9M9U0	Q9SYY3	Q9SVF4

GT35 *(retaining; 3D structure available; starch/glycogen phosphorylases); 2 sequences*
Q9LIB2   Q9SD76

GT37 *(inverting; xyloglucan α-1,2-fucosyltransferase); 11 sequences, ,12 modules*

O81053	Q9SJP2	Q9SWH5	Q9X180
Q9LMF2[a]	Q9SJP4	Q9XI77	Q9XI81
Q9SG64	Q9SJP6	Q9X178	

GT41 *(inverting; glycoprotein β-GlcNAc transferases); 2 sequences*
Q96301   Q9M8Y0

GT43 *(inverting; 3D structure available; β-GluA transferases); 4 sequences*
O23194   Q9SFZ7   Q9SXC4   Q9ZQC6

GT47 *(inverting; β-GluA transferases); 3 sequences*
Q9LY62   Q9LZ75   Q9M8P4

GT48 *(inverting; (1→3)β-D-glucan synthases); 12 sequences*

Q9LR43	Q9LXT9	Q9SFU6	Q9SL03
Q9LTG5	Q9LYS6	Q9SHJ3	Q9SYJ7
Q9LUD7	Q9S9U0	Q9SJM0	Q9ZT82

GT50 *(inverting; α-mannosyltransferases); 1 sequence*
Q9C575

What is the origin of the large number of CAZymes in *Arabidopsis*?

Obviously, higher eukaryotes require a larger number of proteins for differential expression and regulation in tissues, organs and organelles and this is easily accounted for by both duplication of chromosomal domains and extensive duplication of individual genes. Other evolutionary events are probably also involved. Indeed, like most genes, CAZyme genes are mostly passed vertically, but different mechanisms of horizontal or lateral gene transfer can bring new material into genomes (Ochman *et al.*, 2000). For instance, the acquisition of eukaryotic genetic material by bacteria is relatively common, with some well-documented cases (Little *et al.*, 1994). There are also a few reverse examples of acquisition of bacterial genes by horizontal transfer to both lower (Garcia-Vallvé *et al.*, 2000) and higher eukaryotes (Smant *et al.*, 1998) including plants (Aoki and Syno, 1999).

The cell-wall CAZymes

The census presented here demonstrates the expected importance of cellulose- and xylan-based systems in plant cell-wall biochemistry. There are at least 40 NDP-sugar-dependent glycosyltransferases of family GT2, many of which are involved in the synthesis of β-linked polysaccharides, such as cellulose (reviewed in Charnock *et al.*, 2001; Richmond and Somerville, 2001). The probable hydrolytic counterparts of cellulose synthases are found among the 27 GH9 potential cellulases. It is unclear why *Arabidopsis* contains just one of the 11 known sequence families of cellulases. This is in marked contrast to all cellulolytic organisms, which display a battery of different cellulases from many different families. This may reflect the varied roles of these enzymes in the natural *milieu*; microorganisms use a battery of different enzymes for the complete degradation of cellulose as a food source, whereas plants use cellulases for selective modifica-

A. Glycoside hydrolases

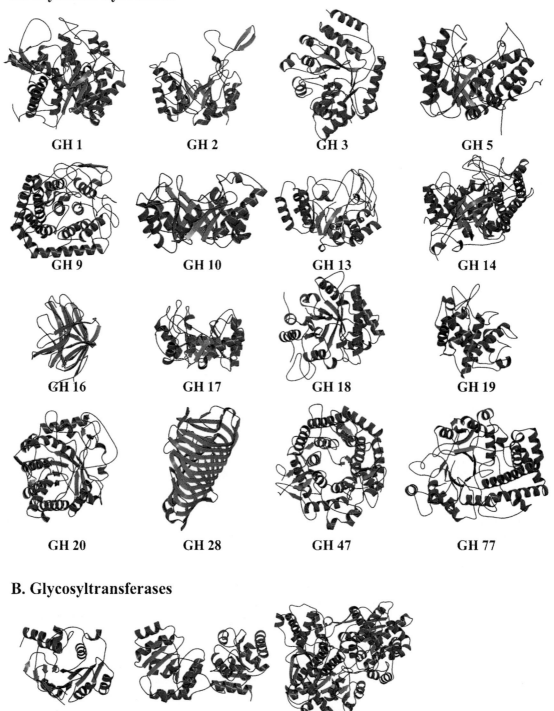

GH 1   **GH 2**   **GH 3**   **GH 5**

GH 9   **GH 10**   **GH 13**   **GH 14**

GH 16   **GH 17**   **GH 18**   **GH 19**

GH 20   **GH 28**   **GH 47**   **GH 77**

B. Glycosyltransferases

GT 2   **GT 28**   **GT 35**

Figure 3. Representative 3D structures of glycoside hydrolases (A) and glycosyltransferases (B) for the families found in *Arabidopsis*. The figure made using the Molscript program (Kraulis, 1991).

tion during growth, expansion and abscission (Nicol *et al.*, 1998).

During the course of analysis of the *Arabidopsis* genes, it became clear that among the many members of glycoside hydrolase family GH17 (Table 1), encoding potential homologues to grass $(1\rightarrow3)$ β- and $(1\rightarrow3)$, $(1\rightarrow4)$ β-glucanases (see Hrmova and Fincher, 2001), more than half were modular, with an extra C-terminal module of 100–120 residues. The multiple occurrences of this module allowed to define a family, which was called X8 (Henrissat and Davies, 2000). This extension was found sometimes in two copies at the C-terminus of a family GH17 protein. More interesting is the observation of 13 occurrences of this module in isolation, just attached to a membrane-spanning region, or even attached to sequences clearly not related to glycoside hydrolases (Figure 1). In several instances, the ORFs corresponding for the isolated module are annotated as '$(1\rightarrow3)\beta$-glucanase-like' pointing to the importance of modularity in ORF annotation. The fact that this domain is found (1) fused to catalytic domains, (2) in isolation and (3) linked to a transmembrane region points to a spectrum of different cellular functions. By analogy to the chitin-binding modules borne by chitinases and the cellulose-binding modules of the cellulases, it is tempting to propose that the X8 modules have $(1\rightarrow3)\beta$-glucan-binding activity. However, for the carbohydrate-binding modules just like for the catalytic domains, substrate specificity is hard to predict due to the frequent 'polyspecificity' found in the sequence-based families.

The enzymes responsible for xylan backbone biosynthesis are still not characterized and possibly missing in our census, as no sequenced protein with xylan synthase activity has been reported. The re-arrangement/modification of xylans involve xyloglucan endotransferases found among the 34 GH16 members (Campbell and Braam, 1999), but also the action of 12 potential xyloglucan fucosyltransferases (family GT37), one of which is already characterized in *Arabidopsis* (Perrin *et al.*, 1999; see Perrin *et al.*, 2001). The degradation of xylans requires a range of CAZymes able to act on xylo- and arabino-glucans (Coughlan and Hazlewood, 1993). For *Arabidopsis* that would correspond to the 12 potential GH10 xylanases, and may be also to two GH43 and two GH51, which encode potential α-L-arabinofuranosidases.

A surprise arising from this study was the huge number of enzymes involved in the degradation of pectin: 66 GH28 (all known members of this family degrade pectin). This number is likely to be significant, as other families of proteins involved in pectin degradation are also largely represented in *Arabidopsis*: 58 potential pectin methylesterases, 13 potential pectin acetylesterases and 28 potential pectate/pectin lyases and 7 potential rhamnogalacturanan lyases (Table 3); in total, probably more than 170 open-reading frames are devoted to pectin degradation in *Arabidopsis*. This abundance of pectin-degrading enzymes probably not only reflect the abundance and structural diversity of pectin, but may also point to other roles for pectin degradation products such as signalling molecules (Spiro *et al.*, 1998). The degree of pectin esterification is known to affect cell division, non-dividing cells presenting higher levels of unesterified pectins (Dolan *et al.*, 1997).

The starch CAZymes and related proteins

Plants utilize starch as a storage material. Starch, the common storage material in plants, is a complex macromolecule made by the concerted action of purely biosynthetic enzymes (starch synthases, starch phosphorylases) and transglycosidases responsible for the creation/hydrolysis of branch points (branching, debranching and disproportionating enzymes) (Myers *et al.*, 2000). Starch hydrolysis is performed by hydrolytic enzymes such as α- and β-amylases and α-glucosidases. In *Arabidopsis*, the complete machinery to synthesize and utilize starch is found among families GT5, GT35, GH13, GH14, GH31 and GH77.

In plants, sucrose constitutes an intermediate storage system alternative to starch. In addition, sucrose can have an extensive role in cellular signalling in plants (Dijkwel *et al.*, 1997; Dejardin *et al.*, 1999). Sucrose synthases, which are among the 22 members of family GT4, are likely to work closely with invertases (Sturm and Tang, 1999) which are found among the 11 GH32 members in the *Arabidopsis* genome.

In *Arabidopsis*, genes encoding functional enzymes for the synthesis (family GT20) and degradation of trehalose (family GH37) have been detected recently and the first investigations suggest that trehalose interferes with carbon allocation to the sink tissues by inducing starch synthesis in the source tissues and that trehalose can modulate sugar-mediated gene expression (Wingler *et al.*, 2000).

Table 3. Listing of other *Arabidopsis thaliana* ORFs possibly involved in pectin degradation (accession numbers of the deduced protein sequences are from SwissProt/TrEMBL or GenBank databases). The availability of a 3D structural representative in each family and the known enzymatic activities in the families are mentioned.

CE8 (3D structure pending; pectin methylesterases): 58 sequences

Q9FHN5	O3038	O1415	Q9XD9	QMAL0
Q9FHN4	O23447	O81516	Q9LXK7	Q9SG77
Q9FKF3	O48711	Q42534	Q9LY17	Q9SG78
Q9FGK7	O48712	Q43867	Q9LY18	Q9SMY6
Q9FM79	O49006	Q9LPX7	Q9LY19	Q9SMY7
Q9FK05	O49298	Q9LPX8	Q9LYT5	Q9QRJ8
Q9FJ21	O64479	Q9LPX9	Q9LZZ0	Q9SRX4
Q9FLF6	O80721	Q9LPZ8	Q9M1Q7	Q9STY3
Q9FF78	O80722	Q9LRN4	Q9M3B0	Q9SX50
O22149	O81300	Q9LSP1	Q9M7Y9	Q9T0P8
O22256	O81301	Q9LUL8	Q9M9W6	
O22902	O81320	Q9LVQ0	Q9M9W7	

CE13 (pectin acetylesterases); 13 sequences

Q9FVU3	O65250	O80537	Q9M9K8	Q9SR23
Q9FF93	O65712	O80731	Q9SFF6	
Q9FH82	O65713	Q9M1R8	Q9SR22	

PL1 (3D structure known; pectate lyases, pectin lyases); 28 sequences

Q9C8G4	O23665	O65456	Q9M3D5	Q9SRH4
Q9FM66	O23666	O65457	Q9M8Z8	Q9SV40
Q9F181	O23667	Q9LFP5	Q9MAF1	Q9SVP1
Q9FMK5	O23668	Q9LJ42	Q9SB71	Q9SVQ6
Q9FY87	O64510	Q9LRM5	Q9SCP2	
O23017	O65388	Q9LTZ0	Q9SF49	

PL4 (rhamnogalacturonan lyases); 7 sequences

O04510	O04512	Q9STV1	Q9ZQ51
O04511	O65620	Q9SZK3	

Other CAZymes

Biosynthesis of the predominant membrane galactolipids involves two types of glycosyl transferases, 1,2-diacylglycerol 3-β-galactosyltransferase (MGDG synthase; Shimojima *et al.*, 1997) and digalactosyldiacylglycerol synthase (DGDG synthase; Dörmann *et al.*, 1999). In *Arabidopsis*, three MGDG synthases, which transfer a β-galactose from UDP-galactose on diacylglycerol, are found in inverting family GT28. Two DGDG synthases, which transfer an α-galactose to the O-6 position of monogalactosylglycerol, are found in retaining family GT4.

Protein glycosylation in plants shares common features with other eukaryotes (Sears and Wong, 1998). Glycosyltransferases and glycosidases, which catal-

yse the stepwise addition and trimming of sugar residues on N-glycans, respectively, are generally considered as working in a co-ordinated and highly ordered fashion (Lerouge *et al.*, 1998). On the hydrolytic side, *Arabidopsis* has two family GH63 members, potentially encoding processing α-glycosidases, and α-mannosidases from both families GH38 (four members) and GH47 (five members). The glycosyltransferases possibly involved in N-glycan biosynthesis are many and probably include members of families GT10, GT13, GT14, GT17, GT22, GT24, GT29, GT31, GT32, GT33, GT34 and GT41. Only a very few of these enzymes have been characterized. These include a family GT13 (1→2)β-*N*-acetylglucosaminyltransferase (Strasser *et al.*, 1999) and a newly described and not yet classified *Ara-*

bidopsis $(1\rightarrow2)\beta$- xylosyltransferase responsible for the plant specific xylosylation activity (Strasser *et al.*, 2000). Although, two potential α-fucosyltransferases are present in *Arabidopsis* (family GT10), no putative fucosidase has been identified in the present survey. This hydrolytic activity, might either be in the last pieces of the genome which were not available at the time this census was performed or simply contained in one, or several, of the many uncharacterized ORFs encoding potential glycosidases.

CAZymes in plant defence

As large groups of organisms (herbivores, pathogens) attack plants, the latter have elaborated a range of defence strategies. Several of these strategies rely on the production of plant CAZymes for a direct action on pathogens (Bishop *et al.*, 2000). The expression of such enzymes can be induced by a variety of factors, such as wounding, the detection of the chemical signature of a pathogen or elicitor, or simply of defence-related signal molecules (Reymond and Farmer, 1998). Among the CAZymes activated in this manner, chitinases from families GH18 and GH19 and endo-$(1\rightarrow3)\beta$-glucanases from family GH17 are abundant in Arabidopsis with 10, 15 and 47 sequences, respectively. By analogy with other plants, the large array of GH17 CAZymes in *Arabidopsis* could be associated with different pathogen-response mechanisms (Jin *et al.*, 1999). While plants do not make chitin, they do synthesize $(1\rightarrow3)\beta$-D-glucan (callose) as part of their defence strategy against microbes (Benhamou, 1995), as well as during initiation of cell plate formation (Hong *et al.*, 2001), pollen tube growth (Doblin *et al.*, 2001), and a certain stages of cotton fibre development (Cui *et al.*, 2001). *Arabidopsis* has at least 12 open-reading frame candidates for this function (Table 2).

Other defence mechanisms involve the release of toxic compounds upon attack of the plant by herbivory. These compounds are synthesized and stored by the plant as non-toxic precursors such as cyanogenic glycosides and glucosinolates. Several glycosyltransferases from family GT1 are known to participate in the synthesis of these precursors. Upon attack by an herbivore, for instance through mastication, the precursors are mixed with hydrolytic enzymes that release the toxic aglycons (Rask *et al.*, 2000). The hydrolytic enzymes, β-glucosidases such

as myrosinase and cyanogenic β-glucosidase, are found in family GH1.

CAZymes in plant signalling

The role of glycosylation in plant defence does not account for the prevalence of family GT1 enzymes, which are found in 116 copies in the *Arabidopsis* genome (Table 2). Because a glycosyltransferase must recognize both the sugar nucleotide and the appropriate acceptor, the multiplicity of the GT1 enzymes most likely reflects a large variety of acceptor specificity. GT1 enzymes typically transfer β-linked monosaccharides such as glucose and xylose. In many higher organisms (*D. melanogaster*, *C. elegans*, *Homo sapiens*), this enzyme family is also very abundant and implicated in detoxification processes by glucuronylation of lipophilic compounds (Tukey and Strassburg, 2000). Because hydrolysis of the resulting conjugates is a potential hazard to the organism (Minton *et al.*, 1986), these higher eukaryotes have very few hydrolytic enzymes to perform the reverse reaction (Coutinho and Henrissat, in preparation). By contrast, *Arabidopsis* contains 49 members of glycoside hydrolase family GH1, the likely hydrolytic counterpart of the 116 potential glycosyltransferases of family GT1. This points to a dominant role for the plant GH1 glycosidases, which are often aglycon-specific enzymes (Cicek *et al.*, 2000), and thereby probably control a range of signalling molecules (hormone, coniferin and flavonol glucosides) by glycosylation/deglycosylation (Mita *et al.*, 1997).

How (in)complete is this census?

The present census covers the entries released from GenBank up to 1 June 2001. Therefore, the final number of entries displayed in Tables 1–3 may vary slightly when the last pieces of genomic sequence are added and when redundancy is completely eliminated. Since our search is based on existing families, the families which have not been previously reported could not be picked in our census. Because glycoside hydrolases are relatively easier to characterize that glycosyltransferases, and because glycoside hydrolases have been studied for a much longer time, the number of glycoside hydrolase families still unknown is probably very small. On the other hand, it is likely that some families of glycosyltransferases are not yet discovered, and in this case, they were not picked in our search. Another limitation with the family classifications is that a family can be defined only when one of its members is characterized at the biochemical level. For example, while cellulose synthases are found in family GT2, several fungal and plant sequences, different from any known glycosyltransferase and thus potentially forming a separate family, have been annotated in sequence databanks as putative $(1\rightarrow3)\beta$-glucan synthases. However, the lack of evidence that the encoded proteins had UDP-glucose glucosyltransferase activity prevented the creation of a family until very recently when such evidence was found for an enzyme from *Pneumocystis carinii* (Kottom and Limper, 2000). It is possible, therefore, that the glycosyltransferase repertoire of *Arabidopsis* is even larger than described here.

Acknowledgements

We would like to thank the EU (TMR grant BIO4-CT97-5077 to P.M.C.) and CNRS. We are grateful to Chris Somerville for useful discussions during a Gordon Research Conference on Plant Cell Walls. The York laboratory is supported by the BBSRC and the Wellcome Trust. G.J.D. is a Royal Society University Research Fellow.

References

Arabidopsis Genome Initiative. 2000. Analysis of the genome sequence of the flowering plant *Arabidopsis thaliana*. Nature 408: 796–815.

Altschul, S.F., Madden, T.L., Schaffer, A.A., Zhang, J., Zhang, Z., Miller, W. and Lipman, D.J. 1997. Gapped BLAST and PSI-BLAST: a new generation of protein database search programs. Nucl. Acids Res. 25: 3389–3402.

Aoki, S. and Syno, K. 1999. Horizontal gene transfer and mutation: *ngrol* genes in the genome of *Nicotiana glauca*. Proc. Natl. Acad. Sci. USA 96: 13229–13234.

Arabidopsis Genome Initiative. 2000. Analysis of the genome sequence of the flowering plant *Arabidopsis thaliana*. Nature 408: 796–815.

Benhamou, N. 1995. Immunocytochemistry of plant defense mechanisms induced upon microbial attack. Microsc. Res. Tech. 31: 63–78.

Bishop, J.G., Dean, A.M. and Mitchell-Olds, T. 2000. Rapid evolution in plant chitinases: molecular targets of selection in plant-pathogen coevolution. Proc. Natl. Acad. Sci. USA 97: 5322–5327.

Burmeister, W.P., Cottaz, S., Rollin, P., Vasella, A. and Henrissat, B. 2000. High resolution X-ray crystallography shows that ascorbate is a cofactor for myrosinase and substitutes for the function of the catalytic base. J. Biol. Chem. 275: 39385–39393.

Callebaut, I., Labesse, G., Durand, P., Poupon, A., Canard, L., Chomilier, J., Henrissat, B. and Mornon, J.P. 1997. Deciphering protein sequence information through hydrophobic cluster analysis (HCA): current status and perspectives. Cell. Mol. Life Sci. 53: 621–645.

Campbell, J.A., Davies, G.J., Bulone, V. and Henrissat, B. 1997. A classification of nucleotide-diphospho-sugar glycosyltransferases based on amino acid sequence similarities. Biochem. J. 326: 929–939.

Campbell, P. and Braam, J. 1999. *In vitro* activities of four xyloglucan endotransglycosylases from *Arabidopsis*. Plant J. 18: 371–382.

Charnock, S.J., Bolam, D.N., Turkenburg, J.P., Gilbert, H.J., Ferreira, L.M., Davies, G.J. and Fontes, C.M. 2000. The X6 'thermostabilizing' domains of xylanases are carbohydrate-binding modules: structure and biochemistry of the *Clostridium thermocellum* X6b domain. Biochemistry 39: 5013–5021.

Charnock, S.J. and Davies, G.J. 1999. Structure of the nucleotide-diphospho-sugar transferase, SpsA from *Bacillus subtilis*, in native and nucleotide-complexed forms. Biochemistry 38: 6380–6385.

Charnock, S.J., Henrissat, B. and Davies, G. 2001. Three-dimensional structures of UDP-sugar glycosyltransferases illuminate the biosynthesis of plant polysaccharides. Plant Physiol. 125: 527–531.

Cicek, M., Blanchard, D., Bevan, D.R. and Esen, A. 2000. The aglycone specificity-determining sites are different in 2,4-dihydroxy-7-methoxy-1,4-benzoxazin-3-one (DIMBOA)-glucosidase (maize β-glucosidase) and dhurrinase (sorghum β-glucosidase). J. Biol. Chem. 275: 20002–20011.

Coughlan, M.P. and Hazlewood, G.P. 1993. β-1,4-D-xylan-degrading enzyme systems: biochemistry, molecular biology and applications. Biotechnol. Appl. Biochem. 17: 259–289.

Coutinho, P. and Henrissat, B. 1999a. Carbohydrate-active enzymes: an integrated database approach. In: H. Gilbert, G. Davies, B. Henrissat and B. Svensson (Eds.) Recent Advances in Carbohydrate Bioengineering, Royal Society of Chemistry, Cambridge, UK, pp. 3–12.

Coutinho, P.M. and Henrissat, B. 1999b. Life with no sugars? J. Mol. Microbiol. Biotechnol. 1: 307–308.

Cui, X., Shin, H., Charlotte Song, C., Laosinchai1, W., Amano, Y. and Brown, R.M. Jr. 2001. A putative plant homolog of the yeast β-1,3-glucan synthase subunit FKS1 from cotton (*Gossypium hirsutum* L.) fibers. Planta, in press.

Davies, G. and Henrissat, B. 1995. Structures and mechanisms of glycosyl hydrolases. Structure 3: 853–859.

Davies, G.J. 1998. Structural studies on cellulases. Biochem. Soc. Transact. 26: 167–173.

Dejardin, A., Sokolov, L.N. and Kleczkowski, L.A. 1999. Sugar/osmoticum levels modulate differential abscisic acid-independent expression of two stress-responsive sucrose synthase genes in *Arabidopsis*. Biochem. J. 344: 503–509.

Dijkwel, P.P., Huijser, C., Weisbeek, P.J., Chua, N.H. and Smeekens, S.C. 1997. Sucrose control of phytochrome A signaling in *Arabidopsis*. Plant Cell 9: 583–595.

Doblin, M.S., De Melis, L., Newbigin, E., Bacic, A. and Read, S.M. 2001. Pollen tubes of *Nicotiana alata* express two genes from different β-glucan synthase families. Plant Physiol., in press.

Dolan, L., Linstead, P. and Roberts, K. 1997. Developmental regulation of pectic polysaccharides in the root meristem of *Arabidopsis*. J. Exp. Bot. 48: 713–720.

Dörmann, P., Balbo, I. and Benning, C. 1999. *Arabidopsis* galactolipid biosynthesis and lipid trafficking mediated by DGD1. Science 284: 2181–2184.

Garcia-Vallvé, S., Romeu, A. and Palau, J. 2000. Horizontal gene transfer of glycosyl hydrolases of the rumen fungi. Mol. Biol. Evol. 17: 352–361.

Gastinel, L.N., Cambillau, C. and Bourne, Y. 1999. Crystal structures of the bovine β4galactosyltransferase catalytic domain and its complex with uridine diphosphogalactose. EMBO J. 18: 3546–3557.

Gebler, J., Gilkes, N.R., Claeyssens, M., Wilson, D.B., Béguin, P., Wakarchuk, W.W., Kilburn, D.G., Miller, R.C. Jr., Warren, R.A. and Withers, S.G. 1992. Stereoselective hydrolysis catalyzed by related β-1,4-glucanases and β-1,4-xylanases. J. Biol. Chem. 267: 12559–12561.

Ha, S., Walker, D., Shi, Y. and Walker, S. 2000. The 1.9 Å crystal structure of Escherichia coli MurG, a membrane-associated glycosyltransferase involved in peptidoglycan biosynthesis. Protein Sci. 9: 1045–1052.

Henrissat, B. 1991. A classification of glycosyl hydrolases based on amino acid sequence similarities. Biochem. J. 280: 309–316.

Henrissat, B. and Bairoch, A. 1993. New families in the classification of glycosyl hydrolases based on amino acid sequence similarities. Biochem. J. 293: 781–788.

Henrissat, B. and Bairoch, A. 1996. Updating the sequence-based classification of glycosyl hydrolases. Biochem. J. 316: 695–696.

Henrissat, B. and Davies, G. 1997. Structural and sequence-based classification of glycoside hydrolases. Curr. Opin. Struct. Biol. 7: 637–644.

Henrissat, B. and Davies, G. 2000. Glycoside hydrolases and glycosyltransferases: families, modules and implications for genomics. Plant Physiol. 124: 1515–1520.

Hong, Z., Delauney, A.J. and Verma, D.P.S. 2001. A cell plate-specific callose synthase and its interaction with phragmoplastin. Plant Cell, in press.

Hrmova, H. and Fincher, G.B. 2001. Three-dimensional structures, substrate specificities and biological functions of β-D-glucan endo- and exohydrolases from higher plants. Plant Mol. Biol., this issue.

Jin, W., Horner, H.T., Palmer, R.G. and Shoemaker, R.C. 1999. Analysis and mapping of gene families encoding β-1,3-glucanases of soybean. Genetics 153: 445–452.

Kaiser, J. 2000. From genome to functional genomics. Science 288: 1715.

Kottom, T.J. and Limper, A.H. 2000. Cell wall assembly by Pneumocystis carinii: evidence for a unique Gsc-1 subunit mediating β-1,3-glucan deposition. J. Biol. Chem. 275: 40628–40634.

Kraulis, P.J. 1991. Molscript: a program to produce both detailed and schematic plots of protein structures. J. Appl. Crystallogr. 24: 946–950.

Lerouge, P., Cabanes-Macheteau, M., Rayon, C., Fischette-Laine, A.C., Gomord, V. and Faye, L. 1998. N-glycoprotein biosynthesis in plants: recent developments and future trends. Plant Mol. Biol. 38: 31–48.

Little, E., Bork, P. and Doolittle, R.F. 1994. Tracing the spread of fibronectin type III domains in bacterial glycohydrolases. J. Mol. Evol. 39: 631–643.

McCarter, J.D. and Withers, S.G. 1994. Mechanisms of enzymatic glycoside hydrolysis. Curr. Opin. Struct. Biol. 4: 885–892.

Minton, J.P., Walaszek, Z., Schooley, W., Hanausek-Walaszek M., and Webb, T.E. 1986. β-Glucuronidase levels in patients with fibrocystic breast disease. Breast Cancer Res. Treatm. 8: 217–222.

Mita, S., Murano, N., Akaike, M. and Nakamura, K. 1997. Mutants of Arabidopsis thaliana with pleiotropic effects on the expression of the gene for β-amylase and on the accumulation of anthocyanin that are inducible by sugars. Plant J. 11: 841–851.

Myers, A.M., Morell, M.K., James, M.G. and Ball, S.G. 2000. Recent progress toward understanding biosynthesis of the amylopectin crystal. Plant Physiol. 122: 989–997.

Nicol, F., His, I., Jauneau, A., Vernhettes, S., Canut, H. and Hofte, H. 1998. A plasma membrane-bound putative endo-1,4-β-D-glucanase is required for normal wall assembly and cell elongation in Arabidopsis. EMBO J. 17: 5563–5576.

Ochman, H., Lawrence, J.G. and Groisman, E.A. 2000. Lateral gene transfer and the nature of bacterial innovation. Nature 405: 299–304.

Pedersen, L.C., Tsuchida, K., Kitagawa, H., Sugahara, K., Darden, T.A. and Negishi, M. 2000. Heparan/chondroitin sulfate biosynthesis: structure and mechanism of human glucuronyltransferase I. J. Biol. Chem. 275: 34580–34585.

Perrin, R.M., DeRocher, A.E., Bar-Peled, M., Zeng, W., Norambuena, L., Orellana, A., Raikhel, N.V. and Keegstra, K. 1999. Xyloglucan fucosyltransferase, an enzyme involved in plant cell wall biosynthesis. Science 284: 1976–1979.

Perrin, R., Wilkerson, C. and Keegstra, K. 2001. Golgi enzymes that synthesize plant cell wall polysaccharides: finding and evaluating candidates in the genomic era. Plant Mol. Biol., this issue.

Rask, L., Andreasson, E., Ekbom, B., Eriksson, S., Pontoppidan, B. and Meijer, J. 2000. Myrosinase: gene family evolution and herbivore defense in Brassicaceae. Plant Mol. Biol. 42: 93–113.

Reymond, P. and Farmer, E.E. 1998. Jasmonate and salicylate as global signals for defense gene expression. Curr. Opin. Plant Biol. 1: 404–411.

Richmond, T.A. and Somerville, C.R. 2001. Integrative approaches to determining Csl function. Plant Mol. Biol., this issue.

Sears, P. and Wong, C.H. 1998. Enzyme action in glycoprotein synthesis. Cell. Mol. Life Sci. 54: 223–252.

Shimojima, M., Ohta, H., Iwamatsu, A., Masuda, T., Shioi, Y. and Takamiya, K. 1997. Cloning of the gene for monogalactosyldiacylglycerol synthase and its evolutionary origin. Proc. Natl. Acad. Sci. USA 94: 333–337.

Sinnott, M.L. 1990. Catalytic mechanisms of enzymic glycosyl transfer. Chem. Rev. 90: 1171–1202.

Smant, G., Stokkermans, J.P., Yan, Y., de Boer, J.M., Baum, T.J., Wang, X., Hussey, R.S., Gommers, F.J., Henrissat, B., Davis, E.L., Helder, J., Schots, A. and Bakker, J. 1998. Endogenous cellulases in animals: isolation of β-1,4-endoglucanase genes from two species of plant-parasitic cyst nematodes. Proc. Natl. Acad. Sci. USA 95: 4906–4911.

Spiro, M.D., Ridley, B.L., Eberhard, S., Kates, K.A., Mathieu, Y., O'Neill, M.A., Mohnen, D., Guern, J., Darvill, A. and Albersheim, P. 1998. Biological activity of reducing-end-derivatized oligogalacturonides in tobacco tissue cultures. Plant Physiol. 116: 1289–1298.

Strasser, R., Mucha, J., Mach, L., Altmann, F., Wilson, I.B., Glossl, J. and Steinkellner, H. 2000. Molecular cloning and functional expression of β-1,2-xylosyltransferase cDNA from Arabidopsis thaliana. FEBS Lett. 472: 105–108.

Strasser, R., Mucha, J., Schwihla, H., Altmann, F., Glossl, J. and Steinkellner, H. 1999. Molecular cloning and characterization of cDNA coding for β-1,2-N-acetylglucosaminyltransferase I (GlcNAc-TI) from Nicotiana tabacum. Glycobiology 9: 779–785.

Sturm, A. and Tang, G.Q. 1999. The sucrose-cleaving enzymes of plants are crucial for development, growth and carbon partitioning. Trends Plant Sci. 4: 401–407.

Tukey, R. and Strassburg, C. 2000. Human UDP-glucuronosyltransferases: metabolism, expression, and disease. Annu. Rev. Pharmacol. Toxicol. 40: 581–616.

Ünligil, U., Zhou, S., Yuwaraj, S., Sarkar, M., Schachter, H. and Rini, J. 2000. X-ray crystal structure of rabbit N-acetylglucosaminyltransferase I: catalytic mechanism and a new protein superfamily. EMBO J. 19: 5269–5280.

Vrielink, A., Ruger, W., Driessen, H.P. and Freemont, P.S. 1994. Crystal structure of the DNA modifying enzyme β-glucosyltransferase in the presence and absence of the substrate uridine diphosphoglucose. EMBO J. 13: 3413–3422.

Wingler, A., Fritzius, T., Wiemken, A., Boller, T. and Aeschbacher, R.A. 2000. Trehalose induces the ADP-glucose pyrophospho-rylase gene, ApL3, and starch synthesis in *Arabidopsis*. Plant Physiol. 124: 105–114.

Zechel, D.L. and Withers, S.G. 2000. Glycosidase mechanisms: anatomy of a finely tuned catalyst. Acc. Chem. Res. 33: 11–18.

Plant Molecular Biology **47**: 73–91, 2001.

Structure-function relationships of β-D-glucan endo- and exohydrolases from higher plants

Maria Hrmova and Geoffrey B. Fincher*

*Department of Plant Science, University of Adelaide, Waite Campus, Glen Osmond, SA 5064, Australia (*author for correspondence; e-mail geoff.fincher@adelaide.edu.au)*

Key words: catalytic mechanism, cell wall hydrolysis, monocotyledons, protein modelling, substrate binding, subsite mapping

Abstract

(1→3),(1→4)-β-D-Glucans represent an important component of cell walls in the Poaceae family of higher plants. A number of glycoside endo- and exohydrolases is required for the depolymerization of (1→3),(1→4)-β-D-glucans in germinated grain or for the partial hydrolysis of the polysaccharide in elongating vegetative tissues. The enzymes include (1→3),(1→4)-β-D-glucan endohydrolases (EC 3.2.1.73), which are classified as family 17 glycoside hydrolases, (1→4)-β-D-glucan glucohydrolases (family 1) and β-D-glucan exohydrolases (family 3). Kinetic analyses of hydrolytic reactions enable the definition of action patterns, the thermodynamics of substrate binding, and the construction of subsite maps. Mechanism-based inhibitors and substrate analogues have been used to study the spatial orientation of the substrate in the active sites of the enzymes, at the atomic level. The inhibitors and substrate analogues also allow us to define the catalytic mechanisms of the enzymes and to identify catalytic amino acid residues. Three-dimensional structures of (1→3),(1→4)-β-D-glucan endohydrolases, (1→4)-β-D-glucan glucohydrolases and β-D-glucan exohydrolases are available or can be reliably modelled from the crystal structures of related enzymes. Substrate analogues have been diffused into crystals for solving of the three-dimensional structures of enzyme-substrate complexes. This information provides valuable insights into potential biological roles of the enzymes in the degradation of the barley (1→3),(1→4)-β-D-glucans during endosperm mobilization and in cell elongation.

Abbreviations: 3D, three-dimensional; HCA, hydrophobic cluster analysis; 4NPG, 4-nitrophenyl β-D-glucoside.

Introduction

Primary cell walls of the Poaceae family of higher plants comprise cellulosic microfibrils embedded in a matrix that consists predominantly of glucuronoarabinoxylans and (1→3),(1→4)-β-D-glucans. Smaller amounts of xyloglucans, pectic polysaccharides, glucomannans and structural proteins may also be present (Bacic *et al.*, 1988; Carpita and Gibeaut, 1993). The relative proportions of these wall components vary between species and between particular tissues in a single species. Furthermore, primary walls are dynamic structures in which the amounts and fine structural characteristics of constituent polysaccharides change at different stages of development, and in response to abiotic and biotic stresses. Clearly, a large number of synthetic and hydrolytic enzymes will be involved in these processes (Henrissat *et al.*, 2001). Here we will focus on enzymes that hydrolyse, or potentially hydrolyse, the (1→3),(1→4)-β-D-glucans during normal cell wall metabolism in the Poaceae.

The (1→3),(1→4)-β-D-glucans are especially abundant in walls of the starchy endosperm, where they represent an important source of stored glucose for the developing seedling (Morrall and Briggs, 1978). In barley they constitute about 75% of cell walls in the starchy endosperm of grain, but are considerably less abundant in walls of other tissues (Fincher 1992). The $(1\rightarrow3),(1\rightarrow4)$-$\beta$-D-glucans are generally found as a molecular family in which fine structure and size vary. The water-soluble $(1\rightarrow3),(1\rightarrow4)$-$\beta$-D-glucans from barley and oat endosperm have been characterized in detail. The water-soluble fraction consists of 1000 or more glucosyl residues connected in linear chains via $(1\rightarrow3)$- and $(1\rightarrow4)$-β-glucosidic linkages. The ratio of $(1\rightarrow4)$ to $(1\rightarrow3)$ linkages is in the range of 2:1 to 3:1, and most of the polysaccharide is composed of groups of two or three contiguous $(1\rightarrow4)$-β-glucosyl residues separated by single $(1\rightarrow3)$-β-glucosyl residues (Woodward et al., 1983). However, in the water-soluble $(1\rightarrow3),(1\rightarrow4)$-$\beta$-D-glucans from barley and other cereals, about 10% of the polysaccharide consists of blocks of up to 10 or more contiguous $(1\rightarrow4)$-β-glucosyl residues (Woodward et al., 1983; Wood et al., 1994). Thus, distinct longer regions of adjacent $(1\rightarrow4)$-β-glucosyl residues are dispersed along the polysaccharide chain.

During normal growth and development a battery of endo- and exohydrolases are used by the plant to degrade or modify cell wall $(1\rightarrow3),(1\rightarrow4)$-$\beta$-D-glucans, or oligosaccharides derived from them. Most of the well characterized enzymes have been isolated from germinated cereal grains or from maize or barley coleoptiles. The extent of hydrolysis of $(1\rightarrow3),(1\rightarrow4)$-$\beta$-D-glucans differs in these tissues. In germinated grain the enzymes are likely to completely depolymerize wall $(1\rightarrow3),(1\rightarrow4)$-$\beta$-D-glucans to glucose (Figures 1 and 2; Briggs, 1992), which can subsequently be translocated to the young seedling as a source of metabolic energy. In the elongating coleoptile the $(1\rightarrow3),(1\rightarrow4)$-$\beta$-D-glucans are believed to be only partially hydrolysed so that wall polysaccharides are 'loosened' during turgor-driven cell elongation (Sakurai and Masuda, 1978). However, conclusive evidence for a role for hydrolytic enzymes in the wall loosening process remains elusive (Cosgrove, 1999). Nevertheless, it is again likely that released $(1\rightarrow3),(1\rightarrow4)$-$\beta$-D-oligoglucosides or glucose are recycled for use in other metabolic pathways in the elongating coleoptile (Gibeaut and Carpita, 1991).

In this review several $(1\rightarrow3),(1\rightarrow4)$-$\beta$-D-glucan endo- and exohydrolases from higher plants are examined, their substrate specificities and action patterns are compared, and molecular data on substrate-binding and catalytic mechanisms are presented. Much of this information has been derived from re-

cently obtained three-dimensional (3D) structures of barley enzymes, or from molecular models based on 3D structures of homologous enzymes. Methods that are used to define the enzymic properties are briefly described, together with associated interpretative limitations. Although the $(1\rightarrow3),(1\rightarrow4)$-$\beta$-D-glucan endo- and exohydrolases catalyse the hydrolysis of $(1\rightarrow3),(1\rightarrow4)$-$\beta$-D-glucans or $(1\rightarrow3),(1\rightarrow4)$-$\beta$-D-oligoglucosides in vitro, their precise biological functions with respect to cell wall metabolism in elongating coleoptiles or germinated grain are not always clear. Available evidence in support of potential biological functions for the various enzymes is evaluated.

Enzymic hydrolysis of $(1\rightarrow3),(1\rightarrow4)$-$\beta$-D-glucans

To effect the complete hydrolysis of wall-bound $(1\rightarrow3),(1\rightarrow4)$-$\beta$-D-glucans to glucose, several categories of enzyme are required (Figure 2). The $(1\rightarrow3),(1\rightarrow4)$-$\beta$-D-glucan endohydrolases of the EC 3.2.1.73 category are widely distributed in commercially important cereal species (Woodward and Fincher, 1982; Stuart et al., 1988) and many have been characterized thoroughly. The $(1\rightarrow3),(1\rightarrow4)$-$\beta$-D-glucan endohydrolases are particularly important in the complete degradation of walls in germinated grain. They are also found in young leaves and roots, but are difficult to detect in elongating coleoptiles (Slakeski and Fincher, 1992). They hydrolyse $(1\rightarrow4)$-β-glucosidic linkages where these linkages are adjacent to a $(1\rightarrow3)$-β-D-glucosyl residue, as follows:

$$\downarrow \qquad\quad \downarrow \qquad\qquad \downarrow$$
$$G\,4\,G\,4\,G\,3\,G\,4\,G\,4\,G\,3\,G\,4\,G\,4\,G\,4\,G\,4\,G\,3\,G\,4\,G\,4..._{red}$$

where G represents a β-D-glucosyl residue, 3 and 4 are $(1\rightarrow3)$ and $(1\rightarrow4)$ linkages, respectively, and red indicates the reducing terminus (Parrish et al., 1960; Anderson and Stone, 1975; Woodward and Fincher, 1982). Thus, the EC 3.2.1.73 enzymes require adjacent $(1\rightarrow3)$- and $(1\rightarrow4)$-β-D-glucosyl residues, they release $(1\rightarrow3),(1\rightarrow4)$-$\beta$-D-tri- and tetrasaccharides (G4G3G_{red} and G4G4G3G_{red}) as major hydrolysis products (Figure 2), but also release higher oligosaccharides of up to 10 or more $(1\rightarrow4)$-β-D-glucosyl residues with a single reducing terminal $(1\rightarrow3)$-β-D-glucosyl residue (e.g. G4G4G4G4G4G4G3G_{red}) from the longer regions of adjacent $(1\rightarrow4)$-linkages alluded to earlier (Woodward et al., 1983; Wood et al., 1994).

In addition to the well characterized $(1\rightarrow3),(1\rightarrow4)$-$\beta$-D-glucan endohydrolases of the EC 3.2.1.73 class,

A

B

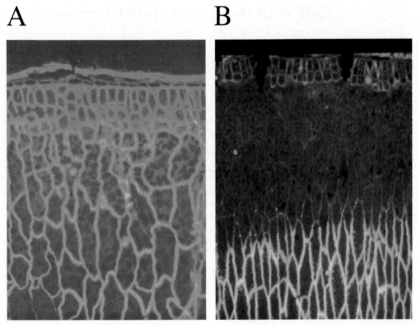

Figure 1. Removal of cell walls in germinated barley grain. Sections of the outer starchy endosperm of ungerminated barley grain (A) and grain germinated for one day (B). Sections were stained with Calcofluor White (Wood and Fulcher, 1978). The starchy endosperm cells range from 80 to 100 μm in diameter. Intact cell wall networks can be seen in ungerminated grain, while the dissolution of the walls is evident in the outer starchy endosperm one day after germination is initiated. The photographs were kindly provided by Meredith Wallwork and Lesley MacLeod.

an unusual $(1\rightarrow3),(1\rightarrow4)$-$\beta$-D-glucan endohydrolase, designated Endo-X in Figure 2, is believed to release larger $(1\rightarrow3),(1\rightarrow4)$-$\beta$-D-glucan molecules from cereal cell walls. This enzyme has been referred to as 'β-D-glucan solubilase' in germinated barley (Bamforth and Martin, 1981) and, although it has not been purified or characterized, a $(1\rightarrow3),(1\rightarrow4)$-$\beta$-D-glucan endohydrolase with similar substrate specificity has been extracted from maize coleoptiles (Inouhe *et al.*, 1999). The maize enzyme releases $(1\rightarrow3),(1\rightarrow4)$-$\beta$-D-glucans with degrees of polymerization of 60 to100 from the isolated polysaccharide substrates (Thomas *et al.*, 2000).

In the absence of a precise definition of the substrate specificity of the barley 'β-D-glucan solubilase' or the maize coleoptile $(1\rightarrow3),(1\rightarrow4)$-$\beta$-D-glucan endohydrolase, one possible explanation for their action patterns is that they are in fact $(1\rightarrow4)$-β-D-glucan endohydrolases of the 'endo-1,4-β-cellulase' (cellulase) group (EC 3.2.1.4). If a cellulase had an extended substrate binding site that required, say, five contiguous $(1\rightarrow4)$-β-D-glucosyl residues, it would only hydrolyse cereal $(1\rightarrow3),(1\rightarrow4)$-$\beta$-D-glucans in the regions of longer blocks of adjacent $(1\rightarrow4)$-β linkages. Because these longer blocks of adjacent $(1\rightarrow4)$-β-D-glucosyl residues represent 10% or less of the total

$(1\rightarrow3),(1\rightarrow4)$-$\beta$-D-glucan (Woodward *et al.*, 1983), very limited hydrolysis would occur and, on a statistical basis, one would expect polysaccharide fragments with degrees of polymerization of about 100 to be released. Balanced against this possibility is the finding that the complete amino acid sequence of the maize coleoptile $(1\rightarrow3),(1\rightarrow4)$-$\beta$-D-glucan endohydrolase (endo-X type) bears no similarity with amino acid sequences of cellulases or other plant glycoside hydrolases (Thomas *et al.*, 1998, 2000).

The $(1\rightarrow3),(1\rightarrow4)$-$\beta$-D-oligoglucosides released by the EC 3.2.1.73 endohydrolases can be further hydrolysed by 'broad-specificity' β-D-glucan exohydrolases or by '$(1\rightarrow4)$-β-D-glucan glucohydrolases/β-D-glucosidases' that are found in germinated grain and in extracts of elongating coleoptiles or young seedlings. There is some difficult nomenclature issue associated with these exohydrolases (Hrmova *et al.*, 1996). Firstly, the 'broad-specificity' β-D-glucan exohydrolases from the family 3 group of glycoside hydrolases (Henrissat, 1991, 1998) have a preference for $(1\rightarrow3)$-β-D-glucans but can also hydrolyse a range of β-D-glucans and β-D-oligoglucosides with $(1\rightarrow2)$, $(1\rightarrow4)$ and $(1\rightarrow6)$ linkages (Hrmova *et al.*, 1996; Hrmova and Fincher, 1998). Furthermore, $(1\rightarrow3),(1\rightarrow4)$-$\beta$-D-glucans and $(1\rightarrow3),(1\rightarrow4)$-$\beta$-D-oligoglucosides are

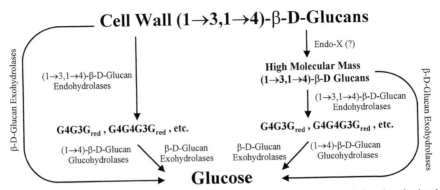

Figure 2. Enzymic hydrolysis of cell wall (1→3),(1→4)-β-D-glucans. In this diagram, enzymes believed to be involved in the release of (1→3),(1→4)-β-D-glucans from cell walls and the complete hydrolysis of the polysaccharide to glucose are shown. In intermediate oligosaccharides, G designates a β-D-glucosyl residue, 3 are (1→3) linkages, 4 are (1→4) linkages, and red denotes the reducing end.

rapidly hydrolysed by the enzymes. As a result, the enzymes cannot be easily classified in any of the existing Enzyme Commission classes. Secondly, the family 1 'β-D-glucosidases' from barley have been so named because they can hydrolyse the synthetic substrate 4-nitrophenyl β-D-glucoside (4NPG) (Simos *et al.*, 1994; Leah *et al.*, 1995; Hrmova *et al.*, 1996). However, their preferred substrates are (1→4)-β-D-oligoglucosides, from which they remove glucose from the non-reducing termini (Hrmova *et al.*, 1996). The rates of hydrolysis of (1→4)-β-D-oligoglucosides by the 'β-D-glucosidases' increase with the degree of polymerization of the substrate. For these reasons Hrmova *et al.* (1998a) suggested that they should be classified as EC 3.2.1.74, (1→4)-β-D-glucan glucohydrolases. Although the 'β-D-glucosidase' designation has been used to avoid confusion since that time, we believe that it is now appropriate to refer to these enzymes as (1→4)-β-D-glucan glucohydrolases; this classification will be used here.

The substrate specificities, action patterns and three-dimensional structures of cereal β-D-glucan endo- and exohydrolases are now well known and will be described further in the following sections. First, however, experimental approaches that allow the precise definition of substrate specificity, action pattern and catalytic mechanisms will be discussed.

Methods for characterizing β-D-glucan endo- and exohydrolases

Substrate specificity

The substrate specificity of a β-D-glucan hydrolase can be readily defined by measuring its activity on a range of polysaccharides, oligosaccharides and alkyl or aryl β-D-glucosides. At the same time, analysis of products generated by hydrolytic action is necessary before the action pattern can be defined and before the enzyme can be correctly classified. Rigorous demonstration of enzyme purity is required before results of such substrate specificity assays can be safely interpreted.

A key determinant of substrate specificity of any enzyme is the complementary shape of the substrate and the binding site on the surface of the enzyme, as originally suggested by Koshland (1958) in his 'induced-fit' model for enzyme-substrate binding. In addition, there must be chemical complementarity between reactive groups on the substrate and the amino acid residues that line the binding site of the enzyme, whether these are ionic interactions, hydrogen bonding, hydrophobic forces, etc. Thus, the enzyme not only has an active site that will physically fit the substrate (but will exclude a differently shaped, non-substrate molecule), but it also aligns the bound substrate in a highly specific spatial orientation. Understanding and defining enzyme-substrate interactions at this level require detailed 3D structural data on the enzyme in complex with substrate or substrate analogues.

Substrate binding to the enzyme can be further analysed by the procedure known as subsite mapping. Polysaccharide hydrolases usually have an extended substrate-binding region that consists of an array of tandemly arranged subsites; each subsite binds a single glycosyl residue of the polymeric substrate (Hiromi, 1970; Thoma *et al.*, 1970; Suganuma *et al.*, 1978). It follows from this arrangement of subsites that kinetic parameters will depend on the degree of polymerization of substrates. The second-order rate kinetic parameter k_{cat}/K_m for substrates of increasing chain length can therefore be used to calculate binding affinities or 'transition state interaction energies' for individual β-D-glucosyl binding subsites on the enzymes (Suganuma et al., 1978; Hrmova et al., 1995, 1998a). Furthermore, concurrent analyses of bond cleavage frequencies of individual substrates with increasing chain lengths allow the position of catalytic amino acids in relation to specific glucosyl-binding subsites to be defined. The position of the linkage that is hydrolysed in relation to the binding subsites is used to distinguish and name the individual subsites. Thus, subsites towards the non-reducing terminus of bound substrate from the point of hydrolysis are consecutively designated -1, -2, -3, etc., while those in the direction of the reducing end of the bound substrate are designated $+1$, $+2$, $+3$, etc. (Biely et al., 1981; Davies et al., 1997).

Catalysis and identification of catalytic amino acid residues

When glycoside hydrolases catalyse the hydrolysis of a glycosidic linkage, the anomeric configuration of the newly generated reducing end of the released product can either be retained in the same configuration as existed in the substrate or it can be inverted (Koshland, 1953; Sinnott, 1990; McCarter and Withers, 1994). This now represents an important property in the classification of glycoside hydrolases (Henrissat and Davies, 1997; Henrissat, 1991, 1998). Proton-NMR can be used to readily determine whether anomeric configuration is retained or inverted during hydrolysis (Withers et al., 1986; Stone and Svensson, 2001). All of the barley β-D-glucan endo- and exohydrolases described here are retaining enzymes (Chen et al., 1995a; Hrmova et al., 1996) and discussion of catalytic mechanisms will therefore be restricted to this class of enzyme.

Once a substrate is bound to a retaining glycoside hydrolase, hydrolysis is initiated by protonation of the glycosidic oxygen atom by an appropriately positioned amino acid, referred to as the catalytic acid/base (White and Rose, 1997; Zechel and Withers, 1999, 2000; Figure 3). This proton donor is usually an unionized carboxylic acid group of an Asp or Glu residue (Legler and Herrchen, 1981). After protonation of the glycosidic oxygen and cleavage of the C1-O bond of the glycosidic linkage, the aglycone portion of the substrate diffuses away from the catalytic site

and is replaced with a water molecule (Heightman and Vasella, 2000). A positively charged oxocarbenium ion-like transition state is formed and this collapses into a covalent glycosyl-enzyme intermediate with inverted configuration at the anomeric centre. The covalent bond in the intermediate is formed with a different, nucleophilic amino acid, which again is usually an Asp or Glu residue (Street et al., 1992; McCarter and Withers, 1994).

Finally, hydrolysis of the covalent glycosyl-enzyme linkage liberates the glycone portion of the hydrolysed substrate and at the same time the catalytic acid on the enzyme is re-protonated. A diagrammatic representation of the likely mechanism of substrate hydrolysis by retaining glycoside hydrolases, occurring by a double displacement mechanism, is shown in Figure 3.

The unequivocal identification of the catalytic nucleophile and the catalytic acid/base of glycoside hydrolases is not a trivial exercise, and should generally be confirmed by as many independent procedures as possible. The first step usually involves multiple sequence alignments and comparisons of hydrophobic cluster analyses (HCA) of enzymes in the same family of glycoside hydrolases (Callebaut et al., 1997). This identifies highly conserved Asp and Glu residues in a conserved local environment. If there are 3D structures available for members of the family, the relative spatial positions of the candidate residues can be assessed. That is, the distance between the putative acid/base and the putative nucleophile can be compared with the 5–6 Å distances that are generally accepted for retaining glycosyl hydrolases (McCarter and Withers, 1994; Davies and Henrissat, 1995; White and Rose, 1997).

There are several chemical methods through which the catalytic nucleophile can be tagged. One tagging procedure that has been applied to barley $(1{\rightarrow}3),(1{\rightarrow}4)$-$\beta$-D-glucan endohydrolases involves the use of epoxyalkyl-β-D-oligoglucosides, which are mechanism-based inhibitors (Legler, 1990; Høj et al., 1989, 1991, 1992; Chen et al., 1993). The β-D-oligoglucoside moiety targets the inhibitor to the substrate-binding site and if the length of the alkyl chain is correct, the epoxide group is brought into the vicinity of the catalytic amino acids. Protonation of the epoxide oxygen opens the epoxide ring and results in the formation of a stable ester linkage between the inhibitor and the catalytic nucleophile (Legler, 1990). The 'tagged' enzyme is subjected to proteolytic hydrolysis and fragments are separated by HPLC. Amino acid sequence analysis of fragments

78

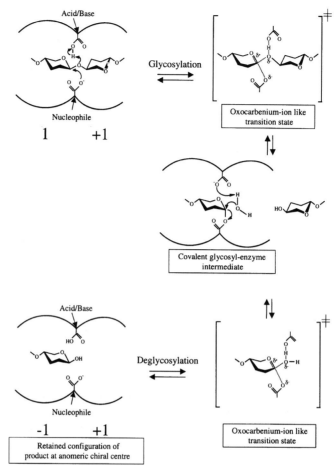

Glycosylation

Oxocarbenium-ion like
transition state

Covalent glycosyl-enzyme
intermediate

Deglycosylation

Oxocarbenium-ion like
transition state

Retained configuration of
product at anomeric chiral centre

Figure 3. Substrate binding and mechanism of catalysis at the active site of retaining plant β-D-glucan endo- and exohydrolases. The double-displacement reaction (Koshland, 1953) at the anomeric chiral centre proceeds through the protonation of the glucosidic oxygen, the formation of an oxocarbenium-ion like transition state, a covalent α-glucosyl-enzyme intermediate, a second oxocarbenium-ion-like transition state and finally, the regeneration of the two catalytic amino acid residues. The anomeric configuration of the released product is retained. Substrate binding subsites −1 and +1 are indicated.

which exhibit altered mobility (compared with the native enzyme), quickly reveals the residue that carries the covalently attached alkyl-β-D-oligoglucoside inhibitor. Conduritol B epoxide (Legler, 1990) or 2′,4′-dinitrophenyl 2-deoxy-2-fluoro-β-D-glucoside (Street *et al.*, 1992) have also been used successfully to label the catalytic nucleophile of a barley β-D-glucan exohydrolase (Hrmova *et al.*, 1998a; M. Hrmova and G.B. Fincher, unpublished data). In some instances the epoxide-based inhibitors do not correctly label the catalytic nucleophile and, at least in one case, a flexible and polar epoxide-bearing xyloside inhibitor 3,4-epoxybutyl-β-xyloside tagged the catalytic acid/base instead of the catalytic nucleophile (Havukainen *et al.*, 1996; Laitinen *et al.*, 2000).

The identification of the catalytic acid/base by chemical tagging procedures is not quite so straightforward. Because the catalytic acid/base residue must carry a proton over the pH range in which the enzyme is active, and because most other acidic amino acid residues on a protein will be in the ionized form in these pH ranges, the catalytic acid can theoretically be labelled with a carbodiimide derivative, which preferentially binds unionized carboxylic acid groups (Hoare and Koshland, 1967). We have used the carbodiimide procedure to label putative catalytic acids in barley (1→3),(1→4)-β-D-glucan endohydrolase (Chen *et al.*, 1993) and in a barley (1→4)-β-D-glucan glucohydrolase (Hrmova *et al.*, 1998a). In the latter case the reagent appeared to label Asp or Glu residues that were close to the catalytic acid/base, but may not have been the actual catalytic acids (Hrmova *et al.*, 1998a). There are numerous other examples, both anecdotal and documented, of incorrect labelling with carbodiimide and one could argue that this method should be abandoned as a procedure for identifying catalytic residues in glycoside hydrolases.

An alternative approach for the identification of the catalytic acid/base employs a mutant enzyme in which the likely catalytic acid/base residue has been changed, linked with careful kinetic analysis of hydrolytic rates of substrates with different leaving groups. Thus, substrates with poor leaving groups such as oligo- and polysaccharides would require protonic assistance from the acid/base, while those with good leaving groups, such as 4NPG and its analogues, would not. As a result, activities of the mutant enzyme should differ considerably between these classes of substrates (Damude *et al.*, 1995; Ly and Withers, 1999).

Site-directed mutagenesis has become an increasingly popular method for defining catalytic amino acids, but it too suffers from interpretative constraints. Altering residues in the substrate-binding region, near to the catalytic site, or indeed at a position remote from the active site, could all lead to subtle changes in enzyme conformation or in the electrostatic interactions between enzyme and substrate. These changes could decrease activity and lead to erroneous conclusions with respect to the identity of the true catalytic acid/base or nucleophile. If site-directed mutagenesis of the catalytic nucleophile is coupled with the 'rescue' or restoration of activity with an exogenous nucleophile, such as sodium azide, the mutagenesis procedure can be used with more confidence to define the catalytic nucleophile (Wang *et al.*, 1994; Viladot *et al.*, 1998; Ly and Withers, 1999).

In summary, any single procedure can rarely be used to unequivocally identify amino acid residues that participate in catalysis in glycoside hydrolases. However, a combination of available procedures, especially if complemented by 3D structural information, has been successfully used to identify catalytic amino acid residues in enzymes that hydrolyse $(1\rightarrow3),(1\rightarrow4)$-$\beta$-D-glucans.

Three-dimensional structures

To solve the 3D structures of enzymes that participate in cereal $(1\rightarrow3),(1\rightarrow4)$-$\beta$-D-glucan hydrolysis, X-ray crystallography remains the method of choice, although in-solution NMR holds considerable promise for the future (Johnson *et al.*, 1999; Asensio *et al.*, 2000). The structures of a barley $(1\rightarrow3),(1\rightarrow4)$-$\beta$-D-glucan endohydrolase and a barley β-D-glucan exohydrolase have now been solved by this procedure (Varghese *et al.*, 1994, 1999). The major bottleneck in the technology is associated with difficulties in obtaining high-quality crystals. This process requires milligram quantities of highly purified enzyme, carefully designed crystallization matrices, patience, and a healthy serving of good luck. The crystallization is usually effected in 8–10 μl 'hanging droplets' containing enzyme at concentrations of up to 10 mg/ml, in ammonium sulfate solutions. The droplets adhere to glass cover slips and are suspended over a well of a microtitre plate. Water is gradually removed from the hanging droplets by vapour diffusion into a slightly more concentrated ammonium sulfate solution in the well of the microtitre plate, and the enzyme may then crystallize from the super-saturated enzyme solution. Crystals can take up to several months to grow to the 0.2–1 mm size required by most X-ray crystallographers (Blundell and Johnson, 1976). If a good quality native data set can be collected from the X-ray diffraction patterns, together with data sets for heavy metal derivatives of the enzymes, crystallographers can generally solve the structure of the enzyme. Solution of the structure is significantly facilitated if the 3D structure of a related enzyme has previously been solved and the co-ordinates made available through protein structure databases.

The 3D structural data open up many opportunities for defining substrate binding, specificity, mechanisms of catalysis and evolutionary relationships between related enzymes. Substrates, non-hydrolysable substrate analogues or active site-directed inhibitors can be diffused into the crystals for the identification of specific amino acid residues that are involved in substrate binding or catalysis (Chipman *et al.*, 1967; Keitel *et al.*, 1993; Parsiegla *et al.*, 2000).

Molecular modelling

Molecular modelling programs are becoming increasingly important as tools in defining 3D structures of enzymes. Enzymes with as little as 25% amino acid sequence identity over about 100 residues can have very similar 3D conformations. If the structure of one of the related enzymes has been solved, the structures of the others can be determined, in good approximation, by homology modelling based on spatial restraints (Sali and Blundell, 1993). Several mathematical procedures are available to test the reliability of the model (Laskowski *et al.*, 1993).

Modelling procedures have been applied during the structural determination of $(1\rightarrow3),(1\rightarrow4)\beta$-D-glucan hydrolases. For example, a reliable structure of a barley $(1\rightarrow4)$-β-D-glucan glucohydrolase was constructed using the coordinates of a cyanogenic β-D-glucosidase from white clover as a template (Barrett *et al.*, 1995; Hrmova *et al.*, 1998a). Similarly, Harvey *et al.* (2000) have used homology modeling, based on spatial restraints, to build from the 3D structure of a barley β-D-glucan exohydrolase (Varghese *et al.*, 1999) reliable models of representatives of family 3 glycoside hydrolases. Sequence identities were as low as 22% over about 180 amino acid residues in some instances, but only one 3D structure was required to provide structural information on more than 100 family 3 glycoside hydrolases from other higher plants, fungi and bacteria (Harvey *et al.*, 2000). Nevertheless, the results of Chothia and Lesk (1986) indicate that comparisons of proteins with low degrees of sequence identity will allow overall folds to be predicted, but that more detailed conclusions on the shapes of substrate-binding regions in distantly-related enzymes should be viewed with some caution.

$(1\rightarrow3),(1\rightarrow4)$-$\beta$-D-Glucan endohydrolases

Three-dimensional structure

The 3D structure of barley $(1\rightarrow3),(1\rightarrow4)$-$\beta$-D-glucan endohydrolase isoenzyme EII has been defined by X-ray crystallography to 2.2–2.3 Å resolution (Varghese *et al.*, 1994). The enzyme is a family 17 glycoside hydrolase (Henrissat, 1991, 1998) that folds into a $(\beta/\alpha)_8$ barrel (Figure 4A). The substrate-binding region consists of a deep cleft that extends across the surface of the enzyme and is long enough to accommodate 6 to 8 glucosyl-binding subsites (Figure 5A). The cleft across the enzyme's surface is consistent with its endo-action pattern, because the enzyme can bind at most positions along the polysaccharide substrate and hydrolyse internal glycoside linkages.

Substrate binding

Although the shape complementarity between substrate and binding site on the enzyme is clearly evident (Figure 5A), the details of chemical interactions between amino acid residues and reactive groups on the substrate have not yet been defined. Attempts to diffuse polysaccharide and oligosaccharide substrates into crystals have not been successful in generating diffraction data for the enzyme-substrate complex (M. Hrmova, J.N. Varghese and G.B. Fincher, unpublished data), mainly because the substrates are hydrolysed and the products diffuse away from the enzyme's surface. Non-hydrolysable S-glycoside substrate analogues (Moreau and Driguez, 1995) might be useful in this connection. The analogues would be expected to bind to the enzyme and to resist hydrolysis; X-ray diffraction data of the enzyme-substrate complex might thereby be collected. Alternatively, substrates might be diffused into crystals of mutant enzyme in which one or both of the catalytic amino acid residues have been altered. Thus, the substrate should bind to the enzyme, but hydrolysis would not occur.

Catalytic amino acid residues

During hydrolysis of $(1\rightarrow4)$-β-glucosyl linkages in $(1\rightarrow3),(1\rightarrow4)$-$\beta$-D-glucans by EC 3.2.1.73 $(1\rightarrow3)$, $(1\rightarrow4)$-β-D-glucan endohydrolases, anomeric configuration is retained (Chen *et al.*, 1995a). The catalytic nucleophile of the enzyme is almost certainly Glu-232. This residue was tagged with specific epoxyalkyl-β-D-oligoglucoside inhibitors (Chen *et al.*, 1993), and it is highly conserved in family 17 glycoside hydrolases. It is located at the bottom of, and about two-thirds of the way along, the substrate binding cleft. The catalytic acid/base was initially identified as Glu-288 by carbodiimide-mediated labelling procedures (Chen *et al.*, 1993). However, Jenkins *et al.* (1995) and Henrissat *et al.* (1995) subsequently suggested that the catalytic acid/base was more likely to be Glu-93. Both residues are highly conserved in family 17 glycoside hydrolases (Høj and Fincher, 1995), but Glu-288 is about 8 Å from the catalytic nucleophile Glu-232 (Varghese *et al.*, 1994) and this is considered to be too far for retaining glycoside hydrolases. The 5–6 Å distance between Glu-232 and Glu-93 is more 'typical' of retaining enzymes (Jenkins *et al.*, 1995; Henrissat *et al.*, 1995). The relative dispositions of these residues in the catalytic site region are shown in Figure 4A. At this stage it is not absolutely clear whether Glu-93 or Glu-288, or possibly both, contribute to protonation of the glycosidic oxygen during $(1\rightarrow3),(1\rightarrow4)$-$\beta$-D-glucan hydrolysis by this enzyme. There are often several highly conserved acidic amino acids in the catalytic region of glycoside hydrolases, together with conserved basic amino

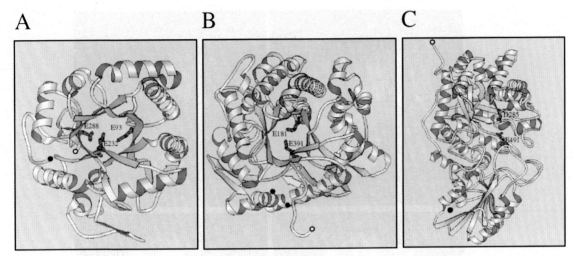

Figure 4. Ribbon representations of barley β-D-glucan endo- and exohydrolases. A. (1→3),(1→4)-β-D-glucan endohydrolase. B. (1→4)-β-D-glucan glucohydrolase. C. β-D-glucan exohydrolase. The secondary structural elements of the 3D fold are shown. The likely catalytic amino acid residues are coloured blue. Modified from Varghese *et al.* (1994, 1999) and Hrmova *et al.* (1998a). The drawings were generated with MOLSCRIPT (Kraulis, 1991).

acids (Chen *et al.*, 1995b) and co-ordinated water molecules. The possibility remains that the proton that eventually hydrolyses the glycosidic linkage of the bound substrate is relatively mobile in this conserved region of acidic and basic amino acid residues.

(1→4)-β-D-Glucan glucohydrolases

Substrate specificity

Plant enzymes have been designated as β-D-glucosidases on the basis of their ability to hydrolyse the synthetic β-D-glucoside, 4-nitrophenyl β-D-glucoside (4NPG). However, β-D-glucosidases can be classified in the family 1 or family 3 groups of glycoside hydrolases and it has become clear that the convenience of the 4NPG assay has created problems in the classification of these enzymes into Enzyme Commission classes (Hrmova *et al.*, 1996).

As mentioned earlier in this review, barley 'β-D-glucosidases' from family 1 have been monitored through their purification by their activity on 4NPG (Simos *et al.*, 1994; Leah *et al.*, 1995; Hrmova *et al.*, 1996; Hrmova *et al.*, 1998a). Closer examination of substrate specificity reveals that the barley enzymes exhibit a marked preference for (1→4)-β-D-oligoglucosides (cellodextrins) and that the rate of hydrolysis increases with the degree of polymerization of the substrate (Figure 6A). Single glucose molecules are released from the non-reducing ends

of substrates, with retention of anomeric configuration (Hrmova *et al.*, 1996). The enzymes do not hydrolyse (1→3),(1→4)-β-D-glucans or (1→3)-β-D-glucans at significant rates. Thus, the substrate specificity and action patterns of the barley β-D-glucosidases are characteristic of polysaccharide exohydrolases of the (1→4)-β-D-glucan glucohydrolase group (EC 3.2.1.74), rather than of an enzyme with a preference for low-molecular-mass β-D-glucosides.

The preference of the barley (1→4)-β-D-glucan glucohydrolases for longer-chain (1→4)-β-D-oligoglucosides (Figure 6A) is consistent with subsite mapping data (Hrmova *et al.*, 1998a), which indicate that the enzymes have 5 to 6 glucosyl-binding subsites (Figure 6B).

Three-dimensional structure

In contrast to the 'open cleft' structure required for hydrolysing internal substrate linkages by endohydrolases, an exohydrolase such as the barley (1→4)-β-D-glucan glucohydrolase aligns its substrate in a dead-end tunnel, slot or funnel so that the non-reducing terminal linkages of the substrate are brought into juxtaposition with catalytic amino acid residues. Although there are no 3D structures for family 1 cereal (1→4)-β-D-glucan glucohydrolases in the databases, models of the barley enzyme were constructed from the co-ordinates of a cyanogenic β-D-glucosidase from white clover, which adopts a (β/α)₈ barrel fold (Barrett *et al.*, 1995; Hrmova *et al.*, 1998a; Figure 4B).

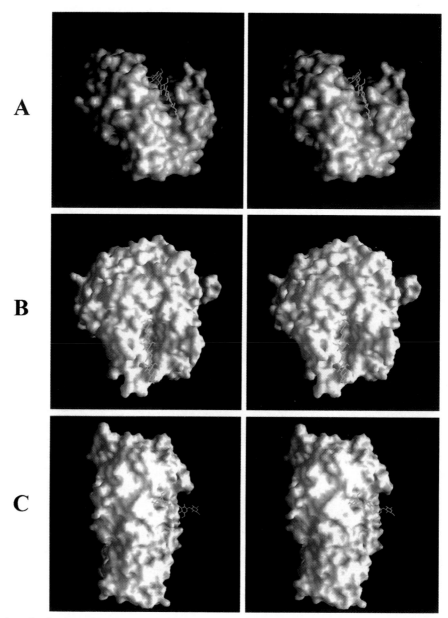

Figure 5. Stereoview of molecular surface representation of barley β-D-glucan endo- and exohydrolases. A. (1→3),(1→4)-β-D-glucan endohydrolase with portion of the (1→3),(1→4)-β-D-glucan substrate bound in a cleft that extends across the surface of the enzyme. B. Model of (1→4)-β-D-glucan glucohydrolase with a (1→4)-β-D-oligoglucoside substrate extended to the bottom of a dead-end funnel. C. β-D-glucan exohydrolase with a linear (1→3)-β-D-oligosaccharide substrate bound in the active site pocket of the enzyme. Modified from Varghese *et al.* (1994, 1999), Jenkins *et al.* (1995) and Hrmova *et al.* (1998a). The drawings were generated by means of GRASP (Nicholls *et al.*, 1991).

The barley (1→4)-β-D-glucan glucohydrolase has a deep, funnel-shaped, dead-end tunnel into which six glucosyl residues of the (1→4)-β-D-oligoglucoside substrate are bound (Hrmova *et al.*, 1998a) and this is again consistent with the subsite mapping data (Figure 6B). As expected, catalytic amino acid residues are located near the bottom of the funnel, close to the glu-

cosidic linkage of the non-reducing terminal residue. It is important to note that the substrate specificities of the template cyanogenic β-D-glucosidase from white clover and the target (1→4)-β-D-glucan glucohydrolase from barley differ. Local variations in parts of the 3D structure of the substrate-binding region of the modelled structure of the barley (1→4)-β-D-glucan

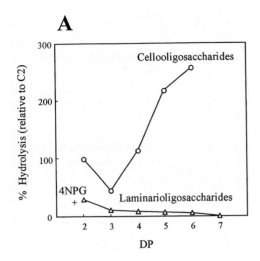

A

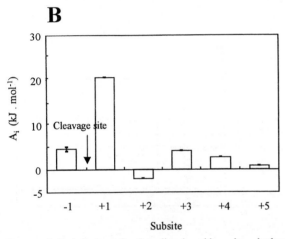

B

Figure 6. Hydrolysis of β-D-oligoglucosides by barley (1→4)-β-D-glucan glucohydrolase. A. Relative hydrolysis rates of a series of (1→4)- (cellooligosaccharides) and (1→3)-linked (laminarioligosaccharides) β-D-oligoglucosides. The rate of hydrolysis of the synthetic substrate 4NPG is shown for comparative purposes. B. Subsite map of the barley (1→4)-β-D-glucan glucohydrolase with a series of cellooligosaccharides. The 'transition state interaction energies' (A_i) are shown for each subsite. Subsite numbering with respect to the cleavage site is as described in the text. Modified from Hrmova *et al.* (1996, 1998a).

glucohydrolase are therefore likely to occur, and the limitations of molecular modelling alluded to earlier need to be taken into account.

The molecular model of the barley (1→4)-β-D-glucan glucohydrolase, with bound substrate, is shown in Figure 5B. It is particularly noteworthy that related substrates such as (1→3)-β-D-oligoglucosidases do not fit into the funnel-shaped pocket on the enzyme surface (data not shown). This can be reconciled with the tight substrate specificity of this class of enzyme

(Hrmova *et al.*, 1996, 1998a; Figure 6A) and clearly reflects the importance of conformational complementarity in enzyme-substrate binding reactions.

Catalytic amino acid residues

The mechanism-based inhibitor conduritol B epoxide was used to identify the catalytic nucleophile of the barley (1→4)-β-D-glucan glucohydrolase (Hrmova *et al.*, 1998a). The inhibitor bound covalently to Glu-391, which is highly conserved in family 1 glycoside hydrolases. The catalytic nucleophile of another family 1 β-D-glucosidase from *Agrobacterium* has been determined previously with 2′,4′-dinitrophenyl-2-deoxy-2-fluoro-β-D-glucoside (Street *et al.*, 1992).

When carbodiimide-based procedures were used in attempts to identify the acid/base in the barley (1→4)-β-D-glucan glucohydrolase, an Asp residue that is located near the entrance to the pocket on the surface of the molecular model, ca. 20 Å from the putative catalytic nucleophile, was tagged. Subsequent sequence alignments, HCAs and examination of the enzyme model indicated that Glu-181 was more likely to be the catalytic acid/base. It was concluded that although the carbodiimide procedure resulted in enzyme inactivation, it did not correctly label the catalytic acid/base of the barley (1→4)-β-D-glucan glucohydrolase (Hrmova *et al.*, 1998a).

The likely catalytic amino acid residues Glu-181 and Glu-391 are located 5–6 Å apart near the bottom of the substrate-binding pocket (Figure 4B). If the non-reducing end of the substrate is pushed to the bottom of the pocket, the Glu-181 and Glu-391 residues are situated close to the oxygen atom of the glycosidic linkage adjacent to the non-reducing end of the substrate. It is not clear whether the non-reducing end of the substrate is 'selected' in preference to the reducing end prior to insertion of the substrate into the active-site funnel, or whether the correct orientation of the substrate is a matter of trial and error that is finally determined to be productive or non-productive by the disposition of amino acid residues at the bottom of the funnel. Furthermore, there does not appear to be enough space at the bottom of the pocket for the released glucose molecule to diffuse out after hydrolysis, and it is likely that the substrate must at least partly dissociate from the enzyme after each hydrolytic event (Hrmova *et al.*, 1998a).

The broad-specificity β-D-glucan exohydrolases

Substrate specificity

The tight specificity of the family 1 barley (1→4)-β-D-glucan glucohydrolases contrasts dramatically with a second group of exohydrolases that are widely distributed in cereals and other higher plants. These have been termed the broad-specificity β-D-glucan exohydrolases, insofar as they can hydrolyse the non-reducing terminal glucosidic linkage in a broad range of polymeric β-D-glucans, β-D-oligoglucosides and aryl β-D-glucosides such as 4NPG, to release glucose (Hrmova and Fincher, 1998). Anomeric configuration is retained (Hrmova *et al.*, 1996). They belong to the family 3 group of glycoside hydrolases (Henrissat, 1991, 1998) but their broad substrate specificity precludes their assignment to existing EC classes (Hrmova and Fincher, 1998). The β-D-glucan exohydrolases can be classified as polysaccharide exohydrolases rather than β-D-glucosidases because they rapidly hydrolyse polysaccharide substrates, such as laminarin and (1→3),(1→4)-β-D-glucan. Indeed, the preferred substrates for the barley β-D-glucan exohydrolases are (1→3)-β-D-glucans, such as laminarins (Hrmova and Fincher, 1998). Similar enzymes have been detected in maize coleoptiles (Labrador and Nevins, 1989; Kim *et al.*, 2000) and in dicotyledonous plants (Cline and Albersheim, 1981; Lienart *et al.*, 1986; Crombie *et al.*, 1998).

Analyses of products released during hydrolysis of laminarin by the barley β-D-glucan exohydrolase show not only the accumulation of glucose and laminaridextrins, as expected, but also the presence of significant levels of the (1→6)-linked glucosyl β-D-disaccharide, gentiobiose (Hrmova and Fincher, 1998). It has been concluded that this product results largely from glycosyl transfer activity and that other higher-molecular-mass glucosyl transfer products are also formed (Hrmova and Fincher, 1998). Whether the glycosyl transfer reactions reflect a real biological function for the enzyme, or merely occur because of high substrate concentrations in *in vitro* hydrolysis reactions, remains to be demonstrated.

Subsite mapping experiments indicate that the barley β-D-glucan exohydrolases have a much shorter substrate-binding region than the (1→4)-β-D-glucan glucohydrolases, and that only 2 to 3 glucosyl binding subsites are present (Figure 7B). Again this can be reconciled with the observation that once a substrate has more than three glucosyl residues, all binding subsites

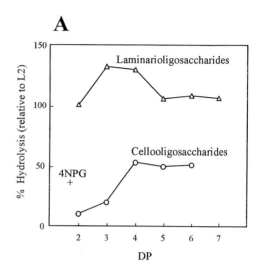

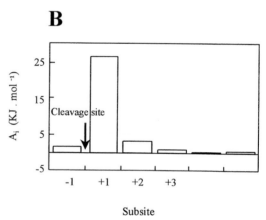

Figure 7. Hydrolysis of β-D-oligoglucosides by barley β-D-glucan exohydrolase. A. Relative hydrolysis rates of a series of (1→4)-(cellooligosaccharides) and (1→3)-linked (laminarioligosaccharides) β-D-oligoglucosides. The rate of hydrolysis of the synthetic substrate 4NPG is shown for comparative purposes. B. Subsite map of the barley β-D-glucan exohydrolase with a series of laminarioligosaccharides. The 'transition state interaction' energies (A_i) are shown for each subsite. Subsite numbering with respect to the cleavage site is as described in the text. Modified from Hrmova *et al.* (1996) and from unpublished data (M. Hrmova and G.B. Fincher).

are occupied and catalytic rates become independent of the degree of polymerization (Figure 7A).

Three-dimensional structure

The 3D structure of barley β-D-glucan exohydrolase isoenzyme ExoI has been determined by X-ray crystallography to 2.2 Å resolution (Hrmova *et al.*, 1998b; Varghese *et al.*, 1999). The enzyme consists of two distinct domains that are connected by a 16-amino acid helix-like linker. The first domain is a $(\beta/\alpha)_8$ barrel of 357 amino acid residues, while the second domain consists of a six-stranded β-sheet flanked on either side by three α-helices (Varghese *et al.*, 1999). A long antiparallel loop of 42 amino acid residues is found at the COOH-terminus of the enzyme. These structural characteristics of the barley β-D-glucan exohydrolase are shown in Figure 4C.

The active site of the barley β-D-glucan exohydrolase consists of a relatively shallow substrate-binding pocket that could accommodate 2 (or at the most 3) glucosyl residues (Varghese *et al.*, 1999), which is consistent with the subsite mapping data (Figure 7B). Because the pocket is only deep enough to accommodate about two glucosyl residues, the unbound portion of the substrate will project away from the enzyme surface (Figure 5C). One would therefore anticipate that β-D-glucan exohydrolase specificity would be largely independent of substrate conformation, in contrast to the conformational constraints imposed on substrates for the $(1{\rightarrow}4)$-β-D-glucan glucohydrolases, which must have the right conformation to penetrate to the bottom of a much deeper, narrower tunnel (Figure 5B). It follows that if substrate binding to the broad specificity β-D-glucan exohydrolase is largely independent of polysaccharide conformation, it would also be largely independent of linkage positions between adjacent β-D-glucosyl residues. This would explain why the β-D-glucan exohydrolases have broad substrate specificities. Thus, the family 3 β-D-glucan exohydrolases and the family 1 $(1{\rightarrow}4)$-β-D-glucan glucohydrolases represent a contrasting pair of polysaccharide exohydrolases with broad and tight linkage specificity, respectively.

The 3D structural analysis revealed that a glucose molecule remains in the active site pocket of the barley β-D-glucan exohydrolase (Figure 8) and is presumably the product of hydrolysis that is not released after catalysis (Varghese *et al.*, 1999). The unexpected bonus was that details of amino acid interactions with the glucose bound at subsite −1 (Figures 7B and 8) could be defined in precise atomic terms (Varghese *et al.*, 1999). Since then, non-hydrolysable S-glycoside substrate analogues (Moreau and Driguez, 1995) have been diffused into barley β-D-glucan exohydrolase crystals and show that the glucosyl residue at subsite +1 is sandwiched between tryptophan residues Trip-286 and Trip-434 that lie above and below the entrance to the pocket (H. Driguez, M. Hrmova, J.N. Varghese and G.B. Fincher, unpub-

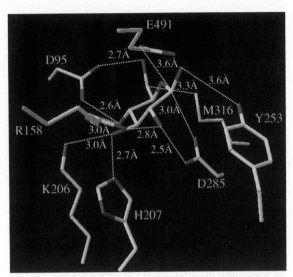

Figure 8. Molecular interactions between amino acid residues that line substrate-binding subsite −1 and the glucose molecule that is trapped in the active site of barley β-D-glucan exohydrolase. Nearest ionic, hydrogen-bonding and hydrophobic interactions between glucose and the contact amino acid residues are shown as dotted lines and with the distances marked. Standard single-letter codes for amino acid residues are used. Modified from Varghese *et al.* (1999). The drawing was generated with Swiss-pdBViewer (Guex and Peitsch, 1997).

lished data). Thus, the chemical binding interactions between the substrate and amino acids at subsites −1 and +1 on the enzyme's surface have now been defined. The fact that about 18 amino acid residues participate in the binding of just two glucosyl residues in the active site attests to the complexity, precision and evolutionary sophistication of chemical and physical complementarity in enzyme-substrate binding.

Catalytic amino acid residues

Labelling with conduritol B epoxide indicates that Asp-285 is the catalytic nucleophile of the two barley β-D-glucan exohydrolases (M. Hrmova and G.B. Fincher, unpublished). Its equivalent (Asp-242) has also been identified as the catalytic nucleophile in a family 3 *N*-acetyl-β-D-glucosaminidase from *Vibrio furnisii* (Vocadlo *et al.* 2000). The Asp-285 residue or its equivalent is highly conserved in family 3 glycoside hydrolases and lies within 3.0 Å of C1 of the glucose molecule that remains bound to the barley enzyme in the substrate-binding pocket (Figure 8). The three other acidic residues, Asp-95, Glu-220 and Glu-491, are also located near C1 of the bound glucose but atomic distances and relative dispositions suggest that Glu-491 is most likely to be the catalytic acid/base

(Varghese *et al.*, 1999). However, Glu-491 is not invariant in family 3 glycoside hydrolases and the role of catalytic acid/base could be assumed by other amino acid residues in more distant members of the family (Harvey *et al.*, 2000).

It is noteworthy that both domains of the barley enzyme contribute amino acid residues that bind the glucose molecule in the active site pocket. Further, the two catalytic residues themselves are located on different domains (Figure 4C). The 16 amino acid residue linker that connects the two domains could act as a molecular hinge that would allow the two domains to move relative to each other. Movement of the second domain away from, or closer to, the active site residues on domain 1 would almost certainly affect activity. This could represent a molecular mechanism whereby enzyme activity is regulated (Varghese *et al.*, 1999). Similarly, movement of the domains relative to each other might be required for the release of the glucose product during its displacement from the active site by the incoming substrate.

Does domain 2 bind (1→3),(1→4)-β-D-glucan?

Kinetic analyses of the barley β-D-glucan exohydrolase indicate that there is positive co-operativity of (1→3),(1→4)-β-D-glucan hydrolysis; this in turn suggests that there is more than one binding site on the enzyme for (1→3),(1→4)-β-D-glucan (Hrmova and Fincher, 1998). A potential non-catalytic (1→3),(1→4)-β-D-glucan-binding site, which is located at the interface of the two domains, has been tentatively assigned to a Rossman fold-like region (Brändén, 1980) on domain 2; this is near the loop between the helices J1 and J2 (Varghese *et al.*, 1999). This relatively small region is rich in hydroxy amino acid residues and low in charge, and therefore shares characteristics with the larger, independently-folding carbohydrate-binding domains found on microbial cellulases (Gilkes *et al.*, 1991). The positive co-operativity observed during (1→3),(1→4)-β-D-glucan hydrolysis (Hrmova and Fincher, 1998) might also be explained by domain movements. If a second (1→3),(1→4)-β-D-glucan molecule bound to domain 2, this might affect binding and hydrolysis of the substrate at the catalytic site.

Biological functions of β-D-glucan endo- and exohydrolases

(1→3),(1→4)-β-D-Glucan endohydrolases

In barley, two (1→3),(1→4)-β-D-glucan endohydrolase isoenzymes are detected in germinated grain and in extracts of young vegetative tissues. Expression patterns of the genes encoding the two isoenzymes are subject to independent, tissue-specific regulation (Slakeski and Fincher, 1992). In germinated grain the primary function of the EC 3.2.1.73 (1→3),(1→4)-β-D-glucan endohydrolases is to participate in cell wall mobilization. The enzymes are expressed in and secreted from the aleurone layer and the scutellar epithelium of germinated grain, where tight temporal and spatial regulation of gene expression is observed (McFadden *et al.*, 1988). Genes encoding both of the barley (1→3),(1→4)-β-D-glucan endohydrolase isoenzymes have been isolated, and nucleotide sequence motifs linked with gibberellic acid induction have been identified in their promoters (Litts *et al.*, 1990; Slakeski *et al.*, 1990; Wolf, 1991). Gibberellic acid induces the transcription of many genes involved in endosperm mobilization (Fincher, 1989).

Although the (1→3),(1→4)-β-D-glucan endohydrolases clearly play a role in wall (1→3),(1→4)-β-D-glucan degradation in germinated barley, the functional significance of isoenzyme EI in young vegetative tissues (Slakeski and Fincher, 1992; Simmons *et al.*, 1992) is not so easy to explain. In barley, transcription of the isoenzyme EI gene is mediated by auxins (Slakeski and Fincher, 1992), as expected for genes involved in tissue elongation, but auxins control many other physiological processes related in wall metabolism in young vegetative tissues, including vascular differentiation, phototropic responses and geotropism. While it might be anticipated that partial hydrolysis of wall polysaccharides would 'loosen' wall structure sufficiently to allow turgor-driven cell elongation in growing tissues, the biochemical evidence for this is not strong (Cosgrove, 1999). Indeed, expression of EC 3.2.1.73 (1→3),(1→4)-β-D-glucan endohydrolases could not be detected, either at the mRNA or enzyme levels, in elongating barley coleoptiles (Slakeski and Fincher, 1992), and we conclude that a role for these enzymes in cell elongation remains unproven.

An alternative possibility has been presented by Roulin *et al.* (2001), who have shown that EC 3.2.1.73 (1→3),(1→4)-β-D-glucan endohydrolase activity in young barley leaves increases 3- to 4-fold when seedlings are transferred into darkness, and that β-D-glucan exohydrolase activity also increases markedly in dark-grown seedlings. They suggest that the endohydrolases mobilize cell wall $(1{\rightarrow}3),(1{\rightarrow}4)$-$\beta$-D-glucans and that glucose is subsequently released from $(1{\rightarrow}3),(1{\rightarrow}4)$-$\beta$-D-oligoglucosides by exohydrolases. The glucose might be used as an energy source after the cessation of photosynthesis in the dark-grown seedlings and might delay the onset of dark-induced senescence (Roulin et al., 2001). A similar function in $(1{\rightarrow}3),(1{\rightarrow}4)$-$\beta$-D-glucan turnover for the $(1{\rightarrow}3),(1{\rightarrow}4)$-$\beta$-D-glucan endohydrolase and the β-D-glucan exohydrolase may occur during normal coleoptile growth (Carpita, 1984; Inouhe and Nevins, 1998).

$(1{\rightarrow}4)$-β-D-Glucan glucohydrolases

In barley, genes encoding the family 1 $(1{\rightarrow}4)$-β-D-glucan glucohydrolases are transcribed in the maturing endosperm, and activity does not increase after germination (Simos et al., 1994; Leah et al., 1995). The enzymes are active in developing grain when $(1{\rightarrow}3),(1{\rightarrow}4)$-$\beta$-D-glucans are being deposited in walls of the starchy endosperm (Leah et al., 1995), and could participate in the trimming or turnover of wall $(1{\rightarrow}3),(1{\rightarrow}4)$-$\beta$-D-glucans during synthesis. Other functions have also been proposed, including roles in the release of active phytohormones from inactive hormone-glucoside conjugates and in the hydrolysis of cyanogenic glucosides, but no evidence for these functions in grain is so far available (Leah et al., 1995).

Another possibility is that the $(1{\rightarrow}4)$-β-D-glucan glucohydrolases hydrolyse the β-D-oligoglucosides released from wall $(1{\rightarrow}3),(1{\rightarrow}4)$-$\beta$-D-glucans by endohydrolases (Figure 2). Substrate specificity studies indicate that the preferred substrates for $(1{\rightarrow}4)$-β-D-glucan glucohydrolases are $(1{\rightarrow}4)$-β-D-oligoglucosides, and that $(1{\rightarrow}3)$-β-D-oligoglucosides are hydrolysed very slowly (Figure 6A). Despite the interpretative limitations associated with molecular modelling, enzyme models provided a rationale for these observations, in that only a straight $(1{\rightarrow}4)$-β-D-oligoglucoside substrate could be threaded all the way to the bottom of the substrate-binding funnel (Figure 5B), where catalytic residues are located. However, the oligosaccharides released by $(1{\rightarrow}3),(1{\rightarrow}4)$-$\beta$-D-glucan endohydrolases have $(1{\rightarrow}4)$-β-D-glucosyl residues at their non-reducing

termini and a single $(1{\rightarrow}3)$-β-D-glucosyl residue at their reducing termini. Thus, an oligosaccharide $G4G4G4G3G_{red}$ released from a block of adjacent $(1{\rightarrow}4)$-β-D-glucosyl residues in the polysaccharide has a very similar shape to cellopentaose $G4G4G4G4G_{red}$, particularly at the non-reducing end. It has been shown that barley $(1{\rightarrow}4)$-β-D-glucan glucohydrolases hydrolyse $(1{\rightarrow}3),(1{\rightarrow}4)$-$\beta$-D-oligoglucosides of this type (Hrmova et al., 1996), and their ability to hydrolyse laminaribiose, albeit slowly (Figure 6A), suggests that these $(1{\rightarrow}3),(1{\rightarrow}4)$-$\beta$-D-oligoglucosides could be completely depolymerized to glucose by this group of enzymes.

β-D-Glucan exohydrolases

The broad specificity of the β-D-glucan exohydrolases has presented problems in assigning a function or functions to these enzymes. The enzymes hydrolyse the $(1{\rightarrow}3),(1{\rightarrow}4)$-$\beta$-D-glucans that are found exclusively in walls of the Poaceae and in this review we have focussed on the potential role of the β-D-glucan exohydrolases in $(1{\rightarrow}3),(1{\rightarrow}4)$-$\beta$-D-glucan metabolism in cereals. However, similar enzymes have been detected in soybean (Glycine max) cultures (Cline and Albersheim, 1981), Acacia verek cells (Lienart et al., 1986) and nasturtium (Tropaeolum majus) (Crombie et al., 1998), where $(1{\rightarrow}3),(1{\rightarrow}4)$-$\beta$-D-glucans are not found.

Crombie et al. (1998) noted that a β-D-glucan exohydrolase (designated a β-D-glucosidase) from nasturtium could hydrolyse oligoglucosides of the type released from wall xyloglucans by endohydrolases, provided the non-reducing glucosyl residue of the oligosaccharide was not substituted. Although xyloglucans from walls of the Poaceae have not been studied in detail, they are present in elongating coleoptiles of maize and barley (Carpita and Gibeaut, 1993; D.M. Gibeaut and G.B. Fincher, unpublished), and in the starchy endosperm of rice (Shibuya and Misaki, 1978). Thus, the β-D-glucan exohydrolases could hydrolyse polymeric $(1{\rightarrow}3),(1{\rightarrow}4)$-$\beta$-D-glucans or xyloglucans, provided substituted monosaccharides were first removed from the xyloglucan by other glycoside hydrolases. They might also hydrolyse cell wall glucomannans, although this does not appear to have been tested at this stage. Further, the β-D-glucan exohydrolases could hydrolyse oligosaccharides released by endohydrolases from both $(1{\rightarrow}3),(1{\rightarrow}4)$-$\beta$-D-glucans and xyloglucans.

The cellular and subcellular locations of the β-D-glucan exohydrolases have been investigated. The enzymes have been extracted from cell wall preparations from maize and barley coleoptiles with high concentrations of LiCl (Kotake *et al.*, 1997; Inouhe *et al.*, 1999; Kim *et al.*, 2000). More recently, Kim *et al.* (2000) showed that two isoforms of the maize β-D-glucan exohydrolases are associated with cell walls, while a third is tightly bound to the plasma membrane. In barley the majority of β-D-glucan exohydrolase activity can be extracted from homogenates of coleoptiles with dilute aqueous buffers, without added LiCl (Harvey *et al.*, 2001), a finding that argues against strong binding to either cell walls or to plasma membranes. Nevertheless, the enzymes are extracellular and could be occluded in the cell wall matrix. At a tissue level, tissue-printing procedures indicate that the maize β-D-glucan exohydrolases are concentrated in the basal portions of the coleoptile and in the elongation zone of the mesocotyl (Kim *et al.*, 2000).

Genes encoding barley β-D-glucan exohydrolases are transcribed in the scutellum of germinated grain, but mRNA levels in aleurone layers are very low; highest levels are detected in elongating coleoptiles (Harvey *et al.*, 2001). High transcription levels in elongating coleoptiles have led to the suggestion that the enzymes participate in auxin-mediated cell elongation (Hoson and Nevins, 1989; Kotake *et al.*, 1997). The $(1\rightarrow3),(1\rightarrow4)$-$\beta$-D-glucan content of coleoptile walls decreases during elongation (Sakurai and Masuda, 1978; Carpita, 1984; Carpita and Gibeaut, 1993) and these collective observations have been connected with wall 'loosening' that is believed to be a prerequisite for cell elongation (Labrador and Nevins, 1989). However, it is not easy to envisage how an exo-acting hydrolase could significantly affect the physical entanglement or chemical cross-linking of polysaccharides in the matrix between cellulosic microfibrils. Kim *et al.* (2000) note that β-D-glucan exohydrolase activity in elongating maize coleoptiles is correlated with $(1\rightarrow3),(1\rightarrow4)$-$\beta$-D-glucan turnover that occurs after growth has ceased, rather than with growth that is in progress.

In the absence of compelling evidence that the β-D-glucan exohydrolases participate in wall 'loosening' and cell elongation (Cosgrove, 1999), what other possible functions might be performed by these enzymes? By focusing on $(1\rightarrow3),(1\rightarrow4)$-$\beta$-D-glucans, have we missed the real substrate for the enzymes? Do they actually participate in the hydrolysis of oligoxyloglucosides during xyloglucan degradation or

turnover, in both monocotyledons and dicotyledons? Are they used in a more general strategy to recover glucose from a range of different classes of polysaccharides or oligosaccharides? They could contribute to the conversion of xyloglucans, glucomannans and $(1\rightarrow3),(1\rightarrow4)$-$\beta$-D-glucans to their certainty constituent monosaccharides. Another possibility is that the β-D-glucan exohydrolases are expressed preemptively to counter potential pathogen attack in young tissues that are particularly vulnerable to fungal infection. The enzymes hydrolyse the $(1\rightarrow3)$-β-D-glucans and $(1\rightarrow3),(1\rightarrow6)$-$\beta$-D-glucans that are found in cell walls of many fungi (Hrmova and Fincher, 1998). They might act in concert with $(1\rightarrow3)$-β-D-glucan endohydrolases to degrade walls of invading fungi.

These considerations and the large amount of speculation regarding β-D-glucan exohydrolase function in the literature point to our inability to describe the biological roles of specific wall-degrading enzymes in expanding plant cells and, indeed, to develop a rigorous model for the role of walls during cell expansion in a more general way. Gene knockout experiments by antisense technology or virus-induced gene silencing (Burton *et al.*, 2000) might now be used to cast further light on the functional roles of these enzymes in higher plants.

Acknowledgements

This work has been supported by grants from the Australian Research Council and the Grains Research and Development Corporation. We are indebted to Andrew Harvey and Jose Varghese for their major contributions to the work described here and to Professor Bruce Stone for critically reading the manuscript.

References

Anderson, M.A. and Stone, B.A. 1975. A new substrate for investigating the specificity of β-D-glucan hydrolases. FEBS Lett. 52: 202–207.

Asensio, J.L., Canada, F.J., Siebert, H.C., Laynez, J., Poveda, A., Nieto, P.M., Soedjanaamadja, U.M., Gabius, H.J. and Jimenez-Barbero, J. 2000. Structural basis for chitin recognition by defense proteins: GlcNAc residues are bound in a multivalent fashion by extended binding sites in hevein domains. Chem. Biol. 7: 529–543.

Bacic, A., Harris, P.J. and Stone, B.A. 1988. Structure and function of plant cell walls. In: J. Preiss (Ed.) The Biochemistry of Plants, Academic Press, New York/London/San Francisco, pp. 297–371.

Bamforth, C.W. and Martin, H.L. 1981. The development of β-D-glucan solubilase during barley germination. J. Inst. Brew. 87: 81–84.

Barrett, T., Suresh, S.G., Tolley, S.P., Dodson, E.J. and Hughes, M.A. 1995. The crystal structure of cyanogenic β-glucosidase from white clover; a family 1 glycosyl hydrolase. Structure 3: 951–960.

Biely, P., Kratky, Z. and Vrsanska, M. 1981. Substrate-binding site of endo-1,4-β-xylanase of the yeast *Cryptococcus albidus*. Eur. J. Biochem. 119: 559–564.

Blundell, T.L. and Johnson, L.N. 1976. Protein Crystallography. Academy Press, New York/London/San Francisco.

Brändén, C.I. 1980. Relation between structure and function of α/β proteins. Q. Rev. Biophys. 13: 317–338.

Briggs, D.E. 1992. Barley germination: biochemical changes and hormonal control. In: P.R. Shewry (Ed.) Barley: Genetics, Biochemistry, Molecular Biology and Biotechnology, CAB International, UK, pp. 369–401.

Burton, R.A., Gibeaut, D.M., Bacic, A., Findlay, K., Roberts, K., Hamilton, A., Baulcombe, D.C. and Fincher, G.B. 2000. Virus-induced silencing of a plant cellulose synthase gene. Plant Cell 12: 691–705.

Callebaut, I., Labesse, G., Durand, P., Poupon, A., Canard, L., Chomolier, J., Henrissat, B. and Mornon, J.P. 1997. Deciphering protein sequence information through hydrophobic cluster analysis (HCA): current status and perspectives. Cell. Mol. Life Sci. 53: 621–645.

Carpita, N.C. 1984. Cell wall development in maize coleoptiles. Plant Physiol 76: 205–212.

Carpita, N.C and Gibeaut, D.M. 1993. Structural models of primary cell walls in flowering plants: consistency of molecular structure with the physical properties of the walls during growth. Plant J. 3: 1–30.

Chen, L., Fincher, G.B. and Høj, P.B. 1993. Evolution of polysaccharide hydrolase substrate specificity. Catalytic amino acids are conserved in barley 1,3;1,4-β-D-glucanase and 1,3-β-D-glucanase. J. Biol. Chem. 268: 13318–13326.

Chen, L., Sadek, M., Stone, B.A., Brownlee, R.T.C., Fincher, G.B. and Høj, P.B. 1995a. Stereochemical course of glucan hydrolysis by barley (1→3)- and (1→3),(1→4)-β-D-glucanases. Biochim. Biophys. Acta 1253: 12–116.

Chen, L., Garrett, T.P.J., Fincher, G.B. and Høj, P.B. 1995b. A tetrad of ionizable amino acids is important for catalysis in barley β-D-glucanases. J. Biol. Chem. 270: 8093–8101.

Chipman, D.M., Grisaro, V. and Sharon, N. 1967. The binding of oligosaccharides containing *N*-acetylglucosamine and *N*-acetylmuramic acid to lysozyme. J. Biol. Chem. 242: 4388–4395.

Chothia, C. and Lesk, A.M. 1986. The relation between the divergence of sequence and structure in proteins. EMBO J. 5: 823–826.

Cline, K. and Albersheim, P. 1981. Host-pathogen interactions. XVI. Purification and characterization of a glucosyl hydrolase/transferase present in the walls of soybean cells. Plant Physiol. 68: 207–220.

Cosgrove, D.J. 1999. Enzymes and other agents that enhance cell wall extensibility. Annu. Rev. Plant Physiol. Plant Mol. Biol. 50: 391–417.

Crombie, H., Chengappa, S., Hellyer, A. and Reid, J.S.G. 1998. A xyloglucan oligosaccharide-active, transglycosylating β-D-glucosidase from the cotyledons of nasturtium (*Tropaeolum majus* L) seedlings: purification, properties and characterization of a cDNA clone. Plant J. 15: 27–38.

Damude, H.G., Withers, S.G., Kilburn, D.G., Miller, R.C. and Warren, R.A.J. 1995. Site-directed mutagenesis of the putative catalytic residues of endoglucanase CenA from *Cellulomonas fimi*. Biochemistry 34: 2220–2224.

Davies, G.J. and Henrissat, B. 1995. Structures and mechanisms of glycosyl hydrolases. Structure 3: 853–859.

Davies, G.J., Wilson, K.S. and Henrissat, B. 1997. Nomenclature for sugar-binding subsites in glycosyl hydrolases. Biochem. J. 321: 557–559.

Fincher, G.B. 1989. Molecular and cellular biology associated with endosperm mobilization in germinating cereal grains. Annu. Rev. Plant Physiol. Plant Mol. Biol. 40: 305–346.

Fincher, G.B. 1992. Cell wall metabolism in barley. In: P.R. Shewry (Ed.) Barley: Genetics, Biochemistry, Molecular Biology and Biotechnology. CAB International, UK, pp. 413–437.

Gibeaut, D.M. and Carpita, N.C. 1991. Tracing cell wall biogenesis in intact cells and plants. Selective turnover and alteration of soluble cell wall polysaccharides in grasses. Plant Physiol. 97: 551–561.

Gilkes, N.R., Henrissat, B., Kilburn, D.G., Miller, J.C. Jr. and Warren, R.A.J. 1991. Domains in microbial β-1,4-glycanases: sequence conservation, function and enzyme families. Microbiol. Rev. 55: 303–315.

Guex, N. and Peitsch, M.C. 1997. SWISS-MODEL and the Swiss-Pdb Viewer: an environment for comparative protein modeling. Electrophoresis 18: 2714–2723.

Harvey, A.J., Hrmova, M., de Gori, R., Varghese, J.N. and Fincher, G.B. 2000. Comparative modeling of the three-dimensional structures of family 3 glycoside hydrolases. Proteins Struct. Funct. Genet. 41: 257–269.

Harvey, A.J., Hrmova, M. and Fincher, G.B. 2001. Regulation of genes encoding β-D-glucan glucohydrolases in barley (*Hordeum vulgare*). Physiol. Plant. 112: in press.

Havukainen, R., Torronen, A., Laitinen, T. and Rouvinen, J. 1996. Covalent binding of three epoxyalkyl xylosides to the active site of endo-1,4-β-xylanase II from *Trichoderma reesei*. Biochemistry 35: 9617–9624.

Heightman, T.D. and Vasella, A.T. 2000. Recent insight into inhibition, structure and mechanism of configuration of retaining glycosidases. Angew. Rev. Int. Ed. 38: 750–770.

Henrissat, B. 1991. A classification of glycosyl hydrolases based on amino acid sequence similarities. Biochem. J. 280: 309–316.

Henrissat, B. 1998. Glycosidase families. Biochem. Soc. Transact. 26: 153–156.

Henrissat, B. and Davies, G. 1997. Structural and sequence-based classification of glycoside hydrolases. Curr. Opin. Struct. Biol. 7: 637–644.

Henrissat, B., Callebaut, I., Fabrega, S., Lehn, P., Mornon, J.P. and Davies, G.J. 1995. Conserved catalytic machinery and the prediction of a common fold for several families of glycosyl hydrolases. Proc. Natl. Acad. Sci. USA 92: 7090–7094.

Henrissat, B., Coutinho, P. M. and Davies, G. J. 2001. A census of carbohydrate-active enzymes in the genome of *Arabidopsis thaliana*. Plant Mol. Biol., this issue.

Hiromi, K. 1970. Interpretation of dependency of rate parameters on the degree of polymerization of substrate in enzyme-catalyzed reactions. Evaluation of subsite affinities of exo-enzyme. Biochem. Biophys. Res. Commun. 40: 1–6.

Hoare, D.G. and Koshland, D.E. Jr. 1967. A method for quantitative modification and estimation of carboxylic acid groups in proteins. J. Biol. Chem. 242: 2447–2453.

Høj, P.B., Rodriguez, E.B., Stick, R.V. and Stone, B.A. 1989. Differences in active site structure in a family of β-D-glucan endohydrolases deduced from the kinetics of inactivation by epoxyalkyl β-oligoglucosides. J. Biol. Chem. 264: 4939–4947.

Høj, P.B., Rodriguez, E.B., Iser, J.R., Stick, R.V. and Stone, B.A. 1991. Active site-directed inhibition by optically pure epoxyalkyl cellobiosides reveals differences in active site geometry of two 1,3-1,4-β-D-glucan 4-D-glucanohydrolases. The importance of epoxide stereochemistry for enzyme inactivation. J. Biol. Chem. 266: 11628–11631.

Høj, P.B., Condron, R., Traeger, J.C., McAuliffe, J.C. and Stone, B.A. 1992. Identification of glutamic acid 105 at the active site of Bacillus amyloliquefaciens 1,3-1,4-β-D-glucan 4-D-glucanohydrolase using epoxide-based inhibitors. J. Biol. Chem. 267: 25059–25066.

Høj, P.B. and Fincher, G.B. 1995. Molecular evolution of plant β-D-glucan endohydrolases. Plant J. 7: 367–379.

Hoson, T. and Nevins, D.J. 1989. β-D-glucan antibodies inhibit auxin-induced cell elongation and changes in cell wall of Zea coleoptile segments. Plant Physiol. 90: 1353–1358.

Hrmova, M. and Fincher, G.B. 1998. Barley β-D-glucan exohydrolases. Substrate specificity and kinetic properties. Carbohydrate Res. 305: 209–221.

Hrmova, M., Garrett, T.P.J. and Fincher, G.B. 1995. Subsite affinities and disposition of catalytic amino acids in the substrate-binding region of barley 1,3-β-D-glucanases. Implications in plant-pathogen interactions. J. Biol. Chem. 270: 14556–14563.

Hrmova, M., Harvey, A.J., Wang, J., Shirley, N.J., Jones, G.P., Stone, B.A., Høj, P.B. and Fincher, G.B. 1996. Barley β-D-glucan exohydrolases with β-D-glucosidase activity. Purification, characterization, and determination of primary structure from a cDNA clone. J. Biol. Chem. 271: 5277–5286.

Hrmova, M., MacGregor, E.A., Biely, P., Stewart, R.J. and Fincher, G.B. 1998a. Substrate binding and catalytic mechanism of a barley β-D-glucosidase/(1,4)-β-D-glucan exohydrolase. J. Biol. Chem. 273: 11134–11143.

Hrmova, M., Varghese, J.N., Høj, P.B. and Fincher, G.B. 1998b. Crystallization and preliminary X-ray analysis of β-D-glucan exohydrolase isoenzyme ExoI from barley (Hordeum vulgare). Acta Crystallogr. D54: 687–689.

Inouhe, M. and Nevins, D.J. 1998. Changes in the activities and polypetide levels of exo- and endoglucanases in cell walls during developmental growth of Zea mays coleoptiles. Plant Cell Physiol. 39: 762–768.

Inouhe, M., Hayashi, K. and Nevins, D.J. 1999. Polypeptide characteristics and immunological properties of exo- and endoglucanases purified from maize coleoptile cell walls. J. Plant Physiol. 154: 334–340.

Jenkins, J., Lo Leggio, L., Harris, G. and Pickersgill, R. 1995. β-Glucosidase, β-galactosidase, family A cellulases, family F xylanases and two barley glycanases for a superfamily of enzymes with 8-fold β/α architecture and with two conserved glutamates near the carboxy-terminal ends of β-strands four and seven. FEBS Lett. 362: 281–285.

Johnson, P.E., Brun, E., Mackenzie, L.F., Withers, S.G. and McIntosh, L.P. 1999. The cellulose-binding domains from Cellulomonas fimi β-1,4-D-glucanase CenC bind nitroxide spin-labeled cellooligosaccharides in multiple orientations. J. Mol. Biol. 287: 609–625.

Keitel, T., Simon, O., Borriss, R. and Heinemann, U. 1993. Molecular and active-site structure of a Bacillus 1,3-1,4-β-D-glucanase. Proc. Natl. Acad. Sci. USA 90: 5287–5291.

Kim, J.B., Olek, A.T. and Carpita, N.C. 2000. Cell wall and membrane-associated exo-β-D-glucanases from developing maize seedlings. Plant Physiol. 123: 471–485.

Koshland, D.E. Jr. 1953. Stereochemistry and the mechanism of enzymatic reactions. Biol. Rev. 28: 416–436.

Koshland, D.E. Jr. 1958. Application of a theory of enzyme specificity to protein synthesis. Proc. Natl. Acad. Sci. USA 44: 98–104.

Kotake, T., Nakagawa, N., Takeda, K. and Sakurai, N. 1997. Purification and characterization of wall-bound exo-1,3-β-D-glucanase from barley (Hordeum vulgare L) seedlings. Plant Cell Physiol. 38: 194–200.

Kraulis, P. 1991. MOLSCRIPT: a program to produce both detailed and schematic plots of protein structures. J. Appl. Crystallogr. 24: 946–950.

Labrador, E. and Nevins, D.J. 1989. An exo-β-D-glucanase derived from Zea coleoptile walls with a capacity to elicit cell elongation. Physiol. Plant. 77: 479–486.

Laitinen, T., Rouvinen, J. and Perakyla, M. 2000. Inversion of the roles of the nucleophile and acid/base catalysts in the covalent binding of epoxyalkyl xyloside inhibitor to the catalytic glutamates of endo-1,4-β-xylanase (XYLII): a molecular dynamics study. Protein Eng. 13: 247–252.

Laskowski, RA., MacArthur, M.W., Moss, D.S. and Thornton, J.M. 1993. PROCHECK: a program to check the stereochemical quality of protein structures. J. Appl. Crystallogr. 26: 283–291.

Leah, R., Kigel, J., Svendsen, I., Mundy, J. 1995. Biochemical and molecular characterization of a barley seed β-glucosidase. J. Biol. Chem. 270: 15789–15797.

Legler, G. 1990. Glycoside hydrolases: mechanistic information from studies with reversible and irreversible inhibitors. Adv. Carbohydrate Chem. Biochem. 48: 319–384.

Legler, G. and Herrchen, M. 1981. Active site directed inhibition of galactosidase by conduritol C epoxides (1,2-anhydro-epi- and neo-inositol). FEBS Lett. 135: 139–144.

Lienart, Y., Comtat, J. and Barnoud, F. 1986. A wall-bound exo-β-D-glucanase from Acacia cultured cells. Biochim. Biophys. Acta 883: 353–360.

Litts, J.C, Simmons, C.R., Karrer, K.E., Huang, N. and Rodriguez, R.L. 1990. The isolation and characterization of a barley (1→3),(1→4)-β-D-glucanase gene. Eur. J. Biochem. 194: 831–838.

Ly, H.D. and Withers, S.G. 1999. Mutagenesis of glycosidases. Annu. Rev. Biochem. 68: 487–522.

McCarter, J.D. and Withers, S.G. 1994. Mechanism of enzymatic glycoside hydrolysis. Curr. Opin. Struct. Biol. 4: 885–892.

McFadden, G.I., Ahluwalia, B., Clarke, A.E. and Fincher, G.B. 1988. Expression sites and developmental regulation of genes encoding (1→3),(1→4)-β-D-glucanase in germinated barley. Planta 173: 500–508.

Morrall, P. and Briggs, D.E. 1978. Changes in cell wall polysaccharides of germinating barley grains. Phytochemistry 17: 1495–1502.

Moreau, V. and Driguez, H. 1995. Enzymic synthesis of hemithiocellodextrins. J. Chem. Soc., Perkin Transact. 1: 525–527.

Nicholls, A., Sharp, K. and Honig, B. 1991. Protein folding and association: insights from the interfacial and thermodynamic properties of hydrocarbons. Proteins 4: 281–296.

Parsiegla, G., Reverbel-Leroy, C., Tardif, C., Belaich, P., Driguez, H. and Haser, R. 2000. Crystal structures of the cellulase Cel48F in complex with inhibitors and substrates give insight into its processive action. Biochemistry 39: 11238–11246.

Parrish, F.W., Perlin, A.S. and Reese, E.T. 1960. Selective enzymolysis of poly-β-D-glucans, and the structure of the polymers. Can. J. Chem. 38: 2094–2104.

Roulin, S., Buchala, A.J. and Fincher, G.B. 2001. Induction of $(1\rightarrow3),(1\rightarrow4)$-$\beta$-D-glucan hydrolases in leaves of dark-incubated barley seedlings. Submitted for publication.

Sakurai, N. and Masuda, Y. 1978. Auxin-induced changes in barley coleoptile cell wall composition. Plant Cell Physiol 19: 1217–1223.

Sali, A. and Blundell, T.L. 1993. Comparative protein modelling by satisfaction of spatial restraints. J. Mol. Biol. 234: 779–815.

Shibuya, N. and Misaki, A. 1978. Structure of hemicellulose isolated from rice endosperm cell wall: mode of linkages and sequences in xyloglucan, β-D-glucan and arabinoxylan. Agric. Biol. Chem. 42: 2267–2274.

Simmons, C.R., Litts, J.C., Huang, N. and Rodriguez, R.L. 1992. Structure of a rice β-D-glucanase gene regulated by ethylene, cytokinin, wounding, salicylic acid and fungal elicitors. Plant Mol. Biol. 18: 33–45.

Simos, G., Panagiotidis, C.A., Skoumbas, A., Choli, D., Ouzounis, C. and Georgatsos, J.G. 1994. Barley β-glucosidase-expression during seed germination and maturation and partial amino acid sequences. Biochim. Biophys. Acta 1199: 52–58.

Sinnott, M.L. 1990. Catalytic mechanisms of enzymic glycosyl transfer. Chem. Rev. 90: 1171–1202.

Slakeski, N., Baulcombe, D.C., Devos, K.M., Ahluwalia, B., Doan, D.N.P. and Fincher, G.B. 1990. Structure and tissue-specific regulation of genes encoding barley $(1\rightarrow3),(1\rightarrow4)$-$\beta$-D-glucan endohydrolases. Mol. Gen. Genet. 224: 437–449.

Slakeski, N. and Fincher, G.B. 1992. Developmental regulation of $(1\rightarrow3),(1\rightarrow4)$-$\beta$-D-glucanase gene expression in barley. I. Tissue specific expression of individual isoenzymes. Plant Physiol. 99: 1226–1231.

Stone, B.A. and Svensson, B. 2001. Enzymic depolymerisation and synthesis of polysaccharides. In: Glycoscience: Chemistry and Chemical Biology, Springer-Verlag, Berlin/Heidelberg/new York, in press.

Street, I.P., Kempton, J.B. and Withers, S.G. 1992. Inactivation of a β-glucosidase through the accumulation of a stable 2-deoxy-2-fluoro-α-D-glucopyranosyl-enzyme intermediate: a detailed investigation. Biochemistry 31: 9970–9978.

Stuart, I.M., Loi, L. and Fincher, G.B. 1988. Varietal and environmental variations in $(1\rightarrow3),(1\rightarrow4)$-$\beta$-D-glucan levels and $(1\rightarrow3),(1\rightarrow4)$-$\beta$-D-glucanase potential in barley: relationships to malting quality. J. Cereal Sci. 7: 61–71.

Suganuma, T., Matsuno, R., Ohnishi, M. and Hiromi, K. 1978. A study of the mechanism of action of Taka-amylase A on linear oligosaccharides by product analysis and computer simulation. J. Biochem. 84: 293–316.

Thoma, J.A., Brothers, C. and Spradlin, J. 1970. Subsite mapping of enzymes. Studies on *Bacillus subtilis* amylase. Biochemistry 9: 1768–1775.

Thomas, B.R., Simmons, C.R. Inouhe. M. and Nevins, D.J. 1998. Maize coleoptile endoglucanase is encoded by a novel gene family (Accession No. AF072326) (PGR98-143). Plant Physiol. 117: 1525.

Thomas, B.R., Inouhe, M., Simmons, C.R. and Nevins, D.J. 2000. Endo-1,3;1,4-β-D-glucanase from coleoptiles of rice and maize:

role in the regulation of plant growth. Int. J. Biol. Macromol. 27: 145–149.

Varghese, J.N., Garrett, T.P.J., Colman, P.M., Chen, L., Høj, P.B. and Fincher, G.B. 1994. Three-dimensional structures of two plant β-D-glucan endohydrolases with distinct substrate specificities. Proc. Natl. Acad. Sci. USA 91: 2785–2789.

Varghese, J.N., Hrmova, M. and Fincher, G.B. 1999. Three-dimensional structure of a barley β-D-glucan exohydrolase; a family 3 glycosyl hydrolase. Structure 7: 179–190.

Viladot, J.L., de Ramon, E., Durany, O. and Planas, A. 1998. Probing the mechanism of *Bacillus* 1,3-1,4-β-D-glucan 4-D-glucanohydrolases by chemical rescue of inactive mutants at catalytically essential residues. Biochemistry 37: 11332–11342.

Vocadlo, D.J., Mayer, C., He, S. and Withers, S.G. 2000. Mechanism of action and identification of Asp242 as the catalytic nucleophile in *Vibrio furnisii* N-acetyl-β-D-glucosaminidase using 2-acetamido-2-deoxy-5-fluoro-α-L-idopyranosyl fluoride. Biochemistry 39: 117–126.

Wang, Q., Graham, R.W., Trimbur, D., Warren, R.A.J. and Withers, S.G. 1994. Changing enzymatic reaction mechanisms by mutagenesis: conversion of a retaining glucosidases to an inverting enzyme. J. Am. Chem. Soc. 116: 11594–11595.

White, A. and Rose, D.R. 1997. Mechanism of catalysis by retaining β-glycosyl hydrolases. Curr. Opin. Struct. Biol. 7: 645–651.

Withers, S.G., Dombroski, D., Berven, L.A., Kilburn, D.G., Miller, R.C. Jr., Warren, A.J. and Gilkes, N.R. 1986. Direct [1]H N.M.R. determination of the stereochemical course of hydrolyses catalysed by glucanase components of the cellulase complex. Biochem. Biophys. Res. Commun. 139: 487–494.

Wolf, N. 1991. Complete nucleotide sequence of a *Hordeum vulgare* gene encoding $(1\rightarrow3),(1\rightarrow4)$-$\beta$-D-glucanase isoenzyme II. Plant Physiol. 96: 1382–1384.

Wood, P.J. and Fulcher, R.G. 1978. Interaction of some dyes with cereal β-D-glucans. Cereal Chem. 55: 952–966.

Wood, P.J., Weisz, J. and Blackwell, B.A. 1994. Structural studies of $(1\rightarrow3),(1\rightarrow4)$-$\beta$-D-glucans by ^{13}C-nuclear magnetic resonance spectroscopy and by rapid analysis of cellulose-like regions using high-performance anion-exchange chromatography of oligosaccharides released by lichenase. Cereal Chem. 71: 301–307.

Woodward, J.R. and Fincher, G.B. 1982. Purification and chemical properties of two 1,3;1,4-β-D-glucan endohydrolases from germinated barley. Eur. J. Biochem. 121: 663–669.

Woodward, J.R., Phillips, D.R. and Fincher, G.B. 1983. Water-soluble $(1\rightarrow3),(1\rightarrow4)$-$\beta$-D-glucans from barley (*Hordeum vulgare*) endosperm. I. Physicochemical properties. Carbohydrate Polymers 3: 143–156.

Zechel, D.L. and Withers, S.G. 1999. Glycosyl transferase mechanisms. In: D. Barton, K. Nakanishi, K. and C.D. Poulter (Eds.) Comprehensive Natural Products Chemistry, Elsevier, New York, vol. 5, pp. 279–314.

Zechel, D.L. and Withers, S.G. 2000. Glycosidase mechanisms: anatomy of a finely tuned catalyst. Acc. Chem. Res. 33: 11–18.

SECTION 3

PRIMARY WALL SYNTHESIS

Plant Molecular Biology **47**: 95–113, 2001.
© 2001 *Kluwer Academic Publishers. Printed in the Netherlands.*

Molecular genetics of nucleotide sugar interconversion pathways in plants

Wolf-Dieter Reiter* and Gary F. Vanzin
*Department of Molecular and Cell Biology, University of Connecticut, Box U-125, 75 North Eagleville Road, Storrs, CT 06269-3125, USA (*author for correspondence; e-mail wdreiter@uconnvm.uconn.edu)*

Key words: Arabidopsis thaliana, cell wall, genomics, monosaccharide, mutant

Abstract

Nucleotide sugar interconversion pathways represent a series of enzymatic reactions by which plants synthesize activated monosaccharides for the incorporation into cell wall material. Although biochemical aspects of these metabolic pathways are reasonably well understood, the identification and characterization of genes encoding nucleotide sugar interconversion enzymes is still in its infancy. *Arabidopsis* mutants defective in the activation and interconversion of specific monosaccharides have recently become available, and several genes in these pathways have been cloned and characterized. The sequence determination of the entire *Arabidopsis* genome offers a unique opportunity to identify candidate genes encoding nucleotide sugar interconversion enzymes via sequence comparisons to bacterial homologues. An evaluation of the *Arabidopsis* databases suggests that the majority of these enzymes are encoded by small gene families, and that most of these coding regions are transcribed. Although most of the putative proteins are predicted to be soluble, others contain N-terminal extensions encompassing a transmembrane domain. This suggests that some nucleotide sugar interconversion enzymes are targeted to an endomembrane system, such as the Golgi apparatus, where they may co-localize with glycosyltransferases in cell wall synthesis. The functions of the predicted coding regions can most likely be established via reverse genetic approaches and the expression of proteins in heterologous systems. The genetic characterization of nucleotide sugar interconversion enzymes has the potential to understand the regulation of these complex metabolic pathways and to permit the modification of cell wall material by changing the availability of monosaccharide precursors.

Abbreviations: AUD, membrane-anchored UDP-D-glucuronate decarboxylase; EST, expressed sequence tag; GAE, UDP-D-glucuronate 4-epimerase; GER, GDP-4-keto-6-deoxy-D-mannose 3,5-epimerase-4-reductase; GFP, green fluorescent protein; GUS, β-glucuronidase; RG-I and RG-II, rhamnogalacturonans I and II; SUD, soluble UDP-D-glucuronate decarboxylase; UER, UDP-4-keto-6-deoxy-D-glucose 3,5-epimerase-4-reductase; UGD, UDP-D-glucose dehydrogenase; UGE, UDP-D-glucose 4-epimerase

Introduction

Most of the carbon fixed by higher plants is utilized for the synthesis of cell wall material while smaller amounts are incorporated into a variety of glycoconjugates including glycoproteins, proteoglycans, and glycolipids (Carpita and Gibeaut, 1993; Reiter, 1998). Glycosyltransferases involved in the synthesis of plant glycans utilize nucleoside 5′-diphospho sugars (also referred to as sugar nucleotides, nu-

cleotide sugars or NDP-sugars) as donor substrates (Feingold and Avigad, 1980; Feingold and Barber, 1990; Mohnen, 1999; Gibeaut, 2000). The two major points of direct synthesis of NDP-sugars from phosphorylated monosaccharides are the production of UDP-D-glucose from UTP and glucose-1-phosphate, and the synthesis of GDP-D-mannose from GTP and mannose-1-phosphate (Figure 1). An alternative pathway for synthesis of UDP-D-glucose is catalyzed by sucrose synthase which converts UDP and sucrose into

UDP-D-glucose and fructose. The latter pathway has been proposed to provide high concentrations of UDP-D-glucose for the synthesis of cellulose at the plasma membrane (Amor *et al.*, 1995).

Many of the other nucleotide sugars may also be synthesized by the sequential action of monosaccharide kinases and NDP-sugar pyrophosphorylases via so-called salvage pathways in which sugars from cell wall turnover events are recycled. However, the primary route of synthesis for most nucleotide sugars is the modification of the sugar moiety in UDP-D-glucose or GDP-D-mannose by oxidation, reduction, epimerization, and/or decarboxylation reactions leading to substrates that can be used directly by glycosyltransferases (Figure 1). D-Glucoronate, the main component of pectic material, can be synthesized as a free monosaccharide via the so-called inositol oxygenation pathway (Loewus *et al.*, 1973), and is then converted into its nucleotide sugar via the sequential action of a monosaccharide kinase and a uridylyltransferase (Figure 1).

The biochemistry of NDP-sugar interconversion reactions and monosaccharide salvage pathways has been studied in a large variety of plant species, and has been extensively reviewed by Feingold and Avigad (1980). Some progress regarding the molecular genetics of these pathways has recently been made by characterizing mutants with altered cell wall composition (Reiter *et al.*, 1993, 1997; Bonin *et al.*, 1997), embryo lethality (Nickle and Meinke, 1998; Lukowitz *et al.*, 2001), increased sensitivity to monosaccharides (Dolezal and Cobbett, 1991; Sherson *et al.*, 1999) or reactive oxygen species (Conklin *et al.*, 1997, 1999). Furthermore, plant genes in NDP-sugar interconversion pathways have been identified via sequence similarity to their bacterial and mammalian counterparts (Dörmann and Benning, 1996; Tenhaken and Thulke, 1996; Kaplan *et al.*, 1997; Bonin and Reiter, 2000) (Table 1). The recent success in deciphering the nucleotide sequence of the entire *Arabidopsis* genome (Arabidopsis Genome Initiative, 2000) permits the identification of candidate genes for nucleotide sugar interconversion enzymes and an assessment of the genetic complexity and redundancy of these biochemical pathways in a plant model organism. In this review we will focus on recent advances in the molecular genetics of NDP-sugar interconversion pathways as they relate to the synthesis of plant cell wall material, and discuss possibilities to identify novel genes in these pathways by analyzing *Arabidopsis* databases.

GDP-sugar interconversion pathways: the biosynthesis of L-fucose, L-galactose and L-ascorbate

Guanosine 5′-diphospho-D-mannose is synthesized from GTP and mannose-1-phosphate via GDP-D-mannose pyrophosphorylase (Feingold and Avigad, 1980). This reaction provides activated D-mannose for incorporation into N-linked glycans and mannose-containing cell wall components such as glucomannans and galactomannans. GDP-D-mannose also serves as the substrate for nucleotide sugar interconversion enzymes yielding GDP-L-fucose and GDP-L-galactose. The latter compound serves as the donor for glycosyltransferases but also plays a key role in the biosynthesis of L-ascorbate (Wheeler *et al.*, 1998; Smirnoff and Wheeler, 2000). Mutations in a gene for GDP-D-mannose pyrophosphorylase of *Arabidopsis* were initially identified by screening for mutants with increased sensitivity to ozone (Conklin *et al.*, 1997). This led to the identification of *vtc1* (*vitamin C locus* 1) plants which are partially deficient in the production of L-ascorbate (vitamin C). Positional cloning of the *VTC1* gene revealed a single point mutation in the GDP-D-mannose pyrophosphorylase gene whereby a highly conserved proline residue is replaced by a serine which leads to a reduction in enzymatic activity by about one-third. The *vtc1* plants contain ca. 25% of the wild-type amount of L-ascorbate suggesting that changes in the availability of GDP-D-mannose have drastic effects on the flux through the L-ascorbate biosynthetic pathway (Conklin *et al.*, 1999).

Keller *et al.* (1999) used an antisense approach to achieve a ca. 50% reduction in GDP-D-mannose pyrophosphorylase activity in potato. The antisense lines showed significant reductions in the mannose content of cell wall material, and in the L-ascorbate content of leaves. This finding confirmed that GDP-D-mannose pyrophosphorylase activity is rate-limiting for the synthesis of vitamin C, and provided evidence that at least some mannosyltransferases appear to be limited by substrate availability *in vivo*. On the other hand, the transgenic plants were not affected in the synthesis of N-linked glycans, and contained essentially normal amounts of L-fucose in their cell wall polymers (Keller *et al.*, 1999). These results suggest that some metabolic pathways are more sensitive to changes in the concentration of GDP-D-mannose than others. The antisense lines with significant reductions in GDP-D-mannose pyrophosphorylase activity showed an early-senescence phenotype which may be related

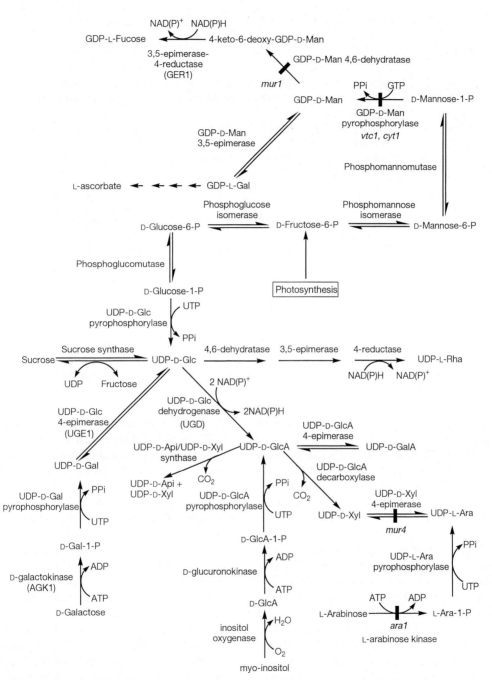

Figure 1. Overview of nucleotide sugar interconversions relevant to the synthesis of cell wall polymers in higher plants. For simplicity, only those monosaccharide salvage pathways discussed in this review are shown. All of the cloned genes and mutant symbols refer to work in *Arabidopsis thaliana*.

Table 1. Cloned plant genes in nucleotide sugar synthesis pathways.

VTC1, CYT1	GDP-D-mannose pyrophosphorylase	AAD04627	A. thaliana	Conklin et al. (1999) Lukowitz et al. (2001)
–	GDP-D-mannose pyrophosphorylase	AAD01737	S. tuberosum	Keller et al. (1999)
AGK1, GAL1	galactokinase	CAA68163	A. thaliana	Kaplan et al. (1997) Sherson et al. (1999)
ARA1	arabinose kinase	CAA74753	A. thaliana	Sherson et al. (1999)
UGE1	UDP-D-glucose 4-epimerase	CAA90941	A. thaliana	Dörmann and Benning (1996)
galE	UDP-D-glucose 4-epimerase	AAA86532	P. sativum	Lake et al. (1998)
–	UDP-D-glucose 4-epimerase	CAA06338	C. tetragonoloba	Joersbo et al. (1999)
–	UDP-D-glucose 4-epimerase	CAA06339	C. tetragonoloba	Joersbo et al. (1999)
–	UDP-D-glucose dehydrogenase	AAB58398	G. max	Tenhaken and Thulke (1996)
UGD	UDP-D-glucose dehydrogenase	BAB11006	A. thaliana	Seitz et al. (2000)
MUR1	GDP-D-mannose 4,6-dehydratase	AAB51505	A. thaliana	Bonin et al. (1997)
GMD1	GDP-D-mannose 4,6-dehydratase	AAF07199	A. thaliana	Bonin et al. (1997)
GER1	GDP-L-fucose synthase	AAC02703	A. thaliana	Bonin and Reiter (2000)

to oxidative damage due to lower concentrations of L-ascorbate (Keller *et al.*, 1999).

Because GDP-D-mannose plays roles in so many cellular processes, complete loss-of-function mutations in GDP-D-mannose pyrophosphorylase are expected to be lethal. The *cyt1* (*cytokinesis 1*) mutation isolated during a screen for embryo lethality causes the formation of highly abnormal cell walls, radial swelling, and an arrest in early embryonic development (Nickle and Meinke, 1998). The *CYT1* gene was positionally cloned and shown to be identical to *VTC1* (Lukowitz *et al.*, 2001). The *cyt1-1* allele contains a Pro-to-Leu amino acid substitution within a conserved region of the protein whereas the *cyt1-2* allele contains a frameshift mutation in the carboxy-terminal part of the protein which is likely to abolish enzyme function entirely. The cytological abnormalities seen in *cyt1-1* embryos are slightly less severe than those observed in the *cyt1-2* genetic background suggesting that the missense mutation in *cyt1-1* does not completely eliminate protein function. Embryos homozygous for *cyt1* mutations show a ca. 50% re-duction in the mannose and fucose content of their cell wall material, and are deficient in the synthesis of N-linked glycans presumably because enzyme activities in the assembly of the mannose-rich core structure are limited by the availability of GDP-D-mannose. The *Arabidopsis* genome contains several sequences similar to the *CYT1*-encoded GDP-D-mannose pyrophosphorylase which offers an explanation why cell walls from *cyt1* embryos contain substantial amounts of mannose and fucose residues (Lukowitz *et al.*, 2001). Alternatively, a maternal contribution may account for residual GDP-D-mannose pyrophosphorylase activity. The cellulose content of *cyt1* embryos is ca. 20% of wild type suggesting that the formation of abnormal and incomplete cell walls is caused by a defect in cellulose synthesis (Lukowitz *et al.*, 2001). Treatment of *Arabidopsis* root tips with tunicamycin (an inhibitor of N-glycosylation) causes radial swelling and callose deposition, a phenotype characteristic of *cyt1* embryos. This suggests that the cell wall defects and embryo lethality caused by the *cyt1* mutations reflect the mutants' inability to properly glycosylate components of the cellulose-synthesizing machinery (Lukowitz *et al.*, 2001).

Eight allelic *Arabidopsis* mutants (*mur1-1* through *mur1-8*; *mur* stands for Latin *murus*, the wall) that are defective in the *de novo* synthesis of L-fucose were isolated as part of a larger effort to isolate cell wall mutants by directly screening for changes in the monosaccharide composition of leaf cell wall material (Reiter *et al.*, 1993, 1997). Mutants carrying tight *mur1* alleles show- 100 to 200-fold lower amounts of L-fucose in all aerial parts of the plants and a ca. 40% reduction in the fucose content of their roots (Reiter *et al.*, 1993). The *MUR1* gene encodes an isoform of GDP-D-mannose 4,6-dehydratase (Bonin *et al.*, 1997), which results in a general fucose deficiency that affects the structure of glycoproteins, xyloglucan, and the pectic polysaccharides rhamnogalacturonan I and II (RG-I and RG-II). Another *Arabidopsis* gene (*GMD1* for GDP-D-mannose 4,6-dehydratase isoform 1) encoding this enzymatic activity is strongly expressed in roots but only weakly in shoot organs which explains the differences in fucose content between roots and shoots of *mur1* plants (Bonin *et al.*, 1997). Plants carrying tight *mur1* alleles are slightly dwarfed, and show a roughly two-fold reduction in the mechanical strength of elongating inflorescence stems (Reiter *et al.*, 1993). The *mur1-3* and *mur1-7* plants show a leaky phenotype (the fucose content in leaf material is ca. 30% and 10% of wild type, respectively) and appear normal in regard to their growth habit and wall strength. All of the phenotypes of *mur1* plants are rescued by addition of L-fucose to the growth medium (Reiter *et al.*, 1993) since the free monosaccharide is incorporated into GDP-L-fucose via a monosaccharide salvage pathway (Feingold and Avigad, 1980). Because the fucose deficiency of *mur1* plants affects a variety of glycans, the altered wall strength and growth habit of the mutants is difficult to interpret unless mutants with lesions in specific fucosylated glycans are available for comparison. The *cgl* (*complex glycan*) mutant of *Arabidopsis* does not contain fucosylated N-linked glycans due to a deficiency in GlcNAc transferase I but shows normal wall strength and morphology (von Schaewen *et al.*, 1993; Reiter *et al.*, 1993). This indicates that the phenotypes observed in *mur1* plants are not caused by an absence of glycoprotein fucosylation but are most likely related to changes in the fucosylation of cell wall polysaccharides.

The fucosylated side chains of xyloglucans are believed to enhance the rate of formation of strong xyloglucan-cellulose interactions (Levy *et al.*, 1991, 1997) and to play an essential role in the modulation of auxin-induced elongation growth by xyloglucan-derived oligosaccharides (York *et al.*, 1984; Mc-Dougall and Fry, 1989). Although the reduced wall strength in *mur1* plants is compatible with changes in the xyloglucan-cellulose network, the shorter plant size is difficult to reconcile with the proposed role of fucosylated xyloglucan fragments. To address this point, Zablackis *et al.* (1996) conducted an in-depth study on the structure of *mur1* xyloglucan. They found that approximately one-third of the positions normally occupied by L-fucose carried the structurally similar monosaccharide L-galactose (L-fucose is 6-deoxy-L-galactose). This study also demonstrated that L-galactosylated xyloglucan oligosaccharides showed an 'anti-auxin' activity comparable to L-fucosylated oligomers suggesting that *mur1* and wild-type plants contain comparable amounts of biologically active 'oligosaccharins'.

The structure of N-linked glycans from *mur1* plants has been investigated by Rayon *et al.* (1999) leading to the conclusion that ca. 5% of the L-fucose residues normally attached to the Asn-linked GlcNAc residue in complex glycans are replaced by L-galactose while the majority of positions remains unsubstituted. The incorporation of L-galactose into xyloglucan and N-linked glycans is presumably catalyzed by fucosyltransferases with different affinities for GDP-L-galactose which leads to varying degrees of substitution depending on the identity of the enzyme and the acceptor molecule. Alternatively, the concentrations of GDP-L-fucose and GDP-L-galactose available to different fucosyltransferases may not be uniform throughout the Golgi stacks.

The screen for mutants with altered monosaccharide composition yielded eight independent lines defective in the *de novo* synthesis of GDP-L-fucose (Reiter *et al.*, 1997) all of which had missense mutations in the *MUR1* gene (Bonin *et al.*, 1997). This raised the question of why no mutations in the subsequent steps of the L-fucose biosynthetic pathway were identified.

Biochemical evidence in bacteria (Ginsburg, 1961), mammals (Chang *et al.*, 1988) and plants (Liao and Barber, 1971) indicated that the intermediate GDP-4-keto-6-deoxy-D-mannose formed by the 4,6-dehydratase reaction undergoes a 3,5 epimerization followed by a 4 reduction yielding GDP-L-fucose. Genetic information on capsule biosynthesis in bacteria (Bastin and Reeves, 1995) permitted the identification of a candidate gene for GDP-4-keto-6-deoxy-

D-mannose 3,5-epimerase-4-reductase (synonymous with GDP-L-fucose synthase), which was then used to identify an *Arabidopsis* homologue (*GER1*) in the database of expressed sequence tags (dbEST) (Bonin and Reiter, 2000). Expression of the intron-less *GER1* gene in *Escherichia coli* yielded a protein with the expected enzymatic activity indicating that the epimerase-reductase in fucose synthesis had indeed been cloned. *GER1* antisense plants showed an up to 50-fold reduction in enzymatic activity but contained normal amounts of fucose in their cell wall material, indicating that the 3,5-epimerase-4-reductase activity is not rate-limiting *in vivo*. The *Arabidopsis* genome contains a sequence predicted to encode a protein highly similar to GER1 (88% amino acid identity over 312 amino acids). According to entries in dbEST, this 'GER2' gene appears to be transcribed both in roots and flower buds suggesting that it may be expressed throughout the plant. If *GER2* were functionally redundant with *GER1*, this would offer an explanation why no mutants defective in either gene were isolated.

The conversion of GDP-D-mannose to GDP-L-galactose is mechanistically similar to the steps catalyzed by the *GER1* gene product (Barber, 1979) since it represents a 3,5 epimerization via *enol* intermediates. Accordingly, GDP-D-mannose 3,5-epimerase may be structurally similar or identical to the 3,5-epimerase-4-reductase in fucose synthesis. GER1 protein expressed in *E. coli* did not convert GDP-D-mannose into detectable products under a variety of reaction conditions speaking against a role of this enzyme in the formation of GDP-L-galactose. Because L-galactose is an intermediate in the synthesis of L-ascorbate, *GER1* antisense plants were assayed for this metabolite but no differences from wild type were observed (Bonin and Reiter, 2000). This leaves GER2 as a candidate for GDP-D-mannose 3,5-epimerase, a hypothesis that remains to be tested.

UDP-sugar interconversion pathways: the biosynthesis of D-galactose, uronic acids, pentoses and L-rhamnose

De novo *and salvage pathways in the formation of* UDP-D-galactose

The activation of D-glucose via UDP-D-glucose pyrophosphorylase or sucrose synthase yields a cytoplasmic pool of UDP-D-glucose that functions as a donor of glucose for the synthesis of cellulose at the plasma membrane (Delmer and Amor, 1995; Amor *et al.*, 1995). UDP-D-glucose also serves as a substrate for the synthesis of other nucleotide sugars required for the synthesis of cell wall matrix components within the Golgi after translocation of the nucleotide sugar across the Golgi membrane (Muñoz *et al.*, 1996). UDP-D-galactose is generated from UDP-D-glucose via a freely reversible 4-epimerization reaction involving an enzyme-bound 4-keto intermediate (Maitra and Ankel, 1971). Most of the UDP-D-galactose is ultimately needed in the Golgi for the synthesis of arabinogalactan-proteins and cell wall polysaccharides including RG-I, RG-II and xyloglucan. In green tissues, substantial amounts of UDP-D-galactose are also needed for the synthesis of chloroplast galactolipids (Joyard *et al.*, 1998). cDNAs encoding UDP-D-glucose 4-epimerase have been cloned from *Arabidopsis* (Dörmann and Benning, 1996), pea (Lake *et al.*, 1998) and the endospermous legume guar (Joersbo *et al.*, 1999). At least two isoforms of this enzyme are expressed in developing seeds of guar to produce UDP-D-galactose required for the synthesis of the storage polysaccharide galactomannan (Joersbo *et al.*, 1999).

The function of a UDP-D-glucose 4-epimerase gene (*UGE1*) on chromosome 1 of *A. thaliana* has been studied in considerable detail via antisense and over-expression approaches (Dörmann and Benning, 1998). Transgenic lines over-expressing the *UGE1* cDNA showed up to three-fold increases in UDP-D-glucose 4-epimerase activity, whereas enzyme function was up to ten-fold lower in antisense lines. These changes in UDP-D-glucose 4-epimerase activity did not significantly alter the ratio between UDP-D-glucose and UDP-D-galactose which is close to the thermodynamic equilibrium value of 3.5 (Wilson and Hogness, 1964) in wild-type *Arabidopsis*. *UGE1* transgenics grown in soil showed a normal growth habit and contained wild-type amounts of D-galactose in their cell wall material and chloroplast lipids. This picture changed considerably upon examination of plants grown axenically in the presence of D-galactose (Dörmann and Benning, 1998). Under these conditions, the ratio between UDP-D-glucose and UDP-D-galactose fell to values as low as 1.9 in wild-type plants, reflecting an increased concentration of UDP-D-galactose. Furthermore, the amount of galactose incorporated into cell wall material increased substantially suggesting that the availability of UDP-D-galactose represents a rate-limiting step in the synthesis of galactose-containing cell wall components. Over-expression of the *UGE1* gene product shifted the UDP-D-glucose to UDP-D-galactose ratio closer to its equilibrium value, and permitted the plants to tolerate otherwise toxic concentrations of D-galactose in the growth medium. Based on this observation, Dörmann and Benning (1998) suggested that over-expression of UDP-D-glucose 4-epimerase may be useful as a selectable marker during plant transformations by 'detoxifying' exogenous D-galactose. The deleterious effects of millimolar concentrations of D-galactose in the growth medium have been observed in a number of plant species (Yamamoto *et al.*, 1988; Maretzki and Thom, 1978), and this may be due to the depletion of phosphate and/or uridine nucleotide pools by the action of galactokinase and UDP-D-galactose pyrophosphorylase; however, more complex metabolic deregulation phenomena may contribute to galactose toxicity in higher plants. Over-expression of UDP-D-glucose 4-epimerase converts excess UDP-D-galactose into UDP-D-glucose, which is rapidly turned over via the action of sucrose phosphate synthase, sucrose synthase or glucosyltransferases leading to the regeneration of UDP. *UGE1* antisense plants were more susceptible to the toxic effects of D-galactose than their wild-type counterparts (Dörmann and Benning, 1998), presumably by exacerbating metabolic deregulation effects caused by the accumulation of UDP-D-galactose.

An evaluation of the *Arabidopsis* genome sequence indicates the presence of four coding regions that are 65–89% identical to UGE1 on the amino acid level. Based on current entries in dbEST, most of these coding regions are transcribed (Table 2), and two of them (*UGE2* and *UGE3*) have been demonstrated to encode functional UDP-D-glucose 4-epimerase (R. Verma and W.-D. Reiter, unpublished results). The up to 10-fold reduction in UDP-D-glucose 4-epimerase activity in *UGE1* antisense lines (Dörmann and Benning, 1998) suggests that this protein represents the predominant UDP-D-glucose 4-epimerase in *Arabidopsis* which is in line with the observation that dbEST contains more ESTs derived from *UGE1* than from homologous coding regions (15 ESTs from UGE1 vs. 10 ESTs from all other isoforms combined). The functional significance of redundant UDP-D-glucose 4-epimerases is difficult to rationalize since the epimerization reaction is freely reversible and does not appear to be rate-limiting under standard growth conditions; however, some of the UGE isoforms may play important roles in specific cell types or under certain environmental conditions. This argument is supported by

the recent finding that the *rhd1* (root hair development locus 1) mutant of *Arabidopsis* (Schiefelbein and Somerville, 1990) carries a defect in the *UGE4* gene (G.J. Seifert, personal communication). *rhd1* plants develop a bulge at the base of the root hairs, and are allelic to the *reb1* (root epidermal cell bulging locus 1) mutants described by Baskin *et al.* (1992). A characterization of *reb1-1* plants indicated a 30% reduction in arabinogalactan-proteins (AGPs) in root material (Ding and Zhu, 1997). Furthermore it was found that the alterations in root cell morphology in *reb1* (=*rhd1*=*uge4*) plants are phenocopied by growth in the presence of (β-D-GLc)$_3$ Yariv reagent (Yariv *et al.*, 1992) which specifically binds to AGPs (Ding and Zhu, 1997; Willats and Knox, 1996). These results suggest that the loss of *UGE-4* function affects the morphology of root epidermal cells by interfering with the synthesis of galactosylated glycans such as AGPs.

The first step in the salvage pathway for re-utilization of free D-galactose is catalyzed by galactokinase, an enzyme for which genes have been isolated from bacterial, fungal and mammalian sources. The first cDNA encoding a plant galactokinase was fortuitously isolated by Kaplan *et al.* (1997) during an attempt to identify *Arabidopsis* homologues of the *Saccharomyces cerevisiae* peroxisome assembly gene *PAS9* via functional complementation of yeast. The *AGK1* cDNA (for *Arabidopsis galactokinase locus* 1) is predicted to encode a soluble protein of 496 amino acids which shows substantial sequence similarities to galactokinases cloned from other sources. Although the *AGK1* cDNA does not complement the yeast *pas1* mutation, it permitted the yeast strain to utilize small amounts of D-galactose (ca. 3 mM) released from agar during autoclaving (Kaplan *et al.*, 1997). Further evidence for the predicted function of *AGK1* was obtained by complementation of a *gal1* strain of yeast which is deficient in galactokinase activity (Kaplan *et al.*, 1997). Sherson *et al.* (1999) demonstrated that an *AGK1* orthologue (termed *GAL1*) from a different ecotype of *Arabidopsis* complemented the *galK* (galactokinase) mutation of *E. coli* and that the complemented strain contained measurable amounts of galactokinase activity.

UDP-D-*glucose dehydrogenase: a key enzyme in the biosynthesis of uronic acids and pentoses*

Most higher plants incorporate large amounts of uronic acids into their cell wall polysaccharides pri-

Table 2. Putative *Arabidopsis* proteins with sequence similarities to UDP-D-glucose 4-epimerases.

Tentative gene name[a]	Protein accession	Length of protein (amino acids)	Chromosomal location	Predicted number of exons	Level of transcription[b]
UGE1	CAA90941	351	Chr. I	8	+++
UGE2	CAB43892	350	Chr. IV	9	+
UGE3	AAG12659	353	Chr. I	9	+
UGE4	AAG21514	348	Chr. I	9	++
UGE5	CAB40064	351	Chr. IV	9	−

[a] *UGE1* has been named and described by Dörmann and Benning (1998).
[b] The level of transcription is based on the number of matching sequences in dbEST: −, none; +, 1–5; ++, 6–10; +++, 10–20.

marily as D-galacturonate residues in the backbones of pectic material, and as D-glucuronate residues in glucuronoarabinoxylans. UDP-D-glucuronate also serves as the precursor for the synthesis of UDP-D-xylose, UDP-L-arabinose, and UDP-D-apiose (Feingold and Avigad, 1980). The first step in this series of nucleotide sugar interconversion reactions is catalyzed by UDP-D-glucose dehydrogenase forming UDP-D-glucuronate in an irreversible and presumably rate-limiting reaction (Dalessandro and Northcote, 1977). An alternate pathway known to exist in many plant species converts *myo*-inositol to D-glucuronate via the action of inositol oxygenase (Loewus *et al.*, 1973). D-Glucuronate is then conjugated to UDP via the sequential action of a monosaccharide kinase and a uridylyltransferase (Feingold and Avigad, 1980). A cDNA encoding UDP-D-glucose dehydrogenase was fortuitously cloned from soybean during an effort to obtain a plant homologue to mammalian NADPH oxidase (Tenhaken and Thulke, 1996). The function of the encoded protein was inferred from its high degree of sequence similarity to the bovine enzyme, and from immunoprecipitation of UDP-D-glucose dehydrogenase activity by an antibody raised against the soybean protein expressed in *E. coli*; however, enzymatic activity of the recombinant protein has not been demonstrated (Tenhaken and Thulke, 1996). Seitz *et al.* (2000) used the soybean cDNA to isolate an *Arabidopsis* orthologue (termed *UGD* for *UDP-D-glucose dehydrogenase*) to conduct detailed expression studies. Although *UGD* was believed to represent a single-copy gene based on Southern blots (Seitz *et al.*, 2000), an evaluation of the *Arabidopsis* database after completion of the genome project indicated the presence of three paralogues (Table 3). The four UGD isoforms are highly similar to each other with amino acid sequence identities between 83% and 93%, and all of them are

transcribed based on the existence of ESTs derived from these genes. Using a variety of procedures including northern blots, promoter-GUS fusions, and promoter-GFP fusions, Seitz *et al.* (2000) demonstrated that *UGD* expression was primarily confined to root tissues in 3-day old *Arabidopsis* seedlings while a more uniform distribution of enzymatic activity was observed in older plants. Furthermore, high UDP-D-glucose dehydrogenase activity could be demonstrated via activity staining in root tips from young seedlings in support of the reporter gene studies. Since UDP-D-glucuronate is required in all plant organs at all stages of development, radiolabeling of young seedlings with *myo*-inositol was used to determine whether the inositol oxygenation pathway contributes toward the synthesis of D-glucuronate in *Arabidopsis*. These experiments showed incorporation of radioactive label preferentially into cotyledons and the hypocotyl where the *UGD* gene was only weakly expressed. This suggests that the UGD and the inositol oxygenation pathways operate at different rates in different organs. Seitz *et al.* (2000) hypothesize that the UGD pathway predominates in roots since anoxic conditions in this organ may render the inositol oxygenation pathway nonfunctional. To analyze the significance of the two pathways toward UDP-D-glucuronate synthesis in more detail, it would be useful to study plant genes encoding *myo*-inositol oxygenase. Currently genes encoding this activity have not been described in any organism making it difficult to identify plant genes encoding this activity via database searches.

Table 3. Putative *Arabidopsis* proteins with sequence similarities to UDP-D-glucose dehydrogenases.

Tentative gene name[a]	Protein accession	Length of protein (amino acids)	Chromosomal location	Predicted number of exons	Level of transcription[b]
UGD1	BAB11006	480	Chr. V	1	+ + ++
UGD2	BAB02581	480	Chr. III	1	+ + ++
UGD3	CAC01748	480	Chr. V	1	+ + +
UGD4	AAF98561	481	Chr. I	1	+

[a]*UGD1* has been described as *UGD* by Seitz *et al.* (2000).
[b]The level of transcription is based on the number of matching sequences in dbEST: +, 1–5, + + +, 10–20; + + ++, >20.

Enzymes acting on UDP-D-glucuronate: the biosynthesis of UDP-D-galacturonate, UDP-D-xylose and UDP-D-apiose

UDP-D-glucuronate represents a major branch-point in the biosynthesis of UDP-sugars which are generated via a series of epimerization, decarboxylation and rearrangement reactions (Figure 1). The interconversion between UDP-D-glucuronate and UDP-D-galacturonate is freely reversible with an equilibrium constant slightly in favor of the latter compound (Feingold and Avigad, 1980). The 4-epimerization reaction proceeds via a 4-keto intermediate presumably generated by an enzyme-bound pyridine nucleotide cofactor similar to the interconversion catalyzed by UDP-D-glucose 4-epimerase. A gene encoding UDP-D-glucuronate 4-epimerase (*cap*1J for *capsular polysaccharide gene* 1J) has recently been cloned from a type 1 strain of *Streptococcus pneumoniae* which requires UDP-D-galacturonate for capsule synthesis (Muños *et al.*, 1999). BLAST searches indicate the presence of *cap*1J homologues in numerous bacterial species; however, the functions of these gene products have not been established. The *Arabidopsis* genome contains six predicted coding regions (tentatively termed *GAE1–GAE6* for *UDP-D-glucuronic acid 4-epimerase* isoforms 1–6) with a high degree of sequence similarity to the bacterial enzymes, and all of these plant sequences are intron-less and transcribed based on the presence of entries in dbEST (Table 4). *GAE1* and *GAE6* appear to be strongly expressed with >25 and >50 ESTs, respectively, whereas the number of ESTs for the other isoforms ranges between 1 and 5. When compared to cap1J, all of the *Arabidopsis* GAE sequences contain N-terminal extensions of ca. 100 amino acids with a hydrophobic segment close to the start of the proteins (Figure 2). This structure is typical of type II membrane proteins such as Golgi-localized glycosyltransferases. We speculate that the GAE proteins are targeted to the Golgi to provide UDP-D-galacturonate at the site of cell wall synthesis. Under this scenario, UDP-D-glucuronate produced in the cytoplasm would be imported into the Golgi to serve as substrate both for glycosyltransferases and for nucleotide sugar interconversion enzymes. TBLASTN searches of dbEST with the *Arabidopsis* GAEs as query sequences identify close homologues in numerous plant species including monocots, dicots, and gymnosperms indicating that these genes are widespread throughout the plant kingdom. cDNA and genomic sequences from fungi and animals do not contain obvious GAE homologues which correlates with the apparent absence of D-galacturonate in these organisms.

UDP-D-glucuronate decarboxylase (also called UDP-D-glucuronate carboxy-lyase or UDP-D-xylose synthase) converts UDP-D-glucuronate into UDP-D-xylose in an essentially irreversible reaction (Feingold and Avigad, 1980). This enzyme may be structurally related to UDP-D-glucoronate 4-epimerase since both proteins act on the same substrate, and catalyze mechanistically similar reactions. Both interconversion enzymes use an NAD(P)$^+$ cofactor to generate a transient 4-keto intermediate which is non-stereospecifically reduced in case of the 4-epimerase yielding an equilibrium mixture of UDP-D-glucuronate and UDP-D-galacturonate. In case of the decarboxylase, the intermediate loses CO_2 in an elimination reaction typical of a α-keto carboxylic acid, and is then stereospecifically reduced to yield UDP-D-xylose (Feingold and Avigad, 1980). These similarities in reaction mechanisms suggest that the 4-epimerase and the decarboxylase have similar primary structures which raises the possibility that one or several of the *GAE* genes of *A. thaliana* encode a decarboxylase rather than a 4-epimerase. A sim-

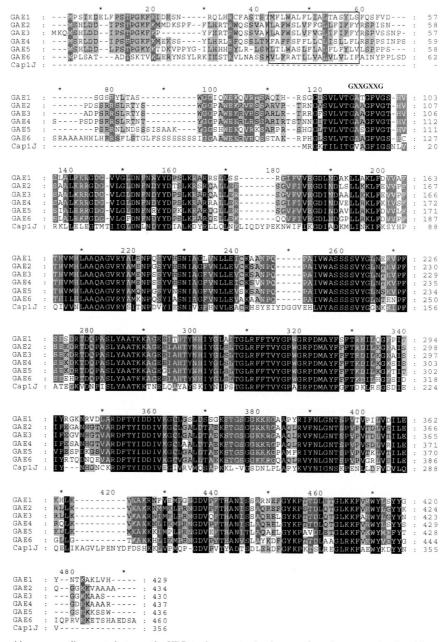

Figure 2. Amino acid sequence alignments between the UDP-D-glucuronate 4-epimerase from *S. pneumoniae* (*cap1J* gene product) and the six related genes in the *Arabidopsis* genome. The predicted transmembrane domains in the *GAE* gene products are boxed, and the GXXGXXG motif predicted to be involved in NAD(P)$^+$ binding is indicated.

ilar argument can be made for the bifunctional enzyme UDP-D-apiose/UDP-D-xylose synthase (also referred to as UDP-D-glucuronate cyclase) which has been purified from duckweed (Wellmann and Grisebach, 1971; Kindel and Watson, 1973) and parsley (Matern and Grisebach, 1977), two plants containing large amounts of D-apiose in the polysaccha-

ride apiogalacturonan (Hart and Kindel, 1970) and the trihydroxyflavone glycoconjugate apiin, respectively. UDP-D-apiose/UDP-D-xylose synthase converts UDP-D-glucuronate into approximately equimolar amounts of UDP-D-apiose and UDP-D-xylose via a common 4-keto intermediate (Matern and Grisebach, 1977). In most higher plants, D-apiose is specifically

Table 4. Putative *Arabidopsis* proteins with sequence similarities to UDP-glucuronate 4-epimerase from *S. pneumoniae.*

Tentative gene name	Protein accession	Length of protein (amino acids)	Chromosal location	Predicted number of exons	Level of transcription[a]
GAE1	CAB79762	429	Chr. IV	1	+ + + +
GAE2	AAF76478	434	Chr. I	1	+
GAE3	CAB80769	430	Chr. IV	1	+
GAE4	AAB82632	437	Chr. II	1	+
GAE5	CAB45972	436	Chr. IV	1	+
GAE6	BAB03000	460	Chr. III	1	+ + + +

[a]The level of transcription is based on the number of matching sequences in dbEST: +, 1–5; + + + +, >20.

found in the complex polysaccharide RG-II where two D-apiose residues serve as attachment points for complex carbohydrate side chains to the homogalacturonan backbone (O'Neill *et al*, 1996; Ishii *et al.*, 1999). The cloning of genes encoding UDP-D-apiose/UDP-D-xylose synthase would be of great interest since down-regulation or elimination of D-apiose synthesis in most plants would specifically affect RG-II by converting a highly complex cell wall component into a structurally simpler polymer. RG-II has recently been shown to become dimerized via a borate tetraester cross-link both *in vitro* and *in vivo* (O'Neill *et al.*, 1996; Ishii and Matsunaga, 1996; Ishii *et al.*, 1999; Kobayashi *et al.*, 1996). The borate is believed to bind tightly to the *cis*-hydroxyls at carbons 2 and 3 of an apiose moiety creating a borate diester which can then react with the apiose moiety of another RG-II molecule to form a tetraester bond. The establishment of this cross-link has recently been shown to reduce wall porosity (Fleischer *et al.*, 1999), a property which may be highly significant for cell wall function and integrity.

If UDP-D-glucuronate 4-epimerase, UDP-D-glucuronate decarboxylase, and UDP-D-apiose/UDP-D-xylose synthase were encoded by members of the *GAE* gene family in *Arabidopsis*, all of these enzymes would be expected to reside in microsomal fractions since the GAE proteins contain a putative transmembrane domain. Membrane localization of the first two proteins has been reported for several dicot species (Feingold and Avigad, 1980; Liljebjelke *et al.*, 1995), and UDP-D-glucuronate decarboxylase activity has been found in the microsome fraction of *Arabidopsis* leaves (Burget and Reiter, 1999). On the other hand, UDP-D-glucuronate decarboxylase from wheat germ (John *et al.*, 1977) and the UDP-D-apiose/UDP-

D-xylose synthases from duckweed (Wellmann and Grisebach, 1971) and parsley (Matern and Grisebach, 1977) are soluble enzymes. The subcellular localization of UDP-D-apiose/UDP-D-xylose synthase in plants such as Arabidopsis which use D-apiose exclusively for the synthesis of RG-II is unknown. UPD-D-glucuronate decarboxylase has recently been purified form pea seedlings which permitted the isolation of a cDNA clone via the N-terminal amino acid sequence (M. Kobayashi and T. Matoh, personal communication; GenBank accession BAB40967). Enzyme assays on recombinant protein from *E. coli* confirmed that the cDNA encodes an enzyme with UDP-D-glucuronate decarboxylase activity. Surprisingly, this protein is much more similar to bacterial dTDP-D-glucose 4,6-dehydratases than to UDP-D-glucuronate 4-epimerases, and represents a close homolog of the '*SUD*' gene family in *Arabidopsis* (see the section on L-rhamnose biosynthesis below). Based on this finding it now appears that at least some UDP-D-glucuronate decarboxylases (and potentially some UDP-D-apiose/UDP-D-xylose synthases) are evolutionary related to 4,6-dehydratases in the biosynthesis of L-rhamnose.

The biosynthesis of UDP-L-arabinose via de novo and salvage pathways

UDP-L-arabinose is synthesized from UDP-D-xylose via a freely reversible 4-epimerization reaction with an equilibrium constant of ca. 0.9, slightly favoring UDP-D-xylose (Salo *et al.*, 1968). UDP-D-xylose 4-epimerase activity from higher plants is membrane-bound, and appears to be specific for the UDP-D-xylose/UDP-L-arabinose pair of nucleotide sugars. A mutant of *A. thaliana* (*mur4*) has been isolated based on a 50% reduction in the amount of L-arabinose in its cell wall material (Reiter *et al.*, 1997) and was recently characterized on the biochemical level (Burget and Reiter, 1999). Membrane preparations from *mur4* plants show a decreased activity of UDP-D-xylose 4-epimerase, and recent genetic data suggest that the MUR4 protein encodes a structural gene for this enzyme (Burget and Reiter, 1999). The residual amount of L-arabinose in *mur4* plants appears to be caused by genetic redundancy rather than leakiness of the mutation (E. Burget and W.-D. Reiter, unpublished results). The partial arabinose deficiency in *mur4* plants is rescued by growth in the presence of 30 mM L-arabinose presumably by exploiting a salvage pathway which converts L-arabinose into its nucleotide sugar via the sequential action of L-arabinose kinase and a UDP-sugar pyrophosphorylase (Feingold and Avigad, 1980). An L-arabinose kinase-deficient mutant (*ara1-1*) of *A. thaliana* was fortuitously isolated by Dolezal and Cobbett (1991) by testing chemically mutagenized plants for growth on media containing various sugars. The *ara1-1* mutation represents a semidominant monogenic Mendelian trait which causes growth retardation and plant death in the presence of millimolar concentrations of L-arabinose in the growth medium. Biochemical assays indicated a tenfold reduction in arabinose kinase activity in crude extracts from the mutant seedlings. Furthermore, the seedlings metabolized free L-arabinose to a much lower extent than wild type (Dolezal and Cobbett, 1991), suggesting a lesion in the structural gene for L-arabinose kinase or in a factor controlling this enzymatic activity. Positional cloning of the *ARA1* gene indicated that it encodes a protein of 988 amino acids which consists of a carboxy-terminal domain with significant sequence similarity to galactokinases, and an N-terminal domain of unknown function (Sherson *et al.*, 1999). The *ara1-1* mutation causes a single nucleotide change leading to a Glu-to-Lys amino acid substitution within the carboxy-terminal domain. Cobbett *et al.* (1992) isolated suppressor mutations of the arabinose sensitivity phenotype of *ara1-1* plants including one line (*ara1-1* sup1) without detectable L-arabinose kinase activity. This line was subsequently shown to contain a nonsense mutation within the carboxy-terminal domain of the *ARA1* gene product (Sherson *et al.*, 1999). It has been suggested that the ARA1 protein plays a role in the uptake of L-arabinose (Cobbett *et al.*, 1992) which may explain why certain *ara1* mutations such as *ara1-1* sup1 are deficient in L-arabinose kinase ac-

tivity but do not cause sensitivity to L-arabinose in the growth medium.

The biosynthesis of L-rhamnose, an essential component of pectic material

L-Rhamnose is the predominant 6-deoxyhexose in most higher plants, and represents a major component of the pectic polysaccharides RG-I and RG-II. L-Rhamnosyl residues are also often conjugated to secondary metabolites yielding soluble rhamnosides for storage in the vacuole.

The biosynthesis of L-rhamnose has been extensively studied in bacteria where this sugar represents a frequent component of capsular polysaccharides (Reeves *et al.*, 1996). The initial activation step is catalyzed by dTDP-D-glucose pyrophosphorylase which transfers a dTMP moiety from dTTP to glucose-1-phosphate. dTDP-D-glucose is then converted to dTDP-L-rhamnose via a succession of three enzymatic steps catalyzed by a 4,6-dehydratase, a 3,5-epimerase, and a 4-reductase (Figure 3). This sequence of reactions is similar to the biosynthetic steps leading from GDP-D-mannose to GDP-L-fucose except that the last two activities in the biosynthesis of dTDP-L-rhamnose reside on separate polypeptide chains, and the configuration at C-4 is inverted in case of dTDP-L-rhamnose, whereas it is retained in case of GDP-L-fucose. The actual interconversion reactions leading to the formation of nucleotide-bound L-rhamnose appear to be catalyzed by enzymes which are evolutionarily conserved between bacteria and plants, but the organization of coding regions differs markedly between these two groups of organisms. As mentioned above, bacteria contain three separate genes encoding the 4,6-dehydratase, 3,5-epimerase, and 4-reductase, respectively. These genes are usually organized as operons which include the coding region for dTDP-D-glucose pyrophosphorylase. For instance, a gene cluster encoding the *O*-antigen biosynthetic proteins of *E. coli* K-12 (Stevensen *et al.*, 1994) contains the genes involved in L-rhamnose biosynthesis in the order *rmlB* (previously called *rfbB* encoding dTDP-D-glucose 4,6-dehydratase), *rmlD* (previously called *rfbD* encoding dTDP-4-keto-L-rhamnose reductase), *rmlA* (previously called *rfbA* encoding dTDP-D-glucose pyrophosphorylase), and *rmlC* (previously called *rfbC* encoding dTDP-4-keto-6-deoxy-D-glucose 3,5-epimerase). As a general rule, plants use UDP-D-glucose rather than dTDP-D-glucose as a precursor for the *de novo* synthesis of L-rhamnose

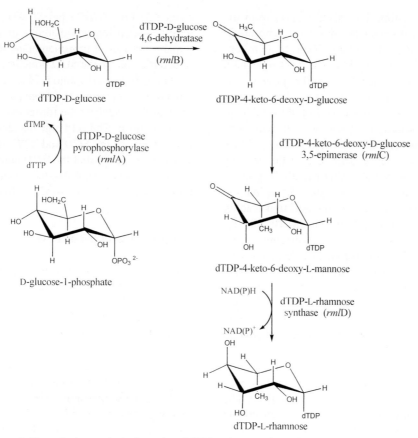

Figure 3. Biosynthetic steps in the formation of dTDP-L-rhamnose from glucose-1-phosphate in bacteria.

(Feingold and Avigad, 1980). In agreement with these biochemical data, the *Arabidopsis* genome does not contain obvious homologues to the bacterial *rml*A gene products which encode dTDP-D-glucose pyrophosphorylase. BLAST searches of the *Arabidopsis* database using dTDP-D-glucose 4,6-dehydratase from *E. coli* identify three coding regions with a high degree of sequence similarity to the bacterial enzymes which we have tentatively designated *RHM1*, *RHM2*, and *RHM3* for rhamnose biosynthetic genes 1, 2 and 3 (Table 5 and Figure 4). The three *RHM*-coding regions are very similar to each other (amino acid identities >84% between isoforms), and consist of two domains: an N-terminal domain of about 340 amino acids with significant sequence similarity to bacterial dTDP-D-glucose 4,6-dehydratases, and a carboxy-terminal domain of about 300 amino acids (Figure 4). When used in BLAST searches against protein databases, the carboxy-terminal domain is most similar to bacterial dTDP-4-keto-L-rhamnose reductases (also known as dTDP-4-keto-6-deoxy-L-

mannose 4-reductase, dTDP-L-rhamnose dehydrogenase and dTDP-L-rhamnose synthase). Although the overall sequence similarity is fairly low with a P value of about 10^{-4}, it is noteworthy that these bacterial sequences are known to catalyze the final step in the *de novo* synthesis of dTDP-L-rhamnose. We speculate that the *Arabidopsis RHM* genes contain all the functions required to convert UDP-D-glucose to UDP-L-rhamnose, and that they contain two NAD(P)$^+$ binding sites. Under this scenario, the first NAD(P)$^+$ moiety would be involved in the 4,6-dehydratase reaction, and may be tightly bound to the enzyme while the second site would be involved in the reduction of the 4-keto intermediate to UDP-L-rhamnose. Both domains of the *Arabidopsis* RHM proteins contain a NAD(P)$^+$ binding motif in their N-terminal regions except that a highly conserved glycine residue in the putative 4,6-dehydratase domain is replaced by an alanine (Figure 4). The same amino acid substitution is found in a human homologue of the bacterial dTDP-D-glucose 4,6-dehydratases (accession number

AAD50061) suggesting that this deviation from the GxxGxxG consensus sequence does not compromise protein function. If the *Arabidopsis* RHM proteins represent a fusion of a 4,6-dehydratase and a 4-reductase, the question arises which part of the protein catalyzes the 3,5-epimerization reaction. We envision a situation where the genes encoding two polypeptide chains representing the bacterial *rmlC* and *rmlD* gene products merged during evolution to produce the type of bifunctional enzyme seen in the *de novo* synthesis of GDP-L-fucose. Alternatively, a primordial 4-reductase may have acquired 3,5-epimerase function via random mutation and selection leading to a bifunctional enzyme. With the PSI-BLAST algorithm during database searches (Altschul *et al.*, 1997), an NAD^+-dependent epimerase domain is predicted within the carboxy-terminal part of the RHM proteins (Figure 4). The same structural domain is present in bacterial 3,5-epimerases catalyzing the second step in the synthesis of dTDP-L-rhamnose from dTDP-D-glucose but is absent from dTDP-L-rhamnose synthases. These findings support the idea that the carboxy-terminal domain of the RHM proteins represent a 3,5-epimerase-4-reductase gene that has been fused to a UDP-D-glucose 4,6-dehydratase gene. The *Arabidopsis* genome contains one predicted coding region (tentatively termed *UER1* for UDP-4-keto-6-deoxy-D-glucose 3,5-epimerase-4-reductase) which is virtually identical to the carboxy-terminal domain of the *RHM* gene products (Figure 4). This putative protein is an obvious candidate for a 3,5-epimerase-4-reductase in L-rhamnose synthesis although other functions cannot be ruled out. With the exception of *RHM3*, all of the coding regions discussed above appear to be heavily transcribed based on the presence of ca. 20 cDNA clones each in dbEST (Table 5).

The possible existence of a separate UDP-4-keto-6-deoxy-D-glucose 3,5-epimerase-4-reductase raises the question whether plant genomes contain coding regions for UDP-D-glucose 4,6-dehydratase which are not linked to other enzymatic activities. The *Arabidopsis* genome encodes six putative proteins with sequence similarities to bacterial dTDP-D-glucose 4,6-dehydratases which appear significant (Figure 5) but are lower than those observed with the *RHM* gene products. Three of these 4,6-dehydratase homologues (tentatively termed AUD1, AUD2, and AUD3) contain N-terminal extensions encompassing a putative transmembrane domain, a protein structure remarkably similar to the *GAE* gene products (Figures 2 and 5). The remaining three coding regions (tentatively

termed *SUD1*, *SUD2*, and *SUD3*) contain small N-terminal extensions compared to their bacterial counterparts but lack a transmembrane domain. As mentioned previously, the SUD proteins are virtually identical in their derived amino acid sequence to a UDP-D-glucuronate decarboxylase from pea strongly suggesting that they catalyze the formation of UDP-D-xylose in *Arabidopsis*. The very high degree of sequence similarity between the *SUD* and *AUD* gene products furthermore suggests that the *AUD* gene family encodes membrane-bound UDP-D-glucuronate decarboxylases rather than UDP-D-glucose 4,6-dehydratases.

Conclusions and perspectives

Nucleotide sugar interconversion pathways produce a variety of activated monosaccharides for incorporation into cell wall material, glycoproteins and low-molecular-weight glycoconjugates. Biochemical aspects of these pathways have been investigated in numerous plant species but the molecular genetics of these reactions has received comparatively little attention. Genes encoding monosaccharide kinases, NDP-sugar pyrophosphorylases and some of the actual interconversion enzymes (i.e. enzymes changing the identity of a sugar moiety) have recently been identified via the characterization of mutants and the identification of coding regions with sequence similarities to functional homologues from other organisms. With the completion of the Arabidopsis Genome Project, numerous candidate NDP-sugar interconversion enzymes have been identified, and can now be used to establish their biochemical function via over-expression or antisense approaches, the isolation of true mutants from tagged populations, or the expression of coding regions in heterologous systems.

Once the function of specific gene products has been established, their subcellular localization, expression pattern, and regulation on the transcriptional and posttranscriptional level can be investigated. It will be of particular interest to determine whether interconversion enzymes with a predicted transmembrane domain will be targeted to a specific membrane system such as the Golgi. Furthermore, it will be instructive to determine the functional significance of genetic redundancy in NDP-sugar interconversion reactions.

The analysis of mutants in the synthesis of activated monosaccharides has already provided some insight into the consequences of reduced precursor

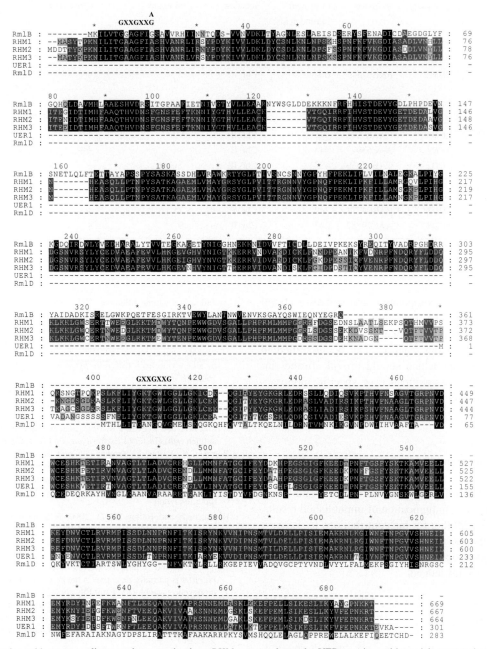

Figure 4. Amino acid sequence alignment between the three *RHM* gene products, the UER protein, and bacterial enzymes involved in the *de novo* synthesis of dTDP-L-rhamnose. The *rmlB* gene from *E. coli* encodes dTDP-D-glucose 4,6-dehydratase, and the *rmlD* gene from *Bacillus halodurans* encodes dTDP-L-rhamnose synthase. The GxxGxxG motifs predicted to be involved in NAD(P)$^+$ binding are indicated. Note that the 4,6-dehydratase domain within the *RHM* gene products contain a G to A amino acid substitution within this sequence.

Table 5. Putative *Arabidopsis* proteins with sequence similarities to dTDP-L-rhamnose biosynthetic enzymes in bacteria.

Tentative gene name	Protein accession	Length of protein (amino acids)	Chromosomal location	Predicted number of exons	Level of transcription[a]
RHM1	AAD30579	669	Chr. I	2	+ + ++
RHM2	AAF78439	667	Chr. I	2	+ + +
RHM3	BAB02645	664	Chr. III	2	+
UER1	AAG48808	301	Chr. I	2	+ + +
AUD1	CAA89205	445	Chr. III	7	+ + +
AUD2	AAC63621	443	Chr. II	7	++
AUD3	CAB67659	433	Chr. III	7	++
SUD1	CAB62035	341	Chr. III	10	+
SUD2	BAB09774	342	Chr. V	11	++
SUD3	AAC79582	343	Chr. II	9	+

[a]The level of transcription is baszed on the number of matching sequences in dbEST: +, 1–5, ++, 6–10, + + +, 11–20; + + ++, >20.

availability for the composition and structure of cell wall material and the formation of an abundant vitamin. In the future, additional mutants in nucleotide sugar interconversion reactions can be identified in tagged collections which should provide further information on the significance of specific monosaccharides for the assembly and integrity of the cell wall.

Acknowledgements

We thank Masaru Kobayashi, Toru Matoh and Georg Seifert for the communication of unpublished results, and Michael Mølhøj for his contributions to database searches. Support of this work by the DOE Energy Biosciences Program (No. DE-FG02-95ER20203) is gratefully acknowledged.

References

Altschul, S.F., Madden, T.L., Schäffer, A.A., Zhang, J., Zhang, Z., Miller, W. and Lipman, D.J. 1997. Gapped BLAST and PSI-BLAST: a new generation of protein database search programs. Nucl. Acids Res. 25: 3389–3402.

Amor, Y., Haigler, C.H., Johnson, S., Wainscott, M. and Delmer, D.P. 1995. A membrane-associated form of sucrose synthase and its potential role in synthesis of cellulose and callose in plants. Proc. Natl. Acad. Sci. USA 92: 9353–9357.

Arabidopsis Genome Initiative. 2000. Analysis of the genome sequence of the flowering plant *Arabidopsis thaliana*. Nature 408: 796–815.

Barber, G.A. 1979. Observations on the mechanism of the reversible epimerization of GDP-D-mannose to GDP-L-galactose by an enzyme from *Chlorella pyrenoidosa*. J. Biol. Chem. 254: 7600–7603.

Baskin, T.I., Betzner, A.S., Hoggart, R., Cork, A. and Williamson, R.E. 1992. Root morphology mutants in *Arabidopsis thaliana*. Aust. J. Plant Physiol. 19: 427–437.

Bastin, D.A. and Reeves, P.R. 1995. Sequence and analysis of the *O* antigen (*rfb*) cluster of *Escherichia coli* O111. Gene 164: 17–23.

Bonin, C.P., Potter, I., Vanzin, G.F. and Reiter, W.-D. 1997. The *MUR1* gene of *Arabidopsis thaliana* encodes an isoform of GDP-D-mannose-4,6-dehydratase, catalyzing the first step in the *de novo* synthesis of GDP-L-fucose. Proc. Natl. Acad. Sci. USA 94: 2085–2090.

Bonin, C.P. and Reiter, W.-D. 2000. A bifunctional epimerase-reductase acts downstream of the *MUR1* gene product and completes the *de novo* synthesis of GDP-L-fucose in *Arabidopsis*. Plant J. 21: 445–454.

Burget, E.G. and Reiter, W.-D. 1999. The *mur4* mutant of *Arabidopsis* is partially defective in the *de novo* synthesis of uridine diphospho L-arabinose. Plant Physiol. 121: 383–389.

Carpita, N.C. and Gibeaut, D.M. 1993. Structural models of primary cell walls in flowering plants: consistency of molecular structure with the physical properties of the walls during growth. Plant J. 3: 1–30.

Chang, S., Duerr, B. and Serif, G. 1988. An epimerase-reductase in L-fucose synthesis. J. Biol. Chem. 263: 1693–1697.

Cobbett, C.S., Medd, J.M. and Dolezal, O. 1992. Suppressors of an arabinose-sensitive mutant of *Arabidopsis thaliana*. Aust. J. Plant Physiol. 19: 367–375.

Conklin, P.L., Pallanca, J.E., Last, R.L. and Smirnoff, N. 1997. L-Ascorbic acid metabolism in the ascorbate-deficient *Arabidopsis* mutant *vtc1*. Plant Physiol. 115: 1277–1285.

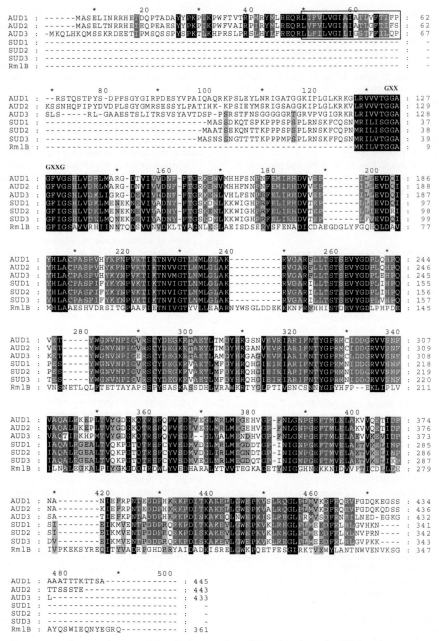

Figure 5. Amino acid sequence alignment between the *rmlB* gene product of *E. coli* and putative UDP-D-glucuronate decarboxylases from *Arabidopsis*. The predicted transmembrane domains in the *AUD* gene products are boxed and the GxxGxxG motif predicted to be involved in NAD(P)⁺ binding is indicated.

Conklin, P.L., Norris, S.R., Wheeler, G.L., Williams, E.H., Smirnoff, N. and Last, R.L. 1999. Genetic evidence for the role of GDP-mannose in plant ascorbic acid (vitamin C) biosynthesis. Proc. Natl. Acad. Sci. USA 96: 4198–4203.

Dalessandro, G. and Northcote, D.H. 1977. Possible control sites of polysaccharide synthesis during cell growth and wall expansion of pea seedlings (*Pisum sativum* L.). Planta 134: 39–44.

Delmer, D.P. and Amor, Y. 1995. Cellulose biosynthesis. Plant Cell 7: 987–1000.

Ding, L. and Zhu, J.-K. 1997. A role for arabinogalactan-proteins in root epidermal cell expansion. Planta 203: 289–294.

Dolezal, O. and Cobbett, C.S. 1991. Arabinose kinase-deficient mutant of *Arabidopsis thaliana*. Plant Physiol. 96: 1255–1260.

Dörmann, P. and Benning, C. 1996. Functional expression of uridine 5′-diphospho-glucose 4-epimerase (EC 5.1.3.2) from *Arabidop-*

112

sis thaliana in *Saccharomyces cerevisiae* and *Escherichia coli*. Arch. Biochem. Biophys. 327: 27–34.

Dörmann, P. and Benning, C. 1998. The role of UDP-glucose epimerase in carbohydrate metabolism of *Arabidopsis*. Plant J. 13: 641–652.

Feingold, D.S. and Avigad, G. 1980. Sugar nucleotide transformations in plants. In: P.K. Stumpf and E.E. Conn (Eds.) The Biochemistry of Plants: A Comprehensive Treatise, Vol. 3, Academic Press, New York, pp. 101–170.

Feingold, D.S. and Barber, G.A. 1990. Nucleotide sugars. In: P.M. Dey (Ed.) Methods in Plant Biochemistry, Vol. 2: Carbohydrates, Academic Press, New York, pp. 39–78.

Fleischer, A., O'Neill, M.A. and Ehwald, R. 1999. The pore size of non-graminaceous plant cell walls is rapidly decreased by borate ester cross-linking of the pectic polysaccharide rhamnogalacturonan II. Plant Physiol. 121: 829–838.

Gibeaut, D.M. 2000. Nucleotide sugars and glycosyltransferases for synthesis of cell wall matrix polysaccharides. Plant Physiol. Biochem. 38: 69–80.

Ginsburg, V. 1961. Studies on the biosynthesis of guanosine diphosphate L-fucose. J. Biol. Chem. 236: 2389–2393.

Hart, D.A. and Kindel, P.K. 1970. Isolation and partial characterization of apiogalacturonans from the cell wall of *Lemna minor*. Biochem. J. 116: 569–579.

Ishii, T. and Matsunaga, T. 1996. Isolation and characterization of a boron-rhamnogalacturonan-II complex from cell walls of sugar beet pulp. Carbohydrate Res. 284: 1–9.

Ishii, T., Matsunaga, T., Pellerin, P., O'Neill, M.A., Darvill, A. and Albersheim, A. 1999. The plant cell wall polysaccharide rhamnogalacturonan II self-assembles into a covalently cross-linked dimer. J. Biol. Chem. 274: 13098–13104.

Joersbo, M., Pedersen, S.G., Nielsen, J.E., Marcussen, J. and Brunstedt, J. 1999. Isolation and expression of two cDNA clones encoding UDP-galactose epimerase expressed in developing seeds of the endospermous legume guar. Plant Sci. 142: 147–154.

John, K.V., Schutzbach, J.S. and Ankel, H. 1977. Separation and allosteric properties of two forms of UDP-glucuronate carboxylyase. J. Biol. Chem. 252: 8013–8017.

Joyard, J., Teyssier, E., Miège, C., Berny-Seigneurin, D., Maréchal, E., Block, M.A., Dorne, A.-J., Rolland, N., Ajlani, G. and Douce, R. 1998. The biochemical machinery of plastid envelope membranes. Plant Physiol. 118: 715–723.

Kaplan, C.P., Tugal, H.B. and Baker, A. 1997. Isolation of a cDNA encoding an *Arabidopsis* galactokinase by functional expression in yeast. Plant Mol. Biol. 34: 497–506.

Keller, R., Springer, F., Renz, A. and Kossmann, J. 1999. Antisense inhibition of the GDP-mannose pyrophosphorylase reduces the ascorbate content in transgenic plants leading to developmental changes during senescence. Plant J. 19: 131–141.

Kindel, P.K. and Watson, R.R. 1973. Synthesis, characterization and properties of uridine 5'-(α-D-apio-D-furanosyl pyrophosphate). Biochem. J. 133: 227–241.

Kobayashi, M., Matoh, T. and Azuma, J. 1996. Two chains of rhamnogalacturonan II are cross-linked by borate-diol ester bonds in higher plant cell walls. Plant Physiol. 110: 1017–1020.

Lake, M.R., Williamson, C.L. and Slocum, R.D. 1998. Molecular cloning and characterization of a UDP-glucose-4-epimerase gene (*galE*) and its expression in pea tissues. Plant Physiol. Biochem. 36: 555–562.

Levy, S., York, W.S., Stuike-Prill, R., Meyer, B. and Staehelin, L.A. 1991. Simulations of the static and dynamic molecular conformations of xyloglucan. The role of the fucosylated sidechain in surface-specific sidechain folding. Plant J. 1: 195–215.

Levy, S., Maclachlan, G. and Staehelin, L.A. 1997. Xyloglucan sidechains modulate binding to cellulose during *in vitro* binding assays as predicted by conformational dynamics simulations. Plant J. 11: 373–386.

Liao, T.-H. and Barber, G.A. 1971. The synthesis of guanosine-5'-diphosphate L-fucose by enzymes of a higher plant. Biochim. Biophys. Acta 230: 64–71.

Liljebjelke, K., Adolphson, R., Baker, K., Doong, R.L. and Mohnen, D. 1995. Enzymatic synthesis and purification of uridine diphosphate [^{14}C]galacturonic acid: a substrate for pectin biosynthesis. Anal. Biochem. 225: 296–304.

Loewus, F., Chen, M.-S. and Loewus, M.W. 1973. The *myo*-inositol oxidation pathway to cell wall polysaccharides. In: F. Loewus (Ed.) Biogenesis of Plant Cell Wall Polysaccharides, Academic Press, New York, pp. 1–27.

Lukowitz, W., Nickle, T.C., Meinke, D.W., Last, R.L., Conklin, P.L. and Somerville, C.R. 2001. *Arabidopsis cyt1* mutants are deficient in a mannose-1-phosphate guanylyltransferase and point to a requirement of *N*-linked glycosylation for cellulose biosynthesis. Proc. Natl. Acad. Sci. USA 98: 2262–2267.

Maitra, U.S. and Ankel, H. 1971. Uridine diphosphate-4-keto-glucose, an intermediate in the uridine diphosphate-galactose 4-epimerase reaction. Proc. Natl. Acad. Sci. USA 68: 2660–2663.

Maretzki, A. and Thom, M. 1978. Characteristics of a galactose-adapted sugarcane cell line grown in suspension culture. Plant Physiol. 61: 544–548.

Matern, U. and Grisebach, H. 1977. UDP-apiose/UDP-xylose synthase. Eur. J. Biochem. 74: 303–312.

McDougall, G.J. and Fry, S.C. 1989. Structure-activity relationships for xyloglucan oligosaccharides with antiauxin activity. Plant Physiol. 89: 883–887.

Mohnen, D. 1999. Biosynthesis of pectins and galactomannans. In: B.M. Pinto (Ed.) Comprehensive Natural Products Chemistry, Vol. 3: Carbohydrates and Their Derivatives Including Tannins, Cellulose, and Related Lignins, Elsevier, Amsterdam, pp. 497–527.

Muños, P., Norambuena, L. and Orellana, A. 1996. Evidence for a UDP-glucose transporter in Golgi apparatus-derived vesicles from pea and its possible role in polysaccharide biosynthesis. Plant Physiol. 112: 1585–1594.

Muños, R., Lópex, R., de Frutos, M. and García, E. 1999. First molecular characterization of a uridine diphosphate galacturonate 4-epimerase: an enzyme required for capsular biosynthesis in *Streptococcus pneumoniae* type 1. Mol. Microbiol. 31: 703–713.

Nickle, T.C. and Meinke, D.W. 1998. A cytokinesis-defective mutant of *Arabidopsis* (*cyt1*) characterized by embryonic lethality, incomplete cell walls, and excessive callose accumulation. Plant J. 15: 321–332.

O'Neill, M.A., Warrenfeltz, D., Kates, K., Pellerin, P., Doco, T., Darvill, A.G. and Albersheim, P. 1996. Rhamnogalacturonan-II, a pectic polysaccharide in the walls of growing plant cell, forms a dimer that is covalently cross-linked by a borate ester. J. Biol. Chem. 271: 22923–22930.

Rayon, C., Cabanes-Macheteau, M., Loutelier-Bourhis, C., Saliot-Maire, I., Lemoine, J., Reiter, W.-D., Lerouge, P. and Faye, L. 1999. Characterization of *N*-glycans from *Arabidopsis thaliana*. Application to a fucose-deficient mutant. Plant Physiol. 119: 725–733.

Reeves, P.R., Hobbs, M., Valvano, M.A., Skurnik, M., Whitfield, C., Coplin, D., Kido, N., Klena, J., Maskell, D., Raetz, C.R.H. and Rick, P.D. 1996. Bacterial polysaccharide synthesis and gene nomenclature. Trends Microbiol. 4: 495–503.

Reiter, W.-D., Chapple, C.C.S. and Somerville, C.R. 1993. Altered growth and cell walls in a fucose-deficient mutant of *Arabidopsis*. Science 261: 1032–1035.

Reiter, W.-D., Chapple, C. and Somerville, C.R. 1997. Mutants of *Arabidopsis thaliana* with altered cell wall polysaccharide composition. Plant J. 12: 335–345.

Reiter, W.-D. 1998. The molecular analysis of cell wall components. Trends Plant Sci. 3: 27–32.

Salo, W.L., Nordin, J.H., Petersen, D.R., Bevill, R.D. and Kirkwood, S. 1968. The specificity of UDP-glucose 4-epimerase from the yeast *Saccharomyces fragilis*. Biochim. Biophys. Acta 151: 484–492.

von Schaewen, A., Sturm, A., O'Neill, J. and Chrispeels, M.J. 1993. Isolation of a mutant Arabidopsis plant that lacks *N*-acetyl glucosaminyl transferase I and is unable to synthesize Golgi-modified complex *N*-glycans. Plant Physiol. 102: 1109–1118.

Schiefelbein, J.W. and Somerville, C. 1990. Genetic control of root hair development in *Arabisopsis thaliana*. Plant Cell 2: 235–243.

Seitz, B., Klos, C., Wurm, M. and Tenhaken, R. 2000. Matrix polysaccharide precursors in *Arabidopsis* cell walls are synthesized by alternate pathways with organ-specific expression patterns. Plant J. 21: 537–546.

Sherson, S., Gy, I., Medd, J., Schmidt, R., Dean, C., Kreis, M., Lecharny, A. and Cobbett, C. 1999. The arabinose kinase, *ARA1*, gene of *Arabidopsis* is a novel member of the galactose kinase gene family. Plant Mol. Biol. 39: 1003–1012.

Smirnoff, N. and Wheeler, G.L. 2000. Ascorbic acid in plants: biosynthesis and function. Crit. Rev. Biochem. Mol. Biol. 35: 291–314.

Stevenson, G., Neal, B., Liu, D., Hobbs, M., Packer, N.H., Batley, M., Redmond, J.W., Lindquist, L. and Reeves, P. 1994. Structure of the *O* antigen of *Escherichia coli* K-12 and the sequence of its *rfb* gene cluster. J. Bact. 176: 4144–4156.

Tenhaken, R. and Thulke, O. 1996. Cloning of an enzyme that synthesizes a key nucleotide sugar precursor of hemicellulose biosynthesis from soybean: UDP-glucose dehydrogenase. Plant Physiol. 112: 1127–1134.

Wellmann, E. and Grisebach, H. 1971. Purification and properties of an enzyme preparation from *Lemna minor* L. catalyzing the synthesis of UDP-apiose and UDP-D-xylose from UDP-D-glucuronic acid. Biochim. Biophys. Acta 235: 389–397.

Wheeler, G.L., Jones, M.A. and Smirnoff, N. 1998. The biosynthetic pathway of vitamin C in higher plants. Nature 393: 365–369.

Willats, W.G.T. and Knox, J.P. 1996. A role for arabinogalactan-proteins in plant cell expansion: evidence from studies on the interaction of *β*-glucosyl Yariv reagent with seedlings of *Arabidopsis thaliana*. Plant J. 9: 919–925.

Wilson, D.B. and Hogness, D.S. 1964. The enzymes of the galactose operon in *Escherichia coli*. I. Purification and characterization of uridine diphosphogalactose 4-epimerase. J. Biol. Chem. 239: 2469–2481.

Yamamoto, R., Inouhe, M. and Masuda, Y. 1998. Galactose inhibition of auxin-induced growth of mono- and dicotyledonous plants. Plant Physiol. 86: 1223–1227.

York, W.S., Darvill, A.G. and Albersheim, P. 1984. Inhibition of 2,4-dichlorophenoxyacetic acid-stimulated elongation of pea stem segments by a xyloglucan oligosaccharide. Plant Physiol. 75: 295–297.

Yariv, J., Rapport, M.M. and Graf, L. 1962. The interaction of glycosides and saccharides with antibody to the corresponding phenylazo glycosides. Biochem. J. 85: 383–388.

Zablackis, E., York, W.S., Pauly, M., Hantus, S., Reiter, W.-D., Chapple, C.C.S., Albersheim, P. and Darvill, A. 1996. Substitution of L-fucose by L-galactose in cell walls of *Arabidopsis mur1*. Science 272: 1808–1810.

Plant Molecular Biology **47**: 115–130, 2001.
© 2001 *Kluwer Academic Publishers. Printed in the Netherlands.*

Golgi enzymes that synthesize plant cell wall polysaccharides: finding and evaluating candidates in the genomic era

Robyn Perrin[1], Curtis Wilkerson[1] and Kenneth Keegstra[2,*]
[1]*Department of Botany and Plant Pathology and* [2]*Department of Biochemistry, MSU-DOE Plant Research Laboratory, Michigan State University, East Lansing, MI 48824, USA* (*author for correspondence; e-mail keegstra@msu.edu*)

Key words: glycan synthase, glycosyltransferase, Golgi, pectin, plant cell wall, polysaccharide biosynthesis, xyloglucang

Abstract

Although the synthesis of cell wall polysaccharides is a critical process during plant cell growth and differentiation, many of the wall biosynthetic genes have not yet been identified. This review focuses on the synthesis of non-cellulosic matrix polysaccharides formed in the Golgi apparatus. Our consideration is limited to two types of plant cell wall biosynthetic enzymes: glycan synthases and glycosyltransferases. Classical means of identifying these enzymes and the genes that encode them rely on biochemical purification of enzyme activity to obtain amino acid sequence data that is then used to identify the corresponding gene. This type of approach is difficult, especially when acceptor substrates for activity assays are unavailable, as is the case for many enzymes. However, bioinformatics and functional genomics provide powerful alternative means of identifying and evaluating candidate genes. Database searches using various strategies and expression profiling can identify candidate genes. The involvement of these genes in wall biosynthesis can be evaluated using genetic, reverse genetic, biochemical, and heterologous expression methods. Recent advances using these methods are considered in this review.

Abbreviations: AtFUT1, *Arabidopsis thaliana* fucosyltransferase 1; EST, expressed sequence tag; FucT, fucosyltransferase; GDP-Fuc, guanosine $5'$-diphospo-α-L-fucose; UDP, uridine $5'$-diphosphate; XyG, xyloglucan

Introduction

Every plant cell is awash in a sea of polysaccharides. This intricate network is important during the growth and differentiation of plant cells and during interactions between plants and other organisms. Despite significant progress in understanding the complex structures of wall components, we are only beginning to understand the process by which they are synthesized.

The biosynthesis of polysaccharides is fundamentally different from the synthesis of other biopoly-

mers. During the biosynthesis of proteins and nucleic acids, a nucleic acid template determines the sequence of monomers within the polymer. In contrast, the sequence of sugars within a polysaccharide is determined by the specificity of enzymes that form the bonds between monosaccharides. Each sugar monomer has several hydroxyl residues that may serve as the attachment site for the next sugar in the chain; consequently, a much greater diversity in structure is possible in carbohydrate polymers than in nucleic acids or proteins. This again emphasizes the importance of enzyme specificity in determining product structure.

When considering the synthesis of a biopolymer, the process can be divided into three stages: initiation, elongation, and termination. For protein and nucleic acid synthesis, these three stages are distinct. Delmer

---

In previous publications (Faik *et al.*, 2000; Perrin *et al.*, 1999) the abbreviation FTase was used to refer to fucosyltransferase. However, the abbreviation FucT is more widely used (for example, see Breton and Imberty, 1999), and, furthermore, the abbreviation FTase has been used by others to refer to farnesyltransferase, an unrelated enzyme.

116

and Stone (1988) have briefly discussed the stages of polysaccharide biosynthesis and, while the details are still not understood, each merits brief consideration. In most cases, initiation of polysaccharide biosynthesis requires a primer. This primer is sometimes a protein, as in the case of glycogen (Smythe and Cohen, 1991), or a lipid, as in the case of some bacterial polysaccharides (Whitfield, 1995; Mengeling and Turco, 1998). At present it is not known whether the initiation of plant cell wall polysaccharide biosynthesis requires a primer. Chain elongation is determined by enzyme specificity. For example, the mixed-linkage glucan synthase creates $(1\rightarrow4)\beta$-linked cellodextrins linked by single $(1\rightarrow3)\beta$-linkages, but with no side-chain substitutions (Gibeaut and Carpita, 1994; Buckeridge et al., 1999). On the other hand, the glucan synthase involved in xyloglucan (XyG) biosynthesis forms a $(1\rightarrow4)\beta$-D-glucan backbone, but it can operate only in concert with a xylosyltransferase that adds xylose side chains (Hayashi, 1989; White et al., 1993). Finally, almost nothing is known about the mechanics of chain termination. It is possible that chain termination occurs when a polysaccharide is transported to the wall via vesicle traffic, thereby separating it from biosynthetic enzymes. Alternatively, it is possible that specific mechanisms determine when a chain is terminated, resulting in removal from the biosynthetic enzymes responsible for elongation. For example, in Acetobacter xylinus and Agrobacterium tumefaciens, cellulose synthesis operons include or are very close to ORFs encoding cellulase homologues. These ORFs are required for cellulose biosynthesis; one possible explanation is that they are involved in chain termination (Standal et al., 1994; Matthysse et al., 1995a, b).

General features of cell wall polysaccharide biosynthesis

Plant cell wall polysaccharides contain significant quantities of nine different sugars (glucose, galactose, mannose, xylose, arabinose, fucose, rhamnose, galacturonic acid, and glucuronic acid) in addition to minor amounts of other sugars (e.g. apiose and aceric acid) (Carpita and Gibeaut, 1993; Brett and Waldron, 1996). Most wall sugars are found in multiple linkages. Thus, given the reasonable assumption that the synthesis of each different linkage requires a different enzyme, it is clear that many different enzymes are required for the synthesis of plant cell wall polysaccharides. Mohnen (1999) calculated that pectin biosynthesis

requires at least 46 different glycosyltransferases. Because pectins constitute but one component of the plant cell wall, it seems likely that more than one hundred different enzymes are required for the synthesis of the various cell wall polysaccharides from activated sugar donors. Indeed, a compilation of known and predicted carbohydrate-active enzymes (http://afmb.cnrs-mrs.fr/~pedro/CAZY/db.html) shows that Arabidopsis has at least 140 Arabidopsis ORFs within families predicted to be involved in polysaccharide and glycoprotein biosynthesis (see also Henrissat et al., 2001, this issue). This does not include the many enzymes predicted or proven to be involved in glycosylation of small molecules (such as hormones) or in starch biosynthesis. It is likely that other genes involved in polysaccharide biosynthesis will be identified as our understanding of these processes increases.

In considering polysaccharide biosynthesis it is useful to distinguish between two general categories of enzymes: glycan synthases and glycosyltransferases. We define glycan synthases as enzymes that link together sugars to make up the backbone of any particular polysaccharide. Very little is known about the glycan synthases located in the Golgi. None of the Golgi glycan synthases have been purified or characterized in detail and none of the genes encoding these enzymes have been identified to date. Based on the assumption that some of the Golgi-localized glycan synthases may be similar to the cellulose synthase or callose synthase from the plasma membrane, some candidate genes have been identified. We will discuss some of these candidates below, but it is important to note that there is currently no published evidence that any of these genes encode Golgi-localized glycan synthases.

Glycosyltransferases are the second category of enzymes involved in polysaccharide biosynthesis. These enzymes transfer a sugar residue from a sugar nucleotide donor to a specific location on an acceptor molecule. They generally have a high degree of specificity with respect to the sugar donor, the range of molecules that will serve as acceptor, and the linkage formed. Glycosyltransferases are usually type II integral membrane proteins containing a short amino-terminal domain facing the cytoplasm, a single transmembrane domain, and a hydrophilic carboxy-terminal domain containing the active site within the lumen of the Golgi apparatus (Kleene and Berger, 1993; Breton and Imberty, 1999). These enzymes have been studied extensively in animal, fungal, and bacterial systems (Breton and Imberty, 1999), but plant cell wall biosynthetic glycosyltransferases have only recently been purified and the genes encoding them identified (Edwards *et al.*, 1999; Perrin *et al.*, 1999). A characteristic feature of the glycosyltransferase family is a lack of extensive sequence identity even among enzymes that catalyze the formation of similar linkages (Breton and Imberty, 1999). However, characteristic motifs have been identified among some enzymes, such as the fucosyltransferases (Breton *et al.*, 1996, 1998a; Sherwood *et al.*, 1998; Oriol *et al.*, 1999; Takahashi *et al.*, 2000) and the galactosyltransferases (Breton *et al.*, 1998b; Imberty *et al.*, 1999). These motifs have been useful for the identification of other candidate genes within the DNA sequences produced by the various genome sequencing projects.

In this review we will focus on polysaccharide biosynthesis within the Golgi apparatus of plant cells. The biosynthesis of cellulose and other polysaccharides at the plasma membrane is considered by other authors (Haigler *et al.*, 2001; Richmond and Somerville, 2001; Turner *et al.*, 2001; Vergara and Carpita, 2001, all in this issue). We will focus on the glycan synthases and glycosyltransferases that link sugars together into non-cellulosic polysaccharides. Other reviews should be consulted for information on the biosynthesis of sugar nucleotide donors and their transport into the Golgi lumen (Butler and Elling, 1999; Gibeaut, 2000; Hirschberg *et al.*, 1998). When appropriate, we will consider advances in other systems that may be relevant to understanding polysaccharide biosynthesis in plants. We will refer to selected biochemical studies, but other more comprehensive reviews should be consulted for details (Gibeaut and Carpita, 1994; Mohnen, 1999). Our major focus will be on recent efforts to identify and characterize the plant glycosyltransferases and glycan synthases involved in polysaccharide biosynthesis, the genes encoding these proteins, and the difficulties associated with biochemical evaluation of candidate gene products.

Strategies for the identification of genes involved in polysaccharide biosynthesis

As emphasized above, elucidation of the plant cell wall polysaccharide biosynthesis process has been slow. In the past, only biochemical strategies were available for studying the enzymes involved in cell wall biosynthesis. A rapid increase in nucleotide and deduced amino acid sequence data in the past few years has made two new strategies possible for identifying genes putatively involved in plant polysaccharide biosynthesis (Figure 1). Each of these strategies is described briefly in the next three paragraphs and illustrated in more detail later in this review. In many cases, candidate genes identified by these three strategies need to be evaluated for their role in cell wall biosynthesis using other techniques (Figure 1). Some methods for confirming the involvement of candidate genes in cell wall biosynthesis will be discussed.

The first strategy, and frequently the most laborious, is to use biochemical assays to follow the purification of an enzyme involved in polysaccharide biosynthesis. Peptide sequence information derived from the purified protein can then be used to identify cDNA clones or genes that encode the protein. The key requirement for this strategy is the assay itself. Developing such assays is a difficult problem for many enzymes in that both acceptor and donor substrates need to be readily available. While many, but not all, donor substrates are commercially available, acceptor substrates are rarely available and usually need to be prepared. For most enzymes the acceptor specificity is unknown, so providing an appropriate acceptor becomes a circular problem. However, when activity assays for specific enzymes are available they also become a valuable tool for proving the function of a candidate gene (Camirand *et al.*, 1987; Farkas and Maclachlan, 1988; Hanna *et al.*, 1991; Maclachlan *et al.*, 1992; Faik *et al.*, 1997; Edwards *et al.*, 1999; Perrin *et al.*, 1999).

A second strategy for identifying candidate genes is to search sequence databases for genes with similarity to genes of known function. As mentioned above, conserved motifs and regions of similarity have been identified for some glycosyltransferases and glycan synthases, despite the overall low level of sequence similarity within these classes of enzymes. Based on these analyses, it has been possible to identify candidate genes in plant genomic databases and propose hypotheses regarding their involvement in cell wall biosynthesis. Some of these hypotheses and ways to evaluate them will be discussed below.

Finally, a third strategy is to use expression profiling to find genes that change in expression levels in concert with other wall biosynthetic genes or in concert with changes in wall biosynthesis. This strategy provides a valuable complement to the second strategy, although one must be aware that regulation may occur at levels other than gene expression. Genes that meet both homology and expression profiling criteria

118

Functional Genomic Strategies

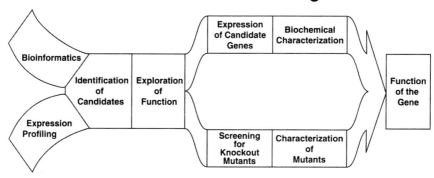

Figure 1. Flow chart of strategies for identifying candidate genes and evaluating their functions. Candidates involved in a particular biological process may be identified by techniques such as bioinformatics-based search strategies or expression profiling. Function of the candidate genes may then be explored via detailed expression analysis, biochemical characterization, and/or genetic analyses. Details of each technique are described in the text.

are the strongest candidates for more detailed investigation. However, it must be emphasized that while these two strategies are valuable for identification of genes and development of hypotheses regarding their function, they cannot be used to prove the role of a gene in cell wall biosynthesis.

Biochemical strategies for gene identification and characterization

Until recently, biochemical strategies were the only method available for identification of candidate proteins and genes involved in cell wall polysaccharide biosynthesis. Thus, most of the emphasis of this short review will be on this strategy. Because of limitations associated with biochemical approaches, described in more detail below, it is expected that genomic strategies will take on enhanced importance in the future. However, it is clear that biochemical analyses remain relevant in the genomic era for evaluation of candidate genes.

Glycosyltransferases

Many cell wall polysaccharides consist of linear backbones with side-chain modifications. Although synthesis of these structures is conducted first by glycan synthases (for the backbones) and later by glycosyltransferases (for the side-chains), we will consider these enzymes in the opposite order. Understanding polysaccharide synthesis from the outside in has proven to be less difficult. For that reason, more literature exists on plant glycosyltransferases than on Golgi

glycan synthases, although studies on the former are still relatively scarce.

Fucosyltransferase genes: a case study

Low homology among glycosyltransferase families, the difficulty of obtaining acceptor substrates, and the challenges of structural analyses of some complex plant polysaccharides all contribute to the difficulty of identifying plant cell wall glycosyltransferases. Identification of a xyloglucan fucosyltransferase (XyG FucT), however, appeared from the outset to be a tractable problem. Understanding XyG biosynthesis is important because XyG interacts with cellulose to form a major load-bearing structure and thus contributes to the structural integrity of the wall (Pauly *et al.*, 1999; Fujino *et al.*, 2000). An amenable activity assay had previously been established for this enzyme (Camirand *et al.*, 1987; Farkas and Maclachlan, 1988; Hanna *et al.*, 1991). Sugar nucleotide (GDP-Fuc) donor is available in both radiolabeled and unlabeled forms. Non-fucosylated xyloglucan acceptor substrate accumulates as seed storage polysaccharides in certain species, namely tamarind (*Tamarindus indica*) or nasturtium (*Tropaeolum majus*); in recent years, this substrate has become commercially available (Megazyme International, Ireland). XyG is composed of a $(1\rightarrow4)\beta$-D-glucan backbone decorated with side chains of varying composition (see Figure 2A). FucT directs the addition of Fuc to Gal in a $(1\rightarrow2)\alpha$ linkage on some side chains. Because Fuc is the terminal modification made to side chains, it seemed unlikely that this enzyme would be dependent upon an interacting protein partner for activity, in contrast to enzymes

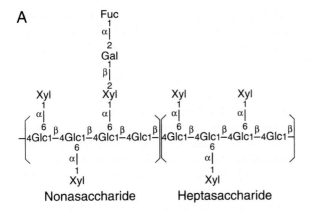

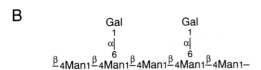

Figure 2. General structures of xyloglucan and galactomannan. A. Xyloglucan consists of a (1→4)-β-D-glucan backbone decorated with side chains of xylose, xylose and galactose, or xylose, galactose and fucose. Variations in side chain structure occur, particularly among the Solanaceae family. From Hayashi (1989) with modifications. B. Galactomannans have a (1→4)β-D-mannan backbone with (1→6)α-D-galactose substitutions. The degree of substitution can vary greatly in different species and affects the physical properties of the polymer. From Mohnen (1999), with modifications.

(such as cellulose synthase) that are part of a larger complex. Carbohydrate analysis of the product could be conducted by methylation analysis to confirm the presence of the Fuc residue after incubation with the enzyme. The pattern of fucosylation in XyG subunits can also be determined by well-established methods (Faik *et al.*, 1997, 2000).

Previous studies had established certain facts about XyG FucT and the FucT gene family as a whole. It had been shown by immuno-electron microscopy using anti-α-L-Fuc-(1→2)-D-Gal antibodies that fucosylated XyG first appears in the lumen of the *trans*-Golgi and *trans*-Golgi network before vesicle-mediated secretion to the cell wall (Moore *et al.*, 1991; Puhlmann *et al.*, 1994). This activity appeared to be spatially distinct from galactosyl- and xylosyltransferase activity (Camirand *et al.*, 1987; Brummell *et al.*, 1990). More recent investigations have shown that XyG FucT is a latent activity in intact Golgi vesicles, thus confirming that the catalytic region of the enzyme is within the lumen of the Golgi apparatus (Wulff *et al.*, 2000). The Golgi localization of activity agreed with similar studies of FucTs in mammals (e.g.

Kimura *et al.*, 1995; Borsig *et al.*, 1996; Narimatsu *et al.*, 1996).

The FucTs received early attention from the medical community because of their role in synthesis of glycoprotein blood group antigens and glycolipids (Field and Wainwright, 1995; Mollicone *et al.*, 1995). Some Fuc-containing antigens and FucT activities also were correlated with tumor formation, serving as clinical indicators of certain types of cancer (Weiser and Wilson, 1981; Shah Reddy *et al.*, 1982; Bolscher *et al.*, 1989; Staudacher *et al.*, 1999). A rash of cloning and characterization studies in mammalian systems in the early to mid-1990s identified a number of FucTs critical to glycobiological processes (e.g. Kukowska-Latallo *et al.*, 1990; Larsen *et al.*, 1990; Weston *et al.*, 1992; Natsuka *et al.*, 1994). Expression-based cloning proved to be a valuable technique to identify candidate genes in many cases (Ernst *et al.*, 1989; Fukuda *et al.*, 1996).

Mammalian studies also established the general structure of FucTs. These enzymes generally have a short N-terminal cytoplasmic tail, a transmembrane helix, an extended luminal stem region, and a globular C-terminal catalytic region (Kleene and Berger, 1993; Breton and Imberty, 1999). The critical nature of the C-terminus in catalysis was illustrated by deletion studies; removal of as little as one amino acid from the C-terminal end could completely abolish all enzymatic activity, while deletion of as much as one third of the N-terminus had little effect (Xu *et al.*, 1996). With time, FucTs were also identified in parasitic organisms such as ulcer-causing bacteria, apparently serving to synthesize mammalian-type antigens so the pathogen can evade detection by the host (Ge *et al.*, 1997; Martin *et al.*, 1997). A FucT involved in synthesis of Nod factors in nitrogen-fixing bacteria was also identified (Quesada *et al.*, 1997; Quinto *et al.*, 1997). None showed high sequence similarity to any plant genomic or cDNA sequences in public databases.

Early studies of plant XyG FucT used rapidly elongating regions of etiolated pea epicotyls as a source of enzyme for partial purification (Hanna *et al.*, 1991). Later studies utilized solubilized, carbonate-washed pea microsomal membrane proteins followed by GDP affinity chromatography and size exclusion chromatography (Faik *et al.*, 2000; Perrin *et al.*, 1999). Tryptic fragments of the purified protein were sequenced. Although none of the amino acid fragments showed similarity to known sequences in the public databases, an *Arabidopsis* EST was identified that encoded all of the pea sequences with only a few differences (Perrin *et al.*, 1999). A full-length cDNA version of this *Arabidopsis* gene was subsequently isolated and its function as a XyG FucT was confirmed by heterologous expression and immunoprecipitation studies (Perrin *et al.*, 1999). The gene locus was named *AtFUT1*. Data from the Arabidopsis Genome Sequencing Initiative indicates that it is located on chromosome 2.

Upon analysis of AtFUT1, it became clear why attempts to identify plant XyG FucT by homology to mammalian enzymes was not successful. AtFUT1 shows very little identity with FucTs from other organisms. AtFUT1 and PsFT1 (the pea XyG FucT homologue) are 62.3% identical (Faik *et al.*, 2000). Interestingly, both enzymes contain motifs that had been identified in other FucTs (Oriol *et al.*, 1999), but combine these motifs in a unique manner (Faik *et al.*, 2000). Three motifs had been identified in $(1{\rightarrow}2)\alpha$- and $(1{\rightarrow}6)\alpha$-FucTs. Motifs I and II had been present in both $(1{\rightarrow}2)\alpha$- and $(1{\rightarrow}6)\alpha$- enzymes but a particular version of motif III had appeared to be characteristic of each group. AtFUT1 and PsFT1, however, contain a hybrid motif III that has features of both the $(1{\rightarrow}2)\alpha$- and $(1{\rightarrow}6)\alpha$-versions. Inclusion of plant FucTs in studies that seek to determine consensus sequences should therefore lead to interesting evolutionary implications. Although site-directed mutagenesis or deletion analyses have not yet been performed with this or any other plant glycosyltransferase, group-specific modifying reagents were used to determine which classes of amino acids might be critical for activity. Other mammalian FucTs have been classified as sensitive or insensitive to some of these modifying reagents, such as N-ethylmaleimide (NEM) (e.g. Britten and Bird, 1997). PsFT1 showed no NEM sensitivity but did show strong sensitivity to reagents that react with Trp, His or Lys residues (Faik *et al.*, 2000). Since conserved Asp residues are thought to perform important roles in catalysis in glycosyltransferases and glycan synthases (Breton *et al.*, 1998b; Wiggins and Munro, 1998), further studies to evaluate these residues as well would be interesting.

In *Arabidopsis*, the presence of many nucleotide sequences that are similar to *AtFUT1* brings up a now common theme: multi-gene families. There are to date ten genes in *Arabidopsis* with identity of encoded amino acid sequences to AtFUT1 ranging from 35% to 73.8% . There is strong evidence that at least four of these genes are expressed (R. Sarria-Millan, unpublished data.) These sequences may represent other FucTs specific to other acceptors, such as Fuc-

containing pectins or glycoproteins. The possibility exists that they may not be FucTs at all, and may indeed represent other classes of glycosyltransferases. This makes it obvious why gene candidates must be evaluated carefully, particularly for enzyme families with disparate sequence homology such as glycosyltransferases. Identification by homology is not sufficient to ensure confidence in biochemical function. Other means of analyses such as heterologous expression and product analysis, identification of null mutants by T-DNA knockout or antisense technology followed by a thorough carbohydrate analysis, or the use of more newly developed methods of virus-mediated gene silencing must be performed (see Evaluation of candidate genes below.)

Galactosyltransferase genes

The gene encoding a $(1{\rightarrow}6)\alpha$-galactosyltransferase (GalT) from fenugreek (*Trigonella foenum-graecum*) that is involved in galactomannan biosynthesis was identified by Edwards *et al.* (1999). Galactomannans are found in seed endosperm of legumes and have many food industrial uses as thickening agents (Reid and Edwards, 1995). The structure of galactomannans is shown in Figure 2B. The strategy used by Edwards *et al.* (1999) to identify galactomannan GalT was similar to that described above for the XyG FucT. An assay for the enzyme had been developed previously (Edwards *et al.*, 1989). When not solubilized the enzyme activity could be detected, presumably due to the presence of an endogenous acceptor, but solubilized enzyme showed an absolute requirement for an exogenous acceptor (Edwards *et al.*, 1999). For this purpose, they used short oligosaccharides of $(1{\rightarrow}4)\beta$-linked mannosyl residues or galactomannan molecules with low levels of galactose. Because galactomannan biosynthesis occurs in the endosperm of developing seeds, the small amounts of available tissue prevented traditional purification. However, high galactomannan synthesis activity in these tissues did allow correlative techniques to be used. Specifically, GalT activity was correlated with a particular 51 kDa polypeptide during isoelectric focusing followed by SDS-PAGE. Amino acid sequence information was obtained from the 51 kDa protein as well as from proteolytic fragments derived from it, and a cDNA clone encoding the protein was isolated by RT-PCR using RNA from developing fenugreek endosperm as template.

Because the GalT had not been purified to homogeneity, it was critical for Edwards *et al.* (1999)

to demonstrate that the cDNA clone encoded the galactomannan GalT. To accomplish this, they expressed the cDNA clone in *Pichia pastoris*. The transformed yeast cells produced a new enzyme activity that could add galactosyl residues onto acceptor molecules, producing galactomannan molecules with the proper structure. Thus, they were able to provide compelling evidence that this cDNA clone encoded the fenugreek galactomannan GalT.

The protein encoded by the fenugreek cDNA clone is predicted to have a membrane-spanning region near the amino terminus, typical of a type II membrane protein. Thus, it has the structure characteristic of other glycosyltransferases. Edwards *et al.* (1999) demonstrated that the carboxyl-terminal hydrophilic domain contains the active site of the enzyme. Presumably this region is located in the lumen of the Golgi apparatus, although this point has not been confirmed experimentally.

Other Golgi glycosyltransferase genes

Biochemical assays have demonstrated the existence of several other glycosyltransferases (for a recent list, see Mohnen, 1999). However, to date, the XyG FucT and the galactomannan GalT are the only two plant enzymes involved in cell wall polysaccharide biosynthesis where the corresponding genes have been cloned. On the other hand, some of the genes involved in glycoprotein biosynthesis have been cloned recently (Leiter *et al.*, 1999; Strasser *et al.*, 1999; Strasser *et al.*, 2000). Because our focus is limited to cell wall polysaccharide biosynthesis, these genes will not be considered in detail. It should however be noted that bioinformatic approaches to identify candidate genes would likely include the genes encoding glycoprotein biosynthetic enzymes in the pool of possibilities, as they share structural features with cell wall biosynthetic glycosyltransferases. Other methods will be needed to distinguish between these two groups of genes (see Figure 1).

Biochemical studies of glycan synthases

In vitro systems have been used to examine the synthesis of polysaccharide backbones by enzymes derived from the Golgi apparatus. In most cases, enzymes contained within membrane vesicles are incubated with radiolabeled sugar nucleotides and the transfer of radiolabeled sugars into larger products is monitored. For such assays to be meaningful, careful product analysis is needed, but such studies are not always performed. These assays generally measure chain elongation, but few studies have investigated the nature of the acceptor molecule, so it is possible that some of the assays result in the initiation of new polysaccharide chains. A limitation on space prevents us from providing a complete review of the biochemical studies on plant polysaccharide biosynthesis; earlier reviews should be consulted for details (Hayashi, 1989; Gibeaut and Carpita, 1994; Delmer, 1999; Mohnen, 1999). We will focus our attention on a few recent reports and on the biosynthesis of XyG, which is probably the best studied of the plant cell wall polysaccharides. Another review in this volume covers the biosynthesis of mixed-linkage glucans in the Golgi of grasses (Vergara and Carpita, 2001).

XyG is the major hemicellulosic polysaccharide in the primary walls of most dicot species and in selected monocots. A great deal is known about its structure (Figure 2). Its biosynthesis has been studied in several plant species (for review, see Hayashi, 1989). Virtually all studies agree that biosynthesis of XyG requires the addition of both UDP-Glc and UDP-Xyl to Golgi vesicles (Hayashi, 1989; White *et al.*, 1993). Thus, it appears that the processes of glucan backbone elongation and xylosyl side chain addition are intimately linked and cannot be physically or temporally separated.

The proteins responsible for the biosynthesis of the XyG backbone have not yet been identified, in part because efforts to solubilize the activities that can be measured in Golgi vesicles result in significant loss of activity. The precise cause of this loss of activity is not clear. Many authors conclude that the enzymes are unstable upon detergent solubilization, but other possibilities exist. For example, detergent solubilization may leave both glucan synthase and XyG in an active form but separated from each other so that they cannot function in the coordinated fashion needed for XyG biosynthesis. Alternatively, perhaps both enzymes are active and in a complex, but the growing chain that functions as an acceptor has been separated from the enzyme complex, or its concentration in the detergent solution is too low for it to be an effective acceptor. Whatever the explanation may be, the loss of activity has prevented purification of the activities responsible for backbone synthesis and has prevented detailed studies of their properties. Consequently, little direct information is available about XyG glucan synthases. However, the recent discovery of a family of cellulose synthase-like (*Csl*) genes has prompted speculation that some of them may encode the Golgi-localized proteins responsible for xyloglucan biosynthesis and possibly other polysaccharides.

The biosynthesis of galactomannan (see Figure 2B) differs from XyG biosynthesis in several important respects. One significant difference is that mannan synthase can operate independently from GalT to create a linear, unsubstituted (1→4)β-D-mannan if UDP-galactose is not added to *in vitro* reactions containing Golgi vesicles capable of galactomannan biosynthesis (Edwards *et al.*, 1989). Moreover, the degree of substitution when UDP-galactose is available is variable and appears to be a stochastic process controlled both by enzyme specificity and the levels of GalT activity (Edwards *et al.*, 1992; Reid *et al.*, 1995). Although the genes and proteins responsible for (1→4)β-D-mannan synthesis have not yet been identified, the studies described above suggest this glycan synthase is likely to have a different relationship with the glycosyltransferases that add side chains than do the glucan synthases involved in XyG biosynthesis.

Other glycan synthases have also been studied. Several research groups have investigated the enzymes involved in biosynthesis of the polygalacturonic acid backbone found in pectic polysaccharides. Earlier studies have been reviewed recently by Mohnen (1999) and will not be covered in detail here. More recently, Doong and Mohnen (1998) were able to solubilize an activity from the microsomal membranes of tobacco cells that could transfer galacturonic acid from UDP-galacturonic acid to acceptors consisting of short oligomers of galacturonic acid. In later studies, they were able to demonstrate that the galacturonic acid residue was added onto the non-reducing end of the acceptor molecule (Scheller *et al.*, 1999). However, the solubilized enzyme added only a single galacturonic acid residue, leading the authors to conclude that the solubilized form of the enzyme had lost the processivity required of the native complex that produces the pectic backbone (Doong and Mohnen, 1998). Further studies will be needed to determine the causes for this change, but the list of possibilities discussed above for explaining the loss of XyG biosynthesis activity after solubilization would also apply to this situation.

Bioinformatics of cell wall biosynthetic genes

Cloning strategies based on obtaining protein sequence data from enzymes of interest are inherently low-throughput in nature because of the difficulties and variability associated with enzyme purification. An alternate approach is to exploit the large amount of available genomic data. Although few plant cell wall biosynthetic genes of known function have been identified, many glycosyltransferase and glycan synthase sequences from other organisms exist in public databases. Bioinformatics tools allow these sequences to be used to mine plant sequence databases, particularly the nearly complete *Arabidopsis* genome. Various techniques may be employed to accomplish this task. In this section, approaches to searching based on overall homology, motifs, and structural features will be described. Expression profiling also represents a valuable correlative technique that may be conducted with microarray or EST library sequence data.

Sequence analysis

Homology searches

In recent years, a large number of glycosyltransferases from various organisms have been sequenced. Most are responsible for the addition of monosaccharide units to glycoproteins. In a number of recent reviews, glycosyltransferases are grouped into families (Campbell *et al.*, 1997; Breton *et al.*, 1998a, b; Kapitonov and Yu, 1999; Oriol *et al.*, 1999; Henrissat *et al.*, 2001), and a curated web site also exists (http://afmb.cnrs-mrs.fr/~pedro/CAZY/db.html). It is clear from these comparisons that there is a great amount of sequence divergence within these families, even between enzymes that have the same donor and acceptor but that catalyze different linkages. Although it will not be possible to easily identify all enzymes that use the same acceptor or donor substrates, homology-based strategies are still useful.

Arabidopsis sequences related to known glycosyltransferases have been identified by sequence similarity search programs; several examples will be given. In the first example, there are twelve sequences in *Arabidopsis* related to the FKS enzyme of *Saccharomyces cerevisiae*. This enzyme synthesizes a (1→3)β-glucan found in the yeast cell wall (Douglas *et al.*, 1994). An obvious possible role for these sequences in *Arabidopsis* is the production of callose. Another example is the presence of over 40 genes in *Arabidopsis* that are related to the cotton cellulose synthase gene. Twelve of these are very closely related to cellulose synthase while the remainder are less well conserved and are referred to as cellulose synthase like (*Csl*) genes. The *Csl* genes may be involved in the synthesis of polymers other than cellulose. More information about FKS homologues, cellulose synthase, and cellulose synthase-like genes in *Arabidopsis* may be found on the web site maintained by Todd Richmond and Chris Somerville at http://cellwall.stanford.edu/cellwall/. In a third example, a somewhat surprising finding is the presence of *Arabidopsis* sequences showing considerable homology (with an expectation score of $5\,e^{-20}$) to a human protein called EXT1 involved in heparan sulfate biosynthesis (C. Wilkerson, unpublished results). EXT1 catalyzes the synthesis of a polymer composed of alternating residues of D-glucuronic acid and *N*-acetyl-D-glucosamine (McCormick *et al.*, 1998). This enzyme is reported to catalyze the addition of both sugars. EXT1 is found as a heterodimer with the protein EXT2 (McCormick *et al.*, 2000). A sequence with significant similarity to EXT2 is also found in *Arabidopsis* (with an expectation score of $1\,e^{-30}$). In total, there are 15 sequences in the *Arabidopsis* genome with significant similarity to EXT1 and EXT2 proteins. As a final example, the *Arabidopsis* genome was searched for possible GalTs. Eight *Arabidopsis* sequences are very similar to the α-GalT from fenugreek involved in galactomannan synthesis described above (Edwards *et al.*, 1999). The relationship between these sequences is illustrated in Figure 3.

Motif searches

A number of short amino acid sequences have been identified that are shared among glycosyltransferases. These motifs are usually found by comparing a number of protein sequences from enzymes that catalyze the same reaction (e.g. Breton *et al.*, 1996, 1998a; Sherwood *et al.*, 1998; Imberty *et al.*, 1999; Oriol *et al.*, 1999; Takahashi *et al.*, 2000). Comparing highly divergent sequences actually increases the likelihood that sequences essential for the enzymatic activity will be identified. The hope is that these motifs will be useful in identifying distantly related proteins that could not be found by standard homology searches.

Conserved residues are identified by several means. A common method is to use a multiple alignment program such as CLUSTAL or PILEUP followed by visual inspection of the resulting alignment for conserved amino acids. Another method that has been used extensively with glycosyltransferases is hydrophobic cluster analysis (HCA) (Geremia *et al.*,

1996; Breton *et al.*, 1998a). This technique uses a graphical representation of the protein sequence that emphasizes secondary structure characteristics. Pairs of sequences are compared to identify motifs. The last commonly used method is to use a computer program to extract patterns from a group of sequences based on some statistical parameter. Examples include MEME (http://meme.sdsc.edu/meme/website/meme-intro.html) from the San Diego Supercomputer Center and BLOCK MAKER (http://blocks.fhcrc.org/blocks/blockmkr/make_blocks.html) from the Fred Hutchinson Cancer Research Center.

The resulting motifs can be validated using a program like the PATTERNFIND server at ISREC (http://www.isrec.isb-sib.ch/software/PATFND_form.html) to search for the occurrence of the putative motif in protein databases. The two desirable aspects of a 'good' motif are selectivity and sensitivity. If a motif is selective, most of the sequences returned from a search with it will be related to the training set used to develop the motif. For example, a motif deduced from glycosyltransferases will return mainly other glycosyltransferases. Conversely, a motif is considered to be sensitive if it is conserved throughout all proteins with a given function. Often motifs are not very selective because they are too short. In these cases the program PHI-BLAST can be used to increase the selectivity of the search (Zhang *et al.*, 1998). This program searches for all occurrences of a given motif in a database and performs a rigorous sequence comparison between the region of the query sequence containing the motif and these sequences. Database sequences that have significant similarity to the query sequence are reported and are available for subsequent searches using the PSI-BLAST algorithm (Altschul *et al.*, 1997).

Several glycosyltransferase motifs have been identified using these techniques. A DxD sequence is common among a large number of glycosyltransferases, particularly in galactosyltransferases (Breton *et al.*, 1998b). The crystal structure of β4Gal-T1, a bovine glycoprotein and glycosphingolipid galactosyltransferase, reveals that the DxD motif is at the bottom of the UDP-Gal binding pocket (Gastinel *et al.*, 1999). Furthermore, these amino acids form a salt bridge with another conserved motif (WG-WGGEDDD) that surrounds this UDP-Gal binding pocket (Gastinel *et al.*, 1999). Useful motifs have also been identified in fucosyltransferases. One such motif, [IV]G[IV][HQ]xRxx[DN], is conserved in all $(1{\rightarrow}2)\alpha$- and $(1{\rightarrow}6)\alpha$-FucTs and has been implicated in donor substrate binding (Takahashi *et al.*, 2000).

124

A

```
            .           :                         :   :   .  :    . ::  ::.       .                        .
AC011809_19  MF-QDGS-------RSSGSGRGLSTTAVSNGGWR-TRG--FLRGWQIQNTLFNNIKFMILCCFVTILILLGTIRVGNLGSS------NADSVNQS-----  78
AC011765_6   MG-QDGSPAHK---RPSGSGGGLPTTTLTNGGGRGGRGGLLPRGRQMQKT-FNNIKITILCGFVTILVLRGTIGVGNLGSS------SADAVNQN-----  84
CAC01676     MGKEDGFRTQKRVSTASSAAAGVLPTTMASGGVR--RP--PPRGRQIQKT-FNNVKMTILCGFVTILVLRGTIGI-NFGTS------DADVVNQN-----  83
T01300       --------------------MIERCLGAYR--------CRRIQRA-LRQLKVTILCLLLTVVVLRSTIGAGKFGTPEQ----DLDEIRQH-FHAR  61
T01099       --------------------MIERCLGAYR--------CRRIQRA-LRQLKVTILCLLLTVVVLRSTIGAGKFGTPEQ----DLDEIRQH-FHAR  61
CAB83122     --------------------MIEKCIGAHR--------FRRLQRF-MRQGKVTILCLVLTVIVLRGTIGAGKFGTPEK----DIEEIREHFFYTR  62
CAC01675     -----------MVSPETSSSHYQSSPMAKYAGTR-----------TRPVVCISDVVLFLGGAFMSLILVWSFFSFSSISPN------LTVKNEES----  67
T04726       --------------------MGKPGGAK----------TRTAVCLSDGVFFLAGAFMSLTLVWSYFSI--FSPS--------FTSLRHDG----  50
CAB52246     -------------------------ATKFGSK-------NKSSPWLSNGCIFLLGAMSALLMIWGLNSFIAPIPNSNPKFNSFTTKLKSLN---  59

             ::                . :
AC011809_19  FIKETIPILAEIPS--DSHSTDLAEPPKADVSPNATYTLEPRIAEIPSDVHSTDLVELPKADISPNATYTLGPRIAEIPSDSHLTDLLEPPKADISPNAT  176
AC011765_6   IIEETNRILAEIRS--DSDPTDLDEPQ-------------------------------------------------------------EGDMNPNAT  118
CAC01676     IIEETNRLLAEIRS--DSDPTDSNEPP-------------------------------------------------------------DSDLDLNMT  117
T01300       KRGEPHRVLEEIQTGGDSSSGDGGGNSGGS----------------------------------------------NNYETFDINKIFVDEGEEEKPD--PNKP  117
T01099       KRGEPHRVLEEIQTGGDSSSGDGGGNSGGS----------------------------------------------NNYETFDINKIFVDEGEEEKPD--PNKP  117
CAB83122     KRGEPHRVLVEVSS--KTTSSEDGGNGG----------------------------------------------NSYETFDINKLFVDEGDEEKSRDRTNKP  116
CAC01675     --SNKCSSGIDMSQ----DPTDPVYYD--------------------------------------------------------------DPD----LT  93
T04726       --KPVQCSGLDMQF----DPSEPGFYD--------------------------------------------------------------DPD----LS  76
CAB52246     FTTNTNFAGPDLLH---DPSDKTFYD---------------------------------------------------------------DPE----TC  87

             *   :    :..**..:*  *:  :*.*            :  :::*** .  *.:*:*:: **: ***:**.*:*. .*.*. : *. :: .:*:* :*
AC011809_19  YT-LGPKITNWDSQRKVWLNQNPEFPNIVN-GKARILLLTGSSPGPCDKPIGNYYLLKAVKNKIDYCRLHGIEIVYNMANLDEELSGYWTKLPMIRTLML  274
AC011765_6   YV-LGPKITDWDSQRKVWLNQNPEFPSTVN-GKARILLLTGSSPPKCDNPIGDHYLLKSVKNKIDYCRLHGIEIVYNMAHLDKELAGYWAKLPMIRRLML  216
CAC01676     YT-LGPKITNWDQKRKLWLTQNPDFPSFIN-GKAKVLLLTGSPPKPCDNPIGDHYLLKSVKNKIDYCRIHGIEIVYNMAHLDKELAGYWAKLPMIRRLML  215
T01300       YT-LGPKISDWDEQRSDWLAKNPSFPNFIGPNKPRVLLVTGSAPKPCENPVGDHYLLKSIKNKIDYCRLHGIEIFYNMALLDAEMAGFWAKLPLIRKLLL  216
T01099       YT-LGPKISDWDEQRSDWLAKNPSFPNFIGPNKPRVLLVTGSAPKPCENPVGDHYLLKSIKNKIDYCRLHGIEIFYNMALLDAEMAGFWAKLPLIRKLLL  216
CAB83122     YS-LGPKISDWDERRDWLKQNPSFPNFVAPNKPRVLLVTGSAPKPCENPVGDHYLLKSIKNKIDYCRIHGIEIFYNMALLDAEMAGFWAKLPLIRKLLL  215
CAC01675     YT-IEKPVKNWDEKRRRWLNLHPSFIPGAE---NRTVMVTGSQSAPCKNPIGDHLLLRFFKNKVDYCRIHGHDIFYSNALLHPKMNSYWAKLPAVKAAMI  189
T04726       YS-IEKPITKWDEKRNQWFESHPSFKPGSE---NRIVMVTGSQSSPCKNPIGDHLLLRCFKNKVDYARIHGHDIFYSNSLLHPKMNSYWAKLPVVKAAML  172
CAB52246     YTMMDKPMKNWDEKRKEWLFHHPSFAAGAT---EKILVITGSQPTKCDNPIGDHLLLRFYKNKVDYCRIHNHDIIYNNALLHPKMDSYWAKYPMVRAAML  184

             :***  *:.:**::****:***:**.  *   .  **..:***:*.    .:.*.:**:* **:** **:::**.* ***       . :           :    .*:: .     .    : *.*
AC011809_19  SHPEVEWIWWMDSDALFTDILFEIPLPRYENHNLVIHGYPDLLFNQKSWVALNTGIFLLRNCQWSLDLLDAWAPMGPKGKIRDETGKILTAYLKGRPAFE  374
AC011765_6   SHPEVEWIWWMDSDALFTDILFQIPLARYQKHNLVIHGYPDLLFDQKSWIALNTGSFLLRNCQWSLDLLDAWAPMGPKGPIRDEAGKVLTAYLKGRPAFE  316
CAC01676     SHPEIEWIWWMDSDALFTDMVFEIPLSRYENHNLVIHGYPDLLFDQKSWIALNTGSFLLRNCQWSLDLLDAWAPMGPKGPIREEAGKILTANLKGRPAFE  315
T01300       SHPEIEFLWWMDSDAMFTDMAFELPWERYKDYNLVMHGWNEMVYDQKNWIGLNTGSFLLRNNQWALDLLDTWAPMGPKGKIREEAGKVLTRELKDRPVFE  316
T01099       SHPEIEFLWWMDSDAMFTDMAFELPWERYKDYNLVMHGWNEMVYDQKNWIGLNTGSFLLRNNQWALDLLDTWAPMGPKGKIREEAGKVLTRELKDRPVFE  316
CAB83122     SHPEIEFLWWMDSDAMFTDMVFELPWERYKDYNLVMHGWNEMVYDQKNWIGLNTGSFLLRNSQWSLDLLDAWAPMGPKGKIREEAGKVLTRELKDRPAFE  315
CAC01675     AHPEAEWIWWVDSDALFTDMDFTPPWRRYKEHNLVVHGWPGVIYNDRSWTALNAGVFLIRNCQWSMELIDTWTGMGPVSPEYAKWGQIQRSIFKDKLFPE  289
T04726       AHPEAEWIWWVDSDAIFTDMEFKPPLHRYRQHNLVVHGWPNIIYEKQSWTALNAGVFLIRNCQWSMDLIDTWKSMGPVSPDYKKWGPIQRSIFKDKLFPE  272
CAB52246     AHPEVEWIWWVDSDAIFTDMEFKLPLWRYKDHNLVIHGWEELVKTEHSWTGLNAGVFLMRNCQWSLDFMDVWASMGPNSPEYEKWGERLRETFKTKVVRD  284

             :***:*: **:        .: : *:::*  *:*.*:*  .        ::  *.:                                             *.  *   .:;***.**
AC011809_19  ADDQSALIYLLLSQKEKWIEKVYVENQYYLHGFWEGLVDRYEEMIEKYHPG--------------------------------LGDERWPFVTHFVGC  440
AC011765_6   ADDQSALIYLLLSQKDTWMEKVFVENQYYLHGFWEGLVDRYEEMIEKYHPG--------------------------------LGDERWPFITHFVGC  382
CAC01676     ADDQSALIYLLLSQKETWMEKVFVENQYYLHGFWEGLVDRYEEMMEKYHPG--------------------------------LGDERWPFITHFVGC  381
T01300       ADDQSAMVYLLATQRDAWGNKVVYLESGYYLHGYWGILVDRYEEMIENYHPG--------------------------------LGDHRWPLVTHFVGC  382
T01099       ADDQSAMVYLLATQRDAWGNKVVYLESGYYLHGYWGILVDRYEEMIENYHPG--------------------------------LGDHRWPLVTHFVGC  382
CAB83122     ADDQSAMVYLLATEREKWGGKVVYLESGYYLHGYWGILVDRYEEMIENHKPG--------------------------------FGDHRWPLVTHFVGC  381
CAC01675     SDDQTALLYLLYKHREVYYPKIYLEGDFYFEGYWLEIVPGLSNVTERYLEMEREDATLRRRHAEKVSERYAAFREERFLKGERGGKGSKRRPFVTHFTGC  389
T04726       SDDQTALIYLLYKHKELYYPKIYLEASYLQGYWIGVFGDFANVTERYLEMEREDDTLRRRHAEKVSERYGAFREERFLKGEFGGRGSRRRAFITHFTGC  372
CAB52246     QDDQTALAYLIAMGEDKWTKKIYMENEYYFEGYWLEISKMYDKMGERYDEIEKRVEGLRRRHAEKVSERYGEMREEYVKN-----LGDMRRPFITHFTGC  379

             :**.        *   . *  . * .  .  :*;:**.***::: :*: *  .*  .                         *:
AC011809_19  KPCG--SYADYAVDRCFKSMERAFNFADNQVLKLYGFSHRGLLSPKIKRIRNETVSPLESVDKFDIRR-MHMETKP-----  513
AC011765_6   KPCG--SYADYAVERCLKSMERAFNFADNQVLKLYGFSHRGLLSPKIKRIRNETVSPLEFVDKFDIRR-TPVETKPQN---  457
CAC01676     KPCG--SYADYAVERCLKSMERAFNFADNQVLKLYGFGHRGLLSPKIKRIRNETTFPLKFVDRFDIRRTTPLKIEARS---  457
T01300       KPCG--KFGDYPVERCLKQMDRAFNFGDNQILQIYGFTHKSLASRKVKRVRNETSNPLEMKDELGLLHPAFKAVKVQTNQV  461
T01099       KPCG--KFGDYPVERCLKQMDRAFNFGDNQILQIYGFTHKSLASRKVKRVRNETSNPLEMKDELGLLHPAFKAVKGEAL--  459
CAB83122     KPCG--KFGDYPVERCLRQMDRAFNFGDNQLQMYGFTHKSLASRKVKPTRNQTDRPLDAKDEFGLLHPPFKAAKLSTTTT  460
CAC01675     QPCSGDHHNKMYDGDTCWNGMIKAINFADNQVMRKYGFVHSDLGKTSP-----LQPVPFDYPDEPW---------------  449
T04726       QPCSGDHHNPSYDGDTCWNEMIRALNFADNQVMRVGYVHSDLSKTSP-----LQPLPFDYPNEAW---------------  432
CAB52246     QPCNGHHNPIYAADDCWNGMERALNFADNQVLRKFGFIHPNLLDKS------VSPLPFGYPAASP---------------  438
```

Figure 3. Comparison of eight protein sequences from *Arabidopsis* with the protein sequence of fenugreek galactomannan GalT. A. Alignment generated using CLUSTALX (Thompson *et al.*, 1997). The fenugreek GalT is the bottommost sequence (accession number CAB52246). Asterisks indicate identical amino acids in all aligned sequences. Semicolons and dots indicate conservative replacements. The residues are colored to represent their physiochemical properties. B (next page). Tree displayed by the program TREEVIEW (Page, 1996). The tree is a representation of a distance matrix produced by CLUSTALX. The numbers below each node represent the number of times this tree was produced out of one thousand attempts using the bootstrap option of CLUSTALX.

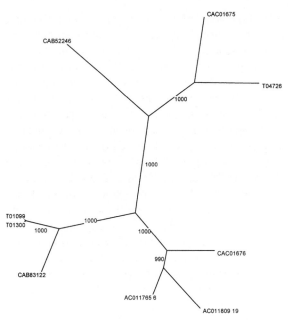

CAC01675

CAB52246

T04726

1000

1000

T01099
T01300

1000

1000

1000

CAB83122

990

CAC01676

AC011765 6

AC011809 19

Figure 3. Continued.

Although short and somewhat ambiguous, this motif is sufficiently selective for searching by methods described above. For example, it identifies 7 proteins out of over 20 000 predicted *Arabidopsis* proteins, and one of the sequences identified is the XyG FucT (C. Wilkerson, unpublished results.) The XyG FucT also shares a KPW motif that was identified as common among GalTs (Breton *et al.*, 1998b; Gibeaut, 2000).

Finally, some glycan synthase motifs have been described. One example is the Dx(33,38)-QxxRW motif, first identified in bacterial and plant cellulose synthases (Saxena *et al.*, 1995; Pear *et al.*, 1996). This motif is present in cellulose synthase, cellulose synthase-like proteins, chitin synthase, NodC and hyaluronan synthase. It has been postulated to be a marker of processive transferases (Saxena and Brown, 1995; see also Richmond and Somerville, 2001, and Vergara and Carpita, 2001, both in this issue).

Protein structural searches

The synthesis of all non-cellulosic cell wall polysaccharides takes place in the Golgi apparatus (Driouich *et al.*, 1993). Many Golgi- retained proteins, including glycosyltransferases, are type II membrane proteins containing a single transmembrane-spanning region that serves as an anchor (for reviews, see Machamer, 1993; Colley, 1997). It has been suggested that the length of the transmembrane region may play a role in segregation of membrane proteins within the en-

domembrane system (Masibay *et al.*, 1993; Colley, 1997). Golgi membrane proteins have a characteristically shorter transmembrane region than proteins localized to other membranes within the cell. Therefore, one approach for identification of glycosyltransferase candidate proteins is to select *Arabidopsis* sequences that have a single, short transmembrane domain. By searching for *Arabidopsis* proteins that have only a single predicted transmembrane helix of a delimited allowable length, it is possible to identify 180 sequences from a total of over 20 000 (C. Wilkerson, unpublished results). This collection of 180 sequences includes the XyG FucT as well as a number of unknown sequences that contain some of the motifs discussed above. Searching for candidate genes on the basis of structural features such as these, combined with the motif-searching strategies outlined above, enhances the possibility of obtaining robust results.

Expression profiling

Microarrays

The Arabidopsis Functional Genomics Consortium has made available the results of a large number (98 as of September 2000) of microarray experiments available to the public. These data are accessible from the AFGC web site at http://genome-www4.stanford.edu/MicroArray/SMD/. It is possible to cluster *Arabidopsis* genes based on their patterns of expression as determined by the microarray data. There are two distinct methods for clustering data: supervised and unsupervised. Using supervised methods, one can discover other genes with expression patterns similar to the genes of interest. The supervised clustering method can then be used to find additional members from a larger group of unclassified objects. One example of such a method applied to microarray data is the support vector machine (Brown *et al.*, 2000). The more common approach is to use unsupervised learning methods, which results in groups of related objects. Unsupervised methods include hierarchical (Eisen *et al.*, 1998) and self-organizing map (Tamayo *et al.*, 1999) algorithms. These techniques are useful when searching for genes related to some physiological state or genetic modification of the organism under study. Examining the annotation of genes in a cluster can reveal the participation of known biochemical or signaling pathways in the process being studied.

Another way to utilize microarrays to identify cell wall biosynthesis candidate genes is to utilize probes

126

diagnostic of cell wall biosynthesis. For example, *Arabidopsis* RNA isolated from stages of cell wall regeneration after preparation of protoplasts might be used. Another alternative would be RNA from root tips versus expanded root sections. The combination of a number of conditions such as these would allow genes involved in the common phenomenon, cell wall biosynthesis, to be distinguished from other processes such as wounding responses and cell division, which are not shared between the two experimental conditions. The same procedure can be used with the *Arabidopsis* chip recently made available from Affymetrix.

EST collections

Recently, many expressed sequence tagged (EST) sequencing projects have been conducted. The data from these projects are available at the NCBI web site (http://www.ncbi.nlm.nih.gov/). It is possible for the user to create personal databases corresponding to each cDNA library using batch ENTREZ at NCBI (http://www.ncbi.nlm.nih.gov/Entrez/batch.html). The libraries can then be searched with a set of genes using BLAST or a comparable similarity search program. If the EST libraries and the query sequences are from the same organism, the comparisons can be made using nucleic acid sequences. If the libraries are from different species, it is necessary to search by means of protein comparisons. If the BLAST algorithm is used, the TBLASTN program will compare input protein sequences to a nucleic acid database. The results of the search are tallies of the number of occurrences of similar sequences for each query sequence in each library searched. These vectors may then be used as input to clustering programs to reveal genes with similar expression profiles (Ewing *et al.*, 1999). However, some drawbacks must be considered. The use of protein comparisons introduces the complication that members of closely related gene families may be scored together. An additional concern is that the libraries do not usually have sufficient numbers to detect changes in genes expressed at low levels. As the number and quality of EST libraries increase, this technique will become more useful.

Evaluation of candidate genes

The methods used to identify candidate genes are generally not sufficiently stringent to allow firm conclusions about the biochemical function or the biological role of the gene product. Particularly when candidate genes have been identified by sequence similarity or by motif searches, further studies are required to confirm hypotheses regarding the biochemical function of the gene. Even when biochemical studies are used to identify candidate genes, independent verification is often needed to ensure that the proper gene has been identified. Expression of the candidate gene in a heterologous system followed by biochemical assay is particularly useful (as shown in the top arm of Figure 1). This approach was utilized in both of the cell wall biosynthetic glycosyltransferase identification studies described to date (Edwards *et al.*, 1999; Perrin *et al.*, 1999). However, although this is a powerful strategy, the lack of biochemical assays for many enzymes will prevent it from being used in many cases. For example, *in vitro* biochemical assays are not available for cellulose synthase, so the activities of the *CesA* gene products have not been confirmed in this manner. Even if *in vitro* assays using plasma membrane preparations become available, it is likely that cellulose synthesis will require the activity of several polypeptides. Consequently, it is unlikely that expression of *CesA* genes in a heterologous system will allow feasible assays for the *CesA* gene products. Similar arguments may apply to many other enzymes involved in polysaccharide biosynthesis.

The other commonly used strategy for evaluating the function of candidate genes is genetics or reverse genetics (shown on the bottom arm of Figure 1). Although a genetic strategy can be very useful, as has been demonstrated for cellulose biosynthesis mutants (Potikha and Delmer, 1995; Turner and Somerville, 1997; Arioli *et al.*, 1998; Taylor *et al.*, 1999), this approach requires an effective and specific screen for mutants in the desired genes. Thus forward genetic strategies may be difficult to apply when one wants to investigate the function of specific genes. However, with the availability of effective methods for the identification of disruptions in a specific gene (Krysan *et al.*, 1996), such as reverse genetics, this strategy will likely become the method of choice for investigating the function of various candidate genes thought to be involved in polysaccharide biosynthesis.

Despite the power of reverse genetic approaches for investigating the function of putative polysaccharide biosynthetic genes, they still have some potentially serious drawbacks. The first is that many important plant genes are present in multiple copies, at least in the case of the *Arabidopsis* genome. Thus, disruption of a single gene may not produce a plant with a visible or a molecular phenotype. Efforts to investigate the functions of the *Csl* genes have encountered this problem (Todd Richmond and Chris Somerville, 2001). On the other hand, disruption of multiple members of an essential gene family may be lethal. For example, if one eliminated all copies of the glucan synthase genes responsible for XyG biosynthesis, one might expect that whole plants could not survive without any XyG. Thus, special efforts using tissue culture cells or inducible promoters for antisense expression may be necessary to overcome these problems.

Concluding remarks

Understanding the molecular details of plant cell wall polysaccharide biosynthesis has been an extremely difficult problem that has experienced limited progress in the past two or three decades. Several obstacles have prevented rapid progress with traditional biochemical strategies. These include the problems associated with obtaining exogenous acceptor substrates needed for assays of the solubilized enzymes and difficulties in purifying non-abundant, membrane-bound enzymes that often become unstable once they have been solubilized. However, the availability of new genomic and genetic strategies offers valuable alternatives for investigating these important problems. Most importantly, they offer new hope for an enhanced rate of progress.

Acknowledgements

We wish to thank members of the Cell Wall group at the MSU-DOE Plant Research Laboratory for helpful comments and Ms Karen Bird for editorial assistance. Work from our laboratory is supported by grants from the Energy Biosciences Program at the U.S. Department of Energy and the Plant Genome Program at the National Science Foundation.

References

Altschul, S.F., Madden, T.L., Schaffer, A.A., Zhang, J., Zhang, Z., Miller, W. and Lipman, D.J. 1997. Gapped BLAST and PSI-BLAST: a new generation of protein database search programs. Nucl. Acids Res. 25: 3389–3402.

Arioli, T., Peng, L., Betzner, A.S., Burn, J., Wittke, W., Herth, W. *et al.* 1998. Molecular analysis of cellulose biosynthesis in *Arabidopsis*. Science 279: 717–720.

Bolscher, J.G., Bruyneel, E.A., Van Rooy, H., Schallier, D.C., Mareel, M.M. and Smets, L.A. 1989. Decreased fucose incorporation in cell surface carbohydrates is associated with inhibition of invasion. Clin. Exp. Metastasis 7: 557–569.

Borsig, L., Kleene, R., Dinter, A. and Berger, E.G. 1996. Immunodetection of α 1-3 fucosyltransferase (FucT-V). Eur. J. Cell Biol. 70: 42–53.

Breton, C. and Imberty, A. 1999. Structure/function studies of glycosyltransferases. Curr. Opin. Struct. Biol. 9: 563–571.

Breton, C., Oriol, R. and Imberty, A. 1996. Sequence alignment and fold recognition of fucosyltransferases. Glycobiology 6: vii-xii.

Breton, C., Oriol, R. and Imberty, A. 1998a. Conserved structural features in eukaryotic and prokaryotic fucosyltransferases. Glycobiology 8: 1–8.

Breton, C., Bettler, E., Joziasse, D. H., Geremia, R. A. and Imberty, A. 1998b. Sequence-function relationships of prokaryotic and eukaryotic galactosyltransferases. J. Biochem. 123: 1000–1009.

Brett, C.T. and Waldron, K. 1996. Physiology and Biochemistry of Plant Cell Walls, 2 ed., Chapman and Hall, London.

Britten, C.J. and Bird, M.I. 1997. Chemical modification of an α 3-fucosyltransferase; definition of amino acid residues essential for enzyme activity. Biochim. Biophys. Acta 1334: 57–64.

Brown, M.P., Grundy, W.N., Lin, D., Cristianini, N., Sugnet, C.W., Furey, T.S., Ares, M. Jr. *et al.* 2000. Knowledge-based analysis of microarray gene expression data by using support vector machines. Proc. Natl. Acad. Sci. USA 97: 262–267.

Brummell, D.A., Camirand, A. and Maclachlan, G.A. 1990. Differential distribution of xyloglucan transferases in pea Golgi dictyosomes and secretory vesicles. J. Cell Sci. 96: 705–710.

Buckeridge, M.S., Vergara, C.E. and Carpita, N. C.1999. The mechanism of synthesis of a mixed-linkage (1→3), (1→4)-β-D-glucan in maize. Evidence for multiple sites of glucosyl transfer in the synthase complex. Plant Physiol. 120: 1105–1116.

Butler, T. and Elling, L. 1999. Enzymatic synthesis of nucleotide sugars. Glycoconjugate J. 16: 147–159.

Camirand, A., Brummell, D. and MacLachlan, G.A. 1987. Fucosylation of xyloglucan: localization of the transferase in dictyosomes of stem stem cells. Plant Physiol. 84: 753–756

Campbell, J.A., Davies, G.J., Bulone, V. and Henrissat, B. 1997. A classification of nucleotide-diphospho-sugar glycosyltransferases based on amino acid sequence similarities. Biochem. J. 326: 929–939.

Carpita, N.C. and Gibeaut, D.M. 1993. Structural models of primary cell walls in flowering plants: consistency of molecular structure with the physical properties of the walls during growth. Plant J. 3: 1–30.

Colley, K.J. 1997. Golgi localization of glycosyltransferases: more questions than answers. Glycobiology 7: 1–13.

Delmer, D. 1999. Cellulose biosynthesis: exciting times for a difficult field of study. Annu. Rev. Plant Physiol. Plant Mol. Biol. 50: 245–276.

Delmer, D.P. and Stone, B.A. 1988. Biosynthesis of plant cell walls. In P.K. Stumpf and E.E. Conn (Eds.) Biochemistry of Plants: A Comprehensive Treatise, vol. 14, Academic Press, New York, pp. 373–419.

Doong, R.L. and Mohnen, D. 1998. Solubilization and characterization of a galacturonosyltransferase that synthesizes the pectic polysaccharide homogalacturonan. Plant J. 13: 363–374.

Douglas, C.M., Foor, F., Marrinan, J.A., Morin, N., Nielsen, J.B., Dahl, A.M. *et al.* 1994. The *Saccharomyces cerevisiae FKS1* (*ETG1*) gene encodes an integral membrane protein which is a subunit of 1,3-β-D-glucan synthase. Proc. Natl. Acad. Sci. USA 91: 12907–12911.

128

Driouich, A., Faye, L. and Staehelin, L.A. 1993. The plant Golgi apparatus: a factory for complex polysaccharides and glycoproteins. Trends Biochem. Sci. 18: 210–214.

Edwards, M., Bulpin, P.V., Dei, I.C.M. and Reid, J.S. 1989. Biosynthesis of legume-seed galactomannans *in vitro*. Planta 178: 41–51.

Edwards, M., Scott, C., Gidley, M.J. and Reid, J.S. 1992. Control of mannose/galactose ratio during galactomannan formation in developing legume seeds. Planta 187: 67–74.

Edwards, M.E., Dickson, C.A., Chengappa, S., Sidebottom, C., Gidley, M.J. and Reid, J.S. 1999. Molecular characterisation of a membrane-bound galactosyltransferase of plant cell wall matrix polysaccharide biosynthesis. Plant J. 19: 691–697.

Eisen, M.B., Spellman, P.T., Brown, P.O. and Botstein, D. 1998. Cluster analysis and display of genome-wide expression patterns. Proc. Natl. Acad. Sci. USA 95: 14863–14868.

Ernst, L.K., Rajan, V.P., Larsen, R.D., Ruff, M.M. and Lowe, J.B. 1989. Stable expression of blood group H determinants and GDP-L-fucose: β-D-galactoside 2-α-L-fucosyltransferase in mouse cells after transfection with human DNA. J. Biol. Chem. 264: 3436–3447.

Ewing, R.M., Kahla, A.B., Poirot, O., Lopez, F., Audic, S. and Claverie, J.M. 1999. Large-scale statistical analyses of rice ESTs reveal correlated patterns of gene expression. Genome Res. 9: 950–959.

Faik, A., Chileshe, C., Sterling, J. and Maclachlan, G. 1997. Xyloglucan galactosyl- and fucosyltransferase activities from pea epicotyl microsomes. Plant Physiol. 114: 245–254.

Faik, A., Bar Peled, M., DeRocher, A.E., Zeng, W., Perrin, R.M., Wilkerson, C. *et al.* 2000. Biochemical characterization and molecular cloning of an α-1,2-fucosyltransferase that catalyzes the last step of cell wall xyloglucan biosynthesis in pea. J. Biol. Chem. 275: 15082–15089.

Farkas, V. and Maclachlan, G. 1988. Fucosylation of exogenous xyloglucans by pea microsomal membranes. Arch. Biochem. Biophys. 264: 48–53.

Field, M.C. and Wainwright, L.J. 1995. Molecular cloning of eukaryotic glycoprotein and glycolipid glycosyltransferases: a survey. Glycobiology 5: 463–472.

Fujino, T., Sone, Y., Mitsuishi, Y. and Itoh, T. 2000. Characterization of cross-links between cellulose microfibrils, and their occurrence during elongation growth in pea epicotyl. Plant Cell Physiol. 41: 486–494.

Fukuda, M., Bierhuizen, M.F. and Nakayama, J. 1996. Expression cloning of glycosyltransferases. Glycobiology 6: 683–689.

Gastinel, L.N., Cambillau, C. and Bourne, Y. 1999. Crystal structures of the bovine β4-galactosyltransferase catalytic domain and its complex with uridine diphosphogalactose. EMBO J. 18: 3546–3557.

Ge, Z., Chan, N.W., Palcic, M.M. and Taylor, D.E. 1997. Cloning and heterologous expression of an α1,3-fucosyltransferase gene from the gastric pathogen *Helicobacter pylori*. J. Biol. Chem. 272: 21357–21363.

Geremia, R.A., Petroni, E.A., Ielpi, L. and Henrissat, B. 1996. Towards a classification of glycosyltransferases based on amino acid sequence similarities: prokaryotic α-mannosyltransferases. Biochem J. 318: 133–138.

Gibeaut, D.M. 2000. Nucleotide sugars and glycosyltransferases for synthesis of cell wall matrix polysaccharides. Plant Physiol. Biochem. 38: 69–80.

Gibeaut, D.M. and Carpita, N.C. 1994. Biosynthesis of plant cell wall polysaccharides. FASEB J. 8: 904–915.

Haigler, C.H., Ivanova-Datcheva, M., Hogan, P.S., Salnikov, V.V., Hwang, S., Martin, L.K. and Delmer, D.P. 2001. Carbon partitioning to cellulose synthesis. Plant Mol. Biol., this issue.

Hanna, R., Brummell, D.A., Camirand, A., Hensel, A., Russell, E.F. and Maclachlan, G.A. 1991. Solubilization and properties of GDP-fucose: xyloglucan 1,2-α-L-fucosyltransferase from pea epicotyl membranes. Arch. Biochem. Biophys. 290: 7–13.

Hayashi, T. 1989. Xyloglucans in the primary cell wall. Annu. Rev. Plant Physiol. Plant Mol. Biol. 40: 139–168.

Henrissat, B., Coutinho, P. and Davies, G.J. 2001. A census of carbohydrate-active enzymes in the genome of *Arabidopsis thaliana*. Plant Mol. Biol., this issue.

Hirschberg, C.B., Robbins, P.W. and Abeijon, C. 1998. Transporters of nucleotide sugars, ATP, and nucleotide sulfate in the endoplasmic reticulum and Golgi apparatus. Annu. Rev. Biochem. 67: 49–69.

Imberty, A., Monier, C., Bettler, E., Morera, S., Freemont, P., Sippl, M. *et al.* 1999. Fold recognition study of α3-galactosyltransferase and molecular modeling of the nucleotide sugar-binding domain. Glycobiology 9: 713–722.

Kapitonov, D. and Yu, R.K. 1999. Conserved domains of glycosyltransferases. Glycobiology 9: 961–978.

Kimura, H., Kudo, T., Nishihara, S., Iwasaki, H., Shinya, N., Watanabe, R. *et al.* 1995. Murine monoclonal antibody recognizing human $\alpha(1,3/1,4)$ fucosyltransferase. Glycoconjugate J. 12: 802–812.

Kleene, R. and Berger, E.G. 1993. The molecular and cell biology of glycosyltransferases. Biochim. Biophys. Acta. 1154: 283–325.

Krysan, P.J., Young, J.C., Tax, F. and Sussman, M.R. 1996. Identification of transferred DNA insertions within *Arabidopsis* genes involved in signal transduction and ion transport. Proc. Natl. Acad. Sci. USA 93: 8145–8150.

Kukowska-Latallo, J.F., Larsen, R.D., Nair, R.P. and Lowe, J.B. 1990. A cloned human cDNA determines expression of a mouse stage-specific embryonic antigen and the Lewis blood group $\alpha(1,3/1,4)$fucosyltransferase. Genes Dev. 4: 1288–1303.

Larsen, R.D., Ernst, L.K., Nair, R.P. and Lowe, J.B. 1990. Molecular cloning, sequence, and expression of a human GDP-L-fucose: β-D-galactoside 2-α-L-fucosyltransferase cDNA that can form the H blood group antigen. Proc. Natl. Acad. Sci. USA 87: 6674–6678.

Leiter, H., Mucha, J., Staudacher, E., Grimm, R., Glossl, J. and Altmann, F. 1999. Purification, cDNA cloning, and expression of GDP-L-Fuc:Asn-linked GlcNAc α1,3-fucosyltransferase from mung beans. J. Biol. Chem. 274: 21830–21839.

Machamer, C.E. 1993. Targeting and retention of Golgi membrane proteins. Curr. Opin. Cell Biol. 5: 606–612.

Maclachlan, G., Levy, B. and Farkas, V. 1992. Acceptor requirements for GDP-fucose:xyloglucan 1,2-α-L-fucosyltransferase activity solubilized from pea epicotyl membranes. Arch. Biochem. Biophys. 294: 200–205.

Martin, S.L., Edbrooke, M.R., Hodgman, T.C., van den Eijnden, D.H. and Bird, M.I. 1997. Lewis X biosynthesis in *Helicobacter pylori*. Molecular cloning of an $\alpha(1,3)$-fucosyltransferase gene. J. Biol. Chem. 272: 21349–21356.

Masibay, A.S., Balaji, P.V., Boeggeman, E.E. and Qasba, P.K. 1993. Mutational analysis of the Golgi retention signal of bovine β-1,4-galactosyltransferase. J. Biol. Chem. 268: 9908–9916.

Matthysse, A.G., Thomas, D.L. and White, A. R.1995a. Mechanism of cellulose synthesis in *Agrobacterium tumefaciens*. J. Bact. 177: 1076–1081.

Matthysse, A.G., White, S. and Lightfoot, R. 1995b. Genes required for cellulose synthesis in *Agrobacterium tumefaciens*. J. Bact. 177: 1069–1075.

McCormick, C., Duncan, G., Goutsos, K.T. and Tufaro, F. 2000. The putative tumor suppressors EXT1 and EXT2 form a stable complex that accumulates in the Golgi apparatus and catalyzes the synthesis of heparan sulfate. Proc. Natl. Acad. Sci. USA 97: 668–673.

McCormick, C., Leduc, Y., Martindale, D., Mattison, K., Esford, L.E., Dyer, A.P. and Tufaro, F. 1998. The putative tumour suppressor EXT1 alters the expression of cell-surface heparan sulfate. Nature Genet. 19: 158–161.

Mengeling, B.J. and Turco, S.J. 1998. Microbial glycoconjugates. Curr. Opin. Struct. Biol. 8: 572–577.

Mohnen, D. 1999. Biosynthesis of pectins and galactomannans. In: P.B. M. (Ed.) Carbohydrates and their Derivatives Including Tannins, Cellulose, and Related Lignins, vol. 3. Elsevier, Amsterdam, pp. 497–527.

Mollicone, R., Cailleau, A. and Oriol, R. 1995. Molecular genetics of H, Se, Lewis and other fucosyltransferase genes. Transfus. Clin. Biol. 2: 235–242.

Moore, P.J., Swords, K.M., Lynch, M.A. and Staehelin, L.A. 1991. Spatial organization of the assembly pathways of glycoproteins and complex polysaccharides in the Golgi apparatus of plants. J. Cell Biol. 112: 589–602.

Narimatsu, H., Iwasaki, H., Nishihara, S., Kimura, H., Kudo, T., Yamauchi, Y. and Hirohashi, S. 1996. Genetic evidence for the Lewis enzyme, which synthesizes type-1 Lewis antigens in colon tissue, and intracellular localization of the enzyme. Cancer Res. 56: 330–338.

Natsuka, S., Gersten, K.M., Zenita, K., Kannagi, R. and Lowe, J.B. 1994. Molecular cloning of a cDNA encoding a novel human leukocyte α-1,3-fucosyltransferase capable of synthesizing the sialyl Lewis x determinant. J. Biol. Chem. 269: 16789–16794.

Oriol, R., Mollicone, R., Cailleau, A., Balanzino, L. and Breton, C. 1999. Divergent evolution of fucosyltransferase genes from vertebrates, invertebrates, and bacteria. Glycobiology 9: 323–334.

Page, R.D.M. 1996. TREEVIEW: an application to display phylogenetic trees on personal computers. Comp. Appl. Biosci. 12: 357–358.

Pauly, M., Albersheim, P., Darvill, A. and York, W.S. 1999. Molecular domains of the cellulose/xyloglucan network in the cell walls of higher plants. Plant J. 20: 629–639.

Pear, J.R., Kawagoe, Y., Schreckengost, W.E., Delmer, D.P. and Stalker, D.M. 1996. Higher plants contain homologs of the bacterial celA genes encoding the catalytic subunit of cellulose synthase. Proc. Natl. Acad. Sci. USA 93: 12637–12642.

Perrin, R.M., DeRocher, A.E., Bar Peled, M., Zeng, W., Norambuena, L., Orellana, A. et al. 1999. Xyloglucan fucosyltransferase, an enzyme involved in plant cell wall biosynthesis. Science 284: 1976–1979.

Potikha, T. and Delmer, D.P. 1995. A mutant of Arabidopsis thaliana displaying altered patterns of cellulose deposition. Plant J. 7: 453–460.

Puhlmann, J., Bucheli, E., Swain, M.J., Dunning, N., Albersheim, P., Darvill, A.G. and Hahn, M.G. 1994. Generation of monoclonal antibodies against plant cell-wall polysaccharides. I. Characterization of a monoclonal antibody to a terminal $\alpha(1\rightarrow2)$-linked fucosyl-containing epitope. Plant Physiol. 104: 699–710.

Quesada, V.D., Fellay, R., Nasim, T., Viprey, V., Burger, U., Prome, J.C. et al. 1997. Rhizobium sp. strain NGR234 NodZ protein is a fucosyltransferase. J. Bact. 179: 5087–5093.

Quinto, C., Wijfjes, A.H.M., Bloemberg, G.V., Blok Tip, L., Lopez Lara, I.M., Lugtenberg, B.J. et al. 1997. Bacterial nodulation protein NodZ is a chitin oligosaccharide fucosyltransferase

which can also recognize related substrates of animal origin. Proc. Natl. Acad. Sci. USA 94: 4336–4341.

Reid, J.S. and Edwards, M. 1995. Galactomannans and other cell wall storage polysaccharides in seeds. In: A.M. Stephen (Ed.) Food Polysaccharides and their Applications, Marcel Dekker, New York, pp. 155–186.

Reid, J.S.G., Edwards, M., Gidley, M.J. and Clark, A.H. 1995. Enzyme specificity in galactomannan biosynthesis. Planta 195: 489–495.

Reiter, W.D., Chapple, C. and Somerville, C.R. 1997. Mutants of Arabidopsis thaliana with altered cell wall polysaccharide composition. Plant J. 12: 335–345.

Richmond, T.A. and Somerville, C.R. 2001. Integrative approaches to determining Csl function. Plant Mol. Biol., this issue.

Saxena, I.M. and Brown, R.M.J. 1995. Identification of a second cellulose synthase gene (acsAII) in Acetobacter xylinum. J. Bact. 177: 5276–5283.

Saxena, I.M., Brown, R.M.J., Fevre, M., Geremia, R.A. and Henrissat, B. 1995. Multidomain architecture of β-glycosyl transferases: implications for mechanism of action. J. Bact. 177: 1419–1424.

Scheller, H.V., Doong, R.L., Ridley, B.L. and Mohnen, D. 1999. Pectin biosynthesis: a solubilized α1,4-galacturonosyltransferase from tobacco catalyzes the transfer of galacturonic acid from UDP-galacturonic acid onto the non-reducing end of homogalacturonan. Planta 207: 512–517.

Shah Reddy, I., Kessel, D.H., Chou, T.H., Mirchandani, I. and Khilanani, U. 1982. Plasma fucosyltransferase as an indicator of imminent blastic crisis. Am. J. Hematol. 12: 29–37.

Sherwood, A.L., Nguyen, A.T., Whitaker, J.M., Macher, B.A., Stroud, M.R. and Holmes, E.H. 1998. Human α 1,3/4-fucosyltransferases. III. A Lys/Arg residue located within the α 1,3-FucT motif is required for activity but not substrate binding. J. Biol. Chem. 273: 25256–25260.

Smythe, C. and Cohen, P. 1991. The discovery of glycogenin and the priming mechanism for glycogen biogenesis. Eur. J. Biochem. 200: 625–631.

Standal, R., Iversen, T.G., Coucheron, D.H., Fjaervik, E., Blatny, J.M. and Valla, S. 1994. A new gene required for cellulose production and a gene encoding cellulolytic activity in Acetobacter xylinum are colocalized with the bcs operon. J. Bact. 176: 665–672.

Staudacher, E., Altmann, F., Wilson, I.B. and Marz, L. 1999. Fucose in N-glycans: from plant to man. Biochim. Biophys. Acta 1473: 216–236.

Strasser, R., Mucha, J., Schwihla, H., Altmann, F., Glossl, J. and Steinkellner, H. 1999. Molecular cloning and characterization of cDNA coding for β 1,2N-acetylglucosaminyltransferase I (GlcNAc-TI) from Nicotiana tabacum. Glycobiology 9: 779–785.

Strasser, R., Mucha, J., Mach, L., Altmann, F., Wilson, I.B., Glossl, J. and Steinkellner, H. 2000. Molecular cloning and functional expression of β 1, 2-xylosyltransferase cDNA from Arabidopsis thaliana. FEBS Lett. 472: 105–108.

Takahashi, T., Ikeda, Y., Tateishi, A., Yamaguchi, Y., Ishikawa, M. and Taniguchi, N. 2000. A sequence motif involved in the donor substrate binding by α 1,6-fucosyltransferase: the role of the conserved arginine residues. Glycobiology 10: 503–510.

Tamayo, P., Slonim, D., Mesirov, J., Zhu, Q., Kitareewan, S., Dmitrovsky, E. et al. 1999. Interpreting patterns of gene expression with self-organizing maps: methods and application to hematopoietic differentiation. Proc. Natl. Acad. Sci. USA 96: 2907–2912.

130

Taylor, N.G., Scheible, W.R., Cutler, S., Somerville, C.R. and Turner, S.R. 1999. The *irregular xylem3* locus of arabidopsis encodes a cellulose synthase required for secondary cell wall synthesis. Plant Cell 11: 769–779.

Thompson, J.D., Gibson, T.J., Plewniak, F., Jeanmougin, F. and Higgins, D.G. 1997. The ClustalX windows interface: flexible strategies for multiple sequence alignment aided by quality analysis tools. Nucl. Acids Res. 24: 4876–4882.

Turner, S.R. and Somerville, C.R. 1997. Collapsed xylem phenotype of *Arabidopsis* identifies mutants deficient in cellulose deposition in the secondary cell wall. Plant Cell 9: 689–701.

Turner, S., Taylor, N. and Jones, L. 2001. Mutations of the secondary wall. Plant Mol. Biol., this issue.

Vergara, C.E. and Carpita, N.C. 2001. Mixed-linkage β-glucan synthase and the *CesA* gene family in cereals. Plant Mol. Biol., this issue.

Weiser, M.M. and Wilson, J.R. 1981. Serum levels of glycosyltransferases and related glycoproteins as indicators of cancer: biological and clinical implications. Crit. Rev. Clin. Lab. Sci. 14: 189–239.

Weston, B.W., Nair, R.P., Larsen, R.D. and Lowe, J.B. 1992. Isolation of a novel human $\alpha(1,3)$fucosyltransferase gene and molecular comparison to the human Lewis blood group α (1,3/1,4)fucosyltransferase gene. Syntenic, homologous, nonallelic genes encoding enzymes with distinct acceptor substrate specificities. J. Biol. Chem. 267: 4152–4160.

White, A.R., Xin, Y. and Pezeshk, V. 1993. Xyloglucan glucosyltransferase in Golgi membranes from *Pisum sativum* (pea). Biochem. J. 294: 231–238.

Whitfield, C. 1995. Biosynthesis of lipopolysaccharide O antigens. Trends Microbiol. 3: 178–185.

Wiggins, C.A. and Munro, S.L. 1998. Activity of the yeast MNN1 α1,3-mannosyltransferase requires a motif conserved in many other families of glycosyltransferases. Proc. Natl. Acad. Sci. USA 95: 7945–7950.

Wulff, C., Norambuena, L. and Orellana, A. 2000. GDP-fucose uptake into the Golgi apparatus during xyloglucan biosynthesis requires the activity of a transporter-like protein other than the UDP-glucose transporter. Plant Physiol. 122: 867–877.

Xu, Z., Vo, L. and Macher, B.A. 1996. Structure-function analysis of human α 1,3-fucosyltransferase. Amino acids involved in acceptor substrate specificity. J. Biol. Chem. 271: 8818-8823.

Zhang, Z., Schaffer, A.A., Miller, W., Madden, T.L., Lipman, D.J., Koonin, E.V. and Altschul, S.F. 1998. Protein sequence similarity searches using patterns as seeds. Nucl. Acids Res. 26: 3986–3990.

Plant Molecular Biology **47**: 131–143, 2001.
© 2001 *Kluwer Academic Publishers. Printed in the Netherlands.*

Integrative approaches to determining Csl function

Todd A. Richmond[1],* and Chris R. Somerville[1,2]
[1]*Carnegie Institution of Washington, Department of Plant Biology, 260 Panama Street, Stanford, CA 94305, USA*
(*author for correspondence; e-mail todd@andrew2.stanford.edu); [2]*Department of Biological Sciences, Stanford University, Stanford, CA 94305, USA*

Key words: Arabidopsis thaliana, cellulose synthase, DNA microarray, glycosyltransferase, *Medicago truncatula*, polysaccharide biosynthesis

Abstract

While there is an ever-increasing amount of information regarding cellulose synthase catalytic subunits (CesA) and their role in the formation of the cell wall, the remainder of the enzymes that synthesize structural cell wall polysaccharides are unknown. The completion of the *Arabidopsis* genome and the wealth of the sequence information from other plant genome projects provide a rich resource for determining the identity of these enzymes. *Arabidopsis* contains six families of genes related to cellulose synthase, the cellulose synthase-like (*Csl*) genes. Our laboratory is taking a multidisciplinary approach to determine the function of the *Csl* genes, incorporating genomic, genetic and biochemical data. Information from expressed sequence tag (EST) projects has revealed the presence of *Csl* genes in all plant species with a significant number of ESTs. Certain *Csl* families appear to be missing from some species. For example, no examples of *Csl*G ESTs have been found in rice or maize. Microarray data and reporter constructs are being used to determine the expression pattern of the *CesA* and *Csl* genes in *Arabidopsis*. Mutations and insertion events have been identified in a majority of the genes in the *Arabidopsis CesA* superfamily and are being characterized by phenotypic and biochemical analysis. While we cannot yet link the function of any of the *Csl* genes to their respective products, the expression and localization of these genes is consistent with the expected expression pattern of polysaccharide synthases that contribute to the primary cell wall.

Abbreviations: CesA, cellulose synthase; Csl, cellulose synthase-like; EST, expressed sequence tag; GAX, glucuronoarabinoxylan; GUS, β-glucuronidase; HG, homogalacturonan; RG-I, rhamnogalacturonan I; RG-II, rhamnogalacturonan II; TM, transmembrane domain; XG, xyloglucan; X-gluc, 5-bromo-4-chloro-3-indoyl glucuronide

Introduction

Relatively little is known about the properties of the enzymes that catalyze the synthesis of the polysaccharides which make up the cell wall, or the regulation of the genes which encode them. These enzymes are thought to be associated with the Golgi or, in the case of cellulose synthase and callose synthase, the plasma membrane (Zhang and Staehelin, 1992; Delmer, 1999). Although it is possible to assay many of the enzymes involved in the synthesis of polymers such as xylan, xyloglucan, callose, arabinan and poly-galacturonic acid in tissue extracts (Robertson *et al.*,

1995), it has not yet been possible to purify enough of the proteins to permit the cloning of the corresponding genes. The cloning of cellulose synthase (*CesA*) genes through genomic and genetic means (Pear *et al.*, 1996; Arioli *et al.*, 1998; Taylor *et al.*, 1999) provided a major breakthrough in plant polysaccharide biosynthesis research.

Comparison of the CesA polypeptide sequences to predicted proteins encoded by the Arabidopsis genome resulted in the identification of six families of about 30 structurally similar genes which encode polypeptides that are sufficiently diverged from the CesA polypeptides that they probably do not participate in cellulose synthesis. These genes have been provisionally termed the cellulose synthase-like (*Csl*) genes (Richmond and Somerville, 2000b). All of the *Csl* genes appear to be integral membrane proteins, with 3 to 6 transmembrane domains in the carboxy-terminal region of the protein and 1 or 2 transmembrane domains in the amino-terminal region. The relatively weak sequence similarity to cellulose synthase raises the possibility that the Csl polypeptides catalyze other processive glycosyltransferase reactions. However, no biological function has been attributed to any of the *Csl* genes as yet. We describe here the results of experiments designed to examine the function of these genes by the use of various genetic, genomic and biochemical criteria.

Genome sequencing

The sequence of the *Arabidopsis* genome is essentially complete and provides a unique and novel opportunity to identify the complete set of the genes necessary for the biosynthesis of a plant cell wall. In addition to the other advantages of using *Arabidopsis* as a model organism, *Arabidopsis* has a cell wall typical of most higher plants (Zablackis *et al.*, 1995), and can be used as a model for most dicots. However, because of substantial differences between the cell walls of dicots and commelinoid monocots (Carpita, 1996), there are a number of specific cell wall-related genes in certain monocots that may not be represented in the *Arabidopsis* genome. Even among monocots, there are substantial differences between the Poales (the family Poaceae, the grasses) and other monocots. In particular, the grasses have relatively small amounts of xyloglucan and instead have an abundance of 'mixed-linkage' glucans, β-D-glucans that contain both $(1{\rightarrow}3)$ and $(1{\rightarrow}4)$ linkages. Many of the world's most important crops belong to the family Poaceae, which includes all of the cereal grains. For these crops, rice will serve as an excellent model. In January 2001, the Torrey Mesa Research Institute (part of the Genomics Unit of Syngenta; http://www.syngenta.com) and Myriad Genetics announced the private completion of the rice genome sequence, which they will make available to academic collaborators. The rice genome is being sequenced for public release by an international consortium (Rice Genome Research Program; http://rgp.dna.affrc.go.jp/index.html), with contributions from Monsanto (http://www.rice-research.org/).

Knowledge of the differences, as well as the similarities, between the complement of genes in *Arabidopsis* and rice will be valuable information for cell wall researchers.

Identification of the cellulose synthase superfamily

We have used computational methods to identify a large family of genes of unknown function that show sequence similarity to cellulose synthase (Richmond and Somerville, 2000b). Reiterative database searches using the *Arabidopsis* Rsw1 (AtCesA1) and the cotton CesA polypeptide sequences as the initial query sequences revealed a large superfamily of at least 40 *CesA* and *CesA*-like genes in *Arabidopsis*. Based on predicted protein sequence, we have grouped these genes into 7 clearly distinguishable families: the *CesA* family, which includes *RSW1* and *IRX3* (*AtCesA7*), and six families of structurally related genes of unknown function designated as the 'cellulose synthase-like' genes (*CslA*, *CslB*, *CslC*, *CslD*, *CslE*, and *CslG*) (Cutler and Somerville, 1997; Richmond and Somerville, 2000b).

While these proteins vary in their degree of sequence similarity to one another, they share several features that have been proposed to be indicative of processive glycosyltransferases (Saxena *et al.*, 1995; Richmond and Somerville, 2000b). The various members of the *Arabidopsis CesA* superfamily appear to belong to family 2 of the inverting nucleotide-diphospho-sugar glycosyltransferases (Campbell *et al.*, 1997) that synthesize repeating β-glycosyl unit structures. They all contain the conserved D,D,D,QxxRW motif thought to define the nucleotide sugar binding domain and the catalytic site of these enzymes. To date, this family of glycosyltransferases has over 670 members, including sequences from viruses, bacteria, fungi, plants and animals (Campbell *et al.*, 1997; Henrissat *et al.*, 2001, in this issue) (see also http://afmb.cnrs-mrs.fr/~pedro/CAZY/gtf_2.html). All of the members of the cellulose synthase superfamily appear to be integral membrane proteins, with three to six transmembrane domains (TMs) in the carboxy-terminal region of the protein and one or two TMs in the amino-terminal region. Intron-exon organization is conserved when comparing the *CesA*, *CslB*, *CslG* and *CslE* gene families to one another but not the *CslA*, *CslC* or *CslD* families (Richmond and Somerville, 2000b).

CesA gene family

Ten full-length *CesA* genes have been sequenced from *Arabidopsis* and an EST represents one additional gene. The genes in the *Arabidopsis* family have 10 to 13 exons and encode proteins ranging in length from 985 to 1088 amino acids. There are 25 full-length CesA proteins in the public databases from 5 different species, though there are only two additional complete genomic sequences, the rice *CesA4* and *CesA7* genes. The proteins in the CesA family range in amino acid identity to each other from 53% to 98%.

Csl gene families

The *CslD* gene family is the most similar of the *Csl* gene families to the *CesA* family (ca. 35% identical at the amino acid level). There are six full-length gene sequences for *Arabidopsis*. Two full-length genes have also been sequenced from rice. The proteins encoded by these genes are as large, or larger than the CesAs (1036–1181 amino acids). The gene structure for this family is unusual in that the eight genes for which complete genomic sequence information is available have five different patterns of intron-exon organization. Based on recent thinking about the evolution of intron/exon structure (de Souza *et al.*, 1998), this would seem to suggest that this gene family is the oldest in the cellulose synthase superfamily and may predate the *CesA* family.

The *Arabidopsis CslA* family contains nine full-length genes, with 7 to 10 exons, encoding the smallest Csl proteins (485–553 amino acids). The closely related *CslC* family contains five full-length genes, each with 4 to 5 exons coding for proteins almost the same size (672–694 amino acids) as the *CslB*, *CslG*, and *CslE* families. There are three full-length or near-full-length rice *CslA* genomic sequences. Two of these proteins (OsCslA2 and OsCslA4) are more similar to one another than to *Arabidopsis* CslA proteins while the third (OsCslA1) is more similar to *Arabidopsis* CslA2. The CslA and CslC proteins are the least similar to the cellulose synthase proteins, showing only ca. 7% sequence identity across the entire length of the proteins.

The *Arabidopsis CslB* gene family is composed of six genes, organized in three clusters of two genes. Four of these genes are located in a single region on chromosome 2; two clusters of two tandem genes separated by ca. 20 kb. The third cluster, on chromosome 4, consists of two genes separated by two cytochrome P450 genes. The predicted proteins range in size from 755 to 759 amino acids. Each member of the *CslB* gene family contains nine predicted exons, with the exception of *AtCslB6*, which is missing the first intron. The individual members of this family share only 23% identity to the CesAs when compared along the entire length of the protein.

The *CslG* family is found as a single cluster of three genes on chromosome 4, suggesting that the three genes have recently evolved by gene duplications. These three genes contain five predicted exons, encoding proteins that range in size from 722 to 760 amino acid residues. The predicted proteins are ca. 22% identical to the CesAs over the entire length of the protein.

The final *Csl* family is represented by a single gene. The *CslE* gene is clearly distinct from the other families described. It is most similar to the *CslG* family to which it is ca. 35% identical at the amino acid level.

EST sequencing

Of the 40 full-length genes in the *Arabidopsis CesA* superfamily, twelve have no ESTs suggesting that they are weakly expressed or are only expressed in a small number of cells or under a limited number of conditions. The percentage of genes without ESTs is similar to that previously reported for the *Arabidopsis* genome as a whole (Lin *et al.*, 1999; Mayer *et al.*, 1999), and this suggests that complete genome sequences of other species will be necessary for a thorough comparison of the *Csl* gene content of *Arabidopsis* to that of other plant species. Presently, the only public data from other plant species that can be compared to the *Arabidopsis* sequences are the seven complete *CesA/Csl* genes from rice. All six are very similar to *Arabidopsis* genes, with high sequence identity (55–84% at the mRNA level, 60–78% at the protein level) and nearly identical intron/exon structure. Given the differences in cell wall composition between *Arabidopsis* and rice, and assuming that the *Csl* genes participate in cell wall synthesis, it is reasonable to expect to find genes in rice that have no equivalent orthologues in *Arabidopsis*, such as those responsible for making mixed-linkage glucans. While there is, as yet, no evidence for additional or missing *Csl* families in rice, only 15% of the rice genomic sequence is currently publicly available as of March, 2001.

While complete genome sequencing is preferable, the genomes of most plant species will not be sequenced in the near future, either for technical reasons or for lack of economic incentive. For example, maize has a genome the size of the human genome (3 billion base pairs) and pine is nearly 7 times larger. Both have a very high percentage of repetitive DNA that is likely to make genome sequencing difficult and costly. New techniques are being developed that utilize the unmethylated state of genes in plant species like maize to selectively clone and sequence genic regions (Rabinowicz *et al.*, 1999). For now, however, expressed sequence tags probably provide the only currently feasible method for large-scale gene identification. EST projects have been initiated for a number of plant species. The Institute for Genomic Research will sequence 150 000 tomato clones (TIGR Tomato Gene Index; http://www.tigr.org/tdb/lgi/index.html) and 300 000 soybean clones will be sequenced as part of the Soybean EST project (Soybean EST Project; http://genome.wustl.edu/est/soybean_esthmpg.html). Extensive EST sequencing projects have also been initiated for cotton, the legume *Medicago truncatula*, and the cereals sorghum, rice, maize and wheat. In the 12-week period between June 2000 and August 2000, these projects produced more plant EST data than was contained in the public databases between 1992 and 1999. There are now over a million plant EST sequences in GenBank, with thousands more being added each week. Besides their usefulness in identifying genes, information regarding tissue specificity and relative expression levels can be obtained. To aid cell wall researchers in utilizing this data, we have established a web site (http://cellwall.stanford.edu) to collect, organize, and summarize *CesA* and *Csl* sequence data for all plant species.

EST expression

Looking at the expression of the *CesA* and *Csl* families as a whole, partial or complete *CesA* sequences constitute the bulk of the sequences in the public databases; 54% of the cellulose synthase superfamily ESTs belong to this family (Table 1). Based on EST information and our own work (S. Cutler, N. Sprenger and C. Somerville, unpublished data), the *CslA* and *CslC* genes are abundantly expressed, both in *Arabidopsis* and in other plant species. They represent 24% of the *CesA* superfamily sequences (15% *CslA*, 9% *CslC*), and most plant species with more than a few thousand ESTs have examples of *CslA* and *CslC* ESTs in the public sequence databases. There are only six *Arabidopsis CslG* EST sequences in the

Table 1. CesA and Csl EST expression.

Family	Percentage of ESTs
CesA	54%
CslA	15%
CslB	2%
CslC	9%
CslD	7%
CslE	4%
CslG	9%

Percentages were calculated from an analysis of over one million plant ESTs available in Genbank (March 2001).

public databases, consistent with the low level of expression that we have observed in *Arabidopsis* by northern blotting (T. Richmond and C. Somerville, unpublished data). They appear to be more abundant in other plant species, representing 9% of all public plant ESTs in the superfamily. About 7% of the cellulose synthase superfamily ESTs in GenBank are *CslD*s. While there is only a single *CslE* gene in *Arabidopsis*, EST data from *Medicago truncatula* and soybean, both members of the legume family, indicates that these species have multiple genes related to the *CslE* family. The *CslB* genes are poorly represented in the EST databases (2%) from all species and are weakly expressed in *Arabidopsis* (T. Richmond and C. Somerville, unpublished data).

In those cases where EST sequences are from tissue-specific cDNA libraries, it is possible to extract some information from EST records about the localization and mRNA abundance of given genes (Ewing and Claverie, 2000; Ewing *et al.*, 1999). The EST information from some plant species is better than others, based on the number of different tissue-specific libraries that have been sampled, and whether or not those libraries have been normalized in some way. In this regard, *Arabidopsis* EST information is less useful than that from other plant species because many of the initial ESTs produced for *Arabidopsis* were selected from mixed tissue libraries. Other EST projects are more informative. Table 2 contains a summary of EST

Table 2. EST distribution for *Medicago truncatula*.

Gene	ESTs	Leaf	Stem	Root	Nodule	Mycorrhiza
*MtCesA*1	37	6	14	12	2	
*MtCesA*2	6	1	3	2		
*MtCesA*3	5	2	3			
*MtCesA*4	3		2			
*MtCesA*5	2		1	1		
*MtCesA*6	3		3			
*MtCesA*7	6	1	2	3		
*MtCesA*8	10		1	5		2
*MtCesA*9	2					
*MtCesA*10	5	1		4		
*MtCesA*11	3			3		
*MtCesA*12	2			2		
*MtCslA*1	70	8		48	3	3
*MtCslB*1	3			3		
*MtCslB*2	2			2		
*MtCslC*1	1			1		
*MtCslC*2	2			2		
*MtCslC*3	2	1				
*MtCslD*1	3			3		
*MtCslE*1	5	1	4			
*MtCslE*2	4	1		1	1	
*MtCslE*3	6	1		1		
*MtCslG*1	29	5	3	19	2	
*MtCslG*2	5			4	1	
*MtCslG*3	3	1	1			
*MtCslG*4	2			1	1	
*MtCslG*5	3	1		1		
*MtCslG*6	3			1		

Expressed sequence tags were organized into sequence groups based on sequence identity (>95%) of overlapping sequences. Sequences groups are not necessarily unique within each *Csl* family, due to the lack of full-length sequences. Tissue localization was extracted from the GenBank (March 2001) annotation for each EST record.

localization and relative abundance from *Medicago truncatula,* for which more than 100 000 ESTs are available from a number of tissue-specific libraries. A notable feature of the data is the very high abundance of ESTs for *MtCesA*1, *MtCslA*1, and *MtCslG*1. Based on this data, we consider it likely that the products of the CslA1 and CslG1 proteins should be relatively abundant in *Medicago* roots and have initiated an examination of the cell wall composition of *Medicago* roots to explore this possibility. More generally, we have noted that there are no *CslG*-like ESTs in any monocot species. This may indicate that the *CslG* gene products produce a polysaccharide found only in dicots.

DNA microarrays

Another useful tool for examining *Csl* gene expression is microarray technology. Two types of microarrays currently exist: DNA-fragment-based and oligonucleotide (oligo)-based microarrays. These differ primarily in the nature of the DNA fixed to the solid support. Both types of arrays are commercially available for *Arabidopsis*, but there are no commercial microarrays available for other plant species at this time. However, there are a number of publicly funded DNA microarray projects which will be producing data for a number of different plant species in the near future (Richmond and Somerville, 2000a). DNA-fragment-based arrays, in which PCR-amplified inserts of partially sequenced cDNA clones are spotted onto glass microscope slides, are the most commonly used. After a large collection of ESTs are sequenced and analyzed, a minimal set is chosen for microarray analysis. While anonymous EST clones can be arrayed, the full power of the technology is not realized unless the identity of all of the elements of the microarray is known (Richmond and Somerville, 2000a).

Presently, the only public microarray data for plants has been provided by the Arabidopsis Functional Genomics Consortium (AFGC) and Stanford University through the Stanford Microarray Database (http://genome-www4.stanford.edu/microarray/smd/). Data on the expression of 12 000 genes, including a number of *Csl* genes, under a variety of experimental conditions is available at SMD. The AFGC project uses DNA-fragment-based arrays, which has one major disadvantage for looking at gene expression. Arrays of this type do not yield quantitative measures of mRNA abundance like oligo-based arrays. DNA-fragment microarrays are best suited to side-by-side comparisons of the two samples hybridized to the same slide. However, by comparing two different tissue types, or by comparing single tissues to a master reference, relative amounts of mRNA abundance can be determined for each gene of interest. Table 3 shows microarray data for a subset of *Arabidopsis CesA* and *Csl* genes. Of the 18 genes shown, 7 exhibit at least a 3-fold difference in expression between tissues. While there is no guarantee that differences in mRNA abundance will equate to differences in cell wall composition, it provides a starting point for conducting detailed phenotypic and chemical analysis. One conclusion from the data is that the genes within a family are not coordinately regulated. For instance, the *CslA*3

Table 3. Differential tissue expression of *CesA* and *Csl* genes.

Gene	F vs. L	L vs. CC	T vs. F	T vs. L	T vs. R	L vs. ES
AtCesA1	−1.2	−1.1	1.3	−1.4	1.2	nc
AtCesA2	–	−1.3	–	–	–	**4.3**
AtCesA3	−1.2	−1.7	1.2	−1.3	1.1	1.7
AtCesA4	−1.5	1.5	−1.2	−1.2	−1.2	**−9.0**
AtCesA5	−1.6	**−2.3**	1.6	−1.8	−1.1	**2.6**
AtCesA6	−1.4	−1.2	1.7	−1.5	1.5	−1.9
AtCesA7	**−2.8**	−1.3	−1.6	−1.7	−1.3	**−2.8**
AtCslA1	**−2.7**	−1.1	**4.9**	−1.4	**6.0**	2.1
AtCslA3	**2.1**	1.5	−1.4	nc	1.5	**−453**
AtCslA7	−1.5	−1.4	1.5	−1.7	1.1	**−28.3**
AtCslC4	nc	−1.1	nc	−1.3	1.7	–
AtCslC5	1.2	1.1	**2.2**	−1.8	1.4	–
AtCslC6	–	1.5	1.8	1.2	1.5	–
AtCslC8	–	–	**2.7**	−1.3	1.9	–
AtCslD2	–	−1.1	1.7	−1.3	**4.0**	–
AtCslD3	−1.2	1.3	1.4	−1.4	**3.0**	–
AtCslG1	1.5	−1.3	−1.6	nc	**2.6**	–
AtCslG3	−1.9	−1.7	nc	1.2	−1.9	–

Positive number: higher expression in first tissue; negative number: higher expression in second tissue. Bold-type face indicates those with >2-fold difference. F, flower; L, leaf; CC, cell culture; T, RNA from all plant tissues; R, root; ES, etiolated seedling. –, no data available; nc, no change.

and *CslA7* genes show strong differential expression between leaf tissues and etiolated seedlings whereas the *CslA1* gene does not.

In addition to tissue specificity, microarray experiments can also suggest treatments and conditions that may be used to reveal or enhance phenotypic changes when examining mutants, or transgenics. In principle, it might be informative to identify conditions that require a large induction of *Csl* family members and then subject loss-of-function mutants to those conditions to reveal or enhance changes in cell wall biosynthesis or structure. Treatments like salt stress and light treatment of etiolated seedlings repress a number of the *CesA* and *Csl* genes, as might be expected of treatments that halt cell expansion and growth (Table 4). Ethylene treatment of mature rosette leaves was the only treatment that caused induction in this limited set of experiments.

Expression data from promoter constructs

One of the possible explanations for the fact that multiple genes represent five of the gene families is that the various genes are expressed in different cell types or are subject to different forms of regulation.

In order to assess this, we have examined where and when the genes from several families are expressed. Reporter-promoter gene constructs have been introduced into transgenic plants for all members of both the *Arabidopsis CslB* and *CslG* families. Examination of β-glucuronidase (GUS) reporter gene expression by histochemical staining with 5-bromo-4-chloro-3-indoyl glucuronide (X-gluc) shows a variety of expression patterns. *AtCslG1* and *AtCslG2* are expressed in an identical pattern (Figure 1). There is strong expression of both constructs in the first true leaves and the vascular tissue of cotyledons and roots in young seedlings, and throughout all mature tissues. Expression is seen primarily in the vascular tissue along the entire length of the root except for the root cap. Since xylem development lags behind the root cap, this is consistent with xylem expression. No data is available yet for *AtCslG3*.

For the *CslB* family, the pattern of expression is varied (Figure 1). Promoter expression data for *AtCslB1*, *AtCslB2*, *AtCslB3*, and *AtCslB6* show expression throughout the entire seedling. *AtCslB2* and *AtCslB6* show stronger expression in the vascular tissue of the root where the secondary roots are attached as well as at the base of the hypocotyl. *AtCslB2* and *AtCslB6*

Table 4. Differential expression of CesA and Csl genes under various treatments.

Gene	Salt stress	Cytokinin	Ethylene (L)	Ethylene (S)	Light
AtCesA1	−3.0	−	+5.0	−	−3.5
AtCesA3	−	−	+5.0	−	−
AtCesA4	−	−	−	−	−4.5
AtCesA6	−	−	−	−	−3.75
AtCslA2	−	−	−	−	−4.5
AtCslA9	−	−	−45	−	−
AtCslC4	−8.0	−	+5.0	−	−4.5
AtCslB1	−	−	−165	−	−4.0
AtCslB2	−	−	−220	−	−
AtCslB3	−	−	−	−5.0	−
AtCslB6	−	−	−220	−	−3.5
AtCslD2	−4.0	−	−	−	−3.75
AtCslD3	−4.0	−75	−	−	−

Positive number indicates induction by treatment, negative number indicates repression by treatment. −, change less than 3-fold. Salt stress: green seedlings grown on filter paper on plates, transferred to liquid MS/sucrose media, with or without 200 mM NaCl. Seedlings were 7 days old at harvest. Ethylene: plants were treated with 50 ppm ethylene in sealed chambers. For seedlings (S), plants were grown in the dark and were 4 days old at harvest. For green tissue (L), leaves were harvested from 4-week old light-grown plants. Cytokinin: green plants grown in liquid culture (MS), treated with 50 μM benzyladenine. Leaves were 3 weeks old at harvest. Light: dark-grown seedlings exposed to 120 μE m^{-2} s^{-1} fluorescent light for 24 h. Seedlings were 4 days old at harvest.

are also expressed throughout all tissues of the plant. *AtCslB5* shows expression only in the root tip, behind the root cap.

Based on these results it appears that at least four members of the *CslB* family are co-expressed in the same cells and may, therefore, be functionally redundant. Similarly, *CslG1* and *CslG2* are co-expressed and may be functionally redundant. It is also clear, however, that even genes that appear to be recently duplicated, like *AtCslB5* and *AtCslB6*, are not necessarily expressed in the same tissues, and may play different functional roles.

Classic and reverse genetic approaches

Classical genetic screens are a powerful technique for unraveling complex processes and have been widely and effectively used in *Arabidopsis*. However, in the area of cell wall biosynthesis, relatively few mutants have been isolated (see Fagard *et al.*, 2000, for review). Direct measurements of the sugar composition of cell walls of plants from a heavily mutagenized population yielded about a dozen classes of mutants (designated *mur*) with relatively slight changes in overall wall composition (Reiter *et al.*, 1993, 1997). The cloning of the genes corresponding to several of the *mur* loci revealed that the mutations affect enzymes involved in nucleotide sugar biosynthesis (Bonin *et al.*, 1997; Burget and Reiter, 1999; see Reiter and Vanzin, 2001, this issue) and one possibly with a defect in xyloglucan side-chain biosynthesis (Perrin *et al.*, 1999). None of the four *mur* mutations characterized to date appears to affect the biosynthesis of polysaccharide backbones.

The failure to identify, as yet, the enzymes responsible for non-cellulosic polysaccharide backbone synthesis may indicate that a defect in these enzymes either results in a lethal phenotype or that multiple genes supply redundant function. The latter possibility is generally consistent with our hypothesis that the *Csl* genes may encode some of the processive glycosyltransferases that catalyze synthesis of the major polysaccharides. The *CslB* and *CslG* gene families consist of sets of tandem genes, with individual members having identical expression patterns (Figure 1). Insertional inactivation of single genes from the *CslC (CslC4)*, *CslG (CslG1)*, or *CslB (CslB5)* families results in no visible phenotype or detectable change in cell wall sugar composition (N. Sprenger, T. Richmond and C. Somerville, unpublished data).

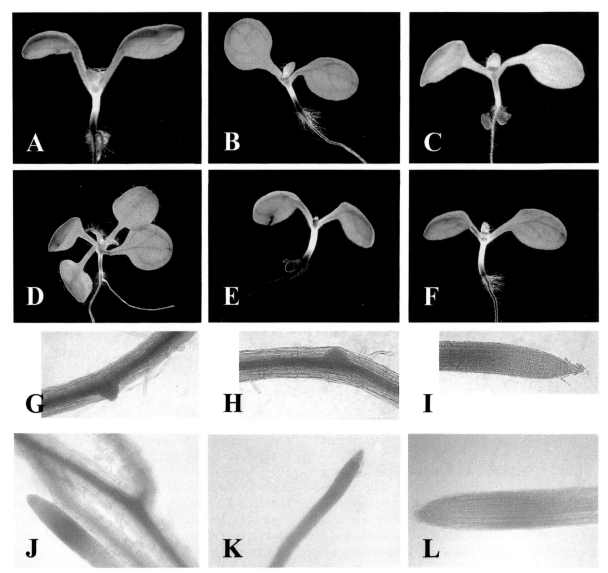

Figure 1. *Csl*::GUS expression data. The promoter regions of the *CslB* and *CslG* genes were placed in front of a GUS/GFP fusion protein in the plant transformation vector pCAMBIA3301. Tissue was harvested from plants at various times in development and expression determined by histochemical staining with 5-bromo-4-chloro-3-indoyl glucuronide (X-gluc). GUS expression is visible as blue staining. A. *CslG*1 seedling. B. *CslG*2 seedling. C. *CslB*1 seedling. D. *CslB*2 seedling. E. *CslB*3 seedling. F. *CslB*6 seedling. G. *CslG*1 seedling root. H. *CslG*2 seedling root. I. *CslG*2 seedling root tip. J. *CslB*2 seedling roots. K. *CslB*3 seedling root tip. L. *CslB*5 seedling root tip.

There are only two mutations in *Arabidopsis Csl* genes that have an identifiable phenotype. The *kojak* (*kjk*) mutant, which is a null mutation in the *Arabidopsis CslD*3 gene, results in seedlings that have no root hairs (Favery *et al.*, 2001). It appears that root hair initials have weakened cell walls that burst, resulting in lack of root hair formation. Since *AtCslD*3 is expressed throughout the plant, the lack of a phenotype in other tissues of the plant suggests the presence of redundant function in tissues other than roots. The

*rat*4 mutation, a dominant T-DNA insertion mutation in the *Arabidopsis CslA*9 gene, confers resistance to *Agrobacterium* root transformation. It is likely that the mutation causes some subtle change in cell wall structure, although *Agrobacterium* binds normally to the root (Nam *et al.*, 1999; Gelvin, 2000). The unpredictable phenotype of the loss of function of *AtCslD*3 and *AtCslA*9 underscores the challenge of designing a phenotypic screen to find mutations in *Csl* genes. Overall plant morphology and growth habit may be affected; stunted plants may arise from the disruption of genes responsible for biosynthesis of the primary cell wall, while larger, weaker plants may indicate defects in secondary cell wall biosynthesis (Turner and Somerville, 1997). Since the cell wall may play a role in plant morphogenesis (Fowler and Quatrano, 1997), tissue organization or cell size may be affected in mutants with changes in the primary cell wall. However, it is impossible to predict the exact phenotype. Multiple mutations can affect cell growth and expansion, and the overall growth habit of the plant. It would be inefficient to search for mutations in the *Csl* genes based on this criterion. Thus, for the *Csl* families a reverse genetic approach will be required in order to test the function of the *Csl* genes by genetic criteria.

Arabidopsis has a number of tools for reverse genetics: large T-DNA and transposon collections (McKinney *et al.*, 1995; Krysan *et al.*, 1996), anti-sense/RNAi methods (Chuang and Meyerowitz, 2000), and virus-induced gene silencing (Dalmay *et al.*, 2000). Presently, we are pursuing a dual reverse genetics approach: identifying loss-of-function mutations in each member of the *CesA* superfamily, and attempting to down-regulate expression of entire *Csl* families via antisense methods. Figure 2 details our progress, and that of other labs, in identifying mutations in the *CesA* superfamily. As mentioned previously, mutations in at least three of the *Csl* genes, from three different families, have revealed no visible phenotypic changes. It seems clear from these initial results that either loss-of-function in several genes in the same family will be necessary to see any phenotypic effects or that we are measuring the wrong cellular constituents.

In order to accomplish down-regulation of entire families, we have created 35S-driven anti-sense lines for several of the families. At present, this approach has not yielded consistent results.

Discussion

The function of the various *Csl* families is not known, but we speculate that they are responsible for producing some of the non-cellulosic polysaccharides found in plant cell walls or in secretions such as root cap or stylar mucilage (Cutler and Somerville, 1997). *Arabidopsis* has a type I wall typical of most higher plants (Zablackis *et al.*, 1995) making it a good genetic model for understanding cell wall synthesis, structure and function. The primary cell walls of all

higher plants that have been studied contain cellulose, the two cross-linking glycans, xyloglucan (XG) and glucuronoarabinoxylan (GAX), the three pectic polysaccharides homogalacturonan, rhamnogalacturonan I (RG-I), and rhamnogalacturonan II (RG-II), and structural glycoproteins. Cellulose, a linear polymer of $(1\rightarrow4)\beta$-linked glucopyranosyl residues, is the major load-bearing polymer in the cell wall, constituting 14% of the total cell wall mass in *Arabidopsis* leaves (Zablackis *et al.*, 1995). Xyloglucan and GAX are the other two polysaccharides that form the basic framework for the primary cell wall. XG can bind tightly to cellulose, and is thought to contribute to the cross-linking of the cellulose microfibril networks(Hayashi, 1989). XG accounts for 20% of the cell wall in *Arabidopsis* leaves while GAX accounts for an additional 4% (Zablackis *et al.*, 1995). XGs are composed of a branched $(1\rightarrow4)\beta$-D-glucan backbone with ca. 75% of the backbone glucosyl residues substituted at the O-6 position with short side-chains containing xylose, galactose and, often, a terminal fucose (McNeil *et al.*, 1984). GAX is an acid polysaccharide consisting of a backbone of $(1\rightarrow4)\beta$-linked D-xylosyl residues. About 25% of the xylosyl residues are substituted at the O-2 position with arabinosyl, glucosyluronic acid, or 4-*O*-methyl glucosyluronic acid residues (Darvill *et al.*, 1980). XG and GAX, along with cellulose, are embedded in matrix of pectic polysaccharides.

There are three pectic polysaccharides found in *Arabidopsis* leaves: homogalacturonan (23%), RG-I (11%) and RG-II (8%) (Zablackis *et al.*, 1995). Homogalacturonan is a linear polymer $(1\rightarrow4)\alpha$-D-galacturonic acid, occasionally interrupted by rhamnosyl residues. The backbone of RG-I consists of repeating $(1\rightarrow2)\alpha$-L-rhamnosyl-$(1\rightarrow4)\alpha$-D-galactosyluronic acid disaccharide units, with long side chains of arabinans and arabinogalactans (Lau *et al.*, 1985). RG-II is a highly substituted $(1\rightarrow4)\alpha$-linked homogalacturonan containing diverse sugars in varying linkages (Darvill *et al.*, 1978). One final polysaccharide, callose, while not found in normal plant cell walls, is involved in cell plate formation and is produced in response to wounding or disease.

It is not known whether there are other minor components of *Arabidopsis* cell walls. Only two detailed analyses have been conducted on *Arabidopsis*, one on leaves (Zablackis *et al.*, 1995) and another on seedling root tissue (Peng *et al.*, 2000). Higher plants have over 40 different cell types, each of which can probably be identified by its unique cell wall composition (Carpita and Vergara, 1998). Even a simple analysis of neutral

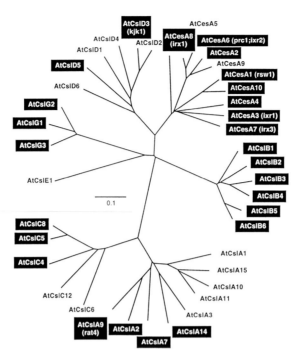

Figure 2. Mutations in the CesA superfamily. Unrooted, bootstrapped tree showing the *Arabidopsis* CesA superfamily. ClustalX (ver. 1.8) was used to create an alignment of the full-length protein sequences which was then bootstrapped (*n* = 5000 trials) to create the final tree. The CesA superfamily can be subdivided into six major families: the cellulose synthases (CesAs) and the cellulose synthase-like gene families (Csls). Names boxed in black indicate genes where T-DNA insertions, transposons, or point mutations have been isolated. Previously known named mutations are shown in parenthesis.

sugars shows that different tissues exhibit major differences in wall composition (Figure 3). Looking at other plant species, other possible polysaccharides might include mixed linkage xylans, arabinans, arabinogalactans, glucomannans, galactoglucomannans, and glucuronomannans (Aspinall, 1980). These polysaccharides might be found in highly specialized cell types, be produced under particular conditions, or be found in very small quantities.

With six families of *Csl* genes and six major non-cellulosic polysaccharides in *Arabidopsis* (i.e., callose, XG, GAX, HG, RG-I, RG-II), it is tempting to propose that each family is responsible for the biosynthesis of one of the principal polysaccharides of the cell wall. However, this is an overly simplistic proposition. First, it may be possible to remove callose from the list of non-cellulosic polysaccharides produced by the *Csl* families. Callose is a linear polymer of $(1{\rightarrow}3)\beta$-linked D-glucopyranosyl residues. Based on similarity to the yeast FKS protein, which synthesizes $(1{\rightarrow}3)\beta$-glucans, a family of putative glucan synthases has been identified in plants. There are twelve of these glucan synthase-like genes in *Arabidopsis* (Cui *et al.*, 2001; Doblin *et al.*, 2001; Hong *et al.*,

2001) (see also http://cellwall.stanford.edu), and callose synthase activity has been demonstrated for one of these genes (Hong *et al.*, 2001).

Second, the six major polysaccharides in *Arabidopsis* have only four different polysaccharide backbones. Cellulose and xyloglucan share a common $(1{\rightarrow}4)\beta$-D-glucan backbone, and homogalacturonan and rhamnogalacturonan II share a $(1{\rightarrow}4)\alpha$-D-galacturonic acid backbone. While xyloglucan and cellulose backbones are likely produced by different enzymes because they are produced at different subcellular locations, there is no reason to expect that the HG and RG-II backbones are produced by different enzymes.

Third, the family 2 glycosyltransferases, to which cellulose synthase belongs, synthesize repeating β-glycosyl unit structures. If the various *Csl* genes do belong in this family and strictly make β-linkages, then the pectins, HG, RG-I and RG-II, as well as arabinans and glucuronomannans must be eliminated from the list of possible products. It is possible that linkage specificity is determined by subtle features in the active site of protein (Stasinopoulos *et al.*, 1999) and the *Arabidopsis CesA* superfamily makes polysac-

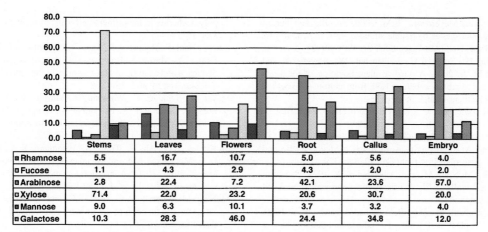

	Stems	Leaves	Flowers	Root	Callus	Embryo
▣ Rhamnose	5.5	16.7	10.7	5.0	5.6	4.0
▢ Fucose	1.1	4.3	2.9	4.3	2.0	2.0
▣ Arabinose	2.8	22.4	7.2	42.1	23.6	57.0
▢ Xylose	71.4	22.0	23.2	20.6	30.7	20.0
▣ Mannose	9.0	6.3	10.1	3.7	3.2	4.0
▢ Galactose	10.3	28.3	46.0	24.4	34.8	12.0

Figure 3. Neutral sugar composition of cell walls from *Arabidopsis* tissues. Neutral sugar composition of cell walls from various *Arabidopsis* tissues, demonstrating the range in cell wall composition possible in wild-type tissues. Analysis was done as previously described (Reiter *et al.*, 1993), excluding glucose due to its presence in non-cell-wall polysaccharides (i.e. starch).

charides with both β- and α-linkages. Recent results concerning the relationship between enzyme structure and function, such as experiments showing that as few as four amino acid changes can alter the catalytic outcome of an enzymatic reaction from desaturation to hydroxylation (Broun *et al.*, 1998), emphasize the need for caution in using sequence similarity to infer function based on sequence (see Vergara and Carpita, 2001, in this issue). However, given the difference in predicted reaction mechanisms for retaining versus inverting processive glycosyltransferases (Carpita *et al.*, 1996), it is unlikely that the members of this family make α-linkages.

Currently we have no direct evidence to link any of the *Csl* genes to their respective products. Preliminary subcellular localization data indicate that the *Csl* proteins are found in the Golgi, as expected for polysaccharide synthases (T. Richmond, N. Sprenger, and C. Somerville, unpublished data). The expression patterns that we see for some of the families (i.e. *CslG* and *CslB*), indicate that they are expressed in all cell types at all times during development. This is consistent with our expectations for the expression pattern of polysaccharide synthases that contribute to the primary cell wall. We are gathering data from DNA microarrays from a variety of tissues and treatments to allow us to focus our attempts to find phenotypes for loss-of-function mutations. As more loss-of-function mutants are isolated, and double and triple mutants are constructed, we are confident that changes in cellular composition and/or structure will be seen, allowing us to unequivocally assign a function to each of the *Csl* families.·

Acknowledgements

This work was supported in part by a grant from the U.S. Department of Energy (DOE-FG02-00ER20133).

References

Arioli, T., Peng, L., Betzner, A.S., Burn, J., Wittke, W., Herth, W. *et al.* 1998. Molecular analysis of cellulose biosynthesis in *Arabidopsis*. Science 279: 717–720.

Aspinall, G.O. 1980. Chemistry of cell wall polysaccharides. In: P.K. Stumpf and E.E. Conn (Eds.) The Biochemistry of Plants: A Comprehensive Treatise, Academic Press, New York, New York, pp. 473–500.

Bonin, C.P., Potter, I., Vanzin, G.F. and Reiter, W.D. 1997. The *MUR1* gene of *Arabidopsis thaliana* encodes an isoform of GDP-D-mannose-4,6-dehydratase, catalyzing the first step in the *de novo* synthesis of GDP-L-fucose. Proc. Natl. Acad. Sci. USA 94: 2085–2090.

Broun, P., Shanklin, J., Whittle, E. and Somerville, C. 1998. Catalytic plasticity of fatty acid modification enzymes underlying chemical diversity of plant lipids. Science 282: 1315–1317.

Burget, E.G. and Reiter, W.D. 1999. The *mur4* mutant of arabidopsis is partially defective in the *de novo* synthesis of uridine diphospho L-arabinose. Plant Physiol. 121: 383–389.

Campbell, J.A., Davies, G.J., Bulone, V. and Henrissat, B. 1997. A classification of nucleotide-diphospho-sugar glycosyltransferases based on amino acid sequence similarities. Biochem. J. 326: 929–939.

Carpita, N., McCann, M. and Griffing, L.R. 1996. The plant extracellular matrix: news from the cell's frontier. Plant Cell 8: 1451–1463.

Carpita, N. and Vergara, C. 1998. A recipe for cellulose. Science 279: 672–673.

Carpita, N.C. 1996. Structure and biogenesis of the cell walls of grasses. Annu. Rev. Plant Physiol. Plant Mol. Biol. 47: 445–476.

142

Chuang, C.F. and Meyerowitz, E.M. 2000. Specific and heritable genetic interference by double-stranded RNA in *Arabidopsis thaliana*. Proc. Natl. Acad. Sci. USA 97: 4985–4990.

Cui, X., Shin, H., Song, C.C., Laosinchai, W., Amano, Y. and Brown, R.M. Jr. 2001. A putative plant homolog of the yeast β-1,3-glucan synthase subunit FKS1 from cotton (*Gossypium hirsutum* L.) fibers. Planta, in press.

Cutler, S. and Somerville, C. 1997. Cloning in silico. Curr. Biol. 7: R108–R111.

Dalmay, T., Hamilton, A., Mueller, E. and Baulcombe, D.C. 2000. Potato virus X amplicons in *Arabidopsis* mediate genetic and epigenetic gene silencing. Plant Cell 12: 369–379.

Darvill, A.G., McNeil, M. and Albersheim, P. 1978. Structure of plant cell walls. VIII. A new pectic polysaccharide. Plant Physiol. 62: 418–422.

Darvill, J.E., McNeil, M., Darvill, A.G. and Albersheim, P. 1980. Structure of plant cell walls. 11. Glucuronoarabinoxylan, a second hemicellulose in the primary cell walls of suspension-cultured sycamore (*Acer pseudoplatanus*) cells. Plant Physiol. 66: 1135–1139.

de Souza, S.J., Long, M., Klein, R.J., Roy, S., Lin, S. and Gilbert, W. 1998. Toward a resolution of the introns early/late debate: only phase zero introns are correlated with the structure of ancient proteins. Proc. Natl. Acad. Sci. USA 95: 5094–5099.

Delmer, D.P. 1999. Cellulose biosynthesis: exciting times for a difficult field of study. Annu. Rev. Plant Physiol. Plant Mol. Biol. 50: 245–276.

Doblin, M.S., De Melis, L., Newbigin, E., Bacic, A. and Read, S.M. 2001. Pollen tubes of *Nicotiana alata* express two genes from different β-glucan synthase families. Plant Physiol, in press.

Ewing, R.M. and Claverie, J.M. 2000. EST databases as multiconditional gene expression data sets. Pac. Symp. Biocomput. 430–442.

Ewing, R.M., Kahla, A.B., Poirot, O., Lopez, F., Audic, S. and Claverie, J.M. 1999. Large-scale statistical analyses of rice ESTs reveal correlated patterns of gene expression. Genome Res. 9: 950–959.

Fagard, M., Hofte, H. and Vernhettes, S. 2000. Cell wall mutants. Plant Physiol. Biochem. 38: 15–25.

Favery, B., Ryan, E., Foreman, J., Linstead, P., Boudonck, K., Steer, M., Shaw, P. and Dolan, L. 2001. *KOJAK* encodes a cellulose synthase-like protein required for root hair cell morphogenesis in *Arabidopsis*. Genes Dev. 15: 79–89.

Fowler, J.E. and Quatrano, R.S. 1997. Plant cell morphogenesis: plasma membrane interactions with the cytoskeleton and cell wall. Annu. Rev. Cell Dev. Biol. 13: 697–743.

Gelvin, S.B. 2000. *Agrobacterium* and plant genes involved in T-DNA transfer and integration. Annu. Rev. Plant Physiol. Plant Mol. Biol. 51: 223–256.

Hayashi, T. 1989. Xyloglucans in the primary cell wall. Annu. Rev. Plant Physiol. Plant Mol. Biol. 40: 139–168.

Henrissat, B., Coutinho, P.M. and Davies, G.J. 2001. A census of carbohydrate-active enzymes in the genome of *Arabidopsis thaliana*. Plant Mol. Biol., this issue.

Hong, Z., Delauney, A.J. and Verma, D.P.S. 2001. A cell plate-specific callose synthase and its interaction with phragmoplastin. Plant Cell, in press.

Krysan, P.J., Young, J.C., Tax, F. and Sussman, M.R. 1996. Identification of transferred DNA insertions within *Arabidopsis* genes involved in signal transduction and ion transport. Proc. Natl. Acad. Sci. USA 93: 8145–8150.

Lau, J.M., McNeil, M., Darvill, A.G. and Albersheim, P. 1985. Structure of the backbone of rhamnogalacturonan I, a pectic

polysaccharide in the primary cell walls of plants. Carbohydrate Res. 137: 111–126.

Lin, X., Kaul, S., Rounsley, S., Shea, T.P., Benito, M.I., Town, C.D., Fujii, C.Y., Mason, T., Bowman, C.L., Barnstead, M. *et al.* 1999. Sequence and analysis of chromosome 2 of the plant *Arabidopsis thaliana*. Nature 402: 761–768.

Mayer, K., Schuller, C., Wambutt, R., Murphy, G., Volckaert, G., Pohl, T., Dusterhoft, A., Stiekema, W., Entian, K.D., Terryn, N. *et al.* 1999. Sequence and analysis of chromosome 4 of the plant *Arabidopsis thaliana*. Nature 402: 769–777.

McKinney, E.C., Ali, N., Traut, A., Feldmann, K.A., Belostotsky, D.A., McDowell, J.M. and Meagher, R.B. 1995. Sequence-based identification of T-DNA insertion mutations in *Arabidopsis*: actin mutants *act2-1* and *act4-1*. Plant J. 8: 613–622.

McNeil, M., Darvill, A.G., Fry, S.C. and Albersheim, P. 1984. Structure and function of the primary cell walls of plants. Annu. Rev. Biochem. 53: 625–663.

Nam, J., Mysore, K.S., Zheng, C., Knue, M.K., Matthysse, A.C. and Gelvin, S.B. 1999. Identification of T-DNA tagged *Arabidopsis* mutants that are resistant to transformation by *Agrobacterium*. Mol. Gen. Genet. 261: 429–438.

Pear, J.R., Kawagoe, Y., Schreckengost, W.E., Delmer, D.P. and Stalker, D.M. 1996. Higher plants contain homologs of the bacterial *celA* genes encoding the catalytic subunit of cellulose synthase. Proc. Natl. Acad. Sci. USA 93: 12637–12642.

Peng, L.C., Hocart, C.H., Redmond, J.W. and Williamson, R.E. 2000. Fractionation of carbohydrates in *Arabidopsis* root cell walls shows that three radial swelling loci are specifically involved in cellulose production. Planta 211: 406–414.

Perrin, R.M., DeRocher, A.E., BarPeled, M., Zeng, W.Q., Norambuena, L., Orellana, A., Raikhel, N.V. and Keegstra, K. 1999. Xyloglucan fucosyltransferase: an enzyme involved in plant cell wall biosynthesis. Science 284: 1976–1979.

Rabinowicz, P.D., Schutz, K., Dedhia, N., Yordan, C., Parnell, L.D., Stein, L., McCombie, W.R. and Martienssen, R.A. 1999. Differential methylation of genes and retrotransposons facilitates shotgun sequencing of the maize genome. Nature Genet. 23: 305–308.

Reiter, W., Chapple, C. and Somerville, C. 1993. Altered growth and cell walls in a fucose-deficient mutant of *Arabidopsis*. Science 261: 1032–1035.

Reiter, W.D., Chapple, C. and Somerville, C.R. 1997. Mutants of *Arabidopsis thaliana* with altered cell wall polysaccharide composition. Plant J. 12: 335–345.

Reiter, W.-D. and Vanzin, G. 2001. Molecular genetics of nucleotide sugar interconversion pathways. Plant Mol. Biol., this issue.

Richmond, T. and Somerville, S. 2000a. Chasing the dream: plant EST microarrays. Curr. Opin. Plant Biol. 3: 108–116.

Richmond, T.A. and Somerville, C.R. 2000b. The cellulose synthase superfamily. Plant Physiol. 124: 495–498.

Robertson, D., McCormack, B.A. and Bolwell, G.P. 1995. Cell-wall polysaccharide biosynthesis and related metabolism in elicitor-stressed cells of French bean (*Phaseolus vulgaris* L.). Biochem. J. 306: 745–750.

Saxena, I.M., Brown, R.M. Jr., Fevre, M., Geremia, R.A. and Henrissat, B. 1995. Multidomain architecture of β-glycosyl transferases: implications for mechanism of action. J. Bact. 177: 1419–1424.

Stasinopoulos, S.J., Fisher, P.R., Stone, B.A. and Stanisich, V.A. 1999. Detection of two loci involved in $(1\rightarrow 3)$-β-glucan (curdlan) biosynthesis by *Agrobacterium* sp. ATCC31749, and comparative sequence analysis of the putative curdlan synthase gene. Glycobiology 9: 31–41.

Taylor, N.G., Scheible, W.R., Cutler, S., Somerville, C.R. and Turner, S.R. 1999. The *irregular xylem3* locus of *Arabidopsis* encodes a cellulose synthase required for secondary cell wall synthesis. Plant Cell 11: 769–780.

Turner, S.R. and Somerville, C.R. 1997. Collapsed xylem phenotype of *Arabidopsis* identifies mutants deficient in cellulose deposition in the secondary cell wall. Plant Cell 9: 689–701.

Vergara, C.E. and Carpita, N.C. 2001. Mixed-linkage β-glucan synthase and the *CesA* gene family in cereals. Plant Mol. Biol., this issue.

Zablackis, E., Huang, J., Muller, B., Darvill, A.G. and Albersheim, P. 1995. Characterization of the cell-wall polysaccharides of *Arabidopsis thaliana* leaves. Plant Physiol. 107: 1129–1138.

Zhang, G.F. and Staehelin, L.A. 1992. Functional compartmentalization of the Golgi-apparatus of plant cells: immunocytochemical analysis of high-pressure frozen-substituted and freeze-substituted sycamore maple suspension-culture cells. Plant Physiol. 99: 1070–1083.

Plant Molecular Biology **47**: 145–160, 2001.
© 2001 *Kluwer Academic Publishers. Printed in the Netherlands.*

145

β-D-Glycan synthases and the *CesA* gene family: lessons to be learned from the mixed-linkage (1→3),(1→4)β-D-glucan synthase

Claudia E. Vergara and Nicholas C. Carpita*
*Department of Botany and Plant Pathology, Purdue University, West Lafayette, IN 47907-1155, USA (*author for correspondence; e-mail carpita@btny.purdue.edu)*

Key words: Arabidopsis thaliana, cellulose synthase, cell walls, *CesA* genes, β-glucan synthase, *Oryza sativa,* phylogeny, polysaccharide synthesis, *Zea mays*

Abstract

Cellulose synthase genes (*CesA*s) encode a broad range of processive glycosyltransferases that synthesize (1→4)β-D-glycosyl units. The proteins predicted to be encoded by these genes contain up to eight membrane-spanning domains and four 'U-motifs' with conserved aspartate residues and a QxxRW motif that are essential for substrate binding and catalysis. In higher plants, the domain structure includes two plant-specific regions, one that is relatively conserved and a second, so-called 'hypervariable region' (HVR). Analysis of the phylogenetic relationships among members of the *CesA* multi-gene families from two grass species, *Oryza sativa* and *Zea mays*, with *Arabidopsis thaliana* and other dicotyledonous species reveals that the *CesA* genes cluster into several distinct sub-classes. Whereas some sub-classes are populated by *CesA*s from all species, two sub-classes are populated solely by *CesA*s from grass species. The sub-class identity is primarily defined by the HVR, and the sequence in this region does not vary substantially among members of the same sub-class. Hence, we suggest that the region is more aptly termed a 'class-specific region' (CSR). Several motifs containing cysteine, basic, acidic and aromatic residues indicate that the CSR may function in substrate binding specificity and catalysis. Similar motifs are conserved in bacterial cellulose synthases, the *Dictyostelium discoideum* cellulose synthase, and other processive glycosyltransferases involved in the synthesis of non-cellulosic polymers with (1→4)β-linked backbones, including chitin, heparan, and hyaluronan. These analyses re-open the question whether all the *CesA* genes encode cellulose synthases or whether some of the sub-class members may encode other non-cellulosic (1→4)β-glycan synthases in plants. For example, the mixed-linkage (1→3)(1→4)β-D-glucan synthase is found specifically in grasses and possesses many features more similar to those of cellulose synthase than to those of other β-linked cross-linking glycans. In this respect, the enzymatic properties of the mixed-linkage β-glucan synthases not only provide special insight into the mechanisms of (1→4)β-glycan synthesis but may also uncover the genes that encode the synthases themselves.

Abbreviations: CSR, class-specific region; GAX, glucuronoarabinoxylan;β-glucan, mixed-linkage (1→3),(1→4)β-D-glucan; HGA, homogalacturonan; HVR, hypervariable region; P-CR, plant-conserved region; (RT)-PCR, (reverse transcriptase)-polymerase chain reaction; RG I, rhamnogalacturonan I; SuSy, sucrose synthase

Introduction

The discovery by Pear *et al.* (1996) of a cotton gene suspected to encode the cellulose synthase catalytic subunit (*CesA*) brought the field of plant cell wall biogenesis fully into the genomic era. The identifi-

cation of bacterial cellulose synthase genes several years earlier (Saxena *et al.*, 1990; Wong *et al.*, 1990; Matthysse *et al.*, 1995) ultimately proved instrumental in the discovery of their homologues in higher plants. Screens of plant cDNA libraries with *Acetobacter xylinus* genes as probes were unsuccessful, suggesting that bacterial and higher-plant cellulose synthase genes are not homologous. The finding that led eventually to the discovery of the higher-plant cellulose synthase was in a report by Saxena *et al.* (1995), which showed that well conserved domains associated with processivity of glycosyl transfer include four regions surrounding conserved aspartyl residues and a QxxRW motif. Those regions are highly conserved among all processive glycosyltransferases in which repeating $(1\rightarrow4)\beta$-glycosyl units are synthesized (Saxena *et al.*, 1995).

Random sequencing of a cotton cDNA library, which was prepared from fibers at the onset of secondary wall synthesis, resulted in the isolation of two *CesA* genes (Pear *et al.*, 1996). The deduced amino acid sequence shares regions of similarity with the bacterial CesA proteins. However, two plant-specific regions are not present in the bacterial counterpart; one shows high sequence conservation, the 'plant-conserved region' (P-CR), and the other shows an apparently high sequence divergence, which was termed a 'hypervariable region' (HVR) (Pear *et al.*, 1996). The higher-plant *CesA* genes are predicted to encode ca. 110 kDa polypeptides, each with a large, cytoplasmic N-terminal region containing two Zn-finger domains (Kawagoe and Delmer, 1997), and a C-terminal region containing up to eight predicted membrane spans (Delmer, 1999). The presumed active site resides between predicted membrane spans 2 and 3, and it comprises a large cytosolic domain with four highly conserved 'U-motifs'. These U-motifs are also found in hyaluronan synthases (DeAngelis *et al.*, 1993), chitin synthases (Bulawa *et al.*, 1986), and the NodC synthase (Geremia *et al.*, 1994), all of which make consecutive β-glycosidic linkages in which one sugar is oriented nearly 180 °C with respect to each neighboring sugar. These and other 'inverting'-type processive synthases comprise the family 2 NDP-sugar glycosyltransferases (Henrissat *et al.*, 2001, this issue). Hyaluronan is a repeating disaccharide unit of alternating $(1\rightarrow4)\beta$ and $(1\rightarrow3)\beta$ residues of glucuronic acid and *N*-acetyl-glucosamine, whereas chitin and the Nod factor are composed of $(1\rightarrow4$ β-D-*N*-acetyl-glucosamine units. Hyaluronan synthase is a bifunctional enzyme, with both glycosyltransferase activities contained within a single polypeptide (Spicer *et al.*, 1997). Another example of a bifunctional synthase is that of heparan sulfate, which catalyzes the formation of a polymer composed of alternating residues of β-D-glucuronic acid (GlcA) and α-*N*-acetyl-D-glucosamine (GlcNAc) joined by

$(1\rightarrow4)$ linkages (Lind *et al.*, 1993). The crystal structure of a single member of the family 2 glycosyltransferases, the *Bacillus subtilis* SpsA polysaccharide synthase, provides the first conformational view of the active site and the role of the aspartyl residues in nucleotide-sugar positioning (Charnock and Davies, 1999).

The secondary cell wall of cotton fibers is almost pure cellulose (Meinert and Delmer, 1977). Hence, the expression of the *GhCesA1* and *GhCesA2* genes in cotton fibers at the onset of secondary wall synthesis is reasonable evidence of function (Pear *et al.*, 1996). However, the first confirmation of the function of plant *CesA* genes in cellulose synthesis came from Arioli *et al.* (1998), who reported the cloning of the *Arabidopsis AtCesA1* gene that restored synthesis of cellulose microfibrils in a temperature-sensitive *root-tip swelling* mutant (*rsw*1). A severe dwarf-hypocotyl mutant, *procuste*, was also traced to a defective *CesA6* (Fagard *et al.*, 2000). Virus-induced suppression of a *CesA* gene specifically lowered cellulose content in cells of tobacco (Burton *et al.*, 2000). The *irx*1 and *irx*3 mutants, both displaying a phenotype of collapsed mature xylem cells and reduced content of secondary cell wall cellulose (Turner and Somerville, 1997), were also determined to be *CesA* homologues (Taylor *et al.*, 1999, 2000; see Turner *et al.*, 2001, this issue). To date, many other *CesA* genes that share similarity with the cotton genes have been identified in higher plants. Furthermore, surveys of *Arabidopsis* sequences reveal several groups of genes related to the *CesA* genes but lack the plant-inserted regions found in the *CesAs* (Richmond and Somerville, 2000). These genes are named cellulose-synthase-like (*Csl*), and their function is unknown. The *CesA* and the *Csl* sequences in higher plants are maintained and updated in the web site (http://cellwall.stanford.edu/cellwall). Clearly, the *CesA* and *Csl* genes represent a large multi-gene family in higher plants (see Richmond and Somerville, 2001, this issue). The prevailing perception is that the *CesA* genes encode glycosyltransferases strictly involved in cellulose biosynthesis at the plasma membrane, and the *Csl* genes function in the synthesis of non-cellulosic β-glycans at the Golgi apparatus. Although it is clear that the *CesA* and *Csl* genes function in glycosyl transfer reactions, the evidence to date is insufficient to conclude that the *CesA*-encoded synthases are restricted to the synthesis of cellulose.

We present a comparative analysis of CesA polypeptides and show that the HVR is not variable, but defines sub-classes among the plant *CesA* genes and contains conserved motifs that might be implicated in the catalytic mechanism. This hypothesis, when considered along with theoretical interpretations about the linkage structure in β-linked polymers, raises the possibility that some members of the *CesA* gene family are involved in the synthesis of polysaccharides other than cellulose. The experimental results of studies to determine the mechanisms of maize mixed-linkage β-glucan synthesis are consistent with the interpretation that this synthase might be encoded by a *CesA* gene (Buckeridge *et al.*, 1999, 2001).

An overview of plant cell-wall $(1\rightarrow4)\beta$-linked polysaccharides

To evaluate the proteins encoded by *CesA* and *Csl* genes with respect to their possible synthases, it is important to review the principal non-cellulosic polysaccharides of the primary and secondary walls. Two structural models of primary cell walls can be distinguished in flowering plants: the Type I wall, which is present in all Dicotyledonae that have been examined and in some Monocotyledonae (Carpita and Gibeaut, 1993), and the Type II wall of commelinoid monocots, which include the grasses, palms, gingers, and bromeliads (Carpita, 1996). The principal crosslinking polysaccharides in Type I primary cell walls are xyloglucans, which are linear chains of $(1\rightarrow4)\beta$-D-glucan with xylosyl units attached at contiguous locations at the O-6 positions of the glucosyl units (McCann and Roberts, 1991). The Type I wall is composed of several other β-linked glycans: the glucuronoarabinoxylans, which are $(1\rightarrow4)\beta$-D-xylan chains substituted with single non-reducing terminal arabinofuranosyl and glucuronosyl units at the O-2 positions, the glucomannans, which are generally unbranched and unsubstituted $(1\rightarrow4)\beta$-D-linked polymers of glucose and mannose, and the galactomannans, which are $(1\rightarrow4)\beta$-D-linked mannans with single α-D-galactosyl residues attached to the O-6 of the mannan backbone. Additional α- and β-linked polymers include the pectic $(1\rightarrow5)\alpha$-L-arabinans, $(1\rightarrow4)\beta$-D-galactans, $(1\rightarrow4)\alpha$-D-galacturonans (HGAs), and $(1\rightarrow2)\alpha$-L-rhamnosyl-$(1\rightarrow4)\alpha$-D-galacturonosyl repeating units of rhamnogalacturonans (RG I). Although each is synthesized by processive glycosyltransferases involving nucleotide-sugar substrates, it is not yet known whether their synthases also possess the D, D, D, QxxRW-containing motifs characteristic of the *CesA* and *Csl* family involved in the 'inverting' synthesis mechanism of polymers that are strictly $(1\rightarrow4)\beta$-D-linked glucose, xylose, or mannose.

The Type II cell wall is composed of cellulose microfibrils crosslinked primarily by glucuronoarabinoxylans (GAXs) instead of xyloglucan (Carpita, 1996). Simplified xyloglucan structures are found, along with glucomannans, arabinans, galactans, HGA and RG I. A remarkable characteristic of the Type II cell wall is that the constituents change during cell expansion. The mixed-linkage $(1\rightarrow3),(1\rightarrow4)\beta$-D-glucan (hereafter called β-glucan) is found only in the Poales (Smith and Harris, 1999) and not in any other commelinoid species or any species with a Type I cell wall. β-Glucan is scarcely detected in meristematic cells but increases markedly in the walls of elongating cells to a maximum that is coincident with the maximum rate of elongation (Kim *et al.*, 2000). When growth ceases, the β-glucan is hydrolyzed extensively and is no longer a major cell wall polymer in mature cells. β-Glucan accumulates in the walls of the endosperm of the developing caryopsis and surrounding maternal tissues (Carpita, 1996). β-Glucan is not a random mixture of $(1\rightarrow3)$- and $(1\rightarrow4)\beta$-D-glucosyl linkages but has a defined structure that was determined with a sequence-dependent endoglucanase from *Bacillus subtilis*, which hydrolyzes $(1\rightarrow4)\beta$-D-glucosyl units preceded by $(1\rightarrow3)\beta$ units (Anderson and Stone, 1975). Hydrolysis of the native polymer reveals that almost 70% of the polysaccharide consists of cellotriosyl units connected by single $(1\rightarrow3)\beta$ linkages (Wood *et al.*, 1994). The remainder of the polymer consists of longer cellodextrin runs also connected by $(1\rightarrow3)\beta$ linkages. The relative amounts of the cellodextrin-$(1\rightarrow3)\beta$-D-glucose oligosaccharides are not randomly generated but constitute a defined structural fingerprint.

Biosynthesis of $(1\rightarrow4)\beta$-linked cell wall polysaccharides

Despite some nominal success (for review, see Brown and Saxena, 2000), the *in vitro* synthesis of cellulose has not been attained with isolated membranes from plants (Delmer, 1999). The lack of progress in identifying polysaccharide synthase polypeptides has been attributed to the difficulties most investigators have encountered in maintaining synthase activities *in vitro*. These include solubilizing the synthases from the membrane in an active form so that they can be purified and characterizing the polysaccharide product to confirm the synthesis of an authentic cellulosic glucan polymer. The majority of cellulose synthase activity is lost during homogenization of plant tissues, but callose synthase, which makes the $(1\rightarrow3)\beta$-glucan wound polymer, is also located at the plasma membrane and possibly arises through disruption of cellulose synthase components (Delmer, 1977). The requirement for an intact membrane indicated that a membrane potential might be part of the reaction mechanism for cellulose (Carpita and Delmer, 1980). The activation of polymer synthesis by artificially induced potentials in plasma membrane vesicles (Bacic and Delmer, 1981) and the activation of *Acetobacter* cellulose synthase *in vitro* (Delmer *et al.*, 1982) provided biochemical evidence for this participation.

Callose synthase is stable after detergent solubilization, and several polypeptides are associated with synthesis (Dhugga and Ray, 1994). Callose is also a native component of some specialized walls, most notably that of the pollen tube, and these calloses are most likely made by unique synthases (Schlüpmann *et al.*, 1993; Turner *et al.*, 1998). More recently, Cui *et al.* (2001), Doblin *et al.* (2001) and Hong *et al.* (2001) reported that a gene expressed in cotton fibers exhibits amino acid similarity with yeast FKS1, which encodes a wall-associated $(1\rightarrow3)\beta$-glucan synthase. However, the existence of a distinct callose synthase synthesizing this polymer at specific developmental stages does not preclude the potential 'default' of cellulose synthase to a callose synthase *in vitro*. Amor *et al.* (1995) proposed that the association of sucrose synthase (SuSy) with cellulose synthase provided a UDP-Glc delivery mechanism, and they provided immunocytochemical evidence that placed a population of SuSy at the plasma membrane of cotton fibers engaged in secondary wall cellulose synthesis. Biochemical support for this idea came from the work of Hirai *et al.* (1998), who demonstrated the synthesis of callose fibrils from sucrose and UDP on immobilized tobacco plasma membrane sheets. A more detailed discussion of the potential interaction of SuSy with cellulose synthase and its significance is found in the article by Haigler *et al.* (2001, this issue).

In contrast to the problems encountered with stabilizing cellulose synthesis *in vitro*, several non-cellulosic polysaccharide $(1\rightarrow4)\beta$-linkage synthases have been stabilized *in vitro* in isolated Golgi membranes. In addition to the mixed-linked β-glucan synthase (Henry and Stone, 1982; Gibeaut and Carpita, 1993; Becker *et al.*, 1995), these activities include the synthases of xyloglucan (Gordon and Maclachlan, 1989), glucuronoxylans (Hobbs *et al.*, 1991), glucomannan (Piro *et al.*, 1993), and galactomannan (Reid *et al.*, 1995). Like cellulose, all of the major cross-linking glycans, xyloglucan, arabinoxylan, and glucomannans, contain $(1\rightarrow4)\beta$ linkages. Each of these synthases has to overcome the steric constraint inherent to this inverting type of linkage. The challenge ahead is to define which *CesA* or *Csl* genes encode these specific synthases.

Whereas the processive glycosyltransferases that generate the backbones of polymers are called synthases, these synthases are associated with several specific glycosyltransferases that add side-chains of single sugars to large oligomers. The thorough discussion of these kinds of glycosyltransferases and the search for the genes that encode them are explored in detail by Perrin *et al.* (2001, this issue) and Gibeaut (2000).

Special features in the biosynthesis of the maize β-glucan

β-Glucan is an unbranched polymer with predominantly $(1\rightarrow4)\beta$ linkages, and its synthase has features that resemble cellulose synthase more than other Golgi-associated synthases (Figure 1). The β-glucan synthase requires Mg^{2+} or Mn^{2+} and has a high apparent K_m for UDP-Glc; maintenance of a high pH external to the Golgi lumen promotes prolonged synthesis of the glucan (Gibeaut and Carpita, 1993). The synthase preserved *in vitro* makes a product with a distribution of cellotriosyl and cellotetraosyl units similar to that of the native β-glucan. Unlike non-gramineous plants, a callose synthase is associated with the Golgi apparatus as well as the plasma membrane, and increases in its activity coincide with loss of β-glucan synthase by action of protonophores or detergent solubilization of the Golgi membranes (Gibeaut and Carpita, 1993, 1994b). The stimulation of β-glucan synthase by addition of small amounts of ATP or steepening of the pH gradient across the Golgi membranes (Gibeaut and Carpita, 1993) provides further evidence that the physiological state of the membrane is important for both cellulose and β-glucan synthesis (Figure 1).

Further investigation of the synthase activity *in vitro* showed that the unit structure could be altered drastically depending on the UDP-Glc concentration (Buckeridge *et al.*, 1999, 2001). Sub-optimal UDP-

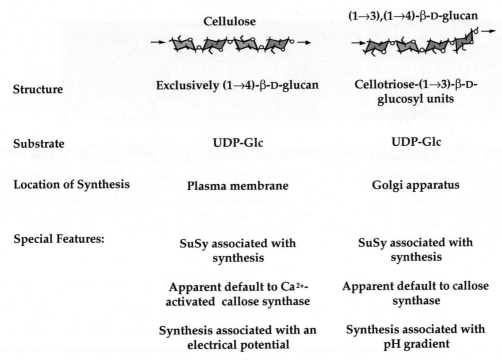

	Cellulose	(1→3),(1→4)-β-D-glucan
Structure	Exclusively (1→4)-β-D-glucan	Cellotriose-(1→3)-β-D-glucosyl units
Substrate	UDP-Glc	UDP-Glc
Location of Synthesis	Plasma membrane	Golgi apparatus
Special Features:	SuSy associated with synthesis	SuSy associated with synthesis
	Apparent default to Ca²⁺-activated callose synthase	Apparent default to callose synthase
	Synthesis associated with an electrical potential	Synthesis associated with pH gradient

Figure 1. Comparison of the fundamental repeating unit structure of the glucan chains in cellulose and mixed-linkage β-glucan with features of their respective biochemical synthesis.

Glc concentrations favor the synthesis of the longer cellodextrin units in β-glucan, particularly the cellote-traosyl unit, whereas at the highest UDP-Glc concentrations tested, the cellotriose accounts for about 80% of the total polymer synthesized, and the cellopentaose unit becomes the next most abundant oligomer (Buckeridge *et al.*, 2001). Thus, as saturation with substrate is approached, the polymer is mostly cellotriose and odd-numbered cellodextrin units connected by single (1→3)β linkages. By comparison, cellulose glucan chains are cellobiose units connected by single (1→4)β linkages (Figure 1). We proposed that this subunit structure is fixed because of the constraints imposed by the position of the non-reducing terminal receptor hydroxyl group when an even or odd number of glycosyltransferase modes are used in the synthase (Buckeridge *et al.*, 1999, 2001).

Early work on the synthesis of non-cellulosic polysaccharides at the Golgi apparatus led to a common misconception that these synthases are saturated at micromolar concentrations of nucleotide-sugar (for review, see Gibeaut and Carpita, 1994a). The idea that Golgi synthases operate strictly at micromolar concentrations originally was thought to be true for xyloglucan synthase, but Gordon and Maclachlan (1989) demonstrated that the synthesis demanded a precise ratio of UDP-Glc to UDP-Xyl, and formation of heptasaccharide units of true xyloglucan required millimolar, not micromolar, concentrations of substrate. The xyloglucan synthase complex is located completely within the lumen of the Golgi apparatus and is dependent upon UDP-Glc transport (Muñoz *et al.*, 1996). The membrane topology of the β-glucan synthase has not been established.

In maize, SuSy was detected immunocytochemically in fractions enriched with both plasma membrane and Golgi membrane, whereas in soybeans, the enzyme was detected only in plasma membranes (Buckeridge *et al.*, 1999). Hence, if β-glucan synthase is derived from an ancestral cellulose synthase and possesses a similar mechanism of synthesis, then part of this mechanism may include the association of SuSy to control the supply of UDP-Glc to the active site(s) of the glucan synthase. For this mechanism to function, the catalytic domain of β-glucan synthase must reside on the cytosolic face of the Golgi membranes. Cytosolic and membrane-associated SuSys appear to be the same gene product but the location is dependent on phosphorylation state (Winter and Huber, 2000). Hence, the membrane to which unphosphorylated SuSy binds appears to depend on the presence of an interacting protein, i.e. cellulose syn-

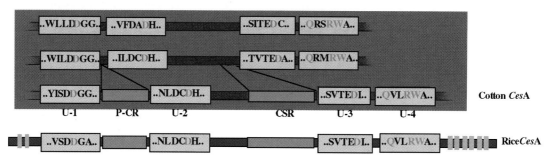

Cotton *CesA*

U-1   P-CR   U-2   CSR   U-3   U-4

Rice*CesA*

Figure 2. Schematic comparison of the predicted domain structures of bacterial and cotton *CesA*s with those of rice *CesA*s. We used the BLAST tool (Altschul *et al.*, 1997) to identify and sequence five rice cDNA clones based on sequence similarity to 'U-motifs' important in catalysis and binding of the substrate UDP-Glc (Pear *et al.*, 1996). GenBank accession numbers for the genes are *OsCesA*1 (D40691), *OsCesA*2 (D48636), *OsCesA*3 (D47622), *OsCesA*11 (D41261), *OsCesA*13 (D46824). The deduced amino acid sequences share the characteristic domain structure of the higher plant CesA proteins; there are two plant-inserted regions, a conserved region between U-1 and U-2 called P-CR, and a CSR (formerly, HVR) between U-2 and U-3.

```
AtCesA7   ..................................KGPKRPKMIS G P FGRRRKNKKFSKNDMNGDVAALGGAEGDKEHLMSEMN

AtCesA8   ....................S.MPSFPKSSSSS  S  CCPGKKEP .KDPSELYRDAKREELDAAIFNLREID .NYDEYERSMLISQTS
GhCesA1   ....................SKPRILPQSSSSS ..CCLTKKKQP .QDPSEIYRDAKREELDAAIFNLGDLD .NYDEYDRSMLISQTS
PtCesA3   ....................SMP SLRKRKDSSS CFS CCCPSKKKPAQDPAEVYRDAKREDLNAAIFNLTEID .NYDEHERSMLISQLS

AtCesA3   ...............LKPKHRKTGILSSL GGSRKKSSKS SKKGSDKKKSGKHVDSTVP VFNLEDIEEGVEG .AGFDDDEKSLLMSQMS  ┐
GhCesA3   ...............IKVKHKKPSLLSKL GGSRKKNSK .AKKESDKKKSGRHTDS TVP VFNLDDIEEGVEG .AGFDDDEKALLMSQMS  │
OsCesA2   ...............IKQKKKGSFLSSL GG .RKKASKSKKKSSDKKKSNKHVDSAVP VFNLEDIEEGVEG .AGFDDDEKSLLMSQMS  │
OsCesA13  ...............IKQKKKGSFLSSL GG .RKKASKSKKKSSDKKKSNKHVDSAVP VFNLEDIEEGVEG .AGFDDDEKSLLMSQMS  ├ II
ZmCesA9   ...................IKQ.KKGGFLSSL GG .RKKGSKS .KKGSDKKKSQKHVDSSVP VFNLEDIEEGVEG .AGFDDDEKSLLMSQMS │
ZmCesA5   ...................VKKKKPGFFSSL GG .RKKTSKS .KKSSEKKKS HRHADS SVP VFNLEDIEEGIEG .SQFDDEKSLIMSQMS _┘

ZmCesA1   ...............LTEADLEPNIVVKS CGRRK ..RKNKSYMDSQSRIMKRTESSAP IFNMEDIEEGIEG ...YEDERSVLMSQRK  ┐
ZmCesA2   ...............LTEADLEPNIVIKS CGRRK ..RKNKSYMDSQSRIMKRTESSAP IFNMEDIEEGIEG ...YEDERSVLMSQRK  │
OsCesA1   ...............LTEADLEPNIVVKS CGGRK ..KKSKSYMDSKNRMMKRTESSAP IFNMEDIEEGIEG ...YEDERSVLMSQKR  ├ III
ZmCesA3   ...............LTEADLEPNIIIKS CGGRK ..KKDKSYIDSKNRDMKRTESSAP IFNMEDIEEGFEG ...YEDERSVLMSQKS  │
AtCesA1   ...............LTEEDLEPNIIVKS CFGSRKKGKSRKIPNYEDNRSIKRSDSNVP LFNMEDIDEDVEG ...YEDEMSLLVSQKR  │
AtCesA10  ...............LTEEDLEPNIIVKS CGSRKKGKSSKKYNYEKRRGINRSDSNAP LFNMEDIDEGFEG ...YDDERSILMSQRS  ┘

GhCesA2   VSEKRPKMT DCWP SW CCCCCGGSRKKSKKKGEKKGLLGGLLYGKKKKMMGKNYVKKGSAP VFDLEEIEEGLEG ..YEELEKSTLMSQKN
PtCesA2   VSEKRPKMT DCWP SW CCCFGGSRKKSKKKGQ.RSLLGG .LYPMKKKMMGKKYTRKASAP VFDLEEIEEGLEG ..YEELEKSSLMSQKS

OsCesA3   KTKKPPSRT NCWPKW FCCCCFGNRK.QK ..KTTK ....PKTEKKKLLFFK ..EENQSPAYALGEIDEAA ...PGAENEKAGIVNQQK  ┐
ZmCesA7   KTKKPPSRT NCWPKW ICCCCFGDRKSKK ..KTTK ....PKTEKKKRSFFKR ..AENQSPAYALGEIEEGA ...PGAENEKAGIVNQQK  │
OsCesA11  KTKKPPSRT NCWPKW CCCCCC .GNRHTKK ..KTTK ....LKPEKKRLFFKK ..AENQSPAYALGEIEEGA ...PGAETDKAGIVNQQK  ├ I
ZmCesA8   KTKKPPSRT NCWPKW LSCCCSRN.KNKK ..KTTK ....PKTEKKKRLFFKK ..AENPSPAYALGEIDEGA ...PGADIEKAGIVNQQK  │
ZmCesA6   KTKKPPSRT NCWPKW ICCCCFGNRKTKKKKTKTSK ....PKFEKIKKL.FKK ..KENQAPAYALGEIDEAA ...PGAENEKASIVNQQK  ┘

AtCesA2   KKKQPPGRT NCWPKW CLCGMRKKKTGK.VKDNQ ....RK .........KP ..KETSKQIHALEHIEEG ..L.QVTNAENNSETAQLK
AtCesA9   KKKKPPGKT NCWPKW CLCGLRKKSKTK.AKD.K ....KT .........NT ..KETSKQIHALENVDEGVIV .PVSNVEKRSEATQLK
AtCesA6   KKKKGPRKT NCWPKW CLLCFGSRKNRKAKTVAADK ....KK .........KN ..REASKQIHALENIEEGRGH .KVLNVEQSTEAMQMK
```

Consensus .............K...............K.............P.F.LE.IEEG.EG....E.EKS.LMSQ..

Figure 3. Amino acid sequence comparison of the class-specific regions (CSRs) among *CesA*s from different plant species. Sequence alignments are from the CSRs of rice *OsCesA*1, *OsCesA*2, *OsCesA*3, *OsCesA*11, and *OsCesA*13, maize *ZmCesA*1 (AF200525), *ZmCesA*2 (AF200526), *ZmCesA*3 (AF200527), *ZmCesA*5 (AF200529), *ZmCesA*6 (AF200530), *ZmCesA*7 (AF200531), *ZmCesA*8 (AF200532), and *ZmCesA*9 (AF200533), *Arabidopsis AtCesA*1 (AF027172), *AtCesA*2 (AF027173), *AtCesA*3 (AF027174), *AtCesA*6 (AF062485), *AtCesA*7 (AF088917), *AtCesA*8 (AL035526), *AtCesA*9 (AC007019), and *AtCesA*10 (AC006300), cotton *GhCesA*1 (U58283), *GhCesA*2 (U58284), and *GhCesA*3 (AF150630), and poplar *PtCesA*2 (AF081534) and *PtCesA*3 (AF072131). The different sub-classes are separated by spaces between the sequences, and the Roman numerals to the right indicate the three cereal sub-classes. Bold residues indicate identity within the sub-classes. The shaded sequence at the bottom of the alignment indicates the consensus sequence. The sequence alignment was generated with the PILEUP program (Deveraux *et al.*, 1984) from the Wisconsin package (Genetics Computer Group). This program ranks the sequences in order of similarities. A dot (·) indicates a gap introduced to optimize the sequence alignment.

thase at the plasma membrane and β-glucan synthase at the Golgi apparatus.

In summary, the reaction conditions of the β-glucan synthase are more similar to those of cellulose synthase than to those of other non-cellulosic β-linked polysaccharides (Figure 1). We propose that the β-glucan synthase is related to an ancestral cellulose synthase and, therefore, it shares features of the synthesis mechanism.

CesA genes in rice: evidence for three distinct sub-classes in cereals

The *Arabidopsis CesA* gene family is arranged in two major groups, the *CesA* genes and the *Csl* genes, which comprise six divergent gene clusters (Holland *et al.*, 2000; Richmond and Somerville, 2001). We characterized five rice cDNA clones that share the domain structure of proteins encoded by the group of *CesA* genes. Sequence comparisons revealed a high degree of sequence identity among the rice and cotton *CesA*s. The U-motifs and the plant-specific domains are well conserved in the rice clones, and hydrophobicity analyses indicate that the number and position of the potential transmembrane domains are consistent with the structure of the CesA polypeptides (Figure 2). Nevertheless, distinct amino acid variability is apparent in the HVR and the C-terminus. A detailed sequence analysis of the rice *CesA*s identified three groups: Class I with 94% identity, Class II with 90% identity, and Class III distinct from the other two classes. Differences in amino-acid sequence identity, which can be traced to the HVR, account for the sub-class structure. Comparisons of *CesA* amino acid sequences from other higher plant species revealed that the sub-class structure is defined primarily by the HVR. Furthermore, this distinction is not restricted to the rice genes, but to all the sequences analysed, and the identity among sequences that belong to the same sub-class is higher than 85% (Figure 3). Theses analyses indicate that the HVR is not variable within a sub-class, even across grass and dicot species. We propose that this region be more appropriately termed class-specific region (CSR).

To further examine the sub-class structure of the rice sequences, we designed gene-specific primers from the CSR that specifically amplify members from each sub-class (Figure 4). The sub-class-specific primers were also able to detect sub-class members among putative *CesA* clones isolated from a rice ge-

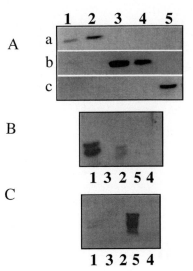

Figure 4. Oligonucleotides from the CSR are specific for each sub-class. Degenerate oligonucleotide primers were synthesized from sub-class-specific amino acid sequences from the CSR. The following forward and reverse primers, respectively, were used in PCR experiments: CSR-I, 5'-G(A/G)AC(C/T)TGCAA(C/T)TGC-3', 5'-GCAGG(C/G/T)GA(C/T)TGATTTT-3'; CSR-II,5'-GAAGTTT CTTGTCATCAC-3', 5'-AAGATTGAAAACTGGCA-3'; CSR-III, 5'-GTACTAACTGAGGCTG-3', 5'-CATGTTAAAGATAGGAG-3'. Of each rice cDNA clone 50 ng was mixed with each oligonucleotide primer set in a final concentration of 0.2 mM and amplified with *Taq* DNA Polymerase (Promega) in the following program: 94 °C, 180 s for a first round of denaturation followed by 30–35 cycles of 94 °C for 60 s denaturing, 50 °C for 90 s annealing, and 72 °C for 90 s synthesis. A final synthesis step of 72 °C for 180 s was included. A. Amplification products were separated by electrophoresis in 2% w/v agarose gels, lanes 1–5: OsCesA3, OsCesA11,OsCesA13, OsCesA2 and OsCesA1 respectively. PCR reactions with sub-class primer pairs I, II, and III are indicated by a, b, and c, respectively. The PCR products were excised from agarose gels, purified by GeneClean (Qiagen) and used as gene-specific probes. Southern blot experiments using the gene-specific probes show sub-class specificity for the CSR probes. B. Southern gel blot analysis with the sub-class I probe. C. Southern gel blot analysis with the sub-class III probe.

nomic library. As expected from the size of other higher-plant *CesA* families, twelve in *Arabidopsis* and nine in maize, the primers did not amplify all of the genomic clones, suggesting that additional sub-classes exist among the rice *CesA* gene family. ased on the primary sequence of the CSR the rice CesA polypeptides can be placed into at least three major sub-classes (Figure 3). This sub-class structure has been also confirmed in alignments of sequences from the 3'-untranslated region (not shown).

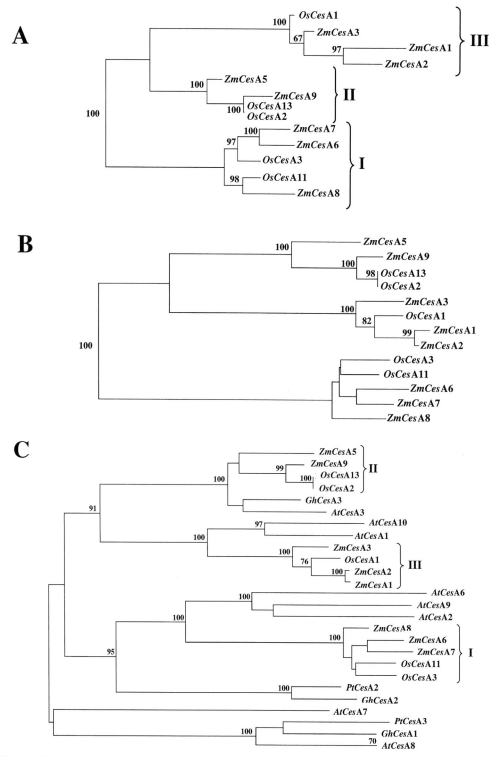

Figure 5. Phylogenetic relationships for the *CesA* gene family. Unrooted phylogenetic trees were generated using Clustal W (Thompson *et al.*, 1994), which uses the Neighbor-Joining method based on a matrix of 'distances' that were constructed according to Kimura's two-parameter method. Phylogenetic trees of the rice CesA proteins based on (A) the cereal complete amino acid sequence and (B) the cereal CSR amino acid sequence. C. Phylogenetic relationships of the *CesA* cereal sequences to *CesA* sequences from other higher plants. The Roman numerals in the brackets indicate the sub-class defined by the CSRs of the cereal sequences. The trees correspond to majority consensus trees from 100 bootstrap replicates; bootstrap values > 70% are shown.

Phylogenetic relationships of the *CesA* gene family

Based on the alignments previously described, phylogenetic relationships for *CesA* genes from rice and other higher plants were determined. The sub-class structure, initially observed in the multiple alignments of the deduced amino acid sequences from the CSR, is supported by unrooted phylogenetic trees. In rice, *OsCesA3* and *OsCesA11* (Class I) form a separate clade from *OsCesA2* and *OsCesA13* (Class II). *OsCesA1* (Class III) is a separate clade that occupies a branch adjacent to the Class II clade. The phylogenetic tree resulting from the analysis was the same whether based on the alignments from complete amino acid sequences or alignments restricted to the CSR sequences (Figure 5).

To determine the evolutionary relationships of the rice genes to other *CesA* genes from cereals and other higher plants, the analysis included full-length and near full-length coding sequences from monocots (rice and maize) and dicots (*Arabidopsis*, cotton and poplar). The rice *CesA* sub-classes are dispersed throughout the tree, and the major lineages include genes from monocots and dicots (Figure 5). This clustering pattern might indicate that the sub-classes evolved prior to speciation events. Similar observations were described by Holland *et al.* (2000) in a comparative analysis of the plant *CesA* gene family. However, a cautionary note should be considered, because designation of orthology based on common function can be misleading as paralogues may show functional convergence. Orthology should be determined by synteny between species and by gene phylogenies, which indicate that genes of different species belong to the same family (Doyle and Gaut, 2000). The resolution of *CesA* gene phylogeny should increase as more *CesA* gene families from different species are characterized.

The *CesA* sequences from cereals are more closely related to each other than to those from dicot species. Whether the trees are produced with alignments restricted to the CSR or the entire coding region, the sub-class I and sub-class II clades cluster *CesA* sequences from maize and rice exclusively. This particular topology is consistent and has strong bootstrap support (Figure 5). This was not the case for the sub-class III clade, which in some instances contains dicot sequences when the complete coding sequence is included (data not shown). When compared to the *Arabidopsis* gene family alone, the rice sequences belong to the *CesA* gene cluster and the three sub-class structure is maintained. In analyses of the entire coding region, the *Arabidopsis Csl* sequences cluster into two distinct clades apart from the *CesA* clades.

Expression of the three cereal *CesA* sub-classes

The high degree of conservation among the distinct *CesA* sub-classes raised the possibility that these gene products might be differentially expressed in a development- or tissue-specific manner. Therefore, the expression patterns of the sub-classes relative to each other were determined with gene-specific probes representing the three sub-classes. The probes were prepared by PCR amplification, taking advantage of unique sequences within the CSR (Figure 4).

Northern gel blot analyses showed that transcripts for the three sub-classes are present in different organs and tissues, such as mature roots, leaves, coleoptiles and endosperm. Quantitative RT-PCR experiments with total RNA revealed some differences in the relative abundance of the *CesA* gene transcripts in the different tissues. For sub-class I the relative expression of the *CesA* genes indicated that this sub-class is predominantly expressed in elongating tissues, root tips and coleoptiles. For sub-class II the relative expression of the *CesA* genes in all tissues analysed is roughly the same. Sub-class III shows a higher level of expression in root tips and low expression in leaves with intermediate expression in mature roots and coleoptiles (Figure 6). These differences in the relative abundance of the transcripts might reflect differential accumulation of transcripts at the cell or tissue level, which can be masked when complex tissues are analyzed. Indeed, this cell-specific expression pattern was observed for some maize *CesA* genes (Holland *et al.*, 2000), as well as *Arabidopsis IRX1* and *IRX3* (Turner *et al.*, 2001) and *PRC1* (Fagard *et al.*, 2001) genes.

As reported by Holland *et al.* (2000), amino acid sequence comparisons may assist in the prediction of expression patterns. Detailed analyses of the tree topologies reveal that the *CesA* sequences, clustered in the most basal clade (Figure 5), represent proteins implicated in secondary wall formation. For example, cotton *GhCesA1* and *GhCesA2* encode proteins expressed during fiber development (Holland *et al.*, 2000), whereas *PtCesA2* and *PtCesA3* were isolated from a xylem-specific cDNA library from poplar. *AtCesA7* (*IRX3*) and *AtCesA8* (*IRX1*) are also associated with secondary cell wall cellulose synthesis (Turner and Somerville, 1997; Taylor *et al.*, 2000). Conversely, the clade formed by proteins from sub-

154

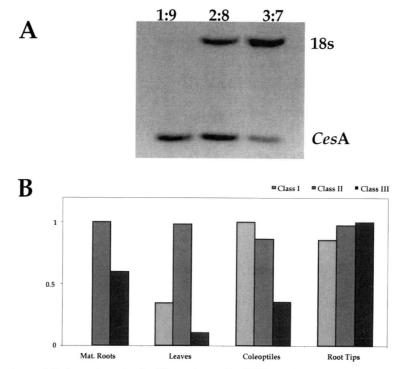

Figure 6. Relative abundance of *CesA* gene transcripts in different tissues. Total RNA was isolated from rice root tips, mature roots, leaves and coleoptiles with the RNeasy Plant Mini Kit (Qiagen). Primer pairs specific for each sub-class were used in multiplex RT-PCR experiments that use two primer sets in a single PCR. One set of sub-class specific primers from the CSR to amplify the *CesA* cDNAs and the other set to amplify an invariant rRNA endogenous control. Primer sets for sub-class II and III were as defined in Figure 4. Forward and reverse primers for sub-class I were: 5'-CGAGGACCTGCAACTGC-3' and 5'-GCAGGGGATTGATTTTCT-3', respectively. The amplification efficiency of the rRNA is modulated by 18S primers (competimers), which have been modified to block extension by DNA polymerase. By mixing 18S primers with increasing amounts of 18S competimers the amplification efficiency of the 18S cDNA can be reduced to a point were both the *CesA* gene and the 18S control give similar yields of product. A. Example of a RT-PCR multiplex reaction. Lanes 1–3: products from root tissue amplified with sub-class II gene-specific primers and different dilutions of primer/competimer mixture. B. The density of each signal is analyzed on one-dimensional profiles with the IP Gel image analysis program (Scanalytics, Vienna, VA). The levels of the *CesA* transcript are normalized against the amplified control, and the relative values can be compared between tissues for estimation of the relative expression of target RNA in the samples.

class III might function in primary wall synthesis. This clade holds *AtCesA*1 (*RSW*1), which is associated with primary cell wall synthesis (Arioli *et al.*, 1998), and *ZmCesA*1, which shows high expression in cells in the root elongation zone that are engaged in primary cell wall deposition (Holland *et al.*, 2000). Consistent with the predictions from phylogenetic comparisons, the relative quantitation studies with RT-PCR of transcripts of this sub-class show high expression in root tips and low expression in leaves (Figure 6).

The sub-class II clade forms a sister clade with genes *AtCesA*3, an RSW1-like sequence with no defined expression pattern, and *GhCesA*3, whose expression pattern is associated with primary wall cellulose synthesis in the elongating trichome. *ZmCesA*8, which is expressed in the differentiating epidermis at the root tip (Holland *et al.*, 2000), resides within the

sub-class I clade. Interestingly, three *Arabidopsis* sequences, *AtCesA*2, *AtCesA*6, and *AtCesA*9, form a sister clade with the sub-class I cereal sequences. *AtCesA*6 (*PRC*1) is involved in primary wall cellulose synthesis in the in roots and dark-grown hypocotyls (Fagard *et al.*, 2000) The function of *AtCesA*2 and *AtCesA*9 genes has yet to be established. Given the specialized functional roles of the cell wall in the different types of plant cells, one expects that the redundancy of the *CesA* genes is a strategy to achieve cell-specific expression as well as specificity of control in developmental expression.

Sequence analyses of the CSR

Early on, the scarcity of *CesA* sequences and the apparent variability within the so-called HVR led to the assumption that this variability denotes a hig rate of evolution because this region is nonessential in the synthase reactions. However, now that a sufficient number of the gene family members have been recognized and their sequences added into an expanding database, it is apparent that these regions are well conserved in paralogues across grass and dicot species (Figure 3). For this reason the HVR is a misnomer, and, as we urged earlier, it should be replaced with the term CSR. A detailed analysis of the CSR reveals several common motifs and conserved amino acid residues, and the number and character of these motifs is characteristic for each sub-class.

Amino acid sequence alignments reveal special features that may be important modulators of function. The regions toward the amino terminus are generally basic, whereas the portions of the sequences towards the C-terminus are acidic. Similarity across all sub-classes is greater towards the C-terminus of the CSR, and a Lys, an Asp, and an aromatic residue are conserved among all class members (Figure 3). More notably, two consensus motifs, E(D/E)x(D/E)E and (D/E)DE ('D-motifs'), are conserved among all the sequences analyzed; although slight variations occur between sub-classes, the substitutions are always conservative (Figure 3). An interesting exception is that of three *Arabidopsis* sequences *AtCesA2*, *AtCesA6*, and *AtCesA9*, which have altered (D/E)DE motif structures with two intervening amino acids and have a CSR motif structure different from the main sub-classes I, II, and III (Figure 3). We suggest that they form a separate sub-class. Analyses including more sequences are necessary to determine if other dicot sequences cluster with these sequences to define additional sub-classes.

Interestingly, these D and DxD motifs closely parallel the structure of the U-2 motif that is conserved in all the family 2 processive glycosyltransferases (Figure 2). This particular motif is iterated in other processive glycosyltransferases, including chitin synthase (Bulawa *et al.*, 1986), hyaluronan synthase (DeAngelis *et al.*, 1993), and NodC synthase (Geremia *et al.* 1994). It is also present in non-processive glycosyl transferases, including those involved in the synthesis of fungal $(1\rightarrow6)\beta$-D-glucans (Roemer *et al.*, 1994), galactosyl transferases (Wiggins and Munro, 1998; Breton and Imberty, 1999; Gastinel *et al.*, 1999), and in the plant 'reversibly glycosylated pro-

tein', which self-glycosylates with nucleotide sugars but whose function in polysaccharide synthesis is unknown (Dhugga *et al.*, 1997; Delgado *et al.*, 1998).

Other motifs within the CSR are associated with sub-class specificity. Two consecutive Cys residues, and a poly-cysteine motif of either six consecutive Cys or five Cys with an Ile insertion add to sub-class distinction (Figure 3). The N-terminal two-thirds of the CSR is enriched in basic amino acid residues and possesses most of the sequence variation that defines the sub-class structure. Two consecutive basic amino acids, Lys or Arg, are found in numbers ranging from 2 to 6, and with the exception of a single sequence, all have at least one motif of three consecutive basic amino acids ('K-motif') and some have as many as four (Figure 3). A chemical effect of having two contiguous Lys residues is that the pK of the ε-amino group is lowered from 10 to 5, which means that it will be uncharged at physiological pH (Fersht, 1999). However, in most instances this chemical effect is more dependent on secondary and tertiary protein structure than on primary structure, and this information will be available only when the 3-dimensional structures of the active sites of the glycosyltransferases are elucidated. Interestingly, the K-motif has been described in other enzymes suspected to be related to plant polysaccharide synthesis. These include the endo-$(1\rightarrow4)\beta$-D-glucanase encoded by the KORRIGAN gene, an enzyme suspected to be involved in polysaccharide synthesis rather than hydrolysis (Nicol *et al.*, 1998; Zuo *et al.*, 2000), and the reversibly glycosylated protein from *Arabidopsis* (AtRGP) (Delgado *et al.*, 1998). Chitin synthase (Bulawa *et al.*, 1986), heparan sulfate synthase (Lind *et al.*, 1993) and hyaluronan synthase (DeAngelis *et al.* 1993) also contain the K-motif. In the hyaluronan synthase family, which includes sequences from prokaryotes as well as eukaryotes, the K-motif is found within a region that is unique to the eukaryotic members of the family (Weigel *et al.*, 1997), as is the case for the CSR of the CesA polypeptides.

Conservation of these motifs implicates the CSR in the function of these synthases. There are several examples where the functions of enzymes in subclasses from a multigene family are altered by limited amino acid substitutions rather than large conformational changes. One such example is a family of human fucosyltransferases in which substrate utilization is determined by a short 'hypervariable segment' composed of basic amino acids (Legault *et al.*, 1995). Substrate binding and catalysis in these fucosyl- and other glycosyltransferases involve hydrogen bond interactions between the reactive hydroxyl group of the acceptor with one or more basic groups on the enzyme. Gene variability at positions encoding charged amino acids has also been documented previously (McLean *et al.*, 1990). A comparative study between *Arabidopsis* and human actin protein families revealed that, unlike the human actin proteins, the *Arabidopsis* sub-classes exhibited variations in non-conservative charged residues, i.e. Lys, Arg, His, Asp, and Glu, or Pro, an amino acid that substantially changes the peptide backbone. These variations are conserved among members of the same sub-class (McDowell *et al.*, 1996). The importance of charged residues in the function of actin is supported by site-directed mutagenesis in the yeast actin protein (Wertman *et al.*, 1992). Comparably, variations in the charged and Pro residues in the CSR are conserved or absent in a sub-class-specific manner (Figure 3).

The features revealed by the amino acid analysis within the CSR are reflected in the topology of the phylogenetic trees. All the sequences forming the sub-class I clade have a conserved Pro residue that is not present in the other CesA polypeptides. In addition, a Pro residue that is present in most of the sequences is absent in the clade formed by *AtCesA2*, *AtCesA6*, and *AtCesA9*. These sequences also differ from the other in the K-motif sequence (Figures 3 and 5). Sub-class II is characterized by having three to four K-motifs; two of these motifs and the DDE motif are conserved within members of this sub-class (Figure 3). Members of sub-class III have one K-motif containing more than three basic amino acids. The second D-motif in this sub-class is EDE, with the exception of *AtCesA10*. Interestingly, the two *Arabidopsis* sequences in this subclass also have an additional D-motif towards the N-terminal region of the CSR (Figure 3).

Speculations about the catalytic domain of the CesA proteins and mechanisms of $(1\rightarrow4)\beta$-glycan synthesis

The catalytic reaction forming a β-glycosidic linkage involves the transfer of a glycosyl residue from a nucleotide-sugar substrate to a growing carbohydrate chain. Evidence to date supports addition to the non-reducing terminal sugar (Koyama *et al.*, 1997). A common thread that links the family 2 $(1\rightarrow4)\beta$-glycan synthases is the 180 °C inversion of the sugar residues with respect to each neighbor. With each sin-

gle addition of a glycosyl residue, the position of the O-4 of the glycosyl acceptor at the non-reducing terminus is moved. We and others proposed that two sites or modes of glycosyl transfer reside within a single catalytic subunit and disaccharide units are added iteratively (Carpita *et al.*, 1996; Koyama *et al.*, 1997; Carpita and Vergara, 1998; Buckeridge *et al.*, 1999, 2001). In this model, glycobiosyl units, or any even-numbered oligomeric units, are added to the non-reducing end to ensure that the $(1\rightarrow4)\beta$ linkages are strictly preserved without inversion of substrate, active site, or terminus of the growing chain.

The crystal structure of the bacterial SpsA synthase provides a first detailed picture of the mechanism of nucleotide-sugar binding and catalysis of glycosyl transfer (Charnock and Davies, 1999). For example, the conserved Asp-39, corresponding to that in the U-1 motif, interacts ionically with the imide of the uracil ring, whereas Asp-98 and Asp-99, which correspond to the DxD or DDx in the U-2 motif, interact with the ribose C-2 hydroxyl group and the Mn^{+2} ion bound to the α- and β-phosphoryl units of UDP, respectively. In addition, two basic residues, Arg-71 and Lys-13, and Thr-9 function in nucleotide binding, and Tyr-11 is predicted to position the uracil by aromatic stacking. Because the protein was co-crystallized with UDP rather than a nucleotide-sugar, the roles of the Asp residue in U-3 and the QxxRW of U-4 remain undefined, although they probably function in the catalytic transfer.

The logic of the two-site mechanism of synthesis notwithstanding, the 3-dimensional structure of the bacterial SpsA synthase argues against more than one glycosyltransferase activity residing in a single polypeptide. Unfortunately, neither the uridine nucleotide-sugar substrate nor the polymer structure made by the SpsA synthase is known. Hence, further comparisons between this synthase and the CesA synthase family are not possible. The conserved residues in the U-1 and U-2 motifs are essential for nucleotide binding in several processive glycosyltransferases (Saxena *et al.*, 1995). However, several features of the biochemical mechanisms of synthesis of $(1\rightarrow4)\beta$-linked cell wall polysaccharides remain to be determined. The information contained within regions that are unique to the eukaryotic family 2 synthases, such as the CSR in the CesA proteins and the conserved motifs, will prove critical in understanding the catalytic and regulatory processes of these proteins.

Our results indicate that conservation of a primary amino acid sequence in one of the plant-specific insertions in the CSR defines 3–5 different sub-classes of *CesA* genes in higher plants. The additional DDx, DxD motifs iterated within the CSR might represent alternating modes of nucleotide-sugar positioning that creates 'toggling' mechanism whereby substrates bind in two conformations to ensure formation of $(1\rightarrow4)\beta$ linkages with respect to the O-4 position of the non-reducing terminal acceptor glycosyl residue. Another possibility is that residues within this region may interact with distant residues in the linear sequence effecting conformational changes that might alter substrate utilization in order to regulate the unit and linkage structure of the polymer being made. This kind of dual effect has been described for the hyaluronan synthases (DeAngelis *et al.*, 1993) and heparan sulfate synthase (Lind *et al.*, 1993).

Whereas the 3D structure of the SpsA synthase argues against more than one catalytic site per subunit, several recent studies provide evidence that cellulose synthase may be a dimer of two different CesA polypeptide (see Haigler *et al.*, 2001, this issue). First, it can be argued on strictly steric grounds that 8 membrane spans are insufficient to provide large enough a channel for the extrusion of a large glucan chain. The discovery of 16 membrane spans in the FKS-like callose synthases (Doblin *et al.*, 2001; Hong *et al.*, 2001) furthers the suggestion that dimerization of CesA products produces an equivalent channel size. In the instance of the latter, two catalytic subunits instead of just one would provide the structural elements of a two-site mechanism of β-glycobiosyl addition. Second, *RSW*1 (*AtCesA*1) and *PRC*1 (*AtCesA*6) are expressed in the same cell yet mutations in either one is sufficient to disrupt cellulose synthesis (Arioli *et al.*, 1998; Fagard *et al.*, 2000). Clearly, their co-expression is not a redundancy in the strict sense. Third, *IRX*1 (*AtCesA*8) and *IRX*3 (*AtCesA*7) are not only expressed in the same cells of the xylem and mutations in either disrupt cellulose synthesis, but also direct evidence for their interaction has been provided (Taylor *et al.*, 2000).

Structurally distinct from cellulose and other strictly $(1\rightarrow4)\beta$-linked glycans, the unique linkage structure of the cereal mixed-linkage β-glucans provides a valuable lesson to be learned about synthase mechanisms of all β-linked polysaccharides (Buckeridge *et al.*, 1999, 2001). The trisaccharide unit of β-glucan can be formed if an ancestral cellulose synthase complex is modified to contain a third mode of glucosyl transfer in addition to the cellobiose generating activity (Carpita *et al.*, 1996). The acquisition of

a mechanism to add primarily an odd number of glucosyl units means that the first and third residues are not inverted 180 °C relative to each other and, instead of the O-4, the O-3 hydroxyl of the non-reducing terminal sugar is positioned for attachment (Buckeridge *et al.*, 2001). The β-glucan synthesis mechanism is the exception that proves the rule: the odd-numbered cellodextrin units of β-glucan will always be linked by single $(1\rightarrow3)\beta$ linkages, whereas the cellobiose units in the even-numbered synthases will always be linked by $(1\rightarrow4)\beta$ linkages. The position and number of K and D motifs within the CSR might function in the regulation of substrate specificity, substrate affinity, and the number of modes of glycosyl transfer in inverting-type synthases. In addition to the practical limits of function predicted on similarity, the relevance of these possibilities will await further experimental investigation in the function of these proteins and the availability of tertiary structures for the enzymes.

With respect to evolutionary relationships, our phylogenetic analyses show that the *CesA* gene family can be divided into at least four sub-classes (Figure 5). Two groups of *CesA* cereal sequences appear to be more closely related among themselves, forming independent clades separated from the dicot species. Nonetheless, the major groups in the tree are built with sequences from dicot and monocot species, suggesting that the sub-classes originated from lineages that split before speciation events (Holland *et al.*, 2000). The *CesA* genes may represent an ancient class or classes specialized for a certain function, and these classes diverged further into sub-classes that have evolved significant functional and regulatory differences. Gene duplications are positively selected when they confer adaptive value to the plant. Redundancy can increase the amount of a gene product, can modify a catalytic function, or can permit a developmental regulation to optimize expression in different tissues and at different stages during the life cycle of a plant. All of these possibilities are certainly feasible with respect to the synthesis of plant cell wall polysaccharides, considering the diverse functional roles these polymers play in plant growth and development. As more *CesA* gene families from higher plants are characterized, the confident designation of orthologues in different species will become possible and the resolution of the *CesA* gene phylogenies should increase. Richmond and Somerville (2000) suggest that *CesA* genes encode only cellulose synthases, and that the six *Csl* gene families encode the synthases of the six major non-cellulosic polysaccharides of the cell walls

158

of all angiosperm species. They caution that sequence divergence does not necessarily mean functional divergence. Conversely, we caution that subtle sequence divergence could generate functional divergence. These considerations are made ever more important when considering that in addition to the six non-cellulosic cell-wall polymers cited by Richmond and Somerville (2000), processive synthases for the type I $(1\rightarrow4)\beta$-D-galactans, the type II $(1\rightarrow3),(1\rightarrow6)\beta$-D-galactans, $(1\rightarrow5)\alpha$-L-arabinans, $(1\rightarrow4)\beta$-D-glucomannans, and galactomannans also remain to be identified. A working hypothesis that emerged from our study suggests that the plant *CesA* sub-classes represent proteins that have evolved significant functional and regulatory differences. Therefore, the *CesA* genes might not be restricted to encoding cellulose synthases but also of non-cellulosic glycan synthases that catalyze the formation of other polymers with backbones of $(1\rightarrow4)\beta$-linked residues. A decade ago, not one single gene encoding a plant cell wall polysaccharide synthase had been identified. Today, the situation is very different, with databases devoted solely to the sequence update of putative glycosyltransferases in great many species, and Web site devoted to the study of these genes (see http://cellwall.stanford.edu/cellwall/; http://afmb.cnrs-mrs.fr/~pedro/CAZY/db.html). The discovery of the *CesA* genes by Pear *et al.* (1996) was instrumental in this transition. A major challenge is to deduce the specific functions of the enormous number of genes and protein sequences we now have accumulated.

Acknowledgements

We thank Deborah P. Delmer (UC Davis) and Kanwarpal Dhugga (Pioneer Hibred, Intl.) for providing maize CesA sequences in advance of publication. We thank Drs Larry Dunkle, Steve Goodwin, Mark Hermodson, Mark Levinthal, Catherine Rayon, Chris Staiger, and Charles Woloshuk for critical review of the manuscript and many helpful suggestions. This work was supported by a grant for the US Department of Energy, Division of Energy Biosciences. Journal paper 16,400 of the Purdue University Agricultural Experiment Station.

References

Altschul, S.F., Madden, T.L., Schaffer, A.A., Zhang, J.H., Zhang, Z., Miller, W. and Lipman, D.J. 1997. Gapped BLAST and PSI-BLAST: a new generation of protein database search programs. Nucl. Acids Res. 25: 3389–3402.

Amor, Y., Haigler, C.H., Johnson, S., Wainscott, M. and Delmer, D.P. 1995. A membrane-associated form of sucrose synthase and its potential role in synthesis of cellulose and callose in plants. Proc. Natl. Acad. Sci. USA 92: 9353–9357.

Anderson, M.A. and Stone, B.A. 1975. A new substrate for investigating the specificity of β-glucan hydrolases. FEBS Lett. 52: 202–207.

Arioli, T., Peng, L.C., Betzner, A.S., Burn, J., Wittke,, W., Herth, W., Camilleri, C., Höfte, H., Plazinski, J., Birch, R., Cork, A., Glover, J., Redmond, J. and Williamson, R.E. 1998. Molecular analysis of cellulose biosynthesis in *Arabidopsis*. Science 279: 717–720.

Bacic, A. and Delmer, D.P. 1981. Stimulation of membrane-associated polysaccharide synthetases by a membrane potential in developing cotton fibers. Planta 152: 346–351.

Becker, M., Vincent, C. and Reid, J.S.G. 1995. Biosynthesis of $(1\rightarrow3),(1\rightarrow4)$-$\beta$-glucan in barley (*Hordeum vulgare* L.). Planta 195: 331–338.

Breton, C. and Imberty, A. 1999. Structure/function studies of glycosyltransferases. Curr. Opin. Struct. Biol. 9: 563–571.

Brown, R.M. Jr. and Saxena, I.M. 2000. Cellulose biosynthesis: a model for understanding the assembly of biopolymers. Plant Physiol. Biochem. 38: 57–67.

Buckeridge, M.S., Vergara, C.E. and Carpita, N.C. 1999. The mechanism of synthesis of a cereal mixed-linkage $(1\rightarrow3),(1\rightarrow4)$-$\beta$-D-glucan: evidence for multiple sites of glucosyl transfer in the synthase complex. Plant Physiol. 120: 1105–1116.

Buckeridge, M.S., Vergara, C.E. and Carpita, N.C. 2001. Insight into multi-site mechanisms of glycosyl transfer in $(1\rightarrow4)\beta$-D-glycan synthases provided by the cereal mixed-linkage $(1\rightarrow3),(1\rightarrow4)\beta$-D-glucan synthase. Phytochemistry, in press.

Bulawa, C.E., Slater, M., Cabib, E., Au-Young, J., Sburlati, A., Adair, W.L. Jr. and Robbins, P.W. 1986. The *S. cerevisiae* structural gene for chitin synthase is not required for chitin synthesis *in vivo*. Cell 46: 213–225.

Burton, R.A, Gibeaut, D.M., Bacic, A., Findlay, K., Roberts, K., Hamilton, A., Baulcombe, D.C. and Fincher, G.B. 2000. Virus-induced silencing of a plant cellulose synthase gene. Plant Cell 12: 691–705.

Carpita, N.C. 1996. Structure and biogenesis of the cell walls of grasses. Annu. Rev. Plant Physiol. Plant Mol. Biol. 47: 445–476.

Carpita, N.C. and Delmer, D.P. 1980. Protection of cellulose synthesis in detached cotton fibers by polyethylene glycol. Plant Physiol 66: 911–916.

Carpita, N.C. and Gibeaut, D.M. 1993. Structural models of the primary cell walls in flowering plants: consistency of molecular structure with the physical properties of the walls during growth. Plant J. 3: 1–30.

Carpita, N.C. and Vergara, C.E. 1998. A recipe for cellulose. Science 279: 672–673.

Carpita, N.C, McCann, M. and Griffing, L.R. 1996. The plant extracellular matrix: news from the cell's frontier. Plant Cell 8: 1451–1463.

Charnock, S.J. and Davies, G.J. 1999. Structure of the nucleotide-diphospho-sugar transferase, SpsA from *Bacillus subtilis*, in native and nucleotide-complexed forms. Biochemistry 38: 6380–6385.

Cui, X., Shin, H., Song, C.C., Laosinchai, W., Amano, Y. and Brown, R.M. Jr. 2001. A putative plant homolog of the yeast β-1,3-glucan synthase subunit FKS1 from cotton (*Gossypium hirsutum* L.) fibers. Planta, in press

DeAngelis, P.L., Papaconstantinou, J. and Weigel, P.H. 1993. Molecular cloning, identification, and sequence of the hyaluronan synthase gene from Group A *Streptococcus pyogenes*. J. Biol. Chem. 268: 19181–19184.

Delgado, I.J., Wang, Z., deRocher, A., Keegstra, K. and Raihkel, N.V. 1998. Cloning and characterization of AtRGP1. A reversibly autoglycosylated *Arabidopsis* protein implicated in cell wall biosynthesis. Plant Physiol. 116: 1339–1349.

Delmer, D.P. 1977. Biosynthesis of cellulose and other plant cell wall polysaccharides. Rec. Adv. Phytochem. 11: 105–153.

Delmer, D.P. 1999. Cellulose biosynthesis: exciting times for a difficult field of study. Annu. Rev. Plant Physiol. Plant Mol. Biol. 50: 245–276.

Delmer, D.P., Benziman, M. and Padan, E. 1982. Requirement for a membrane potential for cellulose synthesis in intact cells of *Acetobacter xylinum*. Proc. Natl. Acad. Sci. USA 79: 5282–5286.

Devereaux, J., Haeberli, P. and Smithies, O.A. 1984. Comprehensive set of sequence analysis programs for the VAX. Nucl. Acids Res. 12: 387–395.

Dhugga, K.S. and Ray, P.M. 1994. Purification of $(1\rightarrow 3)$-β-D-glucan synthase activity from pea tissue. Two polypeptides of 55 kDa and 70 kDa copurify with enzyme activity. Eur. J. Biochem. 220: 943–953.

Dhugga, K.S., Tiwari, S.C. and Ray, P.M. 1997. A reversibly glycosylated polypeptide (RGP1) possibly involved in plant cell wall synthesis: purification, gene cloning, and trans-Golgi localization. Proc. Natl. Acad. Sci. USA 94: 7679–7684.

Doblin, M.S., De Melis, L., Newbigin, E., Bacic, A. and Read, S.M. 2001. Pollen tubes of *Nicotiana alata* express two genes from different β-glucan synthase families. Plant Physiol., in press.

Douglas, C.M., Foor, F., Marrinan, J.A., Morin, N., Nielsen, J.B., Dahl, A.M. *et al.* 1994. The *Saccharomyces cerevisiae* FKS1 (ETG1) gene encodes an integral membrane protein which is a subunit of $(1\rightarrow 3)$-β-D-glucan synthase. Proc. Natl. Acad. Sci. USA 91: 12907–12911.

Doyle, J.J. and Gaut, B.S. 2000. Evolution of genes and taxa: a primer. Plant Mol. Biol. 42: 1–23.

Fagard, M., Desnos, T., Desprez, T., Goubet, F., Refregier, G., Mouille, G. *et al.* 2000. *PROCUSTE1* encodes a cellulose synthase required for normal cell elongation specifically in roots and dark-grown hypocotyls of *Arabidopsis*. Plant Cell 12: 2409–2423.

Fersht, A. 1999. Structure and Mechanism in Protein Science: A Guide to Enzyme Catalysis and Protein Folding. W.H. Freeman, New York.

Gastinel, L.N., Cambillau, C. and Bourne, Y. 1999. Crystal structures of the bovine β-4-galactosyltransferase catalytic domain and its complex with uridine diphosphate galactose. EMBO J. 18: 3546–3557.

Geremia, R.A., Mergaert, P., Geelen, D., Van Montagu, M. and Holsters, M. 1994. The NodC protein of *Azorhizobium caulinodans* is an N-acetylglucosaminyltransferase. Proc. Natl. Acad. Sci. USA 91: 2669–2673.

Gibeaut, D.M. 2000. Nucleotide sugars and glycsoyltransferases for synthesis of cell wall matrix polysaccharides. Plant Physiol. Biochem. 38: 69–80.

Gibeaut, D.M. and Carpita, N.C. 1993. Synthesis of $(1\rightarrow 3),(1\rightarrow 4)\beta$-D-glucan in the Golgi apparatus of maize coleoptiles. Proc. Natl. Acad. Sci. USA 90: 3850–3854.

Gibeaut, D.M. and Carpita, N.C. 1994a. Biosynthesis of plant cell-wall polysaccharides. FASEB J. 8: 904–915.

Gibeaut, D.M. and Carpita, N.C. 1994b. Improved recovery of $(1\rightarrow 3),(1\rightarrow 4)\beta$-D-glucan synthase activity from Golgi apparatus of *Zea mays* (L.) using differential centrifugation. Protoplasma 180: 92–97.

Gordon, R. and Maclachlan, G. 1989. Incorporation of UDP-[^{14}C]glucose into xyloglucan by pea membranes. Plant Physiol. 91: 373–378.

Haigler, C.H., Ivanova-Datcheva, M., Hogan, P.S., Salnikov, V.V., Hwang, S., Martin, L.K. and Delmer, D.P. 2001. Carbon partitioning to cellulose synthesis. Plant Mol. Biol., this issue.

Henrissat, B., Coutinho, P.M. and Davies, G.J. 2001. A census of carbohydrate-active enzymes in the genome of *Arabidopsis thaliana*. Plant Mol. Biol., this issue.

Henry, R.J. and Stone, B.A. 1982. Factors influencing β-glucan synthesis by particulate enzymes from suspension-cultured *Lolium multiflorum* endosperm cells. Plant Physiol. 69: 632–636.

Hirai, N., Sonobe, S. and Hayashi, T. 1998. *In situ* synthesis of β-glucan microfibrils on tobacco plasma membrane sheets. Proc. Natl. Acad. Sci. USA 95: 15102–15106.

Hobbs, M.C., Delange, M.H.P., Baydoun, E.A.-H. and Brett, C.T. 1991. Differential distribution of a glucuronoyltransferase, involved in glucuronoxylan synthesis, with the Golgi apparatus of pea (*Pisum sativum* var. Alaska). Biochem. J. 277: 653–658.

Holland, N., Holland, D., Helentjaris, T., Dhugga, K., Xoconostle-Cazares, B. and Delmer, D.P. 2000. A comparative analysis of the plant cellulose synthase (*Ces*A) gene family. Plant Physiol. 123: 1313–1323.

Hong, Z., Delauney, A.J. and Verma, D.P.S. 2001. A cell plate-specific callose synthase and its interaction with phragmoplastin. Plant Cell, in press.

Kawagoe, Y. and Delmer, D.P. 1997. Pathways and genes involved in cellulose biosynthesis. Genet. Eng. 19: 63–87.

Kim, J.B., Olek, A.T. and Carpita, N.C. 2000. Cell wall and membrane-associate exo-β-D-glucanases from developing maize seedlings. Plant Physiol. 123: 471–485.

Koyama, M., Helbert, W., Imai, T., Sugiyama, J. and Henrissat, B. 1997. Parallel-up structure evidences the molecular directionality during biosynthesis of bacterial cellulose. Proc. Natl. Acad. Sci. USA 94: 9091–9095.

Legault, D.J., Kelly, R.J., Natsuka, Y. and Lowe, J.B. 1995. Human α (1,3/1,4)-fucosyltransferases discriminate between different oligosaccharide acceptor substrates through a discrete peptide fragment. J. Biol. Chem. 270: 20987–20996.

Lind, T., Lindahl, U. and Lidholt, K.J. 1993. Biosynthesis of heparin heparan sulfate. Identification of a 70 kDa protein catalyzing both the D-glucuronosyltransferase and the N-acetyl-D-glucosaminyltransferase reactions. J. Biol. Chem. 268: 20705–20708.

Matthysse, A.G., White, S. and Lightfoot, R. 1995. Genes required for cellulose synthesis in *Agrobacterium tumefasciens*. J. Bact. 177: 1069–1075.

McCann, M.C. and Roberts, K. 1991. Architecture of the primary cell wall. In: C.W. Lloyd (Ed.) The Cytoskeletal Basis of Plant Growth and Form, Academic Press, London, pp. 109–129.

McDowell, J.M., Huang, S., McKinney, E.C., An, Y.Q. and Meagher, R.B. 1996. Structure and evolution of the actin gene family in *Arabidopsis thaliana*. Genetics 142: 587–602.

McLean, B.G., Huang, E.C., McKinney, E.C. and Meagher, R.B. 1990. Plants contain highly divergent actin isovariants. Cell Motil. 17: 276–290.

Meinert, M.C. and Delmer, D.P. 1977. Changes in biochemical composition of cell wall of cotton fiber during development. Plant Physiol. 59: 1088–1097.

Muñoz, P., Norambuena, L. and Orellana, A. 1996. Evidence for a UDP-glucose transporter in Golgi apparatus-derived vesicles from pea and its possible role in polysaccharide biosynthesis. Plant Physiol. 112: 1585–1594.

Nicol, F., His, I., Jauneau, A., Vernhettes, S., Canut, H. and Höfte, H. 1998. A plasma membrane-bound putative endo-1,4-β-D-glucanase is required for normal wall assembly and cell elongation in *Arabidopsis*. EMBO J. 17: 5563–5576.

Pear, J.R., Kawagoe, Y., Schreckengost, W.E., Delmer, D.P. and Stalker, D.M. 1996. Higher plants contain homologs of the bacterial *cel*A genes encoding the catalytic subunit of cellulose synthase. Proc. Natl. Acad. Sci. USA 93: 12637–12642.

Perrin, R., Wilkerson, C. and Keegstra, K. 2001. Golgi enzymes that synthesize plant cell wall polysaccharides: finding and evaluating candidates in the genomic era. Plant Mol. Biol., this issue.

Piro, G., Dalessandro, G. and Northcote, D.H. 1993. Glucomannan synthesis in pea epicotyls: the mannose and glucose transferases. Planta 190: 206–220.

Reid, J.S.G., Edwards, M., Gidley, M.J. and Clark, A.H. 1995. Enzyme specificity in galactomannan biosynthesis. Planta 195: 489–495.

Richmond, T.A. and Somerville, C.R. 2000. The cellulose synthase superfamily. Plant Physiol. 124: 495–498.

Richmond, T.A. and Somerville, C.R. 2001. Integrative approaches to determining *Csl* function. Plant Mol. Biol., this issue.

Roemer, T., Paravicini, G., Payton M.A. and Bussey, H. 1994. Characterization of the yeast (1,6)-β-glucan biosynthetic components, Kre 6p and Skn1p, and genetic interactions between the PKC1 pathway and extracellular matrix assembly. J. Cell Biol. 127: 567–579.

Saxena, I.M., Lin, F.C., Brown, R.M. Jr. 1990. Cloning and sequencing of the cellulose synthase catalytic subunit gene of *Acetobacter xylinum*. Plant Mol. Biol. 15: 673–683.

Saxena, I.M., Brown, R.M. Jr., Fevre, M., Geremia, R.A. and Henrissat, B. 1995. Multidomain architecture of β-glucosyl transferases: implications for mechanism of action. J. Bact. 177: 1419–1424.

Schlüpmann, H., Bacic, A. and Read, S.M. 1993. A novel callose synthase from pollen tubes of *Nicotiana*. Planta 191: 470–481.

Smith, B.G. and Harris, P.J. 1999. The polysaccharide composition of Poales cell walls: Poaceae cell walls are not unique. Biochem. System. Ecol. 27: 33–53.

Spicer, A.P., Olson, J.S. and McDonald, J.A. 1997. Molecular cloning and characterization of a cDNA encoding the third putative mammalian hyaluronan synthase. J. Biol. Chem. 272: 8957–8961.

Stasinopoulis, S.J., Fisher, P.R., Stone, B.A. and Stanisich, V.A. 1999. Detection of two loci involved in (1→3)-β-glucan (curdlan) biosynthesis by *Agrobacterium* sp. ATCC31749, and comparative sequence analysis of the putative curdlan synthase gene. Glycobiology 9: 31–41.

Taylor, N.G., Scheible, W.R., Cutler, S., Somerville, C.R. and Turner, S.R. 1999. The *irregular xylem*3 locus of *Arabidopsis* encodes a cellulose synthase required for secondary cell wall synthesis. Plant Cell 11: 769–779.

Taylor, N.G., Laurie, S. and Turner, S.R. 2001. Multiple cellulose synthase catalytic subunits are required for cellulose synthesis in *Arabidopsis*. Plant Cell 12: 2529–2539.

Thompson, J.D., Higgins, D.G. and Gibson, T.J. 1994. Clustal W: improving the sensitivity of progressive multiple sequence alignment through sequence weighting, position-specific gap penalties and weight matrix choice. Nucl. Acids Res. 22: 4673–4680.

Turner, S.R. and Somerville, C.R. 1997. Collapsed xylem phenotype of *Arabidopsis* identifies mutants deficient in cellulose deposition in the secondary cell wall. Plant Cell 9: 689–701.

Turner, A., Bacic, A., Harris, P.J. and Read, S.M. 1998. Membrane fractionation and enrichment of callose synthase from pollen tubes of *Nicotiana alata* Link et Otto. Planta 205: 380–388.

Turner, S., Taylor, N. and Jones, L. 2001. Mutations of the secondary wall. Plant Mol. Biol., this issue.

Weigel, P.H., Hascall, V.C. and Tammi, M. 1997. Hyaluronan synthases. J. Biol. Chem. 272: 13997–14000.

Wertman, K.F., Drubin, D.G. and Botstein, D. 1992. Systematic mutational analysis of the yeast ACT1 gene. Genetics 132: 208–211.

Wiggins, C.A.R. and Munro, S. 1998. Activity of the yeast MNN1 β-1,3-mannosyltransferase requires a motif conserved in many other families of glycosyltransferases. Proc. Natl. Acad. Sci. USA 95: 7945–7950.

Winter, H. and Huber, S.C. 2000. Regulation of sucrose metabolism in higher plants: localization and regulation of activity of key enzymes. Crit. Rev. Biochem. Mol. Biol. 35: 253–289.

Wong, H.C., Fear, A.L., Calhoon, R.D., Eichinger, G.H., Mayer, R., Amikam, D. *et al.* 1990. Genetic organization of the cellulose synthase operon in *Acetobacter xylinum*. Proc. Natl. Acad. Sci. USA 87: 8130–8134.

Wood, P.J., Weisz, J. and Blackwell, B.A. 1994. Structural studies of (1→3),(1→4)-β-D-glucans by ^{13}C-nuclear magnetic resonance spectroscopy and by rapid analysis of cellulose-like regions using high-performance anion-exchange chromatography of oligosaccharides released by lichenase. Cereal Chem. 71: 301–307.

Zuo, J.R., Niu, Q.W., Nishizawa, N., Wu, Y., Kost, B. and Chua, N.H. 2000. KORRIGAN, an *Arabidopsis* endo-1,4-β-glucanase, localizes to the cell plate by polarized targeting and is essential for cytokinesis. Plant Cell 12: 1137–1152.

Plant Molecular Biology **47**: 161–176, 2001.
© 2001 *Kluwer Academic Publishers. Printed in the Netherlands.*

161

The complex structures of arabinogalactan-proteins and the journey towards understanding function

Yolanda Gaspar[1], Kim L. Johnson[1], James A. McKenna[1], Antony Bacic[1,2,*] and Carolyn J. Schultz[2,3]
[1]*Plant Cell Biology Research Centre, University of Melbourne, 3010, Australia (*author for correspondence; e-mail abacic@unimelb.edu.au);* [2]*Cooperative Research Centre for Bioproducts, University of Melbourne, 3010, Australia;* [3]*Current address: Department of Plant Science, Adelaide University, RMB1 Glen Osmond, SA 5064, Australia*

Key words: *Arabidopsis,* arabinogalactan-protein, cell wall, fasciclin, glycosylphosphatidylinositol-anchor, hydroxyproline

Abstract

Arabinogalactan-proteins (AGPs) are a family of complex proteoglycans found in all higher plants. Although the precise function(s) of any single AGP is unknown, they are implicated in diverse developmental roles such as differentiation, cell-cell recognition, embryogenesis and programmed cell death. DNA sequencing projects have made possible the identification of the genes encoding a large number of putative AGP protein backbones. In contrast, our understanding of how AGPs undergo extensive post-translational modification is poor and it is important to understand these processes since they are likely to be critical for AGP function. Genes believed to be responsible for post-translational modification of an AGP protein backbone, include prolyl hydroxylases, glycosyl transferases, proteases and glycosylphosphatidylinositol-anchor synthesising enzymes. Here we examine models for proteoglycan function in animals and yeast to highlight possible strategies for determining the function(s) of individual AGPs in plants.

Abbreviations: AGP, arabinogalactan-protein; AG, arabinogalactan; GlcY, β-glucosyl Yariv reagent; ER, endoplasmic reticulum; GlcNH$_2$, glucosamine; GPI, glycosylphosphatidylinositol; HRGP, hydroxyproline-rich glycoprotein; PLC, phospholipase C; PLD, phospholipase D; RP-HPLC, reversed-phase high-performance liquid chromatography

Introduction

Plant cell walls play a pivotal role in determining the final shape and function of plant cells and tissues (Bacic *et al.*, 1988; Carpita and Gibeaut, 1993). They consist of complex macromolecules and are principally comprised of carbohydrate building blocks. Small proportions of the macromolecules contain both carbohydrate and protein and are known collectively as proteoglycans or glycoproteins, depending on the relative proportions of carbohydrate and protein. The protein component of these molecules often contain abundant amounts of hydroxyproline (Hyp), an imino acid that is also found in proteins of the extracellular matrix in animals, although in animals Hyp residues are not glycosylated. This family of plant molecules is known as the Hyp-rich glycoprotein (HRGP) family, and includes the extensins, arabinogalactan-proteins (AGPs), Pro/Hyp-rich glycoproteins and the solanaceous lectins (Showalter, 1993; Nothnagel, 1997; Sommer-Knudsen *et al.*, 1997; Bacic *et al.*, 2000). In terms of structure, these sub-classes of HRGPs should not be viewed as discrete families but rather as a continuum of molecules ranging from basic, minimally glycosylated proteins, for example Pro-rich proteins, to acidic highly glycosylated proteoglycans, such as AGPs (Sommer-Knudsen *et al.*, 1998).

AGPs were traditionally differentiated from the other HRGPs by the linkages of their arabinosyl and galactosyl residues and their ability to bind a class of synthetic phenylazo dyes, the β-glycosyl Yariv reagents (for example, βGlcY) (Yariv et al., 1967). The carbohydrate residues in AGPs occur predominantly as type II arabinogalactans (AGs); $(1\rightarrow3)\beta$-linked galactopyranosyl (D-Galp) residues form a backbone that is substituted at $C(O)6$ by side-chains of $(1\rightarrow6)\beta$-D-linked Galp (Figure 1A). The side-chains often terminate in α-L-arabinofuranosyl (L-Araf) and other sugars, such as Fucp, Rhap and GlcpA (with or without 4-O-methyl) are common in some AGPs (reviewed in Nothnagel, 1997; Serpe and Nothnagel, 1999; Bacic et al., 2000). The carbohydrate is generally linked to the protein backbone via Hyp (McNamara and Stone, 1981; Bacic et al., 1988) and some arabinogalactan (AG) chains may consist of as many as 95–120 sugar residues (Gane et al., 1995).

Several groups have demonstrated, based on susceptibility to periodate oxidation, that the $(1\rightarrow3)\beta$-D-Galp backbone comprises a repetitive structure of about seven Galp residues, interspersed by a periodate-sensitive linkage that is postulated to be either $(1\rightarrow5)\alpha$-L-Araf or $(1\rightarrow6)\beta$-D-Galp (Churms et al., 1981; Bacic et al., 1987). Detailed structural analyses of a type II AG from *Larix* (larch) (Prescott et al., 1995) and a type II pectic AG from *Angelica* root cell walls (Kiyohara et al., 1997) are generally consistent with the proposed structures. The gum arabic glycoprotein (GAGP) from *Acacia senegal* contains mono-, oligo- and polysaccharide chains attached to Hyp residues in the protein backbone (Qi et al., 1991; Goodrum et al., 2000). Partial protein sequence of this unusual AGP was recently obtained and the following distribution of oligo- and polysaccharides was inferred based on the Hyp-contiguity hypothesis; the extensin-like short Ara chains are added to Hyp residues in (Ser/Thr)-Hyp3 motifs and larger AG chains are added to the non-contiguous Hyp residues in Ser-Hyp-Thr-Hyp-Thr (Figure 1B) (Goodrum et al., 2000). Support for the Hyp contiguity hypothesis was obtained using synthetic genes (Shpak et al., 1999). One outstanding question is whether a single Hyp residue is sufficient for glycosylation.

During the 1970s, amino acid analyses indicated that AGPs generally consist of less than 10% w/w protein, and are rich in the residues Hyp/Pro, Ala, Ser and Thr (reviewed in Clarke et al., 1979). The motifs Ala-Pro and Pro-Ala have been suggested to

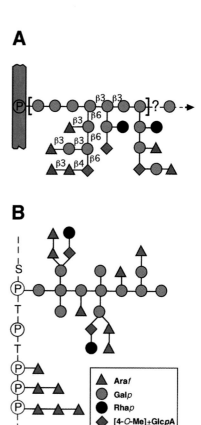

Figure 1. Schematic representation of the carbohydrate modifications of AGPs. A. One Hyp residue (denoted by a circled P) of an AGP protein backbone is shown with a large AG-chain attached to it as proposed in the 'wattle-blossom' model of the structure of an AGP (Fincher et al., 1983). In this model there may be as many as 25 Hyp residues in the protein backbone, each bearing a similar AG-chain. Each AG chain contains repeats of a $(1\rightarrow3)\beta$-D-linked Galp oligosaccharide with a degree of polymerization of about 7 (only one such repeat is shown, within the brackets) (Churms et al., 1981; Bacic et al., 1987). There may be ten or more repeats (indicated by an arrow). The repeats are linked via a periodate-sensitive sugar linkage (indicated by '?') proposed to be either $(1\rightarrow5)\alpha$-L-linked Araf or $(1\rightarrow6)\beta$-D-linked Galp (Bacic et al., 1987). Branching of $(1\rightarrow6)\beta$-D-linked Galp oligosaccharides with non-reducing terminal [4-O-Me]$\pm$GlcpA can occur from any Gal in the backbone chain. Further branching includes a large degree of Araf and Galp oligosaccharides and minor addition of Rhap. B. A revised model of the 'twisted hairy rope' structure (Qi et al., 1991) of an AGP from gum arabic (GAGP) based on recently obtained structural data (Goodrum et al., 2000). Eight amino acid residues (single-letter code) of the protein backbone are shown. The terminal and branching sugars in this model have been positioned arbitrarily. In GAGP, it is proposed that the pattern of glycosylation depends on whether Pro residues are contiguous or non contiguous (Goodrum et al., 2000). Contiguous Pro residues, i.e. two or more Pro/Hyp residues in a row, have short Araf chains attached to them, and non-contiguous Pro/Hyp residues have larger AG chains (containing ca. 23 residues) attached to them. Araf, arabinofuranose; Galp, galactopyranose; [4-O-Me]$\pm$GlcpA, glucuronopyranose acid that may (+) or may not (−) be methylated at $C(O)4$; f, furanose; p, pyranose; Rhaf, rhamnofuranose.

be characteristic of AGPs (Showalter, 1993). Amino acid and carbohydrate analysis of AGPs from different tissues and various plant species suggested that there was heterogeneity in both the protein and carbohydrate components (Nothnagel, 1997). This prediction has certainly proven true for the protein backbones, but less is known about the arrangement of carbohydrates on individual AGPs.

AGP protein backbones

Purification, deglycosylation and protein sequencing of AGP protein backbones by various groups enabled the cloning of the genes (reviewed in Du et al., 1996; Nothnagel, 1997; Schultz et al., 1998; Bacic et al., 2000). This work showed that the deduced AGP protein backbones were organized into domains that led to them being designated as either 'classical' or 'non-classical'.

This classification was useful during the early discovery phase of AGP protein backbone genes. However, the large number of genes with diverse domain structures (Figure 2) suggests that it may be necessary to reconsider this nomenclature. Interestingly, for animal proteoglycans the genes for the protein backbones are designated by name, based on various parameters, for example, size, mutant analysis, domain structures, etc. Subsequently, when the proteoglycan has been isolated and characterized they are assigned to the two major classes, glycosaminoglycans (GAG)-type and mucin-type depending on their sugar chains and/or location (Iozzo, 1998). We have adopted this approach for a subclass of classical AGP protein backbones that contain fasciclin-like domains (see below). However, we will also explain the terms classical and non-classical as they appear in other recent reviews (Du et al., 1996; Nothnagel, 1997; Bacic et al., 2000).

The genes for classical AGPs encode proteins with an N-terminal secretion sequence that is removed from the mature protein, a central domain rich in Pro/Hyp and a C-terminal hydrophobic domain (Figure 2A) (Du et al., 1996). In some cases the Pro/Hyp-rich region is interrupted by a short basic region (Figure 2B) (Pogson and Davies, 1995; Li and Showalter, 1996; Gilson et al., 2001). The C-terminal hydrophobic domain was recently shown to be a signal for the addition of a glycosylphosphatidylinositol (GPI)-anchor (Youl et al., 1998; Oxley and Bacic, 1999; Svetek et al., 1999). In the following discussions, AGPs with a predicted C-terminal hydrophobic/GPI-signal

domain will be assumed to be GPI-anchored, even though in most cases this has not been demonstrated experimentally.

Many genes for AGP protein backbones have been isolated from a single plant species, such as Arabidopsis (Schultz et al., 1998; Sherrier et al., 1999; Schultz et al., 2000) and Pinus (Loopstra et al., 2000; Zhang et al., 2000), and GPI-anchored classical AGPs are the most abundant. In Arabidopsis, the classical AGPs can be divided into at least three subclasses. There are at least ten classical AGPs with mature protein backbones of between 85 and 151 amino acids (Figure 2A) (Schultz et al., 2000), an additional two AGPs containing Lys-rich domains (Gilson et al., 2001) (Figure 2B) and five AG peptides with short protein backbones of about 12 amino acid residues (Figure 2C). AG peptides were first identified in wheat (Fincher et al., 1974). A new class of AGPs was recently identified in Arabidopsis and since they are predicted to be GPI-anchored we will consider them as classical AGPs.

The predicted mature protein backbone of the first member of this gene family to be characterized, At-FLA8 (previously known as AtAGP8), contains two Pro/Hyp-rich regions and two fasciclin-like domains (Figure 2E) (Schultz et al., 2000). A further 16 sequences were identified in the Arabidopsis database with one or two fasciclin-like domains in addition to AGP-like regions (Figure 2E–G). These molecules have been designated fasciclin-like AGPs (FLAs) after the Drosophila cell adhesion molecules, fasciclins (Zinn et al., 1988; Elkins et al., 1990). AtFLA7 is a putative orthologue of the pine AGP, Ptx14A9 (Figure 2F) (Loopstra and Sederoff, 1995). Several extracellular cell matrix proteins with fasciclin domains have been shown to have important roles in development. For example, a Volvox glycoprotein, algal CAM, is required to obtain proper contact between neighbouring cells in this multicellular alga (Huber and Sumper, 1994). As this subclass of AGPs contains domains known to be important in both protein-protein and/or protein-carbohydrate interactions, they may be the most likely AGPs to have developmental roles based on comparisons with other eukaryotic proteoglycans (Selleck, 2000).

Non-classical AGPs are defined as having regions that are atypical of AGPs, for example regions rich in Asn or Cys residues in addition to regions containing Pro/Hyp (Figure 2D) (reviewed in Du et al., 1996). None of the AGPs previously classified as non-classical AGPs are predicted to be

164

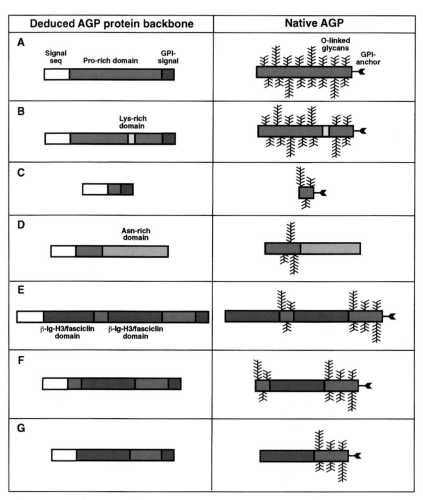

Figure 2. Schematic representation of the different classes of AGPs deduced from DNA sequence (left-hand panels), and the predicted structures of the native AGPs after processing and post-translational modification (right-hand panels). Not drawn to scale. Processing involves removal of the predicted N-terminal signal sequence (white), removal of the predicted C-terminal GPI-signal sequence (if present, in green), followed by the attachment of the GPI-anchor to the C-terminus. Pro residues are hydroxylated to Hyp and *O*-linked glycans (indicated by feathers) are added to most of the Hyp residues. Only *Arabidopsis* AGP genes are described here to highlight the diversity in a single plant species and this list is not exhaustive. A. Classical AGPs contain Pro/Hyp residues throughout the mature protein backbone. There are at least ten classical AGPs in *Arabidopsis* with mature protein backbones 85 to 151 amino acids in length. These AGPs are listed as follows (gene name, GenBank accession number): *AtAGP1*, AF082298; *AtAGP2*, AF082299; *AtAGP3*, AF082300; *AtAGP4*, AF082301; *AtAGP5*, AF082302; *AtAGP6*, AJ012459; *AtAGP7*, AF195888; *AtAGP9*, AF195890; *AtAGP10*, AF195891; *AtAGP11*, AF195892. B. Classical AGPs with a Lys-rich domain (yellow) of 15–20 amino acids in the Hyp-rich region include AtAGP17, AF305939 and AtAGP18, AF305940. C. AG peptides have a predicted backbone size of ca. 12 amino acids and are encoded by *AtAGP12–AtAGP16* (AF195893–AF195897). D. A single putative non-classical AGP containing an Asn-rich domain (orange) at the C-terminus has been identified in *Arabidopsis* (AC003027, 96204–96872). There are three classes of fasciclin-like AGPs (FLAs). The addition of *N*-glycans at up to four sites is predicted for fasciclin domains. E. One class of FLAs have two AGP-like regions (red) and two fasciclin domains (blue): *AtFLA1*, AB009050 (49957–50589, 51333–51974); *AtFLA2*, AL049640 (22114–23325); *AtFLA4*, AL133314 (35714–36976); *AtFLA8* (previously called AtAGP8), AF195889; *AtFLA10*, AL162295 (100831–102099); *AtFLA15*, AL132972 (71060–72311, 73002–73060); *AtFLA16*, AC007017 (25995–27422); *AtFLA17*, AB006700 (61407–62804). F. Two FLAs have a single fasciclin domain in addition to two AGP-like regions: *AtFLA6*, AC007109 (35979–36722); *AtFLA7*, AC006955 (64487–65251); *AtFLA9*, AAD10681 (73512–74255); *AtFLA11*, AB005240 (55910–56650); *AtFLA12*, AB011483 (46416–47165); *AtFLA13*, BAB10980 (19700–20443). G. Another class of the FLAs have one fasciclin domain and one AGP-like region: *AtFLA3*, AC006403 (61009–61851); *AtFLA5*, AL080283 (63600–64436); *AtFLA14*, AB024033 (3972–4739).

GPI-anchored, which is why the FLAs were grouped with the classical AGPs. The non-classical AGP protein backbones are more heterogeneous than those of the classical AGPs and in some instances, such as *PcAGP*2 and *NaAGP*2 (Mau *et al.*, 1995), they undergo post-translational proteolytic processing in plant suspension cultures. In other cases their reactivity to βGlcY reagent is variable or has not been tested. Molecules in this category include the galactose-rich style glycoprotein (GaRSGP) (Sommer-Knudsen *et al.*, 1996), the transmitting-tract-specific (TTS) glycoprotein (Cheung *et al.*, 1995; Wu *et al.*, 2000), the 120 kDa glycoprotein (Lind *et al.*, 1996), and molecules expressed in stages of nodulation. It is likely that the *Arabidopsis* genome contains orthologues for many of the subclasses of AGPs identified in other species. For example, there are at least two genes encoding proteins with similarity to GaRSGP and TTS in *Arabidopsis* (A. van Hengel and K. Roberts, personal communication). There are no obvious orthologues of the Asn-rich containing non-classical AGPs (Mau *et al.*, 1995) in *Arabidopsis,* even though the *Arabidopsis* genome has been sequenced (Martienssen, 2000). However, there are three predicted *Arabidopsis* proteins with C-termini that are similar to NaAGP2 and PcAGP2 (Figure 3). One of these (GenBank accession number AC003027) has a short motif, Ser-Pro-Ala-Pro-Ala-Pro, suggesting it may be an AGP.

NaAGP2	VERQGLSDTRFLENGKYYYD
PcAGP2	SEQQGMSDTRFLENGKYYHD
AtAC003027	EEKQGMSDTRVMENGKYFYD
NtH13	HQQQGMSDTRFLGNGKFYYD
Zinnia TED	AEQQGMSDTRFLENGKYSYD
AtU78721	VERQGMSDTRYMANGKYYYD
Bn Stigma	VERQGMSDTRFMANGKYYYD
AtAC010155	LERQGMSDTRFMEKGSYYYD
Cotton E6	GEKQGMSDTRYLENGKYYYD

Figure 3. A conserved motif within the Asn-rich domains of non-classical AGPs is found in other cell wall proteins. NaAGP2 is a non-classical AGP from *Nicotiana alata* (Mau *et al.*, 1995; S79359); PcAGP2 is a non-classical AGP from *Pyrus communis* (Mau *et al.*, 1995; S79358); AtAC003027 is a genomic clone encoding a putative non-classical AGP from *Arabidopsis*; AtU78721 and AtAC010155 are also from *Arabidopsis*; Bn stigma is a protein that is abundant in *Brassica napus* stigmata (Robert *et al.*, 1994; L34287); cotton E6 is a cell wall protein from developing cotton fibres (John, 1996; U30505); TED3 is a putative cell wall protein expressed in *Zinnia* mesophyll cells in culture prior to differentiation into tracheary elements (Demura and Fukuda, 1994; D30801); and NtH13 is from protoplasts of *Nicotiana tabacum* (Dong *et al.*, 1998; AB012854). The numbers after the reference of each protein are the GenBank accession number of the corresponding clone. Amino acid residues that are identical to NaAGP2 are indicated by a black box with white text and similar amino acid residues are shaded grey with black text.

Post-translational modifications to the protein backbone

The large number and the diversity of AGP protein backbones are now established. The patterns of prolyl hydroxylation and the structure of the carbohydrate moieties attached to the protein backbone are as complex, but are more recalcitrant to analysis.

Prolyl hydroxylation

Prolyl hydroxylation occurs in the endoplasmic reticulum (ER) whereas *O*-glycosylation occurs in both the ER and the Golgi apparatus. The enzyme responsible for the modification of Pro residues, prolyl 4-hydroxylase (procollagen-proline dioxygenase, EC 1.14.11.2), has been extensively studied in animals (Hill *et al.*, 2000). The enzyme is tetrameric and consists of two α- and two β-subunits. The first putative plant cDNA clones for the β-subunit of prolyl 4-hydroxylase were isolated almost ten years ago (Shorrosh and Dixon, 1991). Although no DNA clones have yet been reported for the α-subunit, both protein subunits have now been purified from plants (Wojtaszek *et al.*, 1999) and there are at least seven *Arabidopsis* genes with similarity to the genes for the α-subunit of prolyl 4-hydroxylase from mammals (GenBank accession numbers AC011623, AF024504, AL031986, AB026644, AC004450, AL031032 and AC026234).

For most classical AGPs, 80–90% of Pro residues in the protein backbone are hydroxylated (Youl *et al.*, 1998). Electrospray ionization-mass spectrometry studies of NaAGP1 and PcAGP1 showed that at certain sites, a Pro is often, but not always, hydroxylated (Youl *et al.*, 1998; Oxley and Bacic, 1999). The basis of this heterogeneity is unknown. Predictions have been made about the required amino acid residues surrounding a Pro residue in order for it to be hydroxylated; for example, in extensins and other HRGPs, the Pro residues in Ala-Pro, Pro-Ala, Pro-Pro, Pro-Val and Ser-Pro are always hydroxylated whereas Lys-Pro, Tyr-Pro and Phe-Pro are not hydroxylated (Kieliszewski and Lamport, 1994; Sommer-Knudsen *et al.*, 1998). However, there are not always simple rules and as more sequencing of protein backbones is performed, exceptions to the above rules may be found. For example, in *Arabidopsis*, the central Pro residue in the motif Thr-Pro-Pro is hydroxylated in AtAGP4, but not in AtAGP7 (Schultz *et al.*, 2000). It is not known whether this is due to the nature of other amino acids surrounding these residues, to tissue-specific expression of hydroxylases with varying specificities, or simply a reflection of the flux through the ER during synthesis and secretion.

Glycosylation

Based on the complexity of carbohydrates on several AGPs (Figure 1A), there are likely to be many glycosyl transferases (GTs) for the assembly of the polysaccharide chains. These will include $(1\rightarrow3)\beta$-galactosyltransferases (GalTs) for the $(1\rightarrow3)\beta$-D-galactan framework as well as GTs for the terminal sugars, such as AraTs, FucTs, and RhaTs etc. Most of the GTs are likely to be type II membrane-bound GTs (Breton *et al.*, 1998), but the possibility that the backbone is assembled by family 2 integral membrane proteins, sometimes referred to as polysaccharide synthases cannot be excluded. The genes for GalTs are being characterized in animals (Breton *et al.*, 1998) and plant homologues are being identified (Henrissat and Davies, 2000; Y. Qu, P. Gilson and A. Bacic, unpublished). Identification of the reactions these enzymes catalyse will require a combined biochemical, molecular and genetic approach.

It has been suggested that once assembled, the carbohydrate components of some AGPs are subsequently processed extracellularly by glycosidases (Kreuger and Holst, 1996). This is one mechanism by which AGPs, either as the processed polymer or the released oligosaccharide, could be involved in signalling. Experimental support for this hypothesis has been reviewed (Nothnagel, 1997) and additional support was recently obtained from pine somatic embryogenic cultures where it was demonstrated that sugars were released from βGlcY-precipitated AGPs (Domon *et al.*, 2000).

Addition of the GPI anchor

The other known modification of AGP protein backbones is the addition of a GPI anchor to classical AGPs. The C-terminal signal for addition of a GPI anchor contains small aliphatic amino acids (designated ω, $\omega+1$, and $\omega+2$), followed by a short charged region and terminates in a hydrophobic domain, as found in GPI-anchored proteins from mammals and yeast (Udenfriend and Kodukula 1995). The precise addition site of the GPI anchor has been experimentally

determined for several AGPs from *N. alata* (NaAGP1), *Pyrus communis* (Youl *et al.*, 1998) and *Arabidopsis* (Schultz *et al.*, 2000).

The structure of the lipid moiety of the GPI anchor of two AGPs has been fully characterized, one from *Rosa* (Svetek *et al.*, 1999) and also for PcAGP1 from pear suspension cultures (Oxley and Bacic, 1999). Both AGPs have a phosphoceramide lipid anchor, which is characteristic of yeast and *Dictyostelium* GPI-anchors (Conzelmann *et al.*, 1992; Fankhauser *et al.*, 1993). In both cases the sphingoid base is primarily phytosphingosine and the fatty acid is tetracosanoic. In *Rosa* it was shown that GPI-anchored AGPs could be released from the membrane by the exogenous addition of a phospholipase C (PLC) (Svetek *et al.*, 1999). The only complete primary structure of the glycan moiety of any plant GPI-anchor is for PcAGP1 (Oxley and Bacic, 1999). The oligosaccharides on the anchor of PcAGP1 have a common structure, α-D-Man $(1\rightarrow2)\alpha$-D-Man$(1\rightarrow6)\alpha$-D-Man$(1\rightarrow4)\alpha$-D-GlcNH$_2$-inositol. Slightly more than 50% of the anchors are further substituted at the third mannose with $[\beta$-D-Gal$p$-$(1\rightarrow4)]_{0.54}$ (Figure 4A) (Oxley and Bacic, 1999). The substituted Gal structure is distinct from anchors characterized from mammals, yeast and protozoa.

Processing of the GPI anchor

It is likely that the C-terminal processing of GPI-anchored AGPs occurs in the ER as it does for mammalian, yeast and protozoan GPI-anchored proteins (Udenfriend and Kodukula, 1995; Thompson *et al.*, 2000). As the sequencing of the *Arabidopsis* genome nears completion, most of the putative *Arabidopsis* GPI-anchor biosynthesis genes can be identified (Table 1).

Recently, Takos *et al.* (2000) presented an elegant experimental approach for determining if a C-terminal hydrophobic sequence is a signal for GPI-anchor attachment. They demonstrated that when the predicted GPI-signal sequence of LeAGP-1 from tomato is added to an endoglucanase reporter protein, the endoglucanase is targeted to the cell surface. Addition of a GPI-anchor to the endoglucanase was demonstrated by its release from the cell surface with PLC treatment. When the predicted cleavage site, ω, and the $\omega+1$, and $\omega+2$ sites (Ser-192, Gly-193 and Ala-194 respectively) of the LeAGP-1 GPI-signal sequence were deleted, the hydrophobic GPI-anchor signal sequence was not cleaved. As a consequence, the endoglucanase

Table 1. Mammalian and yeast genes involved in GPI-anchor biosynthesis and an example of a putative plant homologue. Database searches using the mammalian and yeast sequences shown revealed plant genes putatively involved in GPI-anchor biosynthesis (protein identification (PID) numbers are shown for the *Arabidopsis* sequences). '?' indicates that similar sequences have yet to be identified in plants, yeast or mammals. Dol-P-Man, dolichol phosphate mannose; $EtNH_2$-P, ethanolamine phosphate; GAA, GPI-anchor attachment; GlcNAc, N-acetylglucosamine; PIG, phosphatidyl-inositol glycan; SL15, suppressor of Lec15.

Mammals	Yeast	*Arabidopsis*	Function	Ref.
PIG-A	GPI3/SPT14/CWH6	g6911848	GlcNAc transfer	a
PIG-H	?	g3367571	GlcNAc transfer	b
PIG-C	GPI2	g3033393	GlcNAc transfer	c
GPI1	GPI1	g6911887	GlcNAc transfer	d
PIG-L	YM8021.07	g4314376	GlcNAc deacetylation	e
PIG-B	GPI-10	g9755753	Third Man transfer	f
PIG-F	YDR302w	g6587811	$EtNH_2$-P transfer	g
PIG-O	YLL031c	g10177070	$EtNH_2$-P transfer	h
PIG-N	MCD4	g6692267	$EtNH_2$-P transfer	i
?	GPI7/YJL062W	g6598615	$EtNH_2$-P transfer	j
PIG-J	?	?	GlcNAc deacetylation	k
DPM1	DPM1	g8886954	Dol-P-Man synthesis	l
SL15	?	g8885552	Dol-P-Man usage/synthesis	m
PIG-E	?	?	Dol-P-Man synthesis	n
GAA1	GAA1	?	GPI attachment	o
GPI8/PIG-K	GPI8	g3063464	GPI attachment	p

[a]Miyata *et al.*, 1993; Takeda *et al.*, 1993; Leidich *et al.*, 1995; Schönbachler *et al.*, 1995; Vossen *et al.*, 1995; Kawagoe *et al.*, 1996; Vossen *et al.*, 1997; [b]Kamitani *et al.*, 1993; [c]Leidich *et al.*, 1995; Inoue *et al.*, 1996; [d]Leidich *et al.*, 1994; Leidich and Orlean, 1996; Watanabe *et al.*, 1998; [e]Finnegan *et al.*, 1991; Nakamura *et al.*, 1997; [f]Takahashi *et al.*, 1996; Sütterlin *et al.*, 1998; [g]Inoue *et al.*, 1993; Puoti and Conzelmann, 1993; [h]Flury *et al.*, 2000; Hong *et al.*, 2000; [i]Gaynor *et al.*, 1999; Hong *et al.*, 1999; [j]Benachour *et al.*, 1999; [k]Stevens, 1993; [l]Orlean *et al.*, 1988; DeGasperi *et al.*, 1990; [m]Ware and Lehrman, 1996; [n]Chapman *et al.*, 1980; [o]Hamburger *et al.*, 1995; Hiroi *et al.*, 1998; [p]Benghezal *et al.*, 1996; Chen *et al.*, 1996; Yu *et al.*, 1997.

remained bound to the plasma membrane after PLC treatment (Takos *et al.*, 2000). Whilst not precisely determining the ω-site, this technique is considerably more rapid and technically less demanding than the direct structural approach which requires considerable amounts of material and sophisticated instrumentation. These approaches are experimental confirmation of computer prediction programs, such as PSORT (Nakai and Horton, 1999) and big-PI predictor (Eisenhaber *et al.*, 2000).

As has been demonstrated in animal cells (see below), GPI-anchored AGPs presumably proceed to the cell surface via vesicular transport, where they may either be released into the extracellular space through the action of a phospholipase or alternatively may diffuse intact through the cell wall to attach to adjacent cell surfaces via a mechanism of passive diffusion. In pear suspension culture cells, PcAGP1, is apparently transiently attached to the plasma membrane with only

ca. 0.2% of total AGP membrane-bound at any time (Oxley and Bacic, 1999).

It is unclear whether AGPs are released at the cell surface by the action of PLC or phospholipase D (PLD) (Oxley and Bacic, 1999; Svetek *et al.*, 1999). This issue will be difficult to resolve, since the remnant anchor left after cleavage by PLC is the same as the remnant left after a PLD cleavage, followed by a phosphatase (Figure 4A). As yet, no enzymes with PLD activity have been purified from plants; however, a GPI protein hydrolyzing PLC from peanut has been identified (Butikofer and Brodbeck, 1993).

Cross-linking of GPI-anchored proteins to the yeast cell wall

In the yeast *Saccharomyces cerevisiae* some GPI-anchored mannoproteins are cross-linked into the cell wall. In 1994, de Nobel and Lipke hypothesized that

168

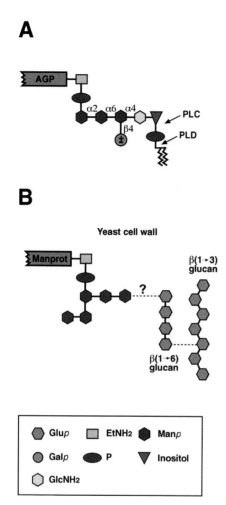

Figure 4. The structure and processing of GPI-anchors in plants and yeast. A. The structure of the GPI anchor added to PcAGP1 (Oxley and Bacic, 1999). There is an ethanolamine (EtNH₂)-phosphate (P) between the anchor and the C-terminus of the protein backbone. This is common to all GPI anchors. The core oligosaccharide of the GPI-anchor of PcAGP1 comprises of 2- and 6-linked Man$p$ residues, a 4-linked GlcNH₂ residue, and a mono-substituted inositol. An additional Gal$p$ residue linked to C(O)4 of the 6-linked Man$p$ residue occurs in some GPI anchors. The lipid moiety is a ceramide composed primarily of a phytosphingosine base and tetracosanoic acid. The arrows show cleavage sites by PLC and PLD, respectively. B. Certain yeast GPI-mannoproteins are cleaved from the plasma membrane and cross-linked into the cell wall via (1→6)β-D-glucan (Kollár et al., 1997). This cleavage is not mediated by phospholipases and likely involves transglycosidation. The mannose residues of the GPI-anchor are cross-linked directly to the (1→6)β-D-glucans in the yeast cell wall. Glu$p$, glucopyranose; EtNH₂, ethanolamine; Gal$p$, galactopyranose; GlcNH₂, glucosamine; Man$p$, mannopyranose; Manprot, mannoprotein; P, phosphate; PLC, phospholipase C; PLD, phospholipase D.

mannoproteins were released from the plasma membrane and cross-linked to (1→6)β-glucan via the GPI remnant glycan core, presumably by transglycosylation (Figure 4B) (de Nobel and Lipke, 1994). This was subsequently confirmed and the nature of the cross-link partially characterized (Lu et al., 1994; Kollár et al., 1997).

In plants, it is likely that the majority of AGPs are not cross-linked into the cell wall because they purify in soluble extracts or are ionically bound to the cell wall (Nothnagel, 1997). This is consistent with the reported amino acid composition of cell walls from Arabidopsis leaves which contain less than 1% Hyp and 5% Pro (Zablackis et al., 1995). The abundant amino acids were Leu, Gln/Glu, Ala, Gly and Asn/Asp, and were each present at levels >9% each (Zablackis et al., 1995). This suggests that classical AGPs and extensins are minor components of Arabidopsis leaf cell walls. However, it is possible that there are cell-type-specific variations in the proportions of 'soluble' versus 'cross-linked' AGPs that would not be detected in such whole-plant/tissue studies.

The implication of cell surface GPI-anchored AGPs

The importance of GPI anchors is unclear although, in animals, it is thought they confer many different properties on their attached proteins such as (1) increased lateral mobility in the lipid bilayer, (2) regulated release from the cell surface, (3) polarized targeting to different cell surfaces, and (4) inclusion in lipid rafts (Hooper, 1997; Muñiz and Riezman, 2000). Another curious feature attributed to the presence of GPI anchors in animals is the passive transfer of GPI-anchored proteins from one membrane to another (reviewed in Kooyman et al., 1998). Whether transfer between membranes is restricted to a few proteins or is a property of all GPI-anchored proteins due to their presence in the outer leaflet of the plasma membrane is unknown. Recent work in this area suggests that GPI-anchored proteins of red blood cells do not transfer spontaneously, but require a catalyst (Suzuki and Okumura, 2000).

Several GPI-anchored glycoproteins from animals activate signaling pathways by binding to growth factors or interacting with extracellular ligands (Selleck, 2000). The GPI-anchored proteoglycan of Drosophila, division abnormally delayed (dally), binds to a number of growth factors in distinct pathways (Nakato et al.,

1995). In another example, GPI-anchored contactin is involved in signal transduction via interactions with other transmembrane proteins, in either the same or adjoining cells (Peles *et al.*, 1997).

In plants, molecules with both extracellular and cytoplasmic domains include wall-associated kinases (WAKs) (He *et al.*, 1999) and the somatic embryogenesis receptor-like kinase (SERK) (Schmidt *et al.*, 1997). WAKs are postulated to be involved in signal transduction as key components of the cytoskeleton-plasma membrane-cell wall continuum (Kohorn, 2000). An intriguing possibility proposed by Gens *et al.*, (2000) suggests that there may be an association at the plasma membrane between AGP epitopes and a WAK, based on their co-localization in immunofluoresence studies (Gens *et al.*, 2000). Other studies however suggest that WAKs are covalently linked to pectin (Kohorn, 2000; Anderson *et al.*, 2001, in this issue). More rigorous biochemical verification is required to determine whether AGPs interact with any of the receptor-like kinases or with other ligands.

Two of the features conferred to GPI-anchored proteins in animals, polarity and signalling, are important in plant embryogenesis (Vroemen *et al.*, 1999; Souter and Lindsey, 2000). AGPs have been implicated in somatic embryogenesis (Kreuger and Holst, 1995; McCabe *et al.*, 1997; Toonen *et al.*, 1997), and in one study an AGP epitope recognized by the monoclonal antibody JIM8 is polarized in cells prior to a division that leads to one embryogenic cell and one non-embryogenic cell (McCabe *et al.*, 1997). Whether this AGP epitope is on a GPI-anchored molecule is unknown. The presence of GPI-anchors on AGPs provides mechanisms by which AGPs could be involved in signalling and polarized targeting processes, known to be important in embryogenesis.

Are non-classical AGPs cross-linked into the cell wall?

There is a possibility that Asn-rich domains could facilitate *in vivo* cross-linking of proteins into the cell wall (see below). Such domains are present in the non-classical AGPs NaAGP2 and PcAGP2. Although these AGPs were purified from the soluble fraction of plant suspension cultures (Mau *et al.*, 1995), this may reflect the prodigious sloughing off of cell wall components that is a special feature of cells in suspension culture (Sims *et al.*, 2000). The basis for suggesting that these two non-classical AGPs are cell

wall-associated is that within the Asn-rich domain is a conserved region of 20 amino acids (70–85% identity) that is found in other cell wall proteins (Figure 3) (John and Crow, 1992; Demura and Fukuda, 1994). For example, the E6 protein from cotton contains an Asn-rich domain and is localized to the primary cell wall and the secondary cell walls of developing fibres (John and Crow, 1992). The two AGPs and E6 from cotton all have Tyr residues, often as Gly-Tyr or Tyr-Gly motifs, that are interspersed between the Asn repeats. It is possible that these Tyr residues cross-link these proteins into the cell wall. Intramolecular cross-linking in the form of iso-dityrosine bridges has been identified in *Chlamydomonas* and higher plants (Epstein and Lamport, 1984; Waffenschmidt *et al.*, 1993). In *Chlamydomonas*, Tyr-Gly-Gly is suggested as the motif involved in the insolubilization of the VSP (vegetative, Ser-Pro-rich) glycoprotein in algal cell walls (Waffenschmidt *et al.*, 1993).

Determining AGP function

A major challenge in AGP research is to establish the function of a single AGP. Determining whether any AGPs have roles in development and which AGPs have overlapping functions will require a variety of different approaches. The finding that there are more than twenty classical AGPs in *Arabidopsis* highlights the problem of assigning AGP function.

The majority of experiments inferring AGP function have used anti-AGP antibodies (reviewed in Knox, 1997), most of which recognize carbohydrate epitopes that are largely uncharacterized (Yates *et al.*, 1996). Although these carbohydrate-directed monoclonal antibodies are useful, such epitopes are likely to be present on many AGPs with different protein backbones, with perhaps only a few epitopes restricted to a specific AGP. However, the possibility that cell type-specific carbohydrate epitopes are involved as the functional elements of an AGP cannot be excluded. In animal proteoglycans both the protein backbone and carbohydrate epitopes have been implicated in function (Selleck, 2000). For example, fibroblast growth factor (FGF) only binds to heparan-sulfate proteoglycans from fibroblasts. This binding is critical for altering the conformation of FGF to permit the appropriate alignment for oligomerization and activation of the secondary messenger cascade by the growth factor receptor complex (Schlessinger *et al.*, 1995).

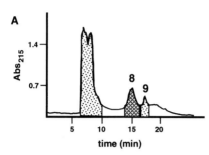

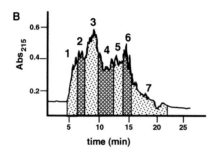

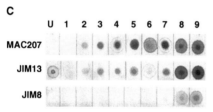

Figure 5. βGlcY-precipitated *Arabidopsis* AGPs react with monoclonal antibodies, MAC207, JIM13 and JIM8. A. Buffer-soluble native AGPs were obtained by precipitation with βGlcY reagent, and separation by RP-HPLC (as in Schultz *et al.*, 2000). AGPs were fractionated by RP-HPLC on a C8 column with a linear gradient from 0% to 100% buffer B (80% acetonitrile in 0.1% trifluoroacetic acid; 25 min; 0.5 ml/min). The two most hydrophobic fractions, 8 (retention time (RT 13–16.5 min) and 9 (RT 17–18 min) were used without further separation. B. The more hydrophilic fractions (RT 6–10 min) were pooled and re-fractionated with a gradient 0–30% buffer B (30 min; 1 ml/min). The X-axis is time (min). The Y-axis is absorbance (Abs) at 215 nm. C. AGP fractions (1 μl; 1 μg carbohydrate, determined by phenol-sulfuric acid; Dubois *et al.*, 1956) were dotted onto nitrocellulose and air-dried. 'U' represents the unbound fraction from the RP-HPLC profile in A, other fractions are as indicated in A and B. Primary antibodies (1:100 dilution) are as indicated at left. Alkaline phosphatase conjugated to goat anti-rat antibody was used as a secondary antibody (1:1000 dilution).

Table 2. Neutral monosaccharide composition of total *Arabidopsis* AGP purified with βGlcY reagent. Neutral sugars were determined by alditol acetate analysis (Harris *et al.*, 1984). tr, trace.

Monosaccharide	Mol (%)	
	leaf	root
Galactose	65	61
Arabinose	32	30
Xylose	2	4
Fucose	1	3
Mannose	tr	tr
Rhamnose	tr	tr

AGP fractions; one dot blot was used for each of the monoclonal antibodies, MAC207, JIM13 and JIM8 (Figure 5C) (Knox *et al.*, 1991; Pennell *et al.*, 1991). JIM13 and MAC207 detect many AGPs, which is consistent with results in other plants (Knox, 1997). JIM8, an antibody that detects an epitope implicated in somatic embryogenesis in carrot suspension cultures (McCabe *et al.*, 1997), only reacted with the two most hydrophobic AGP fractions of *Arabidopsis* (Figure 5C). The number of distinct AGP protein backbones present in each fraction is unknown, but based on SDS-polyacrylamide gel electrophoresis of fractions 8 and 9, there are two distinct broad smears suggesting that each fraction contains at least two different protein backbones (data not shown).

Until the exact number of AGPs that each monoclonal antibody reacts with is established, caution must be exercised in attributing important developmental phenomena to a particular AGP. The development of antibodies against specific protein backbones (Gao and Showalter, 2000) or the use of AGP-fusion proteins (Shpak *et al.*, 1999) will enable the localization of specific protein backbones and should aid in determining AGP function. In these cases specific antibodies will also make it possible to isolate individual AGPs in order to characterize the carbohydrate components.

The identification of the genes for the AGP protein backbones provides an alternative to antibodies for studying gene expression. To fully understand the tissue-specific expression of each gene it will probably be necessary to apply a variety of approaches, as has recently been done for the cellulase synthase (*CesA*) gene family (Holland *et al.*, 2000). Detailed analysis

In a preliminary experiment to investigate the reactivity of several commonly used antibodies to *Arabidopsis* AGPs, native AGPs were separated by reversed-phase high-performance liquid chromatography (RP-HPLC) (Figure 5A and B). Monosaccharide analysis confirmed that the βGlcY precipitated AGPs that were loaded onto the RP-HPLC column contained Ara and Gal as the abundant sugars (Table 2). Three replicate dot blots were made representing ten will be required to look at developmental expression of AGP genes since northern analysis of ten *Arabidopsis* genes showed that many of the AGPs are abundantly expressed in two or more tissue types (Schultz *et al.*, 2000). The use of microarray analysis will also be interesting for determining environmental factors that may control AGP gene expression (Harmer and Kay, 2000).

Reverse genetics approaches

A reverse genetics approach to isolate knockout mutants is a more direct approach than expression studies for determining AGP function. This approach has the advantage that it can be used for all of the genes involved in the assembly of a mature AGP, for example genes for (1) the protein backbone, (2) GPI-anchor synthesis, (3) prolyl hydroxylases and (4) glycosyl transferases. Knockouts in some of these enzymes may have pleotropic effects due to the wide range of enzymatic targets possible. In some cases, redundancy may be a problem and in these cases, RNA interference (RNAi) may be a useful alternative (Burton *et al.*, 2000; Smith *et al.*, 2000). With RNAi, if the DNA identity is over 80% it may be possible to knock out more than one gene at a time. In other cases, knockout mutants may be lethal; for example, all yeast genes known to be involved in GPI-anchor biosynthesis (Table 1) and/or anchoring are essential for viability (Leidich *et al.*, 1994).

A number of groups have shown that null mutants from large gene families rarely show visible phenotypes when grown under standard conditions (Bouchez and Höfte, 1998; Winkler *et al.*, 1998; Meissner *et al.*, 1999). For example, in the MYB transcription factor gene family, individual lines with insertions into 36 distinct MYB genes failed to give any visible phenotypes under normal growth conditions. However, by growing plants under a variety of conditions subtle phenotypes were identified for seven of these mutants (Meissner *et al.*, 1999). Alternatively, in these large gene families it may be necessary to create double, or even triple, mutants to observe a phenotype (Krysan *et al.*, 1999).

An alternative way to analyse mutants for which there is no phenotype is to use microarrays. By comparing RNA from mutant and wild-type plants it should be possible to see if other genes are compensating for the mutation, or if additional genes are affected. The microarray facility at Stanford University, Cali-

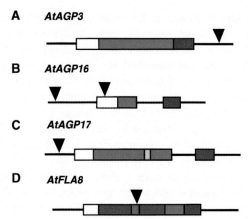

Figure 6. Arabidopsis mutants with insertions in or near AGP genes identified by PCR screening. A. A T-DNA tag is inserted 122 bp downstream of the poly(A) addition site for the gene *AtAGP3*. The tag was identified in the Feldmann line CS10690 (ABRC stock centre number). TAIL-PCR (Liu *et al.*, 1995) (data not shown) showed that this line also has a tag in the coding region of an imbibition protein (AL133248). B. Two independent lines were identified with tags near *AtAGP16*; one line has an insertion in the coding sequence (ABRC stock centre number CS14065) and the other line (ABRC stock centre number CS15474) has an insertion ca. 1.5 kb upstream of the start codon. C. In the Feldmann line CS12955, there is T-DNA tag 1097 bp upstream of *AtAGP17* in the putative promoter region. This line was identified separately on the basis of the phenotype *resistant to Agrobacterium tumefaciens transformation (rat1)* (Nam *et al.*, 1999). D. A mutant for *AtFLA8* (previously known as *AtAGP8*) was identified independently from the *Dissociation* insertion lines (SGT6202), developed by V. Sundaresan (Parinov *et al.*, 1999). Boxes represent coding regions, thin lines represent non-coding regions and triangles represent T-DNA tags. Within the coding regions, white represents secretion signal sequence, blue represents fasciclin domains, red represents Pro/Hyp-rich AGP domains, yellow represents a Lys-rich region and green represent GPI-anchor signal sequences.

fornia (http://afgc.stanford.edu/afgc_html/site2htm) is currently analysing DNA representing ca. 8000 non-redundant expressed sequence tags (ESTs) from *Arabidopsis*, with plans to increase this number to 14000 (ca. 60% of the genome) in 2001.

We currently have several *Arabidopsis* lines with tags in or near AGP genes. Using a PCR-based approach (McKinney *et al.*, 1995) we identified four 'mutants' for three AGP genes from the Feldmann lines (available from the Arabidopsis Biological Resource Center (ABRC), Ohio State University) (Figure 6). In addition, a mutant for *AtFLA8* was identified as one of the 511 *Dissociation* (Ds) insertion lines, for which the flanking sequences were available in a public database (Parinov *et al.*, 1999). We have knockout (null) mutants for *AtAGP16* and *AtFLA8*; one line with a tag in the 3'-untranslated region of *AtAGP3* and

172

two lines with tags in the promoter of *AtAGP16* and *AtAGP17*, respectively (Figure 6). With the exception of *AtAGP17*, preliminary characterization of the AGP mutants has not identified any obvious phenotypes (K. Johnson, C. Schultz and A. Bacic, unpublished results). In the line with the tag in the promoter of *AtAGP17* (ABRC stock center number CS12955), a phenotype was identified by Stanton Gelvin's group (Purdue University) when searching for mutants that were resistant to Agrobacterium tumefaciens transformation (*rat*; Nam *et al.*, 1999). This mutant has a tag 1097 bp upstream of the start codon of *AtAGP17*. We are collaborating with Stanton Gelvin's group to understand the relationship between AGPs and the binding of *Agrobacterium* to roots.

Concluding remarks

The number of genes for AGP protein backbones highlights the diversity of proteoglycans at the cell surface. As with the proteoglycans of other eukaryotes, it is likely that most of the AGPs will have a physical/structural function with only a subset having important developmental roles. It is likely that mutants will provide the first important clues of the function of individual AGPs, but other tools will be equally important. Analysis of post-translational modifications in different cell types will also be necessary to understand the factors controlling the hydroxylation of Pro residues and microheterogeneity of glycosylation of a single protein backbone. This will help determine if this variability is important for function. Should the function reside in the microheterogeneity of *O*-glycosylation, then considerable effort will be needed to elucidate the genes and enzymes involved in glycan chain assembly. This will be a major challenge for the future and will require the efforts of many more groups than are currently involved in AGP research.

Acknowledgements

This work was carried out jointly with funding provided by an Australian Research Council Large Grant (A10020017) and by funding from the Australian Government to the Co-operative Research Centre for Bioproducts. Y.G. and K.J. are recipients of University of Melbourne Research Scholarships. We are grateful to the Arabidopsis Biological Resource Center (Ohio State University and Nottingham) for providing ESTs, DNA pools of the Feldmann Lines and seeds. Many thanks go to Prof. Keith Roberts, John Innes Centre, UK, for the gift of the monoclonal antibodies, MAC207, JIM8 and JIM13. We are grateful to Drs Paul Gilson, Brian Jones and Ed Newbigin for comments on the manuscript.

References

Anderson, C.M., Wagner, T.A., Perret, M., He, Z.-H., Hed, D. and Kohorn, B.D. 2001. WAKs: cell wall-associated kinases linking the cytoplasm to the extracellular matrix. Plant Mol. Biol., this issue.

Bacic, A., Churms, S.C., Stephen, A.M., Cohen, P.B. and Fincher, G.B. 1987. Fine structure of the arabinogalactan-protein from *Lolium multiflorum*. Carbohydrate Res. 162: 85–93.

Bacic, A., Harris, P.J. and Stone, B.A. 1988. Structure and function of plant cell walls. In: J. Priess (Ed.) The Biochemistry of Plants, Academic Press, New York, pp. 297–371.

Bacic, A., Currie, G., Gilson, P., Mau, S.-L., Oxley, D., Schultz, C.J., Sommer-Knudsen, J. and Clarke, A.E. 2000. Structural classes of arabinogalactan-proteins. In: E.A. Nothnagel, A. Bacic and A.E. Clarke (Eds.) Cell and Developmental Biology of Arabinogalactan-Proteins, Kluwer Academic Publishers/Plenum, Dordrecht, Netherlands/New York, pp. 11–23.

Benachour, A., Sipos, G., Flury, I., Reggiori, F., Canivenc-Gansel, E., Vionnet, C., Conzelmann, A. and Benghezal, M. 1999. Deletion of *GPI7*, a yeast gene required for addition of a side chain to the glycosylphosphatidylinositol (GPI) core structure, affects GPI protein transport, remodeling, and cell wall integrity. J. Biol. Chem. 274: 15251–15261.

Benghezal, M., Benachour, A., Rusconi, S., Aebi, M. and Conzelmann, A. 1996. Yeast Gpi8p is essential for GPI anchor attachment onto proteins. EMBO J. 15: 6575–6583.

Bouchez, D. and Höfte, H. 1998. Functional genomics in plants. Plant Physiol. 118: 725–732.

Breton, C., Bettler, E., Joziasse, D.H., Geremia, R.A. and Imberty, A. 1998. Sequence-function relationships of prokaryotic and eukaryotic galactosyltransferases. J. Biol. Chem. 123: 1000–1009.

Burton, R.A., Gibeaut, D.M., Bacic, A., Findlay, K., Roberts, K., Hamilton, A., Baulcombe, D.C. and Fincher, G.B. 2000. Virus-induced silencing of a plant cellulose synthase gene. Plant Cell 12: 691–705.

Butikofer, P. and Brodbeck, U. 1993. Partial purification and characterization of a (glycosyl) inositol phospholipid-specific phospholipase-C from peanut. J. Biol. Chem. 268: 17794–17802.

Carpita, N.C. and Gibeaut, D.M. 1993. Structural models of primary cell walls in flowering plants: consistency of molecular structure with the physical properties of the walls during growth. Plant J. 3: 1–30.

Chapman, A., Fujimoto, K. and Kornfield, S. 1980. The primary glycosylation defect in class-E Thy-1-negative mutant mouse lymphoma cells is an inability to synthesize dolichol-P-mannose. J. Biol. Chem. 255: 4441–4446.

Chen, R., Udenfriend, S., Prince, G.M., Maxwell, S.E., Ramalingam, S., Gerber, L.D., Knez, J. and Medof, M.E. 1996. A defect in glycosylphosphatidylinositol (GPI) transamidase activity in mutant K cells is responsible for their inability to display GPI surface proteins. Proc. Natl. Acad. Sci. USA 93: 2280–2284.

Cheung, A.Y., Wang, H. and Wu, H.-M. 1995. A floral transmitting tissue-specific glycoprotein attracts pollen tubes and stimulates their growth. Cell 82: 383–393.

Churms, S.C., Stephen, A.M. and Siddiqui, I.R. 1981. Evidence for repeating sub-units in the molecular structure of the acidic arabinogalactan from rapeseed (*Brassica campestris*). Carbohydrate Res. 94: 119–122.

Clarke, A.E., Anderson, R.L. and Stone, B.A. 1979. Form and function of arabinogalactans and arabinogalactan-proteins. Phytochemistry 18: 521–540.

Conzelmann, A., Puoti, A., Lester, R.L. and Desponds, C. 1992. Two different types of lipid moieties are present in glycophosphoinositol-anchored membrane proteins of *Saccharomyces cerevisiae*. EMBO J. 11: 457–466.

DeGasperi, R., Thomas, L. J., Sugiyama, E., Chang, H-M., Beck, P. J., Orlean, P. *et al.* 1990. Correction of a defect in mammalian GPI anchor biosynthesis by a transfected yeast gene. Science 250: 988–991.

de Nobel, H. and Lipke, P. N. 1994. Is there a role for GPIs in yeast cell-wall assembly? Trends Cell Biol. 4: 42–45.

Demura, T. and Fukuda, H. 1994. Novel vascular cell-specific genes whose expression is regulated temporally and spatially during vascular system development. Plant Cell 6: 967–981.

Domon, J.-M., Neutelings, G., Roger, D., David, A. and David, H. 2000. A basic chitinase-like protein secreted by embryogenic tissues of *Pinus caribaea* acts on arabinogalactan proteins extracted from the same cell lines. J. Plant Physiol. 156: 33–39.

Dong, L.-Y., Masuda, T., Kawamura, T., Hata, S. and Izui, K. 1998. Cloning, expression, and characterization of a root-form phosphoenolpyruvate carboxylase from *Zea mays*: comparison with the C4-form enzyme. Plant Cell Physiol. 39: 865–873.

Du, H., Clarke, A. E. and Bacic, A. 1996. Arabinogalactan-proteins: a class of extracellular matrix proteoglycans involved in plant growth and development. Trends Cell Biol. 6: 411–414.

Dubois, M., Gillies, K.A., Hamilton, J.K., Rebers, P.A. and Smith, F. 1956. Colorimetric method for the determination of sugars and related substances. Anal. Chem. 28: 350–356.

Eisenhaber, B., Bork, P., Yuan, Y., Löffler, G. and Eisenhaber, F. 2000. Automated annotation of GPI anchor sites: case study *C. elegans*. Trends Biochem. Sci. 25: 340–341.

Elkins, T., Hortsch, M., Bieber, A.J., Snow, P.M. and Goodman, C.S. 1990. *Drosophila* fasciclin I is a novel homophilic adhesion molecule that along with fasciclin III can mediate cell sorting. J. Cell Biol. 110: 1825–1832.

Epstein, L. and Lamport, D.T.A. 1984. An intramolecular linkage involving isodityrosine in extensin. Phytochemistry 23: 1241–1246.

Fankhauser, C., Homans, S.W., Thomas-Oates, J.E., McConville, M.J., Desponds, C., Conzelmann, A. and Ferguson, M.A.J. 1993. Structures of glycosylphosphatidylinositol membrane anchors from *Saccharomyces cerevisiae*. J. Biol. Chem. 268: 26365–26374.

Fincher, G.B., Sawyer, W.H. and Stone, B.A. 1974. Chemical and physical properties of an arabinogalactan-peptide from wheat endosperm. Biochem. J. 139: 535–545.

Fincher, G.B., Stone, B.A. and Clarke, A.E. 1983. Arabinogalactan-proteins: structure, biosynthesis and function. Annu. Rev. Plant Physiol. 34: 47–70.

Finnegan, P.M., Payne, M.J., Keramidaris, E. and Lukins, H.B. 1991. Characterization of a yeast nuclear gene, *AEP2*, required for accumulation of mitochondrial mRNA encoding subunit 9 of the ATP synthase. Curr. Genet. 20: 53–61.

Flury, I., Benachour, A. and Conzelmann, A. 2000. YLL031c belongs to a novel family of membrane proteins involved in the transfer of ethanolaminephosphate onto the core structure of glycosylphosphatidylinositol anchors in yeast. J. Biol. Chem. 275: 24458–24465.

Gane, A.M., Craik, D., Munro, S.L.A., Howlett, G.J., Clarke, A.E. and Bacic, A. 1995. Structural analysis of the carbohydrate moiety of arabinogalactan-proteins from stigmas and styles of *Nicotiana alata*. Carbohydrate Res. 277: 67–85.

Gao, M. and Showalter, A.M. 2000. Immunolocalization of LeAGP-1, a modular arabinogalactan-protein, reveals its developmentally regulated expression in tomato. Planta 210: 865–874.

Gaynor, E.C., Mondesert, G., Grimme, S.J., Reed, S.I., Orlean, P. and Emr, S.D. 1999. MCD4 encodes a conserved endoplasmic reticulum membrane protein essential for glycosylphosphatidylinositol anchor synthesis in yeast. Mol. Biol. Cell 10: 627–648.

Gens, J.S., Fujiki, M., Pickard, B.G. 2000. Arabinogalactan protein and wall-associated kinase in a plasmalemmal reticulum with specialized verticles. Protoplasma 212: 115–134.

Gilson, P., Gaspar, Y., Oxley, D., Youl, J.J. and Bacic, A. 2001. NaAGP4 is an arabinogalactan-protein whose expression is suppressed by wounding and fungal infection in *Nicotiana alata*. Protoplasma 215: 128–139.

Goodrum, L.J., Patel, A., Leykam, J.F. and Kieliszewski, M.J. 2000. Gum arabic glycoprotein contains glycomodules of both extensin and arabinogalactan-glycoproteins. Phytochemistry 54: 99–106.

Hamburger, D., Egerton, M. and Riezman, H. 1995. Yeast Gaa1p is required for attachment of a completed GPI anchor onto proteins. J. Cell Biol. 129: 629–639.

Harmer, S.L. and Kay, S.A. 2000. Microarrays: determining the balance of cellular transcription. Plant Cell 12: 613–615.

Harris, P.J., Henry, R.J., Blakeney, A.B. and Stone, B.A. 1984. An improved procedure for the methylation analysis of oligosaccharides and polysaccharides. Carbohydrate Res. 127: 59–73.

He, Z.-H., Cheeseman, I., He, D.Z. and Kohorn, B.D. 1999. A cluster of five cell wall-associated receptor kinase genes, *Wak1–5*, are expressed in specific organs of *Arabidopsis*. Plant Mol. Biol. 39: 1189–1196.

Henrissat, B. and Davies, G.J. 2000. Glycoside hydrolases and glycosyltransferases. Families, modules, and implications for genomics. Plant Physiol. 124: 1515–1519.

Hill, K.L., Harfe, B.D., Dobbins, C.A. and L'Hernault, S.W. 2000. *dpy*-18 encodes an α-subunit of prolyl-4-hydroxylase in *Caenorhabditis elegans*. Genetics 155: 1139–1148.

Hiroi, Y., Komuro, I., Chen, R., Hosoda, T., Mizuno, T., Kudoh, S., Georgescu, S. P., Medof, M.E. and Yazaki, Y. 1998. Molecular cloning of human homolog of yeast *GAA1* which is required for attachment of glycosylphosphatidylinositols to proteins. FEBS Lett. 421: 252–258.

Holland, N., Holland, D., Helentjaris, T., Dhugga, K.S., Xoconostle-Cazares, B. and Delmer, D.P. 2000. A comparative analysis of the plant cellulose synthase (*CesA*) gene family. Plant Physiol. 123: 1313–1323.

Hong, Y., Maeda, Y., Watanabe, R., Ohishi, K., Mishkind, M., Riezman, H. and Kinoshita, T. 1999. Pig-n, a mammalian homologue of yeast Mcd4p, is involved in transferring phosphoethanolamine to the first mannose of the glycosylphosphatidylinositol. J. Biol. Chem. 274: 35099–35106.

Hong, Y., Maeda, Y., Watanabe, R., Inoue, N., Ohishi, K. and Kinoshita, T. 2000. Requirement of PIG-F and PIG-O for transferring phosphoethanolamine to the third mannose in glycosylphosphatidylinositol. J. Biol. Chem. 275: 20911–20919.

Hooper, N.M. 1997. Glycosyl-phosphatidylinositol anchored membrane enzymes. Clin. Chim. Acta 266: 3–12.

174

Huber, O. and Sumper, M. 1994. Algal-CAMs: isoforms of a cell adhesion molecule in embryos of the alga *Volvox* with homology to *Drosophila* fasciclin I. EMBO J. 13: 4212–4222.

Inoue, N., Kinoshita, T., Orii, T. and Takeda, J. 1993. Cloning of a human gene, PIG-F, a component of glycosylphosphatidylinositol anchor biosynthesis, by a novel expression cloning strategy. J. Biol. Chem. 268: 6882–6885.

Inoue, N., Watanabe, R., Takeda, J. and Kinoshita, T. 1996. PIG-C, one of the three human genes involved in the first step of glycosylphosphatidylinositol biosynthesis is a homologue of *Saccharomyces cerevisiae* GPI2. Biochem. Biophys. Res. Commun. 226: 193–199.

Iozzo, R.V. 1998. Matrix proteoglycans: from molecular design to cellular function. Annu. Rev. Biochem. 67: 609–652.

John, M.E. 1996. Structural characterization of genes corresponding to cotton fiber mRNA, E6: reduced E6 protein in transgenic plants by antisense gene. Plant Mol. Biol. 30: 297–306.

John, M.E. and Crow, L.J. 1992. Gene expression in cotton (*Gossypium hirsutum* L.) fiber: cloning of the mRNAs. Proc. Natl. Acad. Sci. USA 89: 5769–5773.

Kamitani, T., Chang, H.-M., Rollins, C., Waneck, G.L. and Yeh, E.T.H. 1993. Correction of the class H defect in glycosylphosphatidylinositol anchor biosynthesis in Ltk– cells by a human cDNA clone. J. Biol. Chem. 268: 20733–20736.

Kawagoe, K., Kitamura, D., Okabe, M., Taniuchi, I., Ikawa, M., Watanabe, T., Kinoshita, T. and Takeda, J. 1996. Glycosylphosphatidylinositol-anchor-deficient mice: implications for clonal dominance of mutant cells in paroxysmal nocturnal hemoglobinuria. Blood 87: 3600–3606.

Kieliszewski, M.J. and Lamport, D.T.A. 1994. Extensin: repetitive motifs, functional sites, post-translational codes, and phylogeny. Plant J. 5: 157–172.

Kiyohara, H., Zhang, Y.W. and Yamada, H. 1997. Effect of exo-β-D-(1→3)-galactanase digestion on complement activating activity of neutral arabinogalactan unit in a pectic arabinogalactan from roots of *Angelica acutiloba kitagawa*. Carbohydrate Polymers 32: 249–253.

Knox, J.P. 1997. The use of antibodies to study the architecture and developmental regulation of plant cell walls. Int. J. Cytol. 171: 79–120.

Knox, J.P., Linstead, P.J., Peart, J.M., Cooper, C. and Roberts, K. 1991. Developmentally regulated epitopes of cell surface arabinogalactan proteins and their relation to root tissue pattern formation. Plant J. 1: 317–326.

Kohorn, B.D. 2000. Plasma membrane-cell wall contacts. Plant Physiol. 124: 31–38.

Kollár, R., Reinhold, B.B., Petráková, E., Yeh, H.J.C., Ashwell, G., Drgonová, J. *et al.* 1997. Architecture of the yeast cell wall. J. Biol. Chem. 272: 17762–17775.

Kooyman, D.L., Byrne, G.W. and Logan, J.S. 1998. Glycosyl phosphatidylinositol anchor. Exp. Nephrol. 6: 148–151.

Kreuger, M. and van Holst, G.-Jv. 1995. Arabinogalactan-protein epitopes in somatic embryogenesis of *Daucus carota* L. Planta 197: 135–141.

Kreuger, M. and van Holst, G.-Jv. 1996. Arabinogalactan proteins and plant differentiation. Plant Mol. Biol. 30: 1077–1086.

Krysan, P.J., Young, J.C. and Sussman, M.R. 1999. T-DNA as an insertional mutagen in *Arabidopsis*. Proc. Natl. Acad. Sci. USA 11: 2283–2290.

Leidich, S.D., Drapp, D.A. and Orlean, P. 1994. A conditionally lethal yeast mutant blocked at the first step in glycosyl phosphatidylinositol anchor synthesis. J. Biol. Chem. 269: 10193–10196.

Leidich, S.D., Kostova, Z., Latek, R.R., Costello, L.C., Drapp, D.A., Gray, W. *et al.* 1995. Temperature-sensitive yeast GPI anchoring mutants *gpi2* and *gpi3* are defective in the synthesis of *N*-acetylglucosaminyl phosphatidylinositol. J. Biol. Chem. 270: 13029–13035.

Leidich, S.D. and Orlean, P. 1996. Gpi1, a *Saccharomyces cerevisiae* protein that participates in the first step in glycosylphosphatidylinositol anchor synthesis. J. Biol. Chem. 271: 27829–27837.

Li, S.-X. and Showalter, A.M. 1996. Cloning and developmental/stress-regulated expression of a gene encoding a tomato arabinogalactan protein. Plant Mol. Biol. 32: 641–652.

Lind, J.L., Bönig, I., Clarke, A.E. and Anderson, M.A. 1996. A style-specific 120-kDa glycoprotein enters pollen tubes of *Nicotiana alata* in vivo. Sex. Plant Reprod. 9: 75–86.

Liu, Y.-G., Mitsukawa, N., Oosumi, T. and Whittier, R.F. 1995. Efficient isolation and mapping of *Arabidopsis thaliana* T-DNA insert junctions by thermal asymmetric interlaced PCR. Plant J. 8: 457–463.

Loopstra, C.A. and Sederoff, R.R. 1995. Xylem-specific gene expression in loblolly pine. Plant Mol. Biol. 27: 277–291.

Loopstra, C.A., Puryear, J.D. and No, E.-G. 2000. Purification and cloning of an arabinogalactan-protein from xylem of loblolly pine. Planta 210: 686–689.

Lu, C.-F., Kurjan, J. and Lipke, P.N. 1994. A pathway for cell wall anchorage of *Saccharomyces cerevisiae* α-agglutinin. Mol. Cell Biol. 14: 4825–4833.

Martienssen, R.A. 2000. Weeding out the genes: the *Arabidopsis* genome project. Funct. Integr. Genomics 1: 2–11.

Mau, S.-L., Chen, C.-G., Pu, Z.-Y., Moritz, R.L., Simpson, R.J., Bacic, A. and Clarke, A.E. 1995. Molecular cloning of cDNAs encoding the protein backbones of arabinogalactan-proteins from the filtrate of suspension-cultured cells of *Pyrus communis* and *Nicotiana alata*. Plant J. 8: 269–281.

McCabe, P.F., Valentine, T.A., Forsberg, L.S. and Pennell, R.I. 1997. Soluble signals from cells identified at the cell wall establish a developmental pathway in carrot. Plant Cell 9: 2225–2241.

McKinney, E.C., Ali, N., Traut, A., Feldmann, K.A., Belostotsky, D.A., McDowell, J.M. and Meagher, R.B. 1995. Sequence-based identification of T-DNA insertion mutations in *Arabidopsis*: actin mutants *act2-1* and *act4-1*. Plant J. 8: 613–622.

McNamara, M.K. and Stone, B.A. 1981. Isolation, characterization and chemical synthesis of a galactosyl-hydroxyproline linkage compound from wheat endosperm arabinogalactan-peptide. Lebensm. Wiss. Technol. 14: 182–187.

Meissner, R.C., Jin, H., Cominelli, E., Denekamp, M., Fuertes, A., Greco, R. *et al.* 1999. Function search in a large transcription factor gene family in *Arabidopsis*: assessing the potential of reverse genetics to identify insertional mutations in R2R3 *MYB* genes. Plant Cell 11: 1827–1840.

Miyata, T., Takeda, J., Iida, Y., Yamada, N., Inoue, N., Takahashi, M. *et al.* 1993. Cloning of PIG-A, a component in the early step of GPI-anchor biosynthesis. Science 259: 1318–1320.

Muñiz, M. and Riezman, H. 2000. Intracellular transport of GPI-anchored proteins. EMBO J. 19: 10–15.

Nakai, K. and Horton, P. 1999. PSORT: a program for detecting sorting signals in proteins and predicting their subcellular localization. Trends Biochem. Sci. 24: 34–36.

Nakamura, N., Inoue, N., Watanabe, R., Takahashi, M., Takeda, J., Stevens, V.L. and Kinoshita, T. 1997. Expression cloning of PIG-L, a candidate *N*-acetylglucosaminylphosphatidylinositol deacetylase. J. Biol. Chem. 272: 15834–15840.

Nakato, H., Futch, T.A. and Selleck, S.B. 1995. The division abnormally delayed (*dally*) gene: a putative integral membrane proteoglycan required for cell division patterning during postembryonic development of the nervous system in *Drosophila*. Development 121: 3687–3702.

Nam, J., Mysore, K.S., Zheng, C., Knue, M.K., Matthysse, A.G. and Gelvin, S.B. 1999. Identification of T-DNA tagged *Arabidopsis* mutants that are resistant to transformation by *Agrobacterium*. Mol. Gen. Genet. 261: 429–438.

Nothnagel, E.A. 1997. Proteoglycans and related components in plant cells. Int. Rev. Cytol. 174: 195–291.

Orlean, P., Albright, C. and Robbins, P.W. 1988. Cloning and sequencing of the yeast gene for dolichol phosphate mannose synthase, an essential protein. J. Biol. Chem. 263: 17499–17507.

Oxley, D. and Bacic, A. 1999. Structure of the glycosyl-phosphatidylinositol membrane anchor of an arabinogalactan-protein from *Pyrus communis* suspension-cultured cells. Proc. Natl. Acad. Sci. USA 6: 14246–14251.

Parinov, S., Sevugan, M., Ye, D., Yang, W.-C., Kumaran, M. and Sundaresan, V. 1999. Analysis of flanking sequences from *Dissociation* insertion lines: a database for reverse genetics in *Arabidopsis*. Plant Cell 11: 2263–2270.

Peles, E., Nativ, M., Lustig, M., Grumet, M., Schilling, J., Martinez, R. *et al.* 1997. Identification of a novel contactin-associated transmembrane receptor with multiple domains implicated in protein-protein interactions. EMBO J. 16: 978–988.

Pennell, R.I., Janniche, L., Kjellbom, P., Scofield, G.N., Peart, J.M. and Roberts, K. 1991. Developmental regulation of a plasma membrane arabinogalactan protein epitope in oilseed rape flowers. Plant Cell 3: 1317–1326.

Pogson, B.J. and Davies, C. 1995. Characterization of a cDNA encoding the protein moiety of a putative arabinogalactan protein from *Lycopersicon esculentum*. Plant Mol. Biol. 28: 347–352.

Prescott, J.H., Enriquez, P., Jung, C., Menz, E. and Groman, E.V. 1995. Larch arabinogalactan for hepatic drug delivery: isolation and characterization of a 9 kDa arabinogalactan fragment. Carbohydrate Res. 278: 113–128.

Puoti, A. and Conzelmann, A. 1993. Characterization of abnormal free glycosylphosphatidylinositols accumulating in mutant lymphoma cells of classes B, E, F and H. J. Biol. Chem. 268: 7215–7224.

Qi, W., Fong, C. and Lamport, D.T.A. 1991. Gum arabic glycoprotein is a twisted hairy rope: a new model based on *O*-galactosylhydroxyproline as the polysaccharide attachment site. Plant Physiol. 96: 848–855.

Robert, L.S., Allard, S., Gerster, J.L., Cass, L. and Simmonds, J. 1994. Molecular analysis of two *Brassica napus* genes expressed in the stigma. Plant Mol. Biol. 26: 1217–1222.

Schlessinger, J., Lax, I. and Lemmon, M. 1995. Regulation of growth factor activation by proteoglycans: what is the role of the low affinity receptors? Cell 83: 357–360.

Schmidt, E.D.L., Guzzo, F., Toonen, M.A.J. and de Vries, S.C. 1997. A leucine-rich repeat containing receptor-like kinase marks somatic plant cells competent to form embryos. Development 124: 2049–2062.

Schönbachler, M., Horvath, A., Fassler, J.S. and Riezman, H. 1995. The yeast *spt14* gene is homologous to the human PIG-A gene and is required for GPI anchor synthesis. EMBO J. 14: 1637–1645.

Schultz, C.J., Gilson, P., Oxley, D., Youl, J.J. and Bacic, A. 1998. GPI-anchors on arabinogalactan-proteins: implications for signalling in plants. Trends Plant Sci. 3: 426–431.

Schultz, C.J., Johnson, K.L., Currie, G. and Bacic, A. 2000. The classical arabinogalactan protein gene family of *Arabidopsis*. Plant Cell 12: 1751–1767.

Selleck, S.B. 2000. Proteoglycans and pattern formation: sugar biochemistry meets developmental genetics. Trends Genet. 16: 206–212.

Serpe, M.D. and Nothnagel, E.A. 1999. Arabinogalactan-proteins in the multiple domains of the plant cell surface. Adv. Bot. Res. 30: 207–289.

Sherrier, D.J., Prime, T.A. and Dupree, P. 1999. Glycosylphosphatidylinositol-anchored cell-surface proteins from *Arabidopsis*. Electrophoresis 20: 2027–2035.

Shorrosh, B.S. and Dixon, R.A. 1991. Molecular cloning of a putative plant endomembrane protein resembling vertebrate protein disulfide-isomerase and a phosphatidylinositol-specific phospholinase C. Proc. Natl. Acad. Sci USA 88: 10941–10945.

Showalter, A.M. 1993. Structure and function of plant cell wall proteins. Plant Cell 5: 9–23.

Shpak, E., Leykam, J.F. and Kieliszewski, M.J. 1999. Synthetic genes for glycoprotein design and the elucidation of hydroxyproline-*O*-glycosylation codes. Proc. Natl. Acad. Sci. USA 96: 14736–14741.

Sims, I.M., Middleton, K., Lane, A.G., Cairns, A.J. and Bacic, A. 2000. Characterisation of extracellular polysaccharides from suspension cultures of members of the Poaceae. Planta 210: 261–268.

Smith, N.A., Singh, S.P., Wang, M.-B., Stoutjesdijk, P.A., Green, A.G. and Waterhouse, P.M. 2000. Total silencing by intron-spliced hairpin RNAs. Nature 407: 319–320.

Sommer-Knudsen, J., Clarke, A.E. and Bacic, A. 1996. A galactose-rich, cell-wall glycoprotein from styles of *Nicotiana alata*. Plant J. 9: 71–83.

Sommer-Knudsen, J., Clarke, A.E. and Bacic, A. 1997. Proline- and hydroxyproline-rich gene products in the sexual tissues of flowers. Sex. Plant Reprod. 10: 253–260.

Sommer-Knudsen, J., Bacic, A. and Clarke, A.E. 1998. Hydroxyproline-rich plant glycoproteins. Phytochemistry 47: 483–497.

Souter, M. and Lindsey, K. 2000. Polarity and signalling in plant embryogenesis. J. Exp. Bot. 51: 971–983.

Stevens, V.L. 1993. Regulation of glycosylphosphatidylinositol biosynthesis by GTP. Stimulation of *N*-acetylglucosamine-phosphatidylinositol deacylation. J. Biol. Chem. 268: 9718–9724.

Sütterlin, C., Escribano, M.V., Gerold, P., Maeda, Y., Mazon, M.J., Kinoshita, T. *et al.* 1998. *Saccharomyces cerevisiae GPI10*, the functional homologue of human *PIG-B*, is required for glycosylphosphatidylinositol-anchor synthesis. Biochem J. 332: 153–159.

Suzuki, K. and Okumura, Y. 2000. GPI-linked proteins do not transfer spontaneously from erythrocytes to liposomes. New aspects of reorganization of the cell membrane. Biochemistry 39: 9477–9485.

Svetek, J., Yadav, M.P. and Nothnagel, E.A. 1999. Presence of a glycosylphosphatidylinositol lipid anchor on rose arabinogalactan proteins. J. Biol. Chem. 274: 14724–14733.

Takahashi, M., Inoue, N., Ohishi, K., Maeda, Y., Nakamura, N., Endo, Y. *et al.* 1996. PIG-B, a membrane protein of the endoplasmic reticulum with a large lumenal domain, is involved in transferring the third mannose of the GPI anchor. EMBO J. 15: 4254–4261.

Takeda, J., Miyata, T., Kawagoe, K., Iida, Y., Endo, Y., Fujita, T. *et al.* 1993. Deficiency of the GPI anchor caused by a somatic mutation of the *PIG-A* gene in paroxysmal nocturnal hemoglobinuria. Cell 73: 703–711.

176

Takos, A.M., Dry, I.B. and Soole, K.L. 2000. Glycosyl-phosphatidylinositol-anchor addition signals are processed in *Nicotiana tabacum*. Plant J. 21: 43–52.

Thompson, G.A. Jr. and Okuyama, H. 2000. Lipid-linked proteins of plants. Prog. Lipid Res. 39: 19–39.

Toonen, M.A.J., Schmidt, E.D.L., van Kammen, A. and de Vries, S.C. 1997. Promotive and inhibitory effects of diverse arabino-galactan proteins on *Daucus carota* L. somatic embryogenesis. Planta 203: 188–195.

Udenfriend, S. and Kodukula, K. 1995. How glycosyl-phosphatidylinositol-anchored membrane proteins are made. Annu. Rev. Biochem. 64: 563–591.

Vossen, J.H., Ram, A.F.J. and Klis, F.M. 1995. Identification of *SPT14/CWH6* as the yeast homologue of hPIG-A, a gene involved in the biosynthesis of GPI anchors. Biochim. Biophys. Acta 1243: 549–551.

Vossen, J.H., Müller, W.H., Lipke, P.N. and Klis, F.M. 1997. Restrictive glycosylphosphatidylinositol anchor synthesis in *cwh6/gpi3* yeast cells causes aberrant biogenesis of cell wall proteins. J. Bact. 179: 2202–2209.

Vroemen, C., de Vries, S. and Quatrano, R. 1999. Signalling in plant embryos during the establishment of the polar axis. Semin. Cell Dev. Biol. 10: 157–164.

Waffenschmidt, S., Woessner, J.P., Beer, K. and Goodenough, U.W. 1993. Isodityrosine cross-linking mediates insolubilization of cell walls in *Chlamydomonas*. Plant Cell 5: 809–820.

Ware, F.E. and Lehrman, M.A. 1996. Expression cloning of a novel suppressor of the Lec15 and Lec35 glycosylation mutations of Chinese hamster ovary cells. J. Biol. Chem. 271: 13935–13938.

Watanabe, R., Inoue, N., Westfall, B., Taron, C.H., Orlean, P., Takeda, J. and Kinoshita, T. 1998. The first step of glycosylphos-phatidylinositol biosynthesis is mediated by a complex of PIG-A, PIG-H, PIG-C and GPI1. EMBO J. 17: 877–885.

Winkler, R.G., Frank, M.R., Galbraith, D.W., Feyereisen, R. and Feldmann, K.A. 1998. Systematic reverse genetics of transfer-DNA-tagged lines of *Arabidopsis*. Plant Physiol. 118: 743–750.

Wojtaszek, P., Smith, C.G. and Bolwell, G.P. 1999. Ultrastructural localisation and further biochemical characterisation of prolyl 4-hydroxylase from *Phaseolus vulgaris*: comparative analysis. Int. J. Biochem. Cell Biol. 31: 463–477.

Wu, H.-M., Wong, E., Ogdahl, J. and Cheung, A.Y. 2000. A pollen tube growth-promoting arabinogalactan protein from *Nicotiana alata* is similar to the tobacco TTS protein. Plant J. 22: 165–176.

Yariv, J., Lis, H. and Katchalski, E. 1967. Precipitation of arabic acid and some seed polysaccharides by glycosylphenylazo dyes. Biochem. J. 105: 1c–2c.

Yates, E.A., Valdor, J.-F., Haslam, S.M., Morris, H.R., Dell, A., Mackie, W. and Knox, J.P. 1996. Characterization of carbo-hydrate structural features recognized by anti-arabinogalactan-protein monoclonal antibodies. Glycobiology 6: 131–139.

Youl, J.J., Bacic, A. and Oxley, D. 1998. Arabinogalactan-proteins from *Nicotiana alata* and *Pyrus communis* contain glycosylphos-phatidylinositol membrane anchors. Proc. Natl. Acad. Sci. USA 95: 7921–7926.

Yu, J., Nagarajan, S., Knez, J.J., Udenfriend, S., Chen, R. and Medof, M.E. 1997. The affected gene underlying the class K glycosylphosphatidylinositol (GPI) surface protein defect codes for the GPI transamidase. Proc. Natl. Acad. Sci. USA 94: 12580–12585.

Zablackis, E., Huang, J., Muller, B., Darvill, A.G. and Albersheim, P. 1995. Characterization of the cell-wall polysaccharides of *Arabidopsis thaliana* leaves. Plant Physiol. 107: 1129–1138.

Zhang, Y., Sederoff, R.R. and Allona, I. 2000. Differential expression of genes encoding cell wall proteins in vascular tissues from vertical and bent loblolly pine trees. Tree Physiol. 20: 457–466.

Zinn, K., McAllister, L. and Goodman, C.S. 1988. Sequence analysis and neuronal expression of fasciclin I in grasshopper and *Drosophila*. Cell 53: 577–587.

SECTION 4

GROWTH, SIGNALING & DEFENSE

Plant Molecular Biology **47:** 179–195, 2001.
© 2001 *Kluwer Academic Publishers. Printed in the Netherlands.*

The molecular basis of plant cell wall extension

Catherine P. Darley, Andrew M. Forrester and Simon J. McQueen-Mason*
*Plant Laboratory, Department of Biology, University of York, PO Box 373, York YO1 1DD, UK (*author for correspondence)*

Key words: expansion, plant cell wall, rigidification, structural components, wall-modifying enzymes

Abstract

In all terrestrial and aquatic plant species the primary cell wall is a dynamic structure, adjusted to fulfil a diversity of functions. However a universal property is its considerable mechanical and tensile strength, whilst being flexible enough to accommodate turgor and allow for cell elongation. The wall is a composite material consisting of a framework of cellulose microfibrils embedded in a matrix of non-cellulosic polysaccharides, interlaced with structural proteins and pectic polymers. The assembly and modification of these polymers within the growing cell wall has, until recently, been poorly understood. Advances in cytological and genetic techniques have thrown light on these processes and have led to the discovery of a number of wall-modifying enzymes which, either directly or indirectly, play a role in the molecular basis of cell wall expansion.

Abbreviations: CDTA, *trans*-1,2-diaminocyclohexane-N,N,N′,N′-tetraacetic acid; CsExp1, cucumber expansin 1; GAX, glucuronoarabinoxylan; ·OH, hydroxyl radical; IAA, indole-3-acetic acid; PME, pectin methyl-esterase; XET, xyloglucan endotransglycosylases

Introduction

Plant cell growth is driven by internal turgor pressure and restricted by the ability of the cell wall to extend under this pressure. The ability of cell walls to extend depends on at least two different, but connected, levels of events. Firstly, extensibility depends on the composition of the wall and how its components are bound to one another. Secondly, wall extension can be controlled by modification of existing wall structures. In this paper we will review the current state of our understanding of the molecular mechanisms that underlie cell wall extension in plants.

Wall synthesis determines the inherent extensibility of the wall material, but this extensibility is fine-tuned by processes within the wall that alter the structure or binding of existing wall polymers. These processes have been described by Cosgrove (1999) as primary and secondary wall-loosening agents. Primary wall loosening agents directly alter the mechanical properties of existing wall material, and secondary loosening processes influence wall extensibility by altering the composition and structures that primary processes act upon. As both wall synthesis and structure receive extensive review in other parts of this volume the present paper will concentrate solely on aspects of these topics relating directly to wall extension. We will then use this structural background to discuss factors which modulate the extension of existing wall material.

Structural components of the primary cell wall and their assembly

In general, new primary cell wall material must be synthesised in a form that is competent to undergo extension. The primary cell wall is essentially a composite material consisting of a framework of cellulose microfibrils embedded in a matrix of other, mostly polysaccharide, polymers. The matrix polysaccharides can be grouped into two major classes, pectins and cross-linking glycans. Based on the composition of these polysaccharides the primary cell wall can be distinguished into two distinct types: Type I and type II cell walls (Carpita and Gibeaut, 1993). In Type I cell walls the principal polysaccharide interlinking the cellulose microfibrils is xyloglucan, creating a multi-polymer network embedded in pectic polysaccharides. Most dicotyledonous and some monocotyledonous plants possess Type I cell walls and their properties and assembly constitute the major part of the present review. Type II cell walls are characteristic of those found in grasses (Poaceae) and other related 'com-melinoid' species. Although the overall structure of the cellulose-cross linking glycans is similar to Type I cell wall type II cell walls differ in that the major cross-linking polysaccharide is glucuronoarabinoxy-lan (GAX) not xyloglucan. Another discriminating feature of Type I cell walls is a lower content of pectin, especially in growing regions (Knox et al., 1990). This review will now focus on the role of each cell wall component in cell extension.

The role of cellulose

Alterations in cellulose synthesis can have marked effects on plant growth which may be brought about by alterations in wall extensibility. Studies using cellulose biosynthesis inhibitors on cultured plant cells reveal that the cells can survive in the absence of cellulose (Shedletzky et al., 1990). The walls of these cells are effectively devoid of cellulose, and much of this absence appears to be compensated by increased levels of pectin. It is clear that cell growth is greatly restricted in this condition. In addition, the cells adopt a spherical shape, supporting the hypothesis that the cellulose framework may be essential in determining directionality in plant cell growth. In intact plants, cellulose synthesis inhibitors are used as herbicides, implying that correct cellulose synthesis is essential in normal plant life.

In a similar context, alterations in cellulose structure can have effects on cell growth. Shpigel et al. (1998) showed that incubation of tip growing organs (pollen tubes and root hairs) in solutions containing a bacterial cellulose-binding domain would increase their rate of extension. The same protein also inhibited the expansive growth of Arabidopsis root cells. In both cases the protein appeared to disrupt the structure of cellulose deposited in the wall leading to the effect on growth.

Mutations that affect the function of cellulose synthase genes in plants have drastic effects on growth and morphology. A temperature sensitive mutation in the rsw1 gene leads to reduced growth and aberrant cell morphology in Arabidopsis (Arioli et al., 1998). Cells of plants grown at high temperature show growth aberrations with many epidermal cells ballooning out, suggesting they can no longer effectively control their expansion. Analysis of the walls of these cells show that there is a general loss of cellulose crystallinity and this probably results in gross changes in extensibility.

The cellulose/cross-linking glycan network

There is general agreement that the cellulose/matrix glycan network serves to hold the primary cell wall together and is the key determinant of wall extensibility. Glycans, such as xyloglucan in dicots and glucuronoarabinoxylan (GAX) in graminaceous monocots, are thought to bind to cellulose microfibrils via extensive hydrogen bonding between the $(1{\rightarrow}4)\beta$-D-glucan chains in the microfibrils and the glucan or xylan backbone. The matrix glycans are able to bind to more than one microfibril at a time (McCann et al., 1990), which enables them to span between microfibrils and form molecular tethers. An easy assumption to make from this observation is that, by doing this, the tethering matrix glycans would strengthen the wall by endowing interfibril integrity.

Dissociation of cellulose and the tethering polymers from one another requires generally severe extractions such as 4 to 6 M NaOH (Fry, 1988). This suggests that at least a subfraction of the glycans are held to the microfibrils by a more substantial interaction than simple hydrogen bonding to the fibril surface. Indeed, such harsh chemical treatments not only release the matrix glycans but also lead to swelling of the crystalline microfibrils due to partial dissociation of many $(1{\rightarrow}4)\beta$-D-glucan chains from the crystal structure of the microfibrils themselves. This, in turn, may allow the release of xyloglucan trapped within the structure of the microfibrils.

If cellulose and xyloglucan (or other cross-linking glycans) are mixed together in vitro then the amount of glycan that associates with the cellulose is much lower (by as much as a factor of 5) than that found in the wall (Hayashi et al., 1987; Baba et al., 1994). In addition, comparatively mild extractions are sufficient to remove what polymer is bound to the cellulose. Hayashi et al. (1987) showed that if crystalline cellulose and xyloglucans were annealed together by heating and cooling, then more stable associations (requiring 4 M NaOH to disrupt) were formed. This suggested that during heating the crystal structure of the cellulose was partly disrupted allowing the xyloglucan to integrate, in part, with the $(1\rightarrow4)\beta$-D-glucans and then to become entrapped in the crystal structure during annealing.

Pectins

The function of pectins in the wall, with regards to extension, is still not clear. However, it is clear that pectins are mainly found in primary cell walls, indicating that they fulfil an important function during wall extension. In this context, it is worth noting that in Type I walls pectic polymers are in much lower amounts than in Type II walls, suggesting that the role is not necessarily dependent on abundance. In solution, pectins exist as highly hydrated gels and concentrations above 2–5% are difficult to obtain, but in the wall pectins are maintained in a highly condensed form, possibly closer to 20%. How the pectins are maintained in this concentrated form in the wall is unclear, but perhaps it is by purely spatial constraints. This high pectin concentration could provide a high hydration potential to the wall, which in turn may help endow the wall with compressive strength. As will be discussed in a later section, there is gathering evidence that pectins may play a role in wall rigidification at the end of growth.

Bacterial cellulose composites: model structures for the cell wall?

Understanding how cellulose and matrix glycans associate in this network may be key to understanding the molecular basis of wall extension. There is a general assumption that the components of the wall are co-secreted in the apoplast and form the structure of the wall by spontaneous assembly. That is to say that the matrix glycans and pectins secreted into the wall spontaneously assemble themselves around the framework provided by cellulose microfibrils synthesised at the outer face of the plasma membrane. Various investigators have tried to manipulate the components of the wall to assemble correctly *in vitro* (Hayashi *et al.,* 1987; Baba *et al.,* 1994). The major failing in this work was simply in obtaining composites that reflect the correct relative amounts of polymers in them. Type I walls typically contain roughly equal quantities of cellulose, xyloglucan and pectin. However, as pectins exist as highly hydrated gels *in vitro,* it is not possible to dissolve them into the concentrations found in the wall. In addition, it has not been possible to devise conditions for cellulose and xyloglucan to associate in the 1:1 ratio that is found in native walls. Part of this

problem is probably related to the substrates used in these types of experiment. For example, commercially available cellulose is generally derived from whole cell walls such as cotton linters, and this is very different in nature (surface area to volume ratio) than dealing with individual cellulose microfibrils as would occur during wall synthesis.

A more effective means of creating wall-like composites has been recently developed. *Acetobacter xylinus* cultures produce strong cohesive mats of cellulose fibrils. Each bacterial cell secretes a single, long cellulose microfibril. In a culture these come together to form a highly entangled mat known as a pellicle. Whitney *et al.* (1998, 1999) showed that when *Acetobacter* cultures are grown in the presence of xyloglucan or glucomannan, they form a pellicle that reflects many of the characteristics of the cellulose/matrix glycan network in plant cell walls. Not only do these composite materials contain a similar proportion of cellulose and cross-linking glycan to that found in cell walls, but chemical extractions, and NMR studies, showed that the two polymer systems associate in a similar manner (Whitney *et al.,* 1998). Microscopic examination of the structure of the composite showed many similarities to structures observed in pectin-depleted wall materials (McCann *et al.,* 1990, 1993). In this system cellulose synthesis occurs in a medium with high concentrations of matrix glycan. It seems likely that associations can form between matrix glycan molecules and glucan chains in the nascent microfibril as it forms. This probably leads to entrapment of a proportion of the glycans into the crystal structure of the fibril. This is probably very similar to what occurs during plant cell wall synthesis.

At the surface of the plasma membrane in plant cells, cellulose synthesis is thought to occur in areas of high local concentrations of glycans released from Golgi secretions, leading to a similar entrapment in microfibrils during synthesis. In support of this idea, Pauly *et al.,* (1999) demonstrated that there are three domains of xyloglucan within a dicot cell wall. One domain is represented by xyloglucan that is easily extractable, readily digested by endo-β-glucanases, and is probably largely held in the wall by physical entanglements with other polymers. A second, enzyme-inaccessible domain is more tightly held in the wall, probably by hydrogen bonding (along at least part of its length) to the microfibrils. The third domain, also inaccessible to endoglucanases, is very tightly associated with the microfibrils and probably consists of polymer domains trapped within the crystal structure.

Clearly it is possible for any single xyloglucan molecule to have regions of its polymer taking part in all three domains at any time. These different domains of xyloglucan may represent different sites of action for enzymes modulating wall extension.

An interesting observation in the composite pellicle formed between cellulose and matrix glycans by *Acetobacter*, is the drastic difference in mechanical properties between the composite and pure cellulose materials (Whitney *et al.*, 1999). As mentioned above, the cellulose pellicle normally produced by the bacteria is a dense mat of highly entangled microfibrils. Measurement of mechanical properties of this pellicle in uniaxial extension measurements reveal it to be a strong and stiff material. In contrast, the pellicle produced in the presence of xyloglucan is weaker and more extensible. In uniaxial extension measurements, the xyloglucan-containing pellicle exhibited about 20% of the strength of cellulose-only material. As well as being weaker, the composite was more extensible, extending by almost twice as much as the material without xyloglucan (in creep assays). Very similar properties were also reported in composites made with other polysaccharides, such as mannans (Whitney *et al.*, 2000).

The effects of matrix glycans on pellicle structure and material properties might tell us something of the role of these polysaccharides in cell walls. Rather than simply acting as cross-links to strengthen it, the cross-linking glycans may well enhance the extensibility of the cell wall. The presence of matrix glycans during cellulose synthesis probably results in the production of 'fuzzy fibrils' with long trailing ends of the glycans instead of the uncoated, partly crystalline surfaces of fibrils formed in their absence. This 'fuzziness' may help to hold microfibrils apart from one another. Somehow, the xyloglucans may prevent microfibril entanglements giving the wall a much looser structure. In this loose structure, the relatively weak cross-bridges may now also be of significance to the overall extensibility of the wall.

Whitney *et al.* (2000) have also shown that xyloglucan-containing composites made by *Acetobacter* are excellent substrates for the action of the wall-loosening enzyme, 'expansin'. Expansins were the first isolated plant cell wall protein shown to rapidly induce the long-term extension of this material when it is held under a constant physical load. Expansins are discussed in more detail later in this paper. The effects of expansin on *Acetobacter* cellulose composites are very similar to the effects on plant cell walls,

further supporting the validity of these materials for use as models of wall structure. These studies also showed that expansins were much less effective in inducing extension on composites made with shorter xyloglucans than those made with long polymers. In addition, it has been found that endoglucanase treatment of xyloglucan-containing composites results in increasing stiffness of the material, perhaps as the xyloglucan is digested by the enzyme (E. Chanliaud and M. Gidley, personal communication). Taken together, these observations suggest that, at least in this model system, smaller oligosaccharides can lead to stiffer materials. This observation goes against current thinking, which predicts that degradation of xyloglucan or other polymers will lead to more extensible walls.

Structural proteins

In addition to the cellulose-cross-linking glycan network, primary cell walls also contain small amounts of structural proteins that may be important during wall rigidification associated with cessation of cell expansion. Structural proteins, such as extensins, form a network in cell walls, and they were originally conceived to play a role in wall extension (Lamport, 1986). It is now generally accepted that this is unlikely as they are found in mono- and oligomeric forms in growing walls and appear only to be polymerised at the end of growth (Ye *et al.*, 1991). Arabinogalactan proteins (AGPs) have also been considered as potentially playing a role in wall extension. At present these glycoproteins are more generally considered to be signalling molecules (for review, see Gaspar *et al.*, 2001, this issue). There are reports that Yariv reagents, which are phloroglucinol-linked β-glycosides that specifically bind AGPs, can inhibit growth in cultured cells (Serpe and Nothnagel, 1993) and in *Arabidopsis* roots (Willats and Knox, 1996). The authors suggested that the inhibitory effect could be due to a direct effect of the Yariv reagent binding to AGPs and causing wall rigidification. We have investigated whether Yariv reagents can alter growth in cucumber hypocotyls and whether this was due to a direct effect on wall extensibility. The Yariv reagent can almost completely inhibit auxin-induced growth of cucumber hypocotyls (Figure 1A). However, direct application of Yariv reagent does not diminish the extensibility of the hypocotyls (Figure 1B), even though the dye was clearly taken up into the tissues (not shown). This suggests the effects of Yariv reagents, in this instance, are likely to be of a

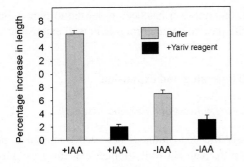

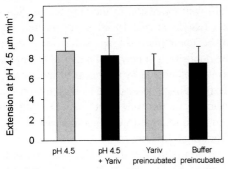

Figure 1. The effects of Yariv Reagent on IAA-induced growth and wall extension of cucumber hypocotyls. A. (top) Percentage increase in length of cucumber hypocotyls incubated in IAA, with and without Yariv reagents. Cucumber hypocotyls (*Cucumis sativus* cv. Burpee Pickler) were grown for 4 days in the dark at 28 °C. Sections (1 cm) were cut from the growing region just below the hook and floated for 1 h on a 5 mM KCl solution. Sections were then transferred to 5 mM KCl containing either 10 μM IAA only, 10 μM IAA with 30 μM Yariv phenylglycoside, no additive, or 30 μM Yariv phenylglycoside only. Sections were incubated in these solutions for 8 h and then re-measured along their axial length. Results are means and standard errors of 10 measurements. B (bottom). Extension of hypocotyl sections at pH 4.5. Apical 1 cm sections of hypocotyl were frozen, thawed and abraded, as described by McQueen-Mason *et al.* (1992). Sections were then either immediately suspended in an extensometer (McQueen-Mason *et al.*, 1992) under a constant load of 20 g in 50 mM sodium acetate pH 4.5 or in the same solution containing 30 μM Yariv phenylglycoside. Alternatively, the sections were pre-incubated for 30 min at room temperature in 50 mM sodium acetate pH 4.5 with or without 30 μM Yariv phenylglycoside, before being placed in the extensometer in 50 mM sodium acetate pH 4.5 as described above. Results are means and standard errors of 6 measurements.

longer-term nature rather than a direct physical result of binding AGPs in the hypocotyl walls.

Cell wall mutants

Genetic approaches to understanding wall structure and extension are becoming increasingly important, and already some important insights into the role of wall structure in normal plant growth have been attained. Reiter *et al.* (1997) screened populations of

EMS-mutagenized *Arabidopsis* for plants with altered cell wall compositions, and found 11 independent loci of '*mur*' mutants (short for *murus*, which is Latin for 'wall') that were affected. Among these are some with clear effects on growth. For example, *mur1*, which has a defective GDP-mannose-4,6-dehydratase required for the synthesis of GDP-fucose, was characterised by a complete absence of fucose from shoot cell walls and was also slightly dwarfed (Zablackis *et al.*, 1996; see Reiter and Vanzin, 2001, this issue). The main effect appears to be on the mechanical properties of the primary cell walls. The terminal-fucosyl residue normally found in xyloglucan in these plants was partly replaced by the rare sugar L-galactose. It was originally believed that this alteration in xyloglucan structure might be responsible for the change in wall extension. However, fucose is found in other polymers such as complex pectins as well as in some glycans from glycoproteins, and the possibility that changes in some of these polymers are involved in the dwarf phenotype cannot be excluded.

A complimentary approach has been adopted by Nicol *et al.* (1998). In this instance, T-DNA-tagged populations of dwarf hypocotyl mutants were screened for defects in cell wall biogenesis. Some of these plants indeed have been shown to have altered wall compositions. One of these mutants, *korrigan* (*kor1*), showed severe cell expansion defects in all tissue types. Microscopic analysis of cell wall material revealed that these defects resulted from incorrect assembly of the cellulose/xyloglucan network (Nicol *et al.*, 1998). The corresponding *KOR* gene has been cloned and identified as a putative membrane-bound endo-$(1\rightarrow4)\beta$-D-glucanase (EGase), a homologue of which was found associated with the ripening of melon fruit (Rose *et al.*, 1998). Although the precise role of this enzyme in cell wall assembly has yet to be elucidated, comparative studies of an EGase (*CelC*) from *A. tumefacians* suggests that this enzyme may play a direct role in the synthesis of cellulose. In *celC* mutants an accumulation of lipid-linked $(1\rightarrow4)\beta$-D-glucan oligosaccharides was detected, indicating that *celC* may be involved with cleavage and assembly of oligosaccharides used in the synthesis of nascent glucan chains (Matthysse *et al.*, 1995a, b). It remains to be established whether EGases are similarly involved in cellulose synthesis in plants.

In addition to mutants showing defects at the whole-plant level, plants have also been identified with organ- or cell-type-specific defects. Perhaps one of the most intriguing of these mutants are those of the *PRO-*

CUSTE (*PRC*) locus in *Arabidopsis*. Disruptions at this locus leads to cellulose defects which manifest as reduced hypocotyl elongation and increased radial expansion of root cells in dark-grown seedlings (Hauser *et al.*, 1995; Desnos *et al.*, 1996; Fagard *et al.*, 2000). The unusual aspect of this mutant is that the aberrant phenotype is light-dependent. Seedlings grown in the light showed no cell wall defects. Studies using double mutants with defective photoreceptors have shown that the light-dependent reversions of phenotype are phytochrome-dependent (Desnos *et al.*, 1996). These observations lead to speculation that PRC is involved in a novel dark-dependent pathway for cellulose synthesis in which, after phytochrome stimulation, *PRC* is down-regulated or its gene products become inactivated.

It seems clear, from the information gathered to date, that studies of these and other cell wall mutants will provide unique insights into the mechanisms underlying the synthesis and assembly of plant cell walls, as well as aspects of structure determining mechanical properties. For an in-depth review of this topic and related issues we recommend the article by Haigler *et al.* (2001, this issue).

The cell wall of grasses

So far, we have largely discussed wall structure and extension in the great majority of dicotyledonous and monocotyledonous cell walls are Type I (Carpita and Gibeaut, 1993). However, the cell walls of grasses and other related 'commelinoid' species that possess Type II walls show substantial differences in wall composition to those of all other angiosperms (Carpita and Gibeaut, 1993). Some of these differences, as well as the broad similarity, may offer insights into the molecular basis of wall extension. Type II cell walls are characterised as having relatively low levels of pectins and different matrix glycans than those of Type I cell walls. In Type II walls the major cross-linking glycan is glucuronoarabinoxylan, and this is generally believed to have a similar role to that played by xyloglucan in Type I walls. In addition to glucuronoarabinoxylan, a second matrix glycan, a mixed-linkage $(1\rightarrow3),(1\rightarrow4)\beta$-D-glucan, appears transiently during cell expansion. Most of this unbranched polymer consists of cellotriose and cellotetraose units connected by single $(1\rightarrow3)\beta$ linkages. This polymer appears at the onset of cell expansion and is rapidly metabolised during growth (Kim *et al.*, 2000). Because of this parallel appearance and metabolism during expan-

sion, this polymer has been proposed to play a direct biochemical and structural role in wall extension.

Wall loosening and expansion

Wall biosynthesis and polymer assembly

Clearly the deposition of new wall material is necessary to maintain wall strength and integrity during extension. Although it has been shown possible to maintain wall extension in the absence of deposition *in vitro* (Rayle *et al.*, 1970) it seems that the two cannot be uncoupled *in vivo*. It has been suggested that the deposition of new wall material might directly lead to wall extension by a process of intersusception. This hypothesis proposes that as new material enters the structure of the wall it permits existing components to slide relative to one another acting as a sort of integrating lubricant.

Little or no direct evidence has been shown for the induction of wall extension by newly secreted polymers. However, a pectin-rich polysaccharide fraction from cucumber hypocotyl cell walls can directly increase the extension of cucumber hypocotyls *in vitro* (Figure 2). These polysaccharides were extracted in CDTA from phenol-extracted walls and are unlikely to contain protein. In support of this effect not being protein based, the activity in these samples survived being heated to 100 °C. The simplest interpretation of these data is that the polysaccharides in this solution are inducing extension by integrating into the wall. An alternative interpretation is that the fraction, although dialysed, might contain residual CDTA, or might themselves serve as chelating agents. However, direct applications of CDTA during extension assays had no stimulating effect. These data support the possibility that wall deposition might itself directly induce wall extension. The molecular mechanisms by which these polysaccharides induce extension are not obvious, and this is an area requiring more work. The majority of this wall fraction is pectinaceous, but it represents a heterogeneous mixture of polymers and is likely to include some matrix glycans as well as pectins. It is worth noting that other wall polysaccharide fractions enriched in matrix glycans did not show this effect (data not shown), but these fractions had been extracted by harsh chemical treatments which might alter structural aspects of the polymers.

Given normal wall assembly there is good evidence that extensibility is directly modulated by activities in

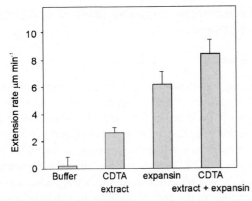

Figure 2. Effects of pectin-rich wall extracts on *in vitro* extension of cucumber hypocotyls. Cucumber seedlings were grown as described in Figure 1. The upper 3 cm of the hypocotyls was excised and homogenised in a Waring blender with an equal volume (to fresh weight) of wash buffer (25 mM Hepes, 2 mM EDTA, 2 mM DTT pH 6.8) and then washed 3 times this buffer. Wall residues were then extracted in wash buffer containing 1 M NaCl for 1 h at room temperature. The residues were then retained on a nylon filter and then suspended and washed in buffer-equilibrated phenol (pH 7.8). Wall residues were pelleted by centrifugation and then washed 3 times in large volumes of 95% ethanol to remove the phenol. Ethanol-washed wall residues were retained on a nylon filter and left to air-dry. These residues were then suspended in 50 mM CDTA and left to stir gently overnight. The extract was filtered through nylon mesh, dialysed against distilled water, and freeze-dried. CDTA extracts were dissolved at 50 mg/ml in 50 mM sodium acetate pH 4.5 for assays. Expansins were prepared from cucumber hypocotyls by ammonium sulfate precipitation of the 1 M NaCl extract as described by McQueen-Mason *et al.* (1992). Extension measurements were carried out as described in the legend to Figure 1. Data are averages and standard errors of 6 measurements.

the wall itself. It is important to bear in mind that most of the polymers in the cell walls of growing cells are not covalently connected to one another, and that the overall integrity of the wall is supported by large collections of non-covalent interactions between these polymers. This indicates that cell wall integrity and mechanical properties will largely be determined by three factors: the size of polymers in the wall, the degree of polymer entanglement, and the interactions among polymers. Changes in any of these three variables could, in theory, be a means of altering wall extensibility.

Most recent studies of cell wall loosening have generally focused on an enzymatic basis for cell wall extension. An exception to this is a novel wall-loosening mechanism proposed by Fry (1998). This mechanism involves the non-enzymatic cleavage of wall polysaccharides by millimolar amounts of L-ascorbate. The proposed mechanism involves the reduction of O_2 and Cu^{2+} by ascorbate in the wall to form H_2O_2 and Cu^+, which then react to form hydroxyl radicals ($\cdot OH$). These radicals then oxidatively cleave polysaccharide chains within the wall. Using an *in vitro* approach, Fry demonstrated that ascorbate-generated $\cdot OH$ could fragment xyloglucan under physiologically relevant conditions. As ascorbate, H_2O_2 and Cu^+ are naturally ocurring wall components it is possible that $\cdot OH$ radicals could be generated *in vivo* and thus lead to wall loosening through the scission of wall polysaccharides. Recently, Schweikert *et al.* (2000), also demonstrated the potential for non-enzymic cleavage of cell wall polysaccharides. In this instance, $\cdot OH$ radicals are generated from O_2 by peroxidases. Whether these processes actually do occur *in vivo* and what contribution they make to cell extension remains to be established.

Wall-modifying enzymes

Expansins

Expansins were first identified in 1992 as mediators of acid-induced wall extension (McQueen-Mason *et al.*, 1992). That is, they catalyse cell wall extension, or 'creep', in response to a unidirectional extensive force. In addition to playing a key role in cell expansion and hence growth, expansins have since been ascribed several diverse roles in plant development, including leaf organogenesis (Fleming *et al.*, 1997), fruit softening (Rose *et al.*, 1997) and wall disassembly (Cho and Cosgrove, 2000). The general properties of expansins have been reviewed in several recent reviews, and we refer the reader to them (McQueen-Mason and Rochange, 1999; Cosgrove, 1999, 2000a, b). The aim of this section of the review is to focus specifically on the molecular aspects of expansins and how these relate to their biochemical mode of action.

The molecular cloning of two expansins from cucumber hypocotyls by Shcherban *et al.* (1995), opened the door to the identification of expansins from a wide-range of higher plants and organs, such as rice internodes (Cho and Kende, 1997a, b), tomato and strawberry fruits (Rose *et al.*, 1997; Civello *et al.*, 1999), cotton fibre (Shimizu *et al.*, 1997) and soybean cell cultures (Downes and Crowell, 1998). Analysis of the accumulating cDNAs has revealed that expansins have distinct and unique structures and can be classified into two subgroups, α and β, on the basis of substrate specificity and sequence similarities.

α-Expansins

α-Expansins were the first expansins to be characterised at both the biochemical and molecular level and, as such, more is understood about their biochemistry and physiology than about β-expansins. They comprise a large multigene family with 58 full-length or partial ESTs from 14 higher-plant species deposited in the GenBank database to date (for current listing, see http://www.bio.psu.edu/expansins/). This large number of isoforms within each gene family may reflect levels of control of wall loosening. In this case each isoform may act on different wall polymers and expression could be regulated by developmental or hormonal signals. Accumulated data support this notion. For example, McQueen-Mason and Cosgrove (1995) showed that two separate expansin isoforms purified from cucumber hypocotyls showed quite distinct effects on stress relaxation properties of hypocotyls and also showed small differences in pH optima. Cho and Kende (1997b, 1998) first demonstrated specific and distinct expression patterns of expansin transcripts in deep water rice. Subsequent studies demonstrated distinctive expression patterns in ripening tomato fruit (Brummell *et al.*, 1997, 1999; Rose *et al.*, 1997, 2000), leaf primordia initiation (Fleming *et al.*, 1997), and in xylem cells (Im *et al.*, 2000). Probably the most convincing evidence for isoform-specific expression and proposed function has recently been demonstrated by Cho and Cosgrove (2000). The authors showed that *AtEXP10* expression was localised to growing leaves and the base of leaf pedicals. Furthermore, morphological analysis of *AtEXP10* antisense inhibited plants indicated a corresponding localised function in these tissues.

Although α-expansin isoforms may have organ-specific expression patterns, they all share a highly conserved amino acid sequence, indicating a corresponding conservation of biochemical function. The primary transcripts of α-expansins characteristically encode a protein with three discrete domains. A signal peptide (typically 22–25 amino acids), which directs the nascent polypeptide into the ER/Golgi and is removed before the mature protein is secreted into the cell wall. A putative polysaccharide-binding domain (ca. 10 kDa) at the carboxyl terminus, which contains a series of conserved tryptophan residues and, resting between these two domains, a cysteine-rich region with limited, but significant, sequence similarity to family-45 endoglucanases (EG45-like domain). The mature protein has a molecular mass of 25–27 kDa and does not appear to possess sites for *N*- or *O*-linked glycosylation.

β-Expansins

The major components of grass pollen are the Group 1 allergens (Esch and Klapper, 1989). These small glycoproteins share a weak sequence similarity to α-expansins (Shcherban *et al.*, 1995) and have been shown to induce cell wall expansion or 'creep' (Cosgrove *et al.*, 1997). It is believed that these proteins soften the stigma and stylar tissue *in vivo*, allowing the penetration of the pollen tube into the maternal tissue. Based on these studies Group 1 allergens and related sequences were designated as β-expansins, referring to their similar rheological effects on the cell wall as α-expansins, whilst indicating their sequence divergence. In common with α-expansins, β-expansins also comprise a multigene family. From sequences so far deposited in the TIGR database (http://www.tigr.org), it appears that β-expansins are more abundant in Type II walls (e.g. rice has 14 α-expansins and 15 β-expansins), than in Type I walls (e.g. *Arabidopsis* has 23 α-expansins and 3 β-expansins). Although it remains to be investigated thoroughly, these distribution patterns are at least consistent with apparent substrate preferences.

Expansin mode of action

Despite considerable knowledge of many aspects of expansin biochemistry, molecular biology and effects on wall rheology, the precise mechanism of how expansin makes the cell wall more extensible has yet to be resolved. Current models of expansin action are based on the interpretation of early studies of expansin biochemistry *in vitro* (McQueen-Mason *et al.*, 1992, 1993; McQueen-Mason and Cosgrove, 1994, 1995) and on similarities in protein sequence to bacterial cellulases (Davies *et al.*, 1995).

Only a limited (ca. 25%) amino acid identity exists between the α and β families. However, there are several conserved motifs within both expansin classes, which presumably have essential catalytic or structural roles, uniting the two expansin classes in a shared function. The most notable of these features are several paired cysteine residues in the N-terminus, a series of tryptophan residues at the C-terminus, and a conserved 'HFD' box in the central domain. We now examine these features in turn and explore how they might be functionally interpreted given the biochemical characteristics of expansin rheology.

The spacing between the C-terminal tryptophan residues shares similarity to that of the cellulose binding domains of some bacterial cellulases (Gilkes *et al.*, 1991; Hamamoto *et al.*, 1992). As α-expansins are observed to bind tightly to cell walls (unpublished data) it has been proposed that the C-terminal domain is involved with cellulose binding and serves to unite the catalytic domain with its substrate (Cosgrove, 1999). However, there are observations which question that tight binding is essential for expansin activity. For instance, the maize β-expansin does not bind strongly to the cell wall but nevertheless exhibits strong wall loosening activity. Alternative roles for the cellulose-binding domain (CBD) may still involve the anchoring of expansin within the cell wall, but in this case the anchor prevents expansins secreted by one cell influencing the growth of neighbouring cells.

The conserved cysteine residues within the N-terminal half of the expansin protein have similar spacing to cysteines in the chitin-binding domain of wheat germ agglutinin (Wright *et al.*, 1991). The cysteine residues in this lectin form a series of intramolecular disulfide bonds that stabilise the protein structure. It has been proposed that expansin may be folded and stabilised in an analogous manner (Shcherban *et al.*, 1995). The limited knowledge of expansin protein chemistry appears to substantiate the presence and importance of disulfide bonds for folding and concomitant activity. Recombinant α-expansins, expressed in either bacterial or bacculovirus systems, require reducing agents in the extraction buffer to produce the same electrophoretic mobility as native expansin. Even under these conditions recombinant proteins consistently show far lower specific activity than native expansin and tend to form insoluble aggregates at higher protein concentrations, suggestive of improper bond formation and incorrectly folded proteins (V. Filatov and S. McQueen-Mason, unpublished data).

Perhaps the most intriguing molecular conservation between expansins is within the central domain. This region shares several characteristics of the catalytic core of family-45 endoglucanases and thus is believed to form the catalytic domain of expansin protein (Cosgrove, 1997). Although sequence similarity to these endoglucanases is slight, the residues which are conserved, the aspartate in the 'HFD' box and the second aspartate close to the N-terminus, are also the key residues at the catalytic site of the endoglucanases (Davies *et al.*, 1995). In addition, conservation of several pairs of cysteine residues within this domain

suggests that disulfide bonds would form and fold the protein enabling the two catalytic aspartates to reside on either side of a substrate binding groove. However, despite similarities in sequence expansins exhibit neither endoglucanase activity (McQueen-Mason and Cosgrove, 1995) nor related transglycosylase activity (McQueen-Mason *et al.*, 1993).

So how does expansin catalyse cell wall extension? Current hypotheses propose that the catalytic domain disrupts non-covalent bonds between wall polysaccharides, thus acting as a kind of 'molecular grease' that allows polymers to slide passed each other. This catalytic action may be facilitated through the cellulose binding domain, which would allow the protein to move along a single cellulose polymer, weakening the tethers of cross-linking glycans as it progresses. This model is consistent with the effect of expansins on the mechanical strength of paper (McQueen-Mason and Cosgrove, 1994), and the reversibility and time-dependence of expansin action (McQueen-Mason and Cosgrove, 1995). However, the question still remains as to how expansin catalyses the disruption of hydrogen bonds between the cellulose/matrix glycan polymers. The exploitation of recombinant proteins harbouring mutations in the putative catalytic residues should throw light on the elusive mechanism of cell wall loosening by expansin.

Substrate specificity has been examined in only one expansin isoform. Whitney *et al.* (1999), examined the activity of recombinant *CsExp1* on cellulose-composite materials made by *Acetobacter* cultures. This expansin displayed high activity on composites containing tamarind xyloglucan, but no activity was detected on similar composites made with glucomannan or galactomannan. This suggests a clear substrate specificity for xyloglucan. Binding studies with the same expansin showed that the protein had little binding affinity for xyloglucans in solution. The expansin showed low affinity for pure crystalline cellulose, but high affinity to complexes involving xyloglucan (or other matrix glycans) and cellulose. Expansin also showed higher binding affinity for swollen cellulose than for crystalline cellulose (McQueen-Mason and Cosgrove, 1995). These data suggest that the protein binds to the junction between crystalline cellulose and soluble glucan chains, whilst the activity data from the composites appears to indicate that the protein is discriminating in the types of polysaccharide it can displace from microfibrils.

Yieldin

There are only two classes of enzymes shown to directly modulate wall mechanical properties. Of these, expansins are the best known and best characterised. However, a new class of enzyme, referred to as 'yieldin', has been reported recently that also modulates wall extension *in vitro* (Okamoto-Nakazato *et al.*, 2000). Yieldins are characterised by their ability to lower the yield threshold for extension in cowpea hypocotyls. The yield threshold is the minimum force necessary to drive the extension of a material. In contrast, expansins catalyse the long-term extension (creep) of cell walls. Yieldins were purified to two bands on SDS-PAGE of around 30 kDa. The coding sequence for a yieldin has been obtained based on information from partial protein sequencing. The identity of the protein encoded by this gene has been confirmed by demonstrating that recombinant protein also has yieldin activity on *Vigna* hypocotyls. The predicted protein sequence of the cloned DNA reveals that yieldin has high amino acid similarity to acidic endochitinases. Although the recombinant protein possessed yieldin activity, it lacked chitinase activity. Given this lack of chitinase activity and the fact that it is generally accepted that plant cell walls do not contain chitin, it seems unlikely that chitin hydrolysis is the mechanism of yieldin action. Whether yieldin proves to be a hydrolytic enzyme or work by a more subtle mechanism remains to be seen.

Xyloglucan endotransglycosylases

The most abundant cross-linking glycan in dicot primary walls is xyloglucan. Given the primary role of this polysaccharide in cell wall architecture it is easy to imagine that factors that modify xyloglucan cross-linking would play a key role in cell wall loosening and expansion. Xyloglucan endotransglycosylases (XETs) were first identified and characterised by Fry *et al.* (1992) and Nishitani and Tominaga (1992). XETs predominantly display true transglycosylase activity, catalysing the cleavage of xyloglucan chains and grafting of the resulting half chain to the non-reducing end of a second xyloglucan chain. However, some species of XET can catalyse the transfer of xyloglucan to a water molecule and, as such, are simply endo-$(1{\rightarrow}4)\beta$-D-glucanases. Although the exact mechanism of XET endotransglycosylase or endoglucanase activity has yet to be elucidated, a stable covalent glycosyl-enzyme intermediate exists prior to transfer of the glycosyl moiety to its donor substrate (Sulova *et al.*, 1998), and this relatively stable intermediate can be exploited for biochemical purification of the enzyme (Steele and Fry, 1999). In addition to its potential ability to alter and loosen the cell wall matrix, studies have shown a strong correlation between XET expression and activity to areas of cell expansion (Fry *et al.*, 1992; Xu *et al.*, 1996; Vissenberg *et al.*, 2000). Given these data it is somewhat surprising that, in contrast to expansins, XETs induce neither unilateral cell wall extension *in vitro* (McQueen-Mason *et al.*, 1993) nor do they have any synergistic interaction with expansins.

Several functions for XETs have been proposed, the most prominent of which is its role in wall biogenesis. As growing plant cells expand the walls do not become thinner, thus new wall material must be incorporated into the extending matrix. Nascent xyloglucan chains are secreted into the wall matrix and covalently linked to existing xyloglucan chains, maintaining the integrity of the wall architecture (Nishitani, 1997). It has been proposed that XETs participate in this process, although as yet the evidence for this is only indirect (Thompson *et al.*, 1997). XETs have also been ascribed roles in fruit ripening (Redgwell and Fry, 1993; Arrowsmith and de Silva, 1995; Schroder *et al.*, 1998; see also Brummell and Harpster, 2001, this issue) and flooding-induced aerenchyma formation (Saab and Sachs, 1996). Whatever the function of XETs probably the key-determining factor surrounding physiological aspects of this enzyme is substrate availability. If XET activity increases as xyloglucan is secreted into the cell walls, polymers will assemble around nascent cellulose microfibrils and wall biogenesis will occur. If xyloglucan is not available, XET activity will probably result in rearrangement and/or degradation of existing xyloglucan polymers. Thus, any presumptive physiological role must therefore take into account the complex interplay between XET activity and the levels of nascent xyloglucan secretion.

Recent advances in the molecular aspects of XETs have given further clues to the function of these enzymes *in vivo*. Systematic sequencing of plant genomes has revealed the existence of large XET and XET-related (XTRs) gene families (reviewed by Campbell and Braam, 1999b). Of the 14 species so far examined, 46 XET sequences have been identified, with 21 XETs or XTRs in the genome of *Arabidopsis* alone. All XETs and XTRs so far reported show a high degree (30–90%) of amino acid identity, suggesting a shared biochemical function. XETs show greatest divergence at the C-terminus, and classification of XETs into 4 groups has been based on the relatedness of sequence in this region. However, all XETs sequences have several conserved features; a hydrophobic N-terminus which most likely acts as a signal peptide (Campbell and Braam, 1998; Rose *et al.*, 2000), a putative catalytic core which is directly preceded by N-linked glycosylation site, and, at the C-terminus, cysteine residues that have the potential to form two disulfide bonds.

The putative catalytic core features the amino acid motif DEIDFEFLG. This sequence shares significant similarity to the active site of *Bacillus* β-glucanases. *Bacillus* β-glucanases, like XETs, cleave $(1\rightarrow4)\beta$ linkages of polysaccharides within glucan backbones but differ from XETs in that their activity is purely hydrolytic. Recently, researchers have exploited the identity between *Bacillus* glucanases and XETs to target specific amino acid residues for site-directed mutagenesis. Campbell and Braam (1999a) have shown that altering the catalytic glutamate to a glutamine eliminates >98% of the enzyme activity, thus demonstrating that this residue is essential for catalysis. To date four proteins have been isolated that do not contain this highly conserved motif. Of these, one which provokes particular interest is the *Nasturtium* seed XET (NXG), which shows mostly hydrolytic activity not transglycosylase activity *in vivo*. This suggests that differences in activity result from slight structural changes in or around the catalytic site. However, in NXG the first phenylalanine residue is replaced with isoleucine and, therefore, substitution would maintain the non-polar, uncharged nature of the residues and is unlikely to affect the structure or catalytic ability of the active site.

Sequence divergence within the carboxyl terminal portions of XETs hints at differences in function of the corresponding isoenzymes. Campbell and Braam (1999a), have also shown subtle differences in the temperature, pH optima, and substrate specificity of four C-terminal divergent *Arabidopsis* XETs when activity was assessed *in vitro*. These subtle differences could translate into specific physiological roles *in vivo*. It has been speculated that the large XET family reflect delegation of a specific XETs to a target organ, tissue, or cell and, as an extension to this, the regulation of certain XET species by developmental or hormonal signals (Xu *et al.*, 1995, 1996).

Future research into XETs will undoubtedly aim to elucidate the precise cellular and physiological func-tions of these enzymes and pinpoint their precise roles in modifying cell walls during plant growth and development.

Endoglucanases

In the 1970s biochemical studies were undertaken to examine the metabolism of wall polymers during extension. Labavitch and Ray (1974), showed that auxin-stimulated growth was accompanied by the release of soluble xyloglucan fragments from pea cell walls. The conclusion of these studies was that the release of xyloglucans resulted from biochemical activities in the wall associated with extension. Thus was born the hypothesis that the action of wall hydrolases, especially β-glucanases, might induce wall extension.

Subsequent studies have also revealed measurable decreases in the molecular mass of xyloglucans during growth in pea (Talbott and Ray, 1992), and azuki bean (Nishitani and Masuda, 1983). Inherent in the general hypothesis is that decreasing the molecular mass of a polymer will also decrease its viscosity. Decreasing the viscosity of major wall polymers will most likely decrease the stiffness of the wall, thus increasing the rate of extension. Considerable indirect support for the role of β-glucanases in wall extension has accumulated. Labrador and Nevins (1989) showed that treatment with β-glucanase could enhance elongation in maize coleoptiles. In addition, Inouhe and Nevins (1991) showed that antibodies raised against a maize glucanase could partially block auxin-induced growth. Similarly, Hoson *et al.* (1991) demonstrated that antibodies that bound xyloglucans could block the decrease in xyloglucan molecular mass, and diminish extension during auxin-induced growth in azuki beans. However, as in the case of XETs, there is no evidence that endoglucanases can directly modulate wall extensibility. In fact, Cosgrove and Durachko (1994) showed that β-glucanase treatment did not induce wall extension *in vitro* although it did weaken the wall.

It is now apparent that $(1\rightarrow4)\beta$-D-glucanases comprise a multi-gene family in higher plants. Studies in tomato (Brummell *et al.*, 1997; Catalá *et al.*, 1997) have examined the expression patterns of various members of this family. Different isoforms have been shown to be expressed in ripening fruit and abscission zones, yet others are associated with cell expansion. Catalá *et al.* (1997) showed that the transcript of a glucanase gene associated with growing hypocotyls was present in both growing and mature tissues, and that auxin induction of expression of this gene was much slower than the effects of auxin on growth. Certainly such an enzyme may be involved in xyloglucan turnover during growth although the pattern of its expression is not suggestive of an enzyme acting as a wall-loosening agent.

In summary, the evidence for a direct role for endo-β-glucanases in cell wall loosening is mixed. As previously discussed, the most convincing evidence for such a role in growth came from the *korrigan* mutant (Nicol *et al.*, 1998), but this particular gene appears involved in wall synthesis rather than loosening. The patterns of activity and expression of glucanases suggest that they certainly are involved in the metabolism xyloglucan and other matrix glycans during growth but a direct role in wall loosening awaits confirmation.

Wall rigidification

Once a cell has reached its final size, wall extension comes to a halt. In parallel with the cessation of growth the cell wall becomes more rigid, preventing turgor-induced extension. Wall rigidification could be brought about by three possible events: (1) the deposition of new wall polymers, (2) an increase in wall cross-linking activity, or (3) a halt in production of wall-loosening proteins, such as expansins. Evidence to date suggests that a combination of these events rather than a single factor leads to rigidification of the wall *in vivo*.

Expansins and wall rigidification

McQueen-Mason *et al.* (1992) showed that expansins extracted from growing regions of cucumber hypocotyls displayed strong expansin activity, whilst those from the base, where growth had stopped, exhibited no extractable activity. Northern blot analyses, with *CsExp*1 used as probe, show the pattern of transcript accumulation in cucumber hypocotyls to be highest in the growing region and then gradually decline along the length of the hypocotyl (Figure 3). This pattern closely mirrors that of cell elongation and acid-induced expansion. Cosgrove and Li (1993), have also shown similar patterns of expansin-induced extension in oat coleoptiles. These data strongly suggest that the cessation of growth involves the progressive decrease in expansin as the wall matures. Contrary to this notion, McQueen-Mason (1995) showed that while expansin activity was probably sufficient to account for wall extension in the growing cells, the loss

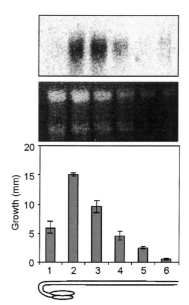

Figure 3. The distribution of *CsExp1* mRNA abundance in etiolated cucumber hypocotyls. A (top). Northern blot probed with *CsExp1*. B (middle). Ethidium bromide-stained denaturing gel. Cucumber (*Cucumis sativa* cv. Burpee Pickler) hypocotyls were germinated and grown in the dark for 4 days at 27 °C. The apical 3 cm including the apical hook was excised and cut into 6 mm × 5 mm sections such that the apical hook comprises the first 5 mm section. Sectioning was carried out under green light conditions and the sections frozen in liquid nitrogen immediately after cutting. Total RNA was extracted from the hypocotyl sections by hot SDS/phenol extraction and lithium chloride precipitation. RNA was separated by formaldehyde denaturing gel electrophoresis and transferred onto nylon membrane by capillary blotting. A DIG-labelled DNA probe was prepared from a full-length clone of *CsExp1* by random prime labelling. Blots were hybridised overnight in 250 mM Na_2HPO_4, 7% SDS at 65 °Cand subsequently washed in 20 mM Na_2HPO_4, 5% SDS at 65 °C, followed by 20 mM Na_2HPO_4, 1% SDS at 65 °C. The probe was detected by a chemiluminescence enzyme immunoassay using a DIG Luminescent Detection Kit according to the manufacturers' instructions (Boeringer-Mannheim). For determination of the distribution of growth in cucumber hypocotyls, 4-day old dark-grown etiolated hypocotyls were marked with spots of oil paint at 5 mm intervals from the hypocotyl tip and grown for further 24 h in the dark; displacement of the marks was measured after 24 h, $n = 30$. All manipulations were carried out under green light conditions.

of expansin activity from the walls significantly lagged behind the cessation of growth. This suggests other events are important in wall rigidification and probably make it unresponsive to the diminishing level of expansins.

Modifications of the polymer networks

If loss of expansins is not on its own responsible for the decrease in wall extensibility, then alternative mechanisms must be considered. Potentially, rigidification could involve modification of any of the polymer networks in the wall. Indeed, Carpita (1984) showed that arabinoxylan polymers become increasingly less branched in maturing maize coleoptiles permitting the polymer to form a tighter complex with cellulose. The removal of mixed-linked $(1\rightarrow4)\beta$-D-glucans also mirrors cessation of growth in these walls. These changes in the polymer network could make the sites at which wall-modifying enzymes bind less accessible and thereby halt further extension of the cell wall.

Although the cellulose/matrix glycan network appears to determine the extensibility of the growing primary wall and to provide the target for wall loosening events, cross-linking of pectins or structural proteins, which form co-extensive networks, could constrain the extension of the cellulose/glycan network preventing cell expansion. Studies have shown that Type I cell walls are extensin and pectin-rich, whilst Type II cell walls are polyphenolic-rich (reviewed by Carpita and Gibeaut (1993). These pronounced differences in wall composition between grasses and dicots suggest that different mechanisms of primary wall rigidification might operate in these plants. There is considerable evidence that pectin cross-linking may be a primary wall rigidifying mechanism in dicots (Yamaoka *et al.*, 1993; McCann *et al.*, 1994), whereas in the grasses phenolic coupling between polysaccharides plays a key role in rigidification (Carpita and Gibeaut, 1993).

In Type I walls polygalacturonan is secreted into the primary wall in a highly methyl-esterified form (Kauss and Hassid, 1967). These methyl esters are introduced during pectin biosynthesis in the Golgi and mask the carboxyl groups of the galacturonyl residues. Subsequent to secretion into the wall, these methyl groups may be removed through the action of pectin methyl-esterases (PMEs) (Knox *et al.*, 1990).

Removal of methyl esters from polygalacturonan has two important effects. Firstly, this exposes the carboxyl group, thus increasing both the net charge of the polymer and increasing its capacity for cross-linking by cations, such as calcium. This can lead to stronger gelling of the pectins, which could in turn, be important both in rigidification as well as in strengthening cell-cell adhesion at the middle lamella. Secondly, many polygalacturonases have low activity against esterified pectins but high activity against the acidic forms. Hence, demethylation can render the pectins more labile to the action of hydrolases. This is clearly the case in fruit ripening, where antisense inhibition of PME leads to pectins that maintain higher degrees of methyl esterification, resulting in decreased depolymerization of pectins by polygalacturonase (Hall *et al.*, 1993).

A number of studies have shown that PME activity shows an inverse correlation with growth rate in expanding tissues, suggesting it may be involved in wall rigidification. More direct evidence for such a role for PME came from some initial studies by McQueen-Mason and Cosgrove (1995), who demonstrated that the loss in sensitivity of cucumber hypocotyl walls to expansin action during wall rigidification probably resulted from PME activity. These authors also showed that incubating cell walls with exogenous PME could result in wall rigidification. Thus, it appears that pectin cross-linking may provide the primary means of wall rigidification in dicotyledons.

Peroxidases

The cross-linking of polymers by oxidative coupling through the action of peroxidases has long been thought to provide a route for wall rigidification. This might occur through the cross-linking/polymerisation of structural proteins, such as hydroxyproline-rich proteins (extensins) leading to a strong network in the wall. Alternatively, it might be through the cross-linking of polysaccharides through phenolic dimerisation. The grass arabinoxylans exist as feruloyl esters and can be cross-linked by peroxidase action (Brett *et al.*, 1999; Rizk *et al.*, 2000). These can subsequently enter free-radical condensation reactions in the wall leading to the formation of ferulate dimers and, hence, polysaccharide cross-links. These cross-links could provide an effective means of wall rigidification catalysed by peroxidases.

Several groups have shown that peroxidase activity in plants increases with the cessation of growth, indicating that the two phenomena are linked in some plants (Goldberg *et al.*, 1987; Tan *et al.*, 1992). However, in other plants there is no clear relationship between peroxidase activity and growth (Swoap *et al.*, 1993; Casal *et al.*, 1994). Schopfer (1996) reported that treatment with H_2O_2 leads to wall rigidification and to reduced growth in maize coleoptiles. He further demonstrated that these tissues contain considerable phenolic compounds in their cell walls and that the level of these was much higher in tissues where growth had ceased (Schindler *et al.*, 1997).

As mentioned earlier, plant cell walls contain considerable quantities of structural proteins. Some classes of these proteins, such as extensins, are polymerised in the wall to form a strong network. Such proteins are highly expressed in response to pathogen ingress and in parts of the plant subject to mechanical stress (Esquerré-Tugayé *et al.*, 1979; Showalter *et al.*, 1985). These proteins are also present as monomers in growing cell walls but cross-linked into the wall by peroxidase action after growth has ceased (Ye *et al.*, 1991), and, as such, this action could provide a means of wall rigidification. In support of this, Brownleader *et al.* (2000) report that treatment of tomato seedlings with extensin peroxidase inhibitors resulted in increased growth.

Concluding remarks

Advances in genetics, molecular biology and biochemistry have provided new tools to investigate both structural and functional components within the growing cell wall. With sophisticated molecular and immunological techniques we are now in a position to unravel the precise relationships of wall-modifying enzymes in the milieu of matrix polysaccharides. The plant genome project is now providing an ever-increasing wealth of data, allowing more efficient schemes for the study of cell wall mutants. In addition, the availability of transgenic technologies now allows us to manipulate the cell wall and shed light on the mechanisms of wall assembly and extension. Taken as a whole, these advances provide very exciting prospects in the study of the molecular basis of cell wall extension.

Acknowledgements

We would like to thank Jeff Patteson for his help in the preparation of the manuscript, and Jennifer Milne for critical reading of this review.

References

Arioli, T., Peng, L.C., Betzner, A.S., Burn, J., Wittke, W., Herth, W. *et al.* 1998. Molecular analysis of cellulose biosynthesis in *Arabidopsis*. Science 279: 717–720.

Arrowsmith, D.A. and de Silva, J. 1995. Characterization of 2 tomato fruit-expressed cDNAs encoding xyloglucan endotransglycosylase. Plant Mol. Biol. 28: 391–403.

Baba, K., Sone, Y., Misaki, A. and Hayashi, T. 1994. Localization of xyloglucan in the macromolecular complex composed of xyloglucan and cellulose in pea stems. Plant Cell Physiol. 35: 439–444.

Brett, C.T., Wende, G., Smith, A.C. and Waldron, K.W. 1999. Biosynthesis of cell-wall ferulate and diferulates. J. Sci. Food Agric. 79: 421–424.

Brownleader, M.D., Hopkins, J., Mobasheri, A., Dey, P.M., Jackson, P. and Trevan, M. 2000. Role of extensin peroxidase in tomato (*Lycopersicon esculentum* Mill) seedling growth. Planta 210: 668–676.

Brummell, D.A., Bird, C.R., Schuch, W. and Bennett, A.B. 1997. An endo-1,4-β-glucanase expressed at high levels in rapidly expanding tissues. Plant Mol. Biol. 33: 87–95.

Brummell, D.A., Harpster, M.H. and Dunsmuir, P. 1999. Differential expression of expansin gene family members during growth and ripening of tomato fruit. Plant Mol. Biol. 39: 161–169.

Campbell, P. and Braam, J. 1998. Co- and/or post-translational modifications are critical for TCH4 XET activity. Plant J. 15: 553–561.

Campbell, P. and Braam, J. 1999a. *In vitro* activities of four xyloglucan endotransglycosylases from *Arabidopsis*. Plant J. 18: 371–382.

Campbell, P. and Braam, J. 1999b. Xyloglucan endotransglycosylases: diversity of genes, enzymes and potential wall-modifying functions. Trends Plant Sci. 4: 361–366.

Carpita, N. 1984. Cell-wall development in maize coleoptiles. Plant Physiol. 76: 205–212.

Carpita, N.C. and Gibeaut, D.M. 1993. Structural models of primary cell walls in flowering plants: consistency of molecular structure with the physical properties of the walls during growth. Plant J. 3: 1–30.

Casal, J.J., Mella, R.A., Ballare, C.L. and Maldonados, S. 1994. Phytochrome-mediated effects on extracellular peroxidase activity, lignin content and bending resistance in etiolated *Vicia faba* epicotyls. Physiol. Plant. 92: 555–562.

Catalá, C., Rose, J.K.C. and Bennett, A.B. 1997. Auxin regulation and spatial localization of an endo-1,4-β-D-glucanase and a xyloglucan endotransglycosylase in expanding tomato hypocotyls. Plant J. 12: 417–426.

Cho, H.T. and Cosgrove, D.J. 2000. Altered expression of expansin modulates leaf growth and pedicel abscission in *Arabidopsis thaliana*. Proc. Natl. Acad. Sci. USA 97: 9783–9788.

Cho, H.T. and Kende, H. 1997a. Expansins and internodal growth of deepwater rice. Plant Physiol. 113: 1145–1151.

Cho, H.T. and Kende, H. 1997b. Expansins in deepwater rice internodes. Plant Physiol. 113: 1137–1143.

Cho, H.T. and Kende, H. 1998. Tissue localization of expansins in deepwater rice. Plant J. 15: 805–812.

Civello, P.M., Powell, A.L.T., Sabehat, A. and Bennett, A.B. 1999. An expansin gene expressed in ripening strawberry fruit. Plant Physiol. 121: 1273–1279.

Cosgrove, D.J. 1997. Assembly and enlargement of the primary cell wall in plants. Annu. Rev. Cell Dev. Biol. 13: 171–201.

Cosgrove, D.J. 1999. Enzymes and other agents that enhance cell wall extensibility. Annu. Rev. Plant Physiol. Plant Mol. Biol. 50: 391–417.

Cosgrove, D.J. 2000a. Expansive growth of plant cell walls. Plant Physiol. Biochem. 38: 109–124.

Cosgrove, D.J. 2000b. New genes and new biological roles for expansins. Curr. Opin. Plant Biol. 3: 73–78.

Cosgrove, D.J. and Durachko, D.M. 1994. Autolysis and extension of isolated walls from growing cucumber hypocotyls. J. Exp. Bot. 45: 1711–1719.

Cosgrove, D.J. and Li, Z.C. 1993. Role of expansin in cell enlargement of oat coleoptiles: analysis of developmental gradients and photocontrol. Plant Physiol. 103: 1321–1328.

Cosgrove, D.J., Bedinger, P. and Durachko, D.M. 1997. Group I allergens of grass pollen as cell wall-loosening agents. Proc. Natl. Acad. Sci. USA 94: 6559–6564.

Davies, G.J., Tolley, S.P., Henrissat, B., Hjort, C. and Schulein, M. 1995. Structures of oligosaccharide-bound forms of the endoglucanase V from *Humicola insolens* at 1.9 Å resolution. Biochemistry 34: 16210–16220.

Desnos, T., Orbovic, V., Bellini, C., Kronenberger, J., Caboche, M., Traas, J. and Hofte, H. 1996. *Procuste*1 mutants identify two distinct genetic pathways controlling hypocotyl cell elongation, respectively in dark and light-grown *Arabidopsis* seedlings. Development 122: 683–693.

Downes, B.P. and Crowell, D.N. 1998. Cytokinin regulates the expression of a soybean β-expansin gene by a post-transcriptional mechanism. Plant Mol. Biol. 37: 437–444.

Esch, R.E. and Klapper, D.G. 1989. Isolation and characterization of a major cross-reactive grass group-I allergenic determinant. Mol. Immunol. 26: 557–561.

Esquerré-Tugayé, M.-T., Lafitte, C., Mazau, D., Toppan, A. and Touze, A. 1979. Cell services in plant-microorganism interactions II. Evidence for the accumulation of hydroxyproline-rich glycoprteins in the cell wall of diseased plants as a defense mechanism. Plant Physiol. 64: 320–326.

Fagard, M., Desnos, T., Desprez, T., Goubet, F., Refregier, G., Mouille, G. *et al.* 2000. *PROCUSTE*1 encodes a cellulose synthase required for normal cell elongation specifically in roots and dark-grown hypocotyls of *Arabidopsis*. Plant Cell 12: 2409–2423.

Fleming, A.J., McQueen-Mason, S., Mandel, T. and Kuhlemeier, C. 1997. Induction of leaf primordia by the cell wall protein expansin. Science 276: 1415–1418.

Fry, S.C. 1988. The Growing Plant Cell Wall: Chemical and Metabolic Analysis. John Wiley, New York.

Fry, S.C. 1998. Oxidative scission of plant cell wall polysaccharides by ascorbate-induced hydroxyl radicals. Biochem. J. 332: 507–515.

Fry, S.C., Smith, R.C., Renwick, K.F., Martin, D.J., Hodge, S.K., and Matthews, K.J. 1992. Xyloglucan endotransglycosylase, a new wall-loosening enzyme activity from plants. Biochem. J. 282: 821–828.

Gilkes, N.R., Henrissat, B., Kilburn, D.G., Miller, R.C. and Warren, R.A.J. 1991. Domains in microbial β-1,4-glycanases: sequence conservation, function, and enzyme families. Microbiol. Rev. 55: 303–315.

Goldberg, R., Liberman, M., Mathieu, C., Pierron, M., and Catesson, A.M. 1987. Development of epidermal-cell wall peroxidases along the mung bean hypocotyl: possible involvement in the cell wall stiffening process. J. Exp. Bot. 38: 1378–1390.

Hall, L.N., Tucker, G.A., Smith, C.J.S., Watson, C.F., Seymour, G.B., Bundick, Y. *et al.* 1993. Antisense inhibition of pectin esterase gene-expression in transgenic tomatoes. Plant J. 3: 121–129.

Hamamoto, T., Foong, F., Shoseyov, O. and Doi, R.H. 1992. Analysis of functional domains of endoglucanases from *Clostridium cellulovorans* by gene cloning, nucleotide sequencing and chimeric protein construction. Mol. Gen. Genet. 231: 472–479.

Hauser, M.T., Morikami, A. and Benfey, P.N. 1995. Conditional root expansion mutants of *Arabidopsis*. Development 121: 1237–1252.

Hayashi, T., Marsden, M.P.F. and Delmer, D.P. 1987. Pea xyloglucan and cellulose 5. Xyloglucan-cellulose interactions *in vitro* and *in vivo*. Plant Physiol 83: 384–389.

Hoson, T., Masuda, Y., Sone, Y. and Misaki, A. 1991. Xyloglucan antibodies inhibit auxin-induced elongation and cell-wall loosening of azuki-bean epicotyls but not of oat coleoptiles. Plant Physiol 96: 551–557.

Im, K.H., Cosgrove, D.T. and Jones, A.M. 2000. Subcellular localization of expansin mRNA in xylem cells. Plant Physiol. 123: 463–470.

Inouhe, M. and Nevins, D.J. 1991. Inhibition of auxin-induced cell elongation in maize coleoptiles by antibodies specific for cell wall glucanases. Plant Physiol. 96: 426–431.

Kauss, H. and Hassid, W.Z. 1967. Enzymic introduction of the methyl ester groups of pectin. J. Biol. Chem. 242: 3449–3453.

Kim, J., Olek, A. and Carpita, N. 2000. Cell wall and membrane-associated exo-β-D-glucanases from developing maize seedlings. Plant Physiol. 123: 471–485.

Knox, J., Linstead, P., King, J., Cooper, C. and Roberts, K. 1990. Pectin esterification is spatially regulated both within cell walls and between developing tissues of root apices. Planta 181: 512–521.

Labavitch, J.M. and Ray, P.M. 1974. The relationship between promotion of xyloglucan metabolism and the induction of elongation by indoleacetic acid. Plant Physiol. ?? (AUTHOR: please mention volume): 499–502.

Labrador, E. and Nevins, D.J. 1989. An exo-β-D-glucanase derived from *Zea* coleoptile walls with a capacity to elicit cell elongation. Physiol. Plant. 77: 479–486.

Lamport, D.T.A. 1986. The primary cell wall: a new model. In: Cellulose: Structure, Modification and Hydrolysis, John Wiley, New York, pp.77–90.

Matthysse, A.G., Thomas, D.L. and White, A.R. 1995a. Mechanism of cellulose synthesis in *Agrobacterium tumefaciens*. J. Bact. 177: 1076–1081.

Matthysse, A.G., White, S. and Lightfoot, R. 1995b. Genes required for cellulose synthesis in *Agrobacterium tumefaciens*. J. Bact. 177: 1069–1075.

McCann, M.C. and Roberts, K. 1994. Changes in cell wall architecture during cell elongation. J. Exp. Bot. 45: 1683–1691.

McCann, M.C., Wells, B. and Roberts, K. 1990. Direct visualization of cross-links in the primary plant cell wall. J. Cell Sci. 96: 323–334.

McCann, M.C., Stacey, N.J., Wilson, R. and Roberts, K. 1993. Orientation of macromolecules in the walls of elongating carrot cells. J. Cell Sci. 106: 1347–1356.

McQueen-Mason, S.J. 1995. Expansins and cell-wall expansion. J. Exp. Bot. 46: 1639–1650.

McQueen-Mason, S.J. and Cosgrove, D.J. 1994. Disruption of hydrogen-bonding between plant cell wall polymers by proteins that induce wall extension. Proc. Natl. Acad. Sci. USA 91: 6574–6578.

McQueen-Mason, S.J. and Cosgrove, D.J. 1995. Expansin mode of action on cell walls: analysis of wall hydrolysis, stress-relaxation, and binding. Plant Physiol. 107: 87–100.

McQueen-Mason, S.J. and Rochange, F. 1999. Expansins in plant growth and development: an update on an emerging topic. Plant Biol. 1: 19–25.

McQueen-Mason, S., Durachko, D.M. and Cosgrove, D.J. 1992. Two endogenous proteins that induce cell-wall extension in plants. Plant Cell 4: 1425–1433.

McQueen-Mason, S.J., Fry, S.C., Durachko, D.M. and Cosgrove, D.J. 1993. The relationship between xyloglucan endotransglycosylase and *in vitro* cell wall extension in cucumber hypocotyls. Planta 190: 327–331.

Nicol, F., His, I., Jauneau, A., Vernhettes, S., Canut, H. and Hofte, H. 1998. A plasma membrane-bound putative endo-1,4-β-D-glucanase is required for normal wall assembly and cell elongation in *Arabidopsis*. EMBO J. 17: 5563–5576.

194

Nishitani, K. 1997. The role of endoxyloglucan transferase in the organization of plant cell walls. Int. Rev. Cytol. 173: 157–206.

Nishitani, K. and Masuda, Y. 1983. Auxin-induced changes in the cell-wall xyloglucans: effects of auxin on the 2 different subfractions of xyloglucans in the epicotyl cell-wall of *Vigna angularis*. Plant Cell Physiol. 24: 345–355.

Nishitani, K. and Tominaga, R. 1992. Endoxyloglucan transferase, a novel class of glycosyltransferase that catalyzes transfer of a segment of xyloglucan molecule to another xyloglucan molecule. J. Biol. Chem. 267: 21058–21064.

Okamoto-Nakazato, A., Nakamura, T. and Okamoto, H. 2000. The isolation of wall-bound proteins regulating yield threshold tension in glycerinated hollow cylinders of cowpea hypocotyl. Plant Cell Environ. 23: 145–154.

Pauly, M., Albersheim, P., Darvill, A. and York, W.S. 1999. Molecular domains of the cellulose/xyloglucan network in the cell walls of higher plants. Plant J 20: 629–639.

Rayle, D.L., Haughton, P.M. and Cleland, R. 1970. An *in vitro* system that simulates plant cell extension growth. Proc. Natl. Acad. Sci. USA 67: 1814–1817.

Redgwell, R.J. and Fry, S.C. 1993. Xyloglucan endotransglycosylase activity increases during kiwifruit (*Actinidia deliciosa*) ripening. Plant Physiol. 103: 1399–1406.

Reiter, W.D., Chapple, C. and Somerville, C.R. 1997. Mutants of *Arabidopsis thaliana* with altered cell wall polysaccharide composition. Plant J. 12: 335–345.

Rizk, S.E., Abdel-Massih, R.M., Baydoun, E.-H. and Brett, C.T. 2000. Protein- and pH-dependent binding of nascent pectin and glucuronoarabinoxylan to xyloglucan in pea. Planta 211: 423–429.

Rose, J.K.C., Lee, H.H. and Bennett, A.B. 1997. Expression of a divergent expansin gene is fruit-specific and ripening-regulated. Proc. Natl. Acad. Sci. USA 94: 5955–5960.

Rose, J., Hadfield, K., Labavitch, J. and Bennett, A. 1998. Temporal sequence of cell wall disassembly in rapidly ripening melon fruit. Plant Physiol. 117: 345–361.

Rose, J.K.C., Cosgrove, D.J., Albersheim, P., Darvill, A.G. and Bennett, A.B. 2000. Detection of expansin proteins and activity during tomato fruit ontogeny. Plant Physiol. 123: 1583–1592.

Saab, I.N. and Sachs, M.M. 1996. A flooding-induced xyloglucan endo-transglycosylase homolog in maize is responsive to ethylene and associated with aerenchyma. Plant Physiol. 112: 385–391.

Schindler, M.G., Bergfeld, R., Ruel, K., Jacquet, G., Lapierre, C., Speth, V. and Schopfer, P. 1997. Structure and distribution of lignin in primary and secondary cell walls of maize coleoptiles analysed by chemical and immunological probes. Planta 201: 146–159.

Schopfer, P. 1996. Hydrogen peroxidase-mediated cell wall stiffening *in vitro* in maize coleoptiles. Planta 199: 43–49.

Schroder, R., Atkinson, R.G., Langenkamper, G. and Redgwell, R.J. 1998. Biochemical and molecular characterisation of xyloglucan endotransglycosylase from ripe kiwifruit. Planta 204: 242–251.

Schweikert, C., Liszkay, A. and Schopfer, P. 2000. Scission of polysaccharides by peroxidase-generated hydroxyl radicals. Phytochemistry 53: 565–570.

Serpe, M.D. and Nothnagel, E.A. 1993. Effects of Yariv phenylglycosides on *Rosa* cell suspensions: evidence for the involvement of arabinogalactan-proteins in cell proliferation. Planta 193: 542–550.

Shcherban, T.Y., Shi, J., Durachko, D.M., Guiltinan, M.J., Mcqueenmason, S.J., Shieh, M. and Cosgrove, D.J. 1995. Molecular cloning and sequence analysis of expansins: a highly conserved, multigene family of proteins that mediate cell-wall extension in plants. Proc. Natl. Acad. Sci. USA 92: 9245–9249.

Shedletzky, E., Shmuel, M., Delmer, D.P. and Lamport, D.T.A. 1990. Adaptation and growth of tomato cells on the herbicide 2,6-dichlorobenzonitrile leads to production of unique cell walls virtually lacking a cellulose-xyloglucan network. Plant Physiol. 94: 980–987.

Shimizu, Y., Aotsuka, S., Hasegawa, O., Kawada, T., Sakuno, T., Sakai, F. and Hayashi, T. 1997. Changes in levels of mRNAs for cell wall-related enzymes in growing cotton fiber cells. Plant Cell Physiol. 38: 375–378.

Showalter, A.M., Bell, J.N., Cramer, C.L., Bailey, J.A., Varner, J.E. and Lamb, C.J. 1985. Accumulation of hydroxyproline-rich glycoprotein messenger RNAs in response to fungal elicitor and infection. Proc. Natl. Acad. Sci. USA 82: 6551–6555.

Shpigel, E., Roiz, L., Goren, R. and Shoseyov, O. 1998. Bacterial cellulose-binding domain modulates *in vitro* elongation of different plant cells. Plant Physiol. 117: 1185–1194.

Steele, N.M. and Fry, S.C. 1999. Purification of xyloglucan endotransglycosylases (XETs): a generally applicable and simple method based on reversible formation of an enzyme-substrate complex. Biochem J. 340: 207–211.

Sulova, Z., Takacova, M., Steele, N.M., Fry, S.C. and Farkas, V. 1998. Xyloglucan endotransglycosylase: evidence for the existence of a relatively stable glycosyl-enzyme intermediate. Biochem. J. 330: 1475–1480.

Swoap, S.J., Sooudi, S.K. and Shinkle, J.R. 1993. Uncorrelated changes in the distribution of stem elongation, tissue extensibilty and cell wall peroxidase-activity along hypocotyl axes of *Cucumis* seedlings exhibiting different patterns of growth. Plant Physiol. Biochem. 31: 361–368.

Talbott, L.D. and Ray, P.M. 1992. Changes in the molecular size of previously deposited and newly synthesised pea cell wall matrix polysaccharides. Plant Physiol. 98: 369–379.

Tan, K.S., Hoson, T., Masuda, Y. and Kamisaka, S. 1992. Involvement of cell wall-bound diferulic acid in light-induced decrease in growth rate and cell wall extensibility of *Oryza* coleoptiles. Plant Cell Physiol. 33: 103–108.

Thompson, J.E., Smith, R.C. and Fry, S.C. 1997. Xyloglucan undergoes interpolymeric transglycosylation during binding to the plant cell wall *in vivo*: evidence from C-13/H-3 dual labelling and isopycnic centrifugation in caesium trifluoroacetate. Biochem J. 327: 699–708.

Vissenberg, K., Martinez-Vilchez, I.M., Verbelen, J.P., Miller, J.G. and Fry, S.C. 2000. *In vivo* colocalization of xyloglucan endotransglycosylase activity and its donor substrate in the elongation zone of *Arabidopsis* roots. Plant Cell 12: 1229–1237.

Whitney, S.E.C., Brigham, J.E., Darke, A.H., Reid, J.S.G. and Gidley, M.J. 1998. Structural aspects of the interaction of mannanbased polysaccharides with bacterial cellulose. Carbohydrate Res. 307: 299–309.

Whitney, S.E.C., Gothard, M.G.E., Mitchell, J.T. and Gidley, M.J. 1999. Roles of cellulose and xyloglucan in determining the mechanical properties of primary plant cell walls. Plant Physiol. 121: 657–663.

Whitney, S.E.C., Gidley, M.J. and McQueen-Mason, S.J. 2000. Probing expansin action using cellulose/hemicellulose composites. Plant J. 22: 327-334.

Willats, W.G.T. and Knox, J.P. 1996. A role for arabinogalactan-proteins in plant cell expansion: evidence from studies on the interaction of β-glucosyl Yariv reagent with seedlings of *Arabidopsis thaliana*. Plant J. 9: 919–925.

Wright, H.T., Sandrasegaram, G., and Wright, C.S. 1991. Evolution of a family of *N*-acetylglucosamine binding proteins containing the disulfide-rich domain of wheat germ agglutinin. J. Mol. Evol. 33: 283–294.

Xu, W., Purugganan, M.M., Polisensky, D.H., Antosiewicz, D.M., Fry, S.C. and Braam, J. 1995. *Arabidopsis tch*4, regulated by hormones and the environment, encodes a xyloglucan endotrans-glycosylase. Plant Cell 7: 1555–1567.

Xu, W., Campbell, P., Vargheese, A.K. and Braam, J. 1996. The *Arabidopsis* XET-related gene family: environmental and hormonal regulation of expression. Plant J. 9: 879–889.

Ye, Z.H., Song, Y.R., Marcus, A., and Varner, J.E. 1991. Comparative localization of 3 classes of cell-wall proteins. Plant J. 1: 175–183.

Zablackis, E., York, W.S., Pauly, M., Hantus, S., Reiter, W.D., Chapple, C.C.S. *et al.* 1996. Substitution of L-fucose by L-galactose in cell walls of *Arabidopsis mur*1. Science 272: 1808–1810.

Plant Molecular Biology **47**: 197–206, 2001.
© 2001 *Kluwer Academic Publishers. Printed in the Netherlands.*

WAKs: cell wall-associated kinases linking the cytoplasm to the extracellular matrix

Catherine M. Anderson[+], Tanya A. Wagner[+], Mireille Perret, Zheng-Hui He[1], Deze He and Bruce D. Kohorn*
*Department of Biology, Rm B353, LSRC, Duke University, Durham, NC 27708, USA (*author for correspondence; e-mail kohorn@duke.edu); [1]present address: Department of Biology, San Francisco State University, San Francisco, CA 94132, USA; [+]these authors contributed equally to this paper*

Key words: cell expansion, GRP, pectin, wall-associated kinase, WAK

Abstract

There are only a few proteins identified at the cell surface that could directly regulate plant cell wall functions. The cell wall-associated kinases (WAKs) of angiosperms physically link the plasma membrane to the carbohydrate matrix and are unique in that they have the potential to directly signal cellular events through their cytoplasmic kinase domain. In *Arabidopsis* there are five WAKs and each has a cytoplasmic serine/threonine protein kinase domain, spans the plasma membrane, and extends a domain into the cell wall. The WAK extracellular domain is variable among the five isoforms, and collectively the family is expressed in most vegetative tissues. *WAK1* and *WAK2* are the most ubiquitously and abundantly expressed of the five tandemly arrayed genes, and their messages are present in vegetative meristems, junctions of organ types, and areas of cell expansion. They are also induced by pathogen infection and wounding. Recent experiments demonstrate that antisense *WAK* expression leads to a reduction in WAK protein levels and the loss of cell expansion. A large amount of WAK is covalently linked to pectin, and most WAK that is bound to pectin is also phosphorylated. In addition, one WAK isoform binds to a secreted glycine-rich protein (GRP). The data support a model where WAK is bound to GRP as a phosphorylated kinase, and also binds to pectin. How WAKs are involved in signaling from the pectin extracellular matrix in coordination with GRPs will be key to our understanding of the cell wall's role in cell growth.

Abbreviations: ECM, extracellular matrix; EGF, epidermal growth factor; GA, gibberellic acid; GRP, glycine-rich protein; GST, glutathione *S*-transferase; GUS, β-glucuronidase; INA, 2,2-dichloroisonicotinic acid; KAPP, protein type 2C phosphatase; SA, salicylic acid; SAR, systemic acquired resistance; SRK, *S*-locus receptor kinase; WAK, wall-associated kinase

Introduction

As plant cells divide and expand they assemble and modify their extracellular matrix (ECM) to permit the subsequent change in cell size and shape. The secretion of an ECM by one cell can also influence neighboring cells, perhaps best exemplified by the interactions between the pollen and stigma. Throughout the plant, the production and regulation of the ECM architecture has the potential to influence many aspects of development. The ECM of plant cells, often referred to as the cell wall, varies in composition throughout development. The primary wall is laid down during cell division and expansion, and the secondary wall is deposited on the primary wall once growth has ceased (Cosgrove, 1997). The cell wall is primarily composed of an ordered array of cellulose microfibrils that are coated with cross-linking glycans and embedded in a gel of pectin. Within this arrangement are proteins with varying amounts of linked carbohydrates (Cosgrove, 1997; Reiter, 1998). This protein and carbohydrate matrix, continuous between cells, provides structure and may define tissues in an organ (Roberts, 1994).

There are many unanswered questions of how cell wall synthesis and deposition is regulated and coordinated with other cellular processes. In particular, our understanding of whether and how communication takes place between the external and internal compartments of the cell is limited. Nevertheless, it is likely that communication requires contact (Carpita and Gibeaut, 1993; Showalter, 1993; Wyatt and Carpita, 1993; Cosgrove, 1997; Lynch and Lintilhac, 1997; Reiter, 1998; Kohorn, 2000). Several classes of proteins may provide this continuum, and these include the arabinogalactan proteins (Oxley and Bacic, 1999; Svetek et al., 1999; Majewska-Sawka and Nothnagel, 2000), cellulose synthases (Pear et al., 1996), a hydrolytic enzyme (Nicol et al., 1998; Zuo et al., 2000), and the wall-associated kinases (He et al., 1996, 1999; Kohorn, 2000). These proteins are bound to both the plasma membrane and the cell wall and have the potential to directly signal between the two compartments. There are many other protein families that are completely secreted by the cell, and thus may only indirectly communicate signals to the cytoplasm. Examples of this class include the hydroxyproline-rich glycoproteins (Showalter, 1993), proline-rich proteins (Fowler et al., 1999), glycine-rich proteins (Cheng et al., 1996), xyloglucan endotransglycosylases (Xu et al., 1996; Vissenberg et al., 2000), endoglucanases (Zuo et al., 2000), and expansins (Cho and Cosgrove, 2000). The completion of the Arabidopsis genome indicates the presence of about 500 receptor-like protein kinases (Arabidopsis Genome Initiative, 2000). Two examples of these, Clavata 2 and the S-locus receptor kinase (SRK), have a demonstrated role in meristem development (Brand et al., 2000; Trotochaud et al., 2000) or fertilization (Schopfer et al., 1999), respectively. As their ligands are extracellular, they may well interact or even form part of the extracellular matrix. It also remains possible that hormones and their receptors, such as BR1 (brassinosteroid; He et al., 2000) and ETR1 (ethylene; Schaller and Bleecker, 1995), have some interactions in the ECM, but this remains to be established. These kinases must eventually be integrated into our understanding of signaling between the extracellular space and the cytoplasm. This review describes recent advances in our understanding of a family of Arabidopsis cell wall-associated kinases (WAKs) that are linked directly to the cell wall, span the membrane, and have a cytoplasmic kinase domain. Recent experiments point to a role of WAKs in cell expansion.

Cell expansion is regulated in a number of ways, but in few cases have the activating stimulant and the resulting mechanistic changes been linked. Plant hormones, including gibberellins (GA) (Phillips, 1998; Silverstone and Sun, 2000), brassinosteroids (Mussig and Altmann, 1999), auxins (Jones, 1998) and ethylene (Bleecker, 1999; Chang and Shockey, 1999), have all been shown to regulate cell expansion in some way. Signal cascades that are influenced by each of the hormones have been characterized biochemically and genetically. In all cases, however, the link between the events that occur just before expansion and those that occur during expansion has not been characterized. For example, it is not clear how auxins alter ion flux at the plasma membrane nor how auxin transport by the PIN proteins (Chen et al., 1998; Galweiler et al., 1998; Luschnig et al., 1998; Muller et al., 1998; Utsuno et al., 1998) is coordinated with expansion. Brassinosteroids have pronounced effects on cell growth (Mussig and Altmann, 1999), and biosynthetic enzymes and a putative receptor have been identified (Li and Chory, 1997; Schumacher and Chory, 2000). The ethylene signal transduction pathway is perhaps the best characterized and uses at least one histidine kinase receptor (ETR1; Chang et al., 1993), a downstream MAP kinase (CTR1; Kieber et al., 1993), activated DNA-binding proteins (EIN3; Chao et al., 1997) and other regulatory components (Chang and Shockey, 1999). Gene activation and repression is frequently the result of hormone-regulated pathways, and this is expected based on the components necessary for cell expansion (Zurek and Clouse, 1994; Caderas et al., 2000; Catala et al., 2000). How the activity of these induced proteins is coordinated with the signaling pathways is not clear. The cytoskeleton also plays an important role in cell expansion and cellulose synthase activity is frequently correlated with the orientation of cortical microtubules (Fisher and Cyr, 1998; Baskin et al., 1999; Kost et al., 1999). However, the mechanisms by which the cytoskeleton and the cell wall communicate remains unclear. Also lacking is an understanding of how the modulation of cell wall polysaccharides, such as pectins, is related to developmental processes. Our work has explored the possibility that WAKs are linked to events that regulate cell expansion during development.

A cytoplasmic protein kinase with a cell wall domain

Since so little is known about communication between the cytoplasm and the cell wall, it was of great interest when we found evidence for a family of five WAKs that each have a cytoplasmic kinase, span the plasma membrane, and have a domain tightly linked to the cell wall (Figure 1). The coding region of the WAK1 kinase domain was expressed in *Escherichia coli* and used to generate antiserum. This WAK serum detected protein on western blots only when plant tissue was boiled in SDS and DTT and the low-speed supernatant was used for analysis. Detergents that solubilize membranes or the cytoskeleton could not extract WAK. Moreover, immunolocalization with transmission electron microscopy showed that the WAK epitope was predominately bound to the cell wall, even in places where the plasma membrane and cell wall had separated due to fixation procedures (He *et al.*, 1996). As the WAK kinase domain was predicted to be cytoplasmic in an intact cell, these data suggested that WAKs could link the cytoplasm and the cell wall. To determine if WAKs indeed spanned the plasma membrane, the cell wall was digested with a crude protoplasting enzyme mix, and a plasma membrane fraction was isolated in the expectation that WAK would remain in the membrane. This was confirmed, demonstrating that WAK can be released from tissue not only by SDS/DTT but also by enzymatic release from the cell wall. Next, protease treatments demonstrated that the kinase domain was on the cytoplasmic face and the amino terminus was extracellular, as expected for a signaling molecule (He *et al.* 1996).

WAKs are bound to pectin

The observation that WAKs could be released into a microsomal fraction after enzymatically digesting the cell wall led to further analysis using enzymes specific for different cell wall carbohydrates. Purified preparations of cellulase, non-cellulosic glycan hydrolases, and pectinase were obtained from a number of sources (including D. Della Penna, Michigan State University). Release from the cell wall was assayed by the ability to detect WAK in the low-speed supernatant of ground, enzymatically treated leaves. Only pectinase treatments released WAK from the cell wall. Antibodies against homogalacturonan (JIM5 and JIM7; Knox *et al.*, 1990; Knox, 1997) both detect this WAK band

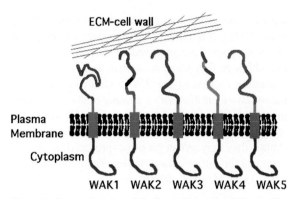

Figure 1. Cartoon depiction of the five wall-associated kinases. The WAK isoforms share the same structure and are 87% identical in their cytoplasmic kinase (blue). Their amino termini are linked to the cell wall and all contain EGF repeats (lightest green) adjacent to the transmembrane domain (red). The extracellular regions are 40–64% identical among isoforms and the differences are depicted by the various colors.

on western blots (Wagner and Kohorn, 2001). Thus, WAKs are bound to the cell wall in part by their covalent association with pectin.

Five clustered *WAK* genes

Initially, we thought that WAK is encoded by a single-copy gene. The *Arabidopsis* genome project had not yet reached this locus, and mapping studies with oligonucleotides specific to the kinase domain identified only one locus on chromosome 1. While isolating the *WAK1* promoter, we realized that there were additional WAKs in this region that were not detected by mapping or Southern blot analysis. Sequence and PCR analysis identified five *WAK* genes arranged in a tandem cluster within 30 kb (He *et al.*, 1999). The kinase domains are 87% identical and all are likely to be recognized by the serum raised to WAK1 kinase. This has been partially confirmed because the WAK1 kinase serum recognizes the WAK2 kinase expressed in *E. coli* (Kohorn, unpublished results). The WAK amino-terminal extracytoplasmic regions are only 40–64% identical. This identity is in clusters, and all WAKs contain conserved EGF-like cysteine repeats (Kohorn *et al.*, 1996; He *et al.*, 1999). The function of these repeats is unknown, and surprisingly remains one of only a few examples (Laval *et al.*, 1999) found in the plant kingdom. In metazoans, EGF repeats aid in the calcium-mediated dimerization of proteins, often receptors. Other cysteine-rich repeats, similar but distinct from EGF repeats, are required for the binding of some spore coat proteins to cellulose (Zhang et al., 1998). It remains to be determined if the WAK cysteine repeats are involved in carbohydrate binding, although they show no such activity *in vitro*. The remaining regions of the WAK extracytoplasmic domains have identities with a number of domains found in proteins associated with the metazoan ECM, but the identities are too weak to be suggestive of function. Rather, they may reflect the use of protein domains appropriate for carbohydrate-rich environments (Kohorn et al., 1996; He et al., 1999).

Expression

To predict how WAKs function, it is necessary to determine the cellular expression patterns for each WAK isoform. If the cellular expression of individual WAKs overlap, WAK isoforms may interact to form heterodimers within a cell. Conversely, if each WAK is expressed only in separate cell types, then the functions for WAKs may be distinct. To address this question, the DNA regions that lie between each of the tandemly arrayed WAK coding regions were predicted to contain the promoters, and these regions were fused to GUS and transformed into *Arabidopsis*. *In situ* hybridization was also used to confirm the expression patterns seen with the promoter GUS fusions.

The five *WAKs* are expressed in a variety of tissues with *WAK1* and *WAK2* messages being the most abundant (He et al., 1999; Wagner and Kohorn, 2001). *WAKs* are expressed in germinating seedlings and at most tissue junctions. Figure 2 shows typical GUS staining patterns seen for plants containing the *WAK2* promoter-GUS fusion. Expression is detected in all layers of the apical meristem and leaf primordia. One WAK isoform is subsequently expressed in cotyledon leaf tips and margins and eventually throughout the entire leaf and cotyledon, with expression decreasing as the leaf matures and expansion stops. Another WAK is found throughout the leaf as well and is particularly abundant in the vasculature. Expression of *WAKs* is induced within minutes of wounding and within hours after pathogen infection. *WAKs* are expressed in the floral tissues, but at far lower levels than in vegetative cells. As *WAKs* can be found in the same but also distinct cell types, the existence of signaling cascades that contain WAK homo- and hetero-dimers is possible.

WAK function

The most direct evidence of function would be to isolate mutant alleles of *WAKs* that have observable phenotypes. We initially identified a plant that had a T-DNA insertion in *WAK1*, but there was also a large rearrangement in the *WAK* locus that made it difficult to ascribe the observed embryo lethal phenotype to *WAK*. Extensive PCR screens of the T-DNA pools from the Ohio stock center and from the Wisconsin collection have only identified one other T-DNA in the *WAK* locus, and this lies in the 3′-untranslated region of the WAK2 transcribed sequence. Plants homozygous for this insertion have no detectable phenotype. Our previous work showed that CaMV 35S promoter-driven *WAK* was lethal, and the 2,2-dichloroisonicotinic acid (INA)-inducible PR1 promoter-driven WAK had limitations due to the role of WAK in pathogenesis (see below and He et al., 1998). We therefore generated WAK antisense lines using a different inducible promoter. When a full-length *WAK* antisense message is induced, the emerging leaves show a loss of cell expansion: the leaves are smaller but have the same number of cells as wild type (Wagner and Kohorn, 2001). The induced antisense *WAK* plants have reduced total WAK protein levels, and this was expected since the kinase sequence of the *WAK* gene used has high identity with all WAKs.

WAKs and pathogenesis

Pathogen infection of angiosperms requires interaction between the ECM and the invading agent, and may be accompanied by signaling between the ECM and cytoplasm (Hammond-Kosack and Jones, 1996). Pathogen infection is often followed by localized cell death and in some cases systemic acquired resistance (SAR). SAR is associated with rises in levels of salicylic acid (SA) and the induction of pathogen-related proteins. We recently found that *WAK* expression is induced when *Arabidopsis* plants are infected with pathogen or stimulated by exogenous SA or its analogue INA. *WAK1* mRNA induction requires the positive regulator NPR1/NIM1 (He et al., 1998). Using antisense *WAK* and a dominant negative allele of WAK1, we also demonstrated that induced expression of WAK1 is required for a plant to survive if stimulated by INA. These experiments suggested that WAK1 expression somehow protects plants from detrimental effects incurred during the pathogenesis response. If so, in-

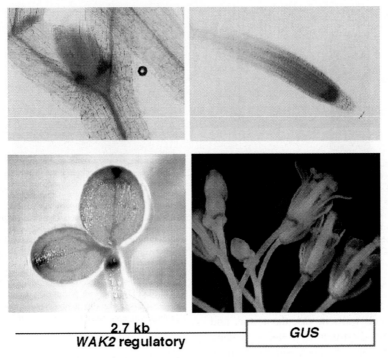

Figure 2. The *WAK2* promoter is active in a variety of tissues. 2.7 kb of the *WAK2* promoter, including the 5′-untranslated region, was fused to the GUS reporter and transformed into *Arabidopsis*. This figure shows representative GUS staining (Silverstone *et al.*, 1997) seen in multiple independently transformed lines. The *WAK2* promoter is active at the cotyledon petiole-hypocotyl junction (top left, thick section), the primary root tip (top right), in the cotyledon blade starting at the tip and margin (bottom left) and at the base of flowers and at the tip of expanding sepals (bottom right).

creased WAK1 expression would protect plants from abnormally high and toxic levels of SA or INA. Indeed, ectopic expression of full-length WAK1 or the kinase domain alone provides resistance to otherwise lethal SA levels (He *et al.*, 1998). These results provide a direct link between a protein kinase that could mediate signals from the ECM to the events that are precipitated by pathogen infection. They also suggest that while pathogen infection induces protective changes in cells, these changes can be detrimental if certain cellular components, such as WAK1, are not present in sufficient amounts.

Seeking WAK ligands (Figure 3)

To understand how WAKs are involved in regulating cell expansion, perhaps in association with pectins, we searched for proteins that might bind the cell wall-associated domain of WAK. The amino-terminal extra-cytoplasmic domain of WAK1 and 2 were used to screen an *Arabidopsis* library in the yeast two-hybrid assay. WAK2 identified two proteins that ap-

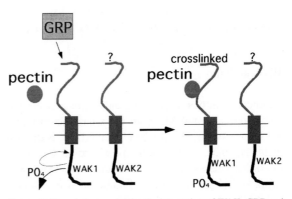

Figure 3. Speculative model for the interaction of WAK, GRP and pectin. GRP binds WAK1 but not WAK2, and some WAKs are bound to pectin. The model suggests that the binding activates the WAK kinase (arrows in cytoplasm), and that WAKs are subsequently found phosphorylated and bound to pectin, cross-linked to the cell wall.

peared specific for WAK2, but these did not contain signal sequences and thus are not predicted to be found in the wall. WAK1 was also used in the two-hybrid screen, and binds to one of a family of glycine-rich proteins, GRP. In pea, GRPs are secreted cell wall

202

GST
GST WAK

Figure 4. WAK has kinase activity *in vitro*. The WAK kinase domain was cloned into the expression vector pGEX-2TK (Amersham), expressed in *E. coli*, and purified using the GST tag according to corporate dogma. The fusion protein (GST WAK) or GST alone (indicated above the lanes) was incubated in the presence of ^{32}P-γATP in kinase buffer (50 mM HEPES pH 7.9, 1 mM EGTA, 2.5 mM MgCl$_2$, 7.5 mM MnCl$_2$, 1 mM DTT, and 50 μM cold ATP) for 30 min at 30 °C, separated by SDS-PAGE, and subjected to autoradiography.

proteins that are often associated with the generation of highly elaborate cell walls in vascular tissue, and hence have been termed 'structural proteins' (Cheng *et al.*, 1996). GRPs are characteristically high in glycine and represented by over 50 genes in *Arabidopsis*. Many of these GRPs are predicted to be nuclear, but at least half have signal sequences that direct them to be secreted. The expression of this GRP overlaps that of *WAK*1. It remains to be determined whether the WAK binding of GRP *in vivo* and the association with pectin is correlated with WAK kinase activity.

WAK activity

To understand how a given ligand might activate the kinase domain of WAK, both the kinase activity and the kinase substrate must be characterized. Based on their sequence, WAKs were predicted to be serine threonine kinases. The kinase domain of WAK1 was expressed in *E. coli* as a glutathione *S*-transferase (GST) fusion protein and purified. When incubated in the presence of ^{32}P-γATP, GST-WAK, but not GST alone, has protein kinase activity *in vitro*, indicating that this domain is indeed a protein kinase (Figure 4).

The GST-WAK1 kinase fusion protein is also active in *E. coli*. Figure 5 shows that WAK serum detects the fusion protein in bacterial extracts on western blots (left panel), and this same protein is detected by serum specific to phosphothreonine (Figure 5, α P^{T}). Site-directed mutagenesis was used to generate a kinase mutant where the lysine at 298 in a conserved region of the kinase domain was changed to a glutamine. This

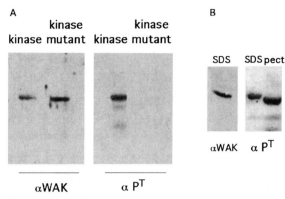

Figure 5. A. WAK kinase is active in *E. coli* and requires lysine residue 298. *E. coli* extracts expressing either GST-WAK-6HIS (kinase) or a similar protein that has a glutamine instead of a lysine at residue 298 (kinase mutant) were western-blotted and probed with WAK serum (left panel) or anti phospho-threonine serum (Zymed Lab; αP^{T}, right panel) as described (He *et al.*, 1996). B. SDS/DTT- and pectinase-released WAK is phosphorylated. Leaves were treated with SDS/DTT or with pectinase (Wagner and Kohorn, 2001), and the extracts were run in a denaturing gel, western-blotted, and probed with anti-phosphothreonine antiserum (Zymed).

protein was also expressed as a GST fusion in *E. coli*, and while it is detected by WAK serum (Figure 5, left panel), anti-phosphothreonine serum does not detect it (Figure 5, right panel kinase mutant α P^{T}), and thus it shows no kinase activity.

The WAK kinase is also phosphorylated in plants. Anti-phosphothreonine serum recognizes WAK protein that is released by SDS/DTT or pectinase treatment of leaf extracts (Figure 5B and Wagner and Kohorn 2001). However, it is not known if the *in vivo* phosphorylation is due to WAK kinase activity or to another kinase. While 'autophosphorylation' is common to tyrosine kinases and in heterodimeric threonine kinases in metazoans, it remains to be established how receptor kinases act in plants (Kohorn, 1999). Thus it is not reasonable to assume that WAK kinase activation by a ligand can be monitored by the phosphorylation state of the WAK kinase. Indeed, while *WAK* mRNA can be rapidly induced by wounding or pathogens, no changes in WAK phosphorylation can be observed (Kohorn, unpublished). A better indicator of WAK activity might be to identify a cytoplasmic substrate and monitor its phosphorylation state.

Seeking WAK kinase substrates

Members of a small GTPase family termed ROPs were initially good candidates for WAK substrates because

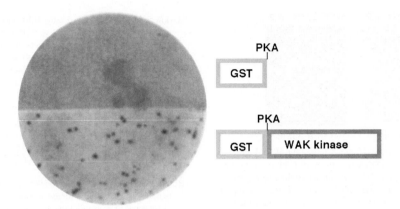

Figure 6. WAK binds a protein phosphatase 2C. Purified GST-WAK or GST was labeled *in vitro* using the cAMP-dependent kinase site (PKA) present in the GST sequence of the pGEX-2TK vector following the manufacturer's directions (Amersham). The ^{32}P-γATP-labeled proteins were used separately to probe a nitrocelluose filter (that was cut in half) containing λGT11 phage expressing the KAPP cDNA. The autoradiographs of the filters are shown. The KAPP cDNA was isolated from the *Arabidopsis* stock center library using the GST-WAK as probe according to established procedures (Blanar and Rutter, 1992).

ROPs appear to regulate cell morphology, perhaps via the cytoskeleton (Franklin-Tong, 1999; Li *et al.*, 1999). In addition, a ROP has been found associated with another receptor protein kinase, CLAVATA 1 (Trotochaud *et al.*, 1999). However, ROPs do not act as a substrate for WAK either as proteins expressed in *E. coli* or in the yeast two-hybrid assay (Wang and Kohorn, unpublished results).

A screen of radioactively labeled and purified WAK1 kinase to probe a bacterial expression library identified a protein type 2C phosphatase (KAPP). KAPP binds to several, but not all, plant receptor kinases (Braun *et al.*, 1997). GST-WAK but not GST binds to KAPP in bacterial extracts displayed on filters (Figure 6). KAPP was expressed in *E. coli* as a GST fusion protein, purified, and labeled with ^{32}P-γATP *in vitro* with a cAMP-dependent protein kinase substrate site in the GST domain. This probe was incubated with various WAK domains expressed and purified from *E. coli*. KAPP binds specifically to the WAK1 kinase domain (Figure 7, W1K) and not to the extracytoplasmic domain (Figure 7, W1N). KAPP also binds to the WAK2 protein (Figure 7, W2), indicating that both of these kinases interact with KAPP. KAPP binding to WAK does not require an active kinase domain as the substitution of Lys-298 for glutamine in a conserved kinase domain abolishes kinase activity but has no effect on KAPP binding *in vitro* (data not shown). This is unlike the binding of KAPP to other receptor kinases, but this difference is not understood.

Summary

WAKs are poised to provide a direct signal from the cell wall to the cytoplasm of many cell types during the development of *Arabidopsis*. Antiserum to the kinase of WAKs identifies a 68 kDa protein that spans the plasma membrane, has a cytoplasmic kinase, and an extracellular domain tightly linked to the cell wall. Enzymatic digestion of the cell wall during protoplast preparation releases WAKs into a plasma membrane fraction. Further enzyme digestions and western analysis show that WAKs are bound to pectin. WAKs can also be extracted from tissue by boiling in SDS/DTT, but not by milder treatments. Immunolocalization detects WAKs on the cell wall-plasma membrane interface. In addition to pectin, WAKs interact with an abundant cell wall protein, GRP.

WAKs are expressed in a variety of organs in a developmentally and environmentally regulated fashion, and *WAK1* and 2 are the most abundantly expressed of the five isoforms. The five WAKs differ most in their cell wall domain, and since WAK1 but not WAK2 binds a GRP and *WAKs* show distinct expression patterns, the different WAK isoforms may sense different wall signals. In contrast, the kinase domains have high identity, suggesting that they signal similar cytoplasmic events. It remains to be determined what are the cytoplasmic Wak substrates, and if indeed multiple cell wall characteristics are funneled into one type of cytoplasmic response.

WAKs have fundamental roles in the regulation of cell expansion. Somehow, the binding of GRP and

204

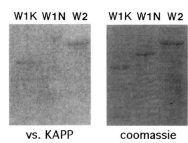

W1K W1N W2   W1K W1N W2

vs. KAPP   coomassie

Figure 7. KAPP binds to WAK1 and 2 but not the WAK1 extracellular domain. The KAPP cDNA was fused to GST (pGEX-2TK),
expressed and purified from *E. coli* according to the manufacturer's
directions (Amersham), and then labeled with ^{32}P-γATP using the
cAMP-dependent kinase substrate sequence present in the GST domain. The WAK kinase domain fused to GST (W1K), the WAK1
extracytoplasmic domain fused to GST (W1N), and a full-length
WAK2 GST fusion (W2), were expressed in *E. coli* and then purified
using a 6HIS tag present at their carboxyl termini (Holzinger *et al.*,
1996). Duplicate samples of these WAK proteins were separated by
SDS-PAGE and either western-blotted (He *et al.*, 1996) for probing with labeled KAPP (left panel; (Blanar and Rutter, 1992) or
Coomassie-stained (right panel).

pectins must be included in an explanation of this
process. It will be important to determine if the GRP
and pectin bound WAK are temporally and spatially
distinct, and if the WAK phosphorylation activity is
related to the N-terminal extracytoplasmic binding activity. The data are consistent with a model in which
GRP negatively regulates WAK1 to control cell expansion. The association of WAK with pectin is especially
intriguing, as chemical modifications and structural
characteristics of pectins have been correlated with developmentally important events. It will also be important to analyze the relationship between WAK activity
and other signaling pathways. Cell wall expansion
must be coordinated with cell growth, and understanding the relationships between these processes
will reveal much about cell physiology and cell wall
biology.

Acknowledgements

We thank Shaun Snyders for invaluable help with protein purification, and Tai Ping Sun and Jim Siedow
for helpful discussions. This work is funded by NSF
MCB9728045.

References

Arabidopsis Genome Initiative. 2000. Analysis of the flowering
plant *Arabidopsis thaliana*. Nature 408: 796–815.

Baskin, T.I., Meekes, H.T., Liang, B.M. and Sharp, R.E. 1999. Regulation of growth anisotropy in well-watered and water-stressed
maize roots. II. Role of cortical microtubules and cellulose
microfibrils. Plant Physiol. 119: 681–692.

Blanar, M.A. and Rutter, W.J. 1992. Interaction cloning: isolation
of a cDNA encoding FIP, a basic-HLH-zip protein that interacts
with cFos. Science 256: 1014–1018.

Bleecker, A.B. 1999. Ethylene perception and signaling: an evolutionary perspective. Trends Plant Sci. 4: 269–274.

Brand, U., Fletcher, J.C., Hobe, M., Meyerowitz, E.M. and Simon, R. 2000. Dependence of stem cell fate in *Arabidopsis*
on a feedback loop regulated by CLV3 activity. Science 289:
617–619.

Braun, D.M., Stone, J.M. and Walker, J.C. 1997. Interaction of the
maize and *Arabidopsis* kinase interaction domains with a subset
of receptor-like protein kinases: implications for transmembrane
signaling in plants. Plant J. 12: 83–95.

Caderas, D., Muster, M., Vogler, H., Mandel, T., Rose, J.K.C.,
McQueen-Mason, S. and Kuhlemeier, C. 2000. Limited correlation between expansin gene expression and elongation growth
rate. Plant Physiol. 123: 1399–1413.

Carpita, N. and Gibeaut, D. 1993. Structural models of primary cell
walls in flowering plants: consistency of molecular structure with
the physical properties of the walls during growth. Plant J. 3:
1–30.

Catalá, C., Rose, J.K.C. and Bennett, A.B. 2000. Auxin-regulated
genes encoding cell wall-modifying proteins are expressed during early tomato fruit growth. Plant Physiol. 122: 527–534.

Chang, C., Kwok, S.F., Bleecker, A.B. and Meyerowitz, E.M. 1993.
Arabidopsis ethylene-response gene *ETR1*: similarity of product
to two-component regulators. Science 262: 539–544.

Chang, C. and Shockey, J.A. 1999. The ethylene-response pathway:
signal perception to gene regulation. Curr. Opin. Plant Biol. 2:
352–358.

Chao, Q., Rothenberg, M., Solano, R., Roman, G., Terzaghi, W.
and Ecker, J.R. 1997. Activation of the ethylene gas response
pathway in *Arabidopsis* by the nuclear protein ETHYLENE-
INSENSITIVE3 and related proteins. Cell 89: 1133–1144.

Chen, R., Hilson, P., Sedbrook, J., Rosen, E., Caspar, T. and
Masson, P.H. 1998. The *Arabidopsis thaliana AGRAVITROPIC1*
gene encodes a component of the polar-auxin-transport efflux
carrier. Proc. Natl. Acad. Sci. USA 95: 15112–15117.

Cheng, S.H., Keller, B. and Condit, C.M. 1996. Common occurrence of homologues of petunia glycine-rich protein-1 among
plants. Plant Mol. Biol. 31: 163–168.

Cho, H.T. and Cosgrove, D.J. 2000. From the cover: altered
expression of expansin modulates leaf growth and pedicel abscission in *Arabidopsis thaliana*. Proc. Natl. Acad. Sci. USA 97:
9783–9788.

Cosgrove, D.J. 1997. Assembly and enlargement of the primary cell
wall in plants. Annu. Rev. Cell Dev. Biol. 13: 171–201.

Fisher, D.D. and Cyr, R.J. 1998. Extending the micro-
tubule/microfibril paradigm. Cellulose synthesis is required for
normal cortical microtubule alignment in elongating cells. Plant
Physiol. 116: 1043–1051.

Fowler, T., Bernhardt, C. and Tierney, M. 1999. Characterization
and expression of four proline-rich cell wall protein genes in
Arabidopsis encoding two distinct subsets of multiple domain
proteins. Plant Physiol. 121: 1081–1092.

Franklin-Tong, V.E. 1999. Signaling and the modulation of pollen
tube growth. Plant Cell 11: 727–738.

Galweiler, L., Guan, C.H., Muller, A., Wisman, E., Mendgen, K.,
Yephremov, A. and Palme, K. 1998. Regulation of polar auxin

transport by *AtPIN1* in *Arabidopsis* vascular tissue. Science 282: 2226–2230.

Hammond-Kosack, K.E. and Jones, J.D.G. 1996. Resistance gene-dependent plant defense responses. Plant Cell 8: 1773–1791.

He, Z., Wang, Z.Y., Li, J., Zhu, Q., Lamb, C., Ronald, P. and Chory, J. 2000. Perception of brassinosteroids by the extracellular domain of the receptor kinase BRI1. Science 288: 2360–2363.

He, Z.-H., Fujiki, M. and Kohorn, B.D. 1996. A cell wall-associated, receptor-like kinase. J. Biol. Chem. 271: 19789–19793.

He, Z.-H., He, D. and Kohorn, B.D. 1998. Requirement for the induced expression of a cell wall associated receptor kinase for survival during the pathogen response. Plant J. 14: 55–63.

He, Z.-H., Cheeseman, I., He, D. and Kohorn, B.D. 1999. A cluster of five cell wall-associated receptor kinase genes, *Wak1–5*, are expressed in specific organs of *Arabidopsis*. Plant Mol. Biol. 39: 1189–1196.

Holzinger, A., Phillips, K.S. and Weaver, T.E. 1996. Single-step purification/solubilization of recombinant proteins: application to Surface Protein B. Biotechniques 20: 804–808.

Jones, A.M. 1998. Auxin transport: down and out and up again. Science 282: 2201–2203.

Kieber, J.J., Rothenberg, M., Roman, G., Feldmann, K.A. and Ecker, J.R. 1993. CTR1, a negative regulator of the ethylene response pathway in *Arabidopsis*, encodes a member of the raf family of protein kinases. Cell 72: 427–441.

Knox, J.P. 1997. The use of antibodies to study the architecture and developmental regulation of plant cell walls. Int. Rev. Cytol. 171: 79–120.

Knox, J.P., Linstead, P.J., King, J., Cooper, C. and Roberts, K. 1990. Pectin esterification is spatially regulated both within cell walls and between developing tissues of root apices. Planta 181: 512–521.

Kohorn, B.D. 1999. Shuffling the deck: plant signaling plays a club. Trends Cell Biol. 10: 381–383.

Kohorn, B.D. 2000. Plasma membrane-cell wall contacts. Plant Physiol. 124: 31–38.

Kohorn, B.D., He, Z.-H. and Fujiki, M. 1996. Elusin: a receptor-like kinase with an EGF domain in the cell wall. In: P.R. Shewry, N.G. Halford and R. Hooley (Eds.) Protein Phosphorylation in Plants, Clarendon Press, Oxford, UK.

Kost, B., Mathur, J. and Chua, N.H. 1999. Cytoskeleton in plant development. Curr. Opin. Plant Biol. 2: 462–470.

Laval, V., Chabannes, M., Carriere, M., Canut, H., Barre, A., Rouge, P., Pont-Lezica, R. and Galaud, J. 1999. A family of *Arabidopsis* plasma membrane receptors presenting animal β-integrin domains. Biochim. Biophys. Acta 1435: 61–70.

Li, J. and Chory, J. 1997. A putative leucine-rich repeat receptor kinase involved in brassinosteroid signal transduction. Cell 90: 929–938.

Li, H., Lin, Y., Heath, R.M., Shu, M.X. and Yang, Z. 1999. Control of pollen tube tip growth by a Rop GTPase-dependent pathway that leads to tip-localized calcium influx. Plant Cell 11: 1731–1742.

Luschnig, C., Gaxiola, R.A., Grisafi, P. and Fink, G.R. 1998. *EIR1*, a root-specific protein involved in auxin transport, is required for gravitropism in *Arabidopsis thaliana*. Genes Dev. 12: 2175–2187.

Lynch, T.M. and Lintilhac, P.M. 1997. Mechanical signals in plant development: a new method for single cell studies. Dev. Biol. 181: 246–256.

Majewska-Sawka, A. and Nothnagel, E.A. 2000. The multiple roles of arabinogalactan proteins in plant development. Plant Physiol. 122: 3–9.

Muller, A., Guan, C., Galweiler, L., Tanzler, P., Huijser, P., Marchant, A., Parry, G., Bennett, M., Wisman, E. and Palme, K. 1998. *AtPIN2* defines a locus of *Arabidopsis* for root gravitropism control. EMBO J. 17: 6903–6911.

Mussig, C. and Altmann, T. 1999. Physiology and molecular mode of action of brassinosteroids. Plant Physiol. Biochem. 37: 363–372.

Nicol, F., His, I., Jauneau, A., Vernhettes, S., Canut, H. and Höfte, H. 1998. A plasma membrane-bound putative endo-1,4-β-D-glucanase is required for normal wall assembly and cell elongation in *Arabidopsis*. EMBO J. 17: 5563–5576.

Oxley, D. and Bacic, A. 1999. Structure of the glycosylphosphatidylinositol anchor of an arabinogalactan protein from *Pyrus communis* suspension-cultured cells. Proc. Natl. Acad. Sci. USA 25: 14246–14251.

Pear, J.R., Kawagoe, Y., Schreckengost, W.E., Delmer, D.P. and Stalker, D.M. 1996. Higher plants contain homologs of the bacterial *celA* genes encoding the catalytic subunit of cellulose synthase. Proc. Natl. Acad. Sci. USA 25: 12637–12642.

Phillips, A.L. 1998. Gibberellins in *Arabidopsis*. Plant Physiol. Biochem. 36: 115–124.

Reiter, W.-D. 1998. The molecular analysis of cell wall components. Trends Plant Sci. 3: 27–32.

Roberts, K. 1994. The plant extracellular matrix: in a new expansive mood. Curr. Opin. Cell Biol. 6: 688–694.

Schaller, G.E. and Bleecker, A.B. 1995. Ethylene-binding sites generated in yeast expressing the *Arabidopsis ETR1* gene. Science 270: 1809–1811.

Schopfer, C.R., Nasrallah, M.E. and Nasrallah, J.B. 1999. The male determinant of self-incompatibility in *Brassica*. Science 286: 1697–1700.

Schumacher, K. and Chory, J. 2000. Brassinosteroid signal transduction: still casting the actors. Curr. Opin. Plant Biol. 3: 79–84.

Showalter, A.M. 1993. Structure and function of plant cell wall proteins. Plant Cell 5: 9–23.

Silverstone, A.L, Chang, C.-W., Krol, E. and Sun, T.-P. 1997. Developmental regulation of the gibberellin biosynthetic gene *GA1* in *Arabidopsis thaliana*. Plant J. 12: 9–19.

Silverstone, A.L. and Sun, T.-P. 2000. Gibberellins and the green revolution. Trends Plant Sci. 5: 1–2.

Svetek, J., Yadav, M.P. and Nothnagel, E.A. 1999. Presence of a glycosylphosphatidylinositol lipid anchor on rose arabinogalactan proteins. J. Biol. Chem. 274: 14724–14733.

Trotochaud, A.E., Hao, T., Wu, G., Yang, Z. and Clark, S.E. 1999. The CLAVATA1 receptor-like kinase requires CLAVATA3 for its assembly into a signaling complex that includes KAPP and a Rho-related protein. Plant Cell 11: 393–406.

Trotochaud, A.E., Jeong, S. and Clark, S.E. 2000. CLAVATA3, a multimeric ligand for the CLAVATA1 receptor-kinase. Science 289: 613–617.

Utsuno, K., Shikanai, T., Yamada, Y. and Hashimoto, T. 1998. *Agr*, an Agravitropic locus of *Arabidopsis thaliana*, encodes a novel membrane-protein family member. Plant Cell Physiol. 39: 1111–1118.

Vissenberg, K., Martinez-Vilchez, I.M., Verbelen, J.-P., Miller, J.G. and Fry, S.C. 2000. *In vivo* colocalization of xyloglucan endotransglycosylase activity and its donor substrate in the elongation zone of *Arabidopsis* roots. Plant Cell 12: 1229–1237.

Wagner, T.A. and Kohorn, B. 2001. Wall associated kinases, WAKs, are expressed throughout plant development and are required for cell expansion. Plant Cell 13: 303–318.

Wyatt, S.E. and Carpita, N.C. 1993. The plant cytoskeleton-cell-wall continuum. Trends Cell Biol. 12: 413–417.

Xu, W., Campbell, P., Vargheese, A.K. and Braam, J. 1996. The *Arabidopsis* XET-related gene family: environmental and hormonal regulation of expression. Plant J. 9: 879–889.

Zhang, Y., Brown, R.J. and West, C. 1998. Two proteins of the *Dictyostelium* spore coat bind to cellulose *in vitro*. Biochemistry 37: 10766–10779.

Zuo, J., Niu, Q.-W., Nishizawa, N., Wu, Y., Kost, B. and Chua, N.-H. 2000. KORRIGAN, an *Arabidopsis* endo-1,4-β-glucanase, localizes to the cell plate by polarized targeting and is essential for cytokinesis. Plant Cell 12: 1137–1152.

Zurek, D.M. and Clouse, S.D. 1994. Molecular cloning and characterization of a brassinosteroid-regulated gene from elongating soybean (*Glycine max* L.) epicotyls. Plant Physiol. 104: 161–170.

SECTION 5

SECONDARY WALL SYNTHESIS

Plant Molecular Biology **47**: 209–219, 2001.
© 2001 *Kluwer Academic Publishers. Printed in the Netherlands.*

Mutations of the secondary cell wall

Simon R. Turner*, Neil Taylor and Louise Jones
*University of Manchester, School of Biological Science, 3.614 Stopford Building, Oxford Road, Manchester M13 9PT, UK (*author for correspondence; e-mail simon.turner@man.ac.uk)*

Key words: cellulose, cell wall, lignin, mutants

Abstract

It has not been possible to isolate a number of crucial enzymes involved in plant cell wall synthesis. Recent progress in identifying some of these steps has been overcome by the isolation of mutants defective in various aspects of cell wall synthesis and the use of these mutants to identify the corresponding genes. Secondary cell walls offer numerous advantages for genetic analysis of plant cell walls. It is possible to recover very severe mutants since the plants remain viable. In addition, although variation in secondary cell wall composition occurs between different species and between different cell types, the composition of the walls is relatively simple compared to primary cell walls. Despite these advantages, relatively few secondary cell wall mutations have been described to date. The only secondary cell wall mutations characterised to date, in which the basis of the abnormality is known, have defects in either the control of secondary cell wall deposition or secondary cell wall cellulose or lignin biosynthesis. These mutants have, however, provided essential information on secondary cell wall biosynthesis.

Abbreviations: C4H, cinnamate 4-hydroxylase; CCR, cinnamoyl-CoA reductase; CCoAOMT, cinnamoyl-CoA *O*-methyltransferase; DP, degree of polymerization; F5H, ferulate 5-hydroxylase; IRX, irregular xylem

Introduction

In contrast to primary cell walls, plant secondary cell walls are deposited once the cell has stopped expanding. The composition of secondary cell walls varies widely among different species and different cell types. In general, however, they are composed of a complex mixture of lignin, carbohydrates and proteins. Consequently, the formation of a secondary cell wall requires the co-ordinate regulation of a number of complex metabolic pathways.

Recent efforts to investigate metabolic pathways involved in synthesising secondary cell wall components have focused on down-regulating individual steps in the biosynthetic pathway using transgenic plants. Although this antisense/sense suppression approach has provided valuable insights into the mechanisms of synthesis of compounds such as lignin, several disadvantages to this approach remain. Such problems include the obvious necessity of protein purification and gene cloning prior to the targeted reg-

ulation of gene expression. In the case of the genes required for cellulose biosynthesis, as well as some steps in the lignin biosynthetic pathway, this has not been possible. Secondly, the sequence of the *Arabidopsis* genome has revealed that many biosynthetic genes are encoded by large multi-gene families and targeting specific genes could prove problematic where significant sequence similarity exists between members of the same family. Finally, a complete knock-out of gene expression is unlikely to be achieved through an antisense/sense approach and hence a definitive analysis of gene function is more difficult.

In order to avoid some of the problems encountered with antisense technology, an alternative approach to investigate the function and synthesis of secondary cell wall components involves the characterisation of naturally occurring and chemically induced mutants. For genetic analysis, the study of secondary cell walls rather than primary walls offers a distinct advantage in that even plants with a very severe alteration in secondary cell wall components still produce viable plants. This is clearly illustrated in the case of cellulose, where severe mutations in primary cell wall cellulose synthesis are often lethal (Arioli *et al.*, 1998), whilst plants with mutations in secondary cell wall cellulose deposition grow relatively normally (Turner and Somerville, 1997). Given this advantage and the complexity of secondary cell wall deposition it is surprising that relatively few mutants have been identified that are specifically defective in some aspect of secondary cell wall deposition.

Secondary cell walls may conveniently be divided as either lignified (woody) or non-lignified. Whilst a number of cell types possess a thick secondary cell wall that does not contain lignin, to our knowledge only two mutants, from cotton fibres (Kohel *et al.*, 1993) and *Arabidopsis* trichomes (Potikha and Delmer, 1995), have been demonstrated to affect non-lignified secondary cell walls. Consequently, the majority of the work covered in this article concerns mutations affecting lignified secondary cell wall formation.

Lignified cell walls contain lignin, cellulose and non-cellulosic glycans together with a variety of proteins and other minor components. In angiosperms the major cross-linking glycan in lignified secondary cell walls is xylans, which tend to have relatively few substitutions on a linear $(1\rightarrow4)\beta$-D-xylose backbone (Hori and Elbein, 1985). In some monocots xylans represent the majority of the secondary cell wall material and are far more abundant than cellulose (Gorshkova *et al.*, 1996). Despite this abundance, no reports of a secondary cell wall defect have yet been attributed to alterations in secondary cell wall xylans. Consequently, this review is divided into three sections that cover mutations in cellulose synthesis, lignin synthesis and the control of secondary cell wall composition. These simple divisions may themselves cause some problems since it is unclear whether some regulatory mutations affect only a single pathway, such as lignin biosynthesis (see below) or all aspects of secondary cell wall synthesis.

Cellulose mutants

Despite the fact that cellulose is the world's most abundant biopolymer, with an estimated 180 billion tonnes being produced annually (Englehardt, 1995), the understanding of the mechanisms involved in cellulose biosynthesis in plants is still incomplete. Early biochemical studies of cellulose synthesis in plants yielded only limited success, and this can be attributed to a number of factors. These include the labile nature of the cellulose synthesising complex or a lack of purification of an essential, associated protein or co-factor. Recent advances have been made, however, using mutants that are deficient in cellulose production. A detailed review of the current knowledge of cellulose synthesis is found elsewhere in this issue.

Barley mutants

Brittle culm lines of barley were first described on the basis of the physical properties of the culm. Although the outer and inner culm diameters of brittle culm lines were not significantly different from normal lines (Kokubo *et al.*, 1989), the maximum bending stress of their culms were found to be less than half the value of normal lines. Cell wall analysis revealed that the maximum bending stress correlated significantly with the cellulose content of the cell walls, but not with non-cellulosic compounds.

A more detailed analysis using isogenic brittle lines demonstrated that the cellulose content of the single-gene brittle mutants was less than half that of the corresponding non-brittle line (Kokubo *et al.*, 1991). No correlation was found between brittleness and lignin content, demonstrating the importance of the cellulose content in determining the physical properties of the culm. Using viscometric techniques to study the degree of polymerisation (DP) of the cellulose microfibrils, no significant differences were found between brittle mutants and the corresponding non-brittle lines. In addition, the number of cellulose molecules per unit length of culm was significantly smaller in the brittle mutants than in the non-brittle line (Kokubo *et al.*, 1991). Thus, the reduction in cellulose content is due to a decrease in the number of the cellulose molecules and not a decrease in the molecular weight of the cellulose molecules. In one of these brittle culm lines, Kobinkatagi 4, the cellulose content of cell walls from other tissues was also determined. The cellulose content in callus and in suspension culture cells, that only possess a primary cell wall, exhibited no differences between brittle and non-brittle lines. This result suggests that the defect in this line is solely affecting secondary cell wall cellulose synthesis (Yeo *et al.*, 1995).

Examination of the number of cellulose synthesising rosettes in the brittle mutant Kobinkatagi 4 demonstrated that number of rosettes per unit area was variable between cells. It was clear, however, that on average there were significantly fewer rosettes in the mutant line. In the normal strain it was found that there were about five times as many rosettes per unit area as in the mutant (Kimura *et al.*, 1999), demonstrating a clear correlation between the number of rosettes and the reduction in cellulose in this brittle culm line. No irregular-shaped rosettes were seen in this mutant. This is in contrast to what has been observed in the *Arabidopsis* mutant *rsw1*. This mutant is deficient in primary cell wall cellulose, thought to be the result of rosette disassembly (Arioli *et al.*, 1998). Consequently, the mutation in Kobinkatagi 4 may result in the decrease in the insertion of terminal complexes in the plasma membrane, rather than disassembly of already existing rosettes (Kimura *et al.* 1999).

Cotton mutants

Developing cotton fibres have been an excellent system for studying cellulose synthesis, since single cells develop synchronously on the boll, and after a period of elongation the synthesis of a nearly pure cellulosic cell wall is initiated. Ligon lintless-1 mutant has drastically shortened cotton fibres, but these fibres have extensively thickened cell walls (Kohel *et al.*, 1993). Experiments that measured the incorporation of [^{14}C]glucose into crystalline cellulose in both the primary and secondary cell walls demonstrated that the rate of cellulose production was reduced in primary walls, correlating with the reduced rate of fibre elongation (Kohel *et al.*, 1993). In secondary walls, however, there was a five-fold increase in crystalline cellulose production per millimetre of fibre compared to wild-type fibres. Thus the Ligon lintless-1 mutation affects both the growth and development of the cotton fibres along with changes in the formation of cellulose in both the primary and secondary cell walls (Kohel *et al.*, 1993). It was thought that the increase in cellulose in the secondary cell walls was due to either the increase in synthetic activity of existing complexes, or an increase in the number of synthetic complexes in the fibre (Kohel *et al.*, 1993).

Arabidopsis *mutants*

Although the brittle culm mutants have aided our understanding of the relationship between secondary cell wall cellulose content and physical properties, barley is not an ideal system for molecular genetic investigations. As a result, more recent work has concentrated on the model plant species *Arabidopsis thaliana*, which is far more tractable for molecular genetic studies.

The first description of a mutant of *Arabidopsis* deficient in secondary cell wall cellulose synthesis was of the *tbr* (*trichome birefringence*) mutant (Potikha and Delmer, 1995). This mutant was initially selected because the leaf and stem trichomes lacked the birefringence under polarised light which is characteristic of plant cell walls containing highly ordered cellulose microfibrils. These *tbr* plants also exhibit reduced birefringence in the xylem of the leaf. Chemical analysis of various tissues, including isolated trichomes, demonstrated that *tbr* was impaired in its ability to deposit secondary cell wall cellulose in specific cell types. Indeed, in the trichomes, the secondary cell wall appeared to be totally absent whilst cellulose deposition in the xylem and other cell types of the stem and root were relatively unaffected in this mutant (Potikha and Delmer, 1995). The *tbr* mutation was considered not to affect primary cell walls, as callus tissue, which consists almost entirely of cells with only primary walls, showed no difference in cellulose content (Potikha and Delmer, 1995). In addition, mature *tbr* trichomes were seen to resemble immature wild-type trichomes, suggesting that primary cell wall deposition occurs as normal but deposition of a thick secondary cell wall is impaired.

A further series of mutants deficient in secondary cell wall cellulose deposition has been described (Turner and Somerville, 1997). These mutations, termed *irregular xylem* (*irx1, 2* and *3*), caused the collapse of mature xylem cells in the inflorescence stems of *Arabidopsis* (Figure 2a). These mutants were identified in a screen involving the microscopic examination of cross sections of stems from a chemically mutagenised population. The collapse of the xylem elements is thought to be due to a weakness in the secondary cell wall of the xylem cells which results in them being unable to withstand the negative pressure generated during water transport up the stem. This collapsed xylem phenotype was also seen in mature hypocotyls and in the primary root and petioles. It was found that the xylem elements initially expanded correctly to attain their normal shape, and these cells only appeared to collapse once they become involved in the transport of water. The *irx* mutants otherwise appeared little different from the wild type, apart from being slightly slower growing and *irx3* showing an inability to maintain an upright growth habit (Turner and Somerville, 1997).

Upon electron microscopic examination of cell walls from the interfascicular region, where no collapse of the cells is evident, it was found that wild-type cells were evenly thickened around the entire cell and stained in a ubiquitous manner. Walls from *irx1* plants were similar in appearance to the wild type, but were slightly thinner. The *irx2* plants, in contrast, showed a preferential deposition of cell wall material at the corners of the cell, with a decrease in the width of the walls at the midpoint. Some darker-staining regions were observed that appeared to divide the wall into different domains. The walls from *irx3* plants also showed different domains, with darker-staining regions and a very uneven pattern of thickening (Turner and Somerville, 1997).

As indicated by the inability of *irx3* plants to maintain an upright growth habit, the *irx* mutants demonstrate altered physical properties of their stems. The stiffness of mature stems, as measured by the bending modulus, was decreased in all *irx* mutants, with *irx3* stems displaying only 10% of the rigidity of wild-type stems. The *irx3* plants were also found to have drastically weaker stems, as determined by the maximum stress at yield. There were no apparent differences in either the deposition of lignin in *irx1*, *2* or *3* stems compared to the wild type, as determined by diagnostic stains for lignin and by chemical analysis, or in the composition of the non-cellulosic carbohydrate fraction of the walls. There was, however, a considerable difference between the cellulose contents of the stems. At all stages of development, all three mutants showed a large decrease in the total cellulose content of the stem. These differences were greatest in mature stems. The *irx1*, *irx2* and *irx3* stems had 40%, 36% and 18% , respectively, of the total cellulose content of the wild type. Mature hypocotyls showed similarly large changes whereas leaves showed very small, if any, changes in cellulose content. This suggested that only secondary cell wall cellulose production was affected in these mutants (Turner and Somerville, 1997).

The *irx3* mutation was mapped to the top arm of chromosome V (Turner and Somerville, 1997). An EST that showed sequence similarity to bacterial cellulose synthase (Cutler and Somerville 1997) was found to be tightly linked to *irx3*, and genomic clones that contained the gene corresponding to this EST were found to complement this mutation (Taylor *et al.*, 1999). Thus, the *irx3* gene encodes a cellulose synthase that is required for cellulose synthesis in secondary cell walls. The IRX3 gene product showed high levels of homology to a previously identified cotton cellulose synthase catalytic subunit (Pear *et al.*, 1996) and a gene, *RSW1* (Arioli *et al.*, 1998), that was identified as being mutated in the *radial swelling1* (*rsw1*) mutation of root tips (Baskin *et al.*, 1992). *RSW1* encodes a cellulose synthase involved in primary cell wall cellulose synthesis (Arioli *et al.*, 1998). Further information on this group of proteins (CesA) may be found elsewhere in this issue.

The *irx3* mutant allele was isolated and found to contain a premature stop codon between the 4th and 5th predicted transmembrane domains, which results in the protein being truncated by 168 amino acids. The expression pattern of *irx3* was determined by northern blotting. No expression was found in the leaves, whereas expression levels were high in the inflorescence stem. RNA levels in *irx3* plants were greatly reduced, and this was thought to be due to the premature stop codon resulting in degradation of the transcript (Taylor *et al.*, 1999). It has recently been shown that there is no protein detectable by an IRX3 specific polyclonal antibody in *irx3* extracts (Taylor *et al.*, 2000).

The *irx1* mutation was also mapped and found to be present in a 140 kb region of chromosome 4. Upon release of the sequence and analysis of the predicted genes present in this region, it was found that another member of the *CesA* (cellulose synthase) family of genes was present (Taylor *et al.*, 2000). A genomic clone carrying this gene was found to complement the *irx1* mutation. Two mutant alleles of *irx1* were cloned and sequenced. *irx1-1* has a mutation that results in the third of the conserved aspartate residues being replaced by an asparagine residue. As this aspartate is present in all processive β-glycosyl transferases, then it is expected that *irx1-1* will be a null allele. The *irx1-2* allele showed a substitution within 4 residues of this third conserved aspartate, which resulted in a serine residue being replaced by a leucine. This serine is conserved between all cellulose synthases. Thus *IRX1* also encodes a cellulose synthase catalytic subunit (Taylor *et al.*, 2000).

It is clear that in xylem cells, using both light microscopy (Turner and Somerville, 1997) and electron microscopy (Taylor *et al.*, 2000), the phenotypes of *irx1* and *irx3* walls are indistinguishable, demonstrating that *irx1* and *irx3* affect exactly the same cells and are required in the same cell types. Confirmation of this proposal was made using polyclonal antibodies raised to the first hypervariable region of each of these proteins, which have been shown to be specific for the individual isoforms (Taylor *et al.*, 2000). Tissue prints

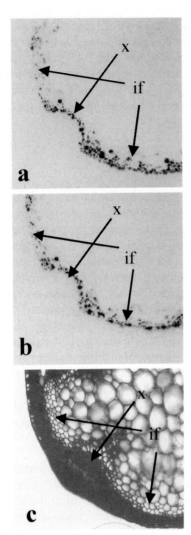

Figure 1. Expression of *IRX*1 and *IRX*3 in *Arabidopsis* inflorescence stems. Sequential tissue prints probed with anti-IRX1 (a) or anti-IRX3 (b) antibody and a toluidine blue section of the material from which the tissue prints were obtained (c). The xylem (x) and interfascicular region (if) are indicated.

from inflorescence stems probed with these antibodies demonstrate that *IRX*1 and *IRX*3 have very similar if not identical expression patterns, consistent with the idea that they are both required in the same cell type (Figure 1).

Since *irx1* and *irx3* appeared to be required in the same cells, it was important to determine whether they interacted directly as part of one complex. An epitope tag, consisting of the recognition sequence for a highly specific monoclonal antibody (RGSHis) and a hexahistidine sequence, was inserted at the amino terminus of the *IRX3* gene. This epitope-tagged *IRX3* was then transformed into *irx3* plants and found to comple-

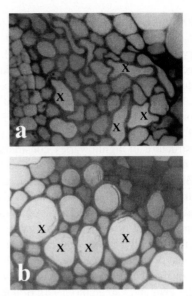

Figure 2. Complementation of the *irx3* mutation with an epitope-tagged *IRX3* gene. Toluidine blue-stained sections of vascular bundles from *irx3* plants (a) and from *irx3* plants transformed with a construct that contained the *IRX3* gene with an epitope inserted close to the N-terminus (b). Xylem elements (x) are indicated.

ment the mutation, confirming that the epitope tag had no effect on the function of the enzyme (Figure 2).

In order to determine whether IRX1 and IRX3 were associated, detergent-solubilised extracts from tagged IRX3 plants were bound to nickel resin to purify the tagged IRX3. A large proportion of the tagged IRX3 bound to the resin, and was specifically eluted by increasing the imidazole concentration, consistent with the binding being due to the hexahistidine sequence. IRX1 was found to follow an almost identical pattern of binding. Binding of IRX1 was absolutely dependent on the presence of the histidine-tagged IRX3 and other plasma membrane proteins did not bind to the resin demonstrating a specific interaction between IRX1 and IRX3 (Taylor *et al.*, 2000).

The identity of the gene defective in *irx2* plants has not been determined. Recently, at least one additional cellulose-deficient complementation group (*irx5*) has been identified. The nature of these mutants awaits further characterisation (N. Taylor and S. Turner, unpublished).

Lignin biosynthesis mutants

The manipulation of the lignin content of plant cell walls, both qualitatively and quantitatively, has received much attention (Baucher *et al.*, 1998; Halpin *et al.*, 2001, this issue). Lignin itself is a complex aromatic polymer and a major component of plant secondary cell walls (Bacic *et al.*, 1988). It provides mechanical strength and support to plant tissue and renders the wall hydrophobic and impermeable to water. Lignin is synthesised by the oxidative polymerisation of three monomers, *p*-coumaryl alcohol, coniferyl alcohol and sinapyl alcohol.

The structure of lignin varies considerably and depends on the tissue and cell type and on the environmental conditions (Campbell and Sederoff, 1996; Campbell and Rogers, 2001, this issue). This heterogeneity is related to the monomeric composition of the polymer and also to the different types of linkages within the polymer and to the presence of non-conventional or unusual phenolic units. Although the structure and biosynthesis of lignin has been well characterised (Lewis and Yamamoto, 1990; Boudet, 1998), more recent work has demonstrated that our understanding is by no means complete (Humphreys *et al.*, 1999). A more extensive review of the lignin biosynthetic pathway may be found elsewhere (Mellerowicz *et al.*, 2001, this issue).

Maize mutants

The earliest lignin mutants were identified by a reddish-brown pigmentation of the leaf midrib. Although these mutants have been found in sorghum and millet, they have been best characterised in maize. This class of mutants have been designated *brown midrib* (*bm*) mutants. Further characterisation revealed these mutants had an altered lignin content and composition and were found to have enhanced digestibility properties. To date, 4 independent *bm* loci have been identified (denoted *bm1–4*) in maize but only 2 have been fully characterised (Barrière and Argillier, 1993). There has been much commercial interest in the use of *bm* mutants as forage crops since lignin is an indigestible component of the cell wall and limits the breakdown of the wall polysaccharides (Cherney *et al.*, 1991).

The *bm3* mutation has been extensively studied. Initial investigations demonstrated a reduction in caffeate *O*-methyltransferase (COMT) activity (Grand *et al.*, 1985). Further analysis revealed the *bm3* mutation resulted from structural changes in the COMT gene (Vignols *et al.*, 1995) and a number of deletion mutations have been identified (Morrow *et al.*, 1997).

The *bm1* mutant displays reduced cinnamyl alcohol dehydrogenase (CAD) activity in lignified tissues and produces lignin with a modified composition. The total lignin content and structure of the polymer is altered in *bm1* plants. The CAD cDNA has been isolated and characterised from maize and found to map close to the *bm1* gene suggesting this mutation directly affects expression of the CAD gene (Halpin *et al.*, 1998).

Pine mutants

The *cad-n1* mutant in loblolly pine displays a characteristic phenotype of wood with a brown coloration (Mackay *et al.*, 1997), similar to the *bm* mutants of maize. Characterisation of the mutant revealed reduced expression of the lignin biosynthetic enzyme CAD whilst the lignin content of the mutant was only slightly decreased. High levels of coniferaldehyde and dihydroconiferyl alcohol, a monomer not normally associated with the lignin biosynthetic pathway, were identified in *cad-n1* plants. Similarly, biphenyl and biphenyl ether bonds were present in large excess in the mutant. Variations in the lignin composition, therefore, do not appear to disrupt the essential functions of lignin in this mutant (Ralph *et al.*, 1997). The *cad-n1* plants utilise non-conventional wall phenolics to construct unusual lignins enriched in resistant interunit bonds to compensate for the shortage in normal lignin precursors (Lapierre *et al.*, 2000).

Commercially, this mutation is particularly useful because it is capable of modulating the lignin composition and, hence, can affect the extractability of the polymer, without compromising the functional properties.

The *cad-n1* mutant has also been used to characterise the quantitative effects associated with this allele. Co-segregation analysis indicated that the *cad* locus itself might represent a gene that governs stem growth in pine (Wu *et al.*, 1999).

Arabidopsis mutants

The *fah-1 Arabidopsis* mutant (initially described as *sin1*) displays an increased transparency to UV light and demonstrates a characteristic red fluorescence under UV. Characterisation of the *fah-1* mutant revealed an absence of sinapic acid esters. The transparency to UV light is the result of a lack of sinapoyl malate in the upper epidermis (Chapple *et al.*, 1992). In the mutant, the conversion of ferulate to 5-hydroxyferulate is blocked, as determined by *in vivo* radiotracer feeding experiments, precursor supplementation studies and enzyme assays. This step in the general phenyl-propanoid pathway is a key reaction in the production of syringyl units. The *fah-1* mutant therefore lacks the sinapic acid-derived components of lignin although the total lignin content remained unchanged. After the isolation of a T-DNA tagged allele, the gene was identified as ferulate 5-hydroxylase (F5H). F5H functions as a cytochrome P450-dependent monooxygenase and exhibits low amino acid sequence homology to other classes of plant P450. F5H was therefore designated as a new family of plant P450, CYP84 (Meyer *et al.*, 1996).

The ectopic over-expression of F5H abolished tissue-specific lignin monomer accumulation (Meyer *et al.*, 1998). Transgenic plants with the cinnamate 4-hydroxylase (C4H) promoter linked to F5H produced syringyl-rich lignin, with a much higher proportion of syringyl units than in any other plant previously reported (Franke *et al.*, 2000; Marita *et al.*, 1999). The specific control of the monomeric composition of lignin is likely to be achieved through the engineering of the F5H gene.

The phenotype demonstrated by *irx4* plants is characteristic of that previously described for other *irx* mutants (Figure 2a) (Turner and Somerville, 1997; Taylor *et al.*, 1999, 2000). The mutants are unable to maintain an upright growth habit. Typically, the xylem vessels in these *irx* plants exhibit a collapsed phenotype, which has been attributed to defective secondary cell walls. Biochemical analysis of *irx4* cell walls revealed a 50% reduction in the total lignin content, as determined by histochemical staining, thioglycolic acid assays and solid-state NMR. In contrast, the cellulose and non-cellulosic glycan content of *irx4* cell walls remained unchanged (Jones *et al.*, 2001).

The most dramatic aspect of the *irx4* phenotype was observed in the ultrastructure of *irx4* secondary cell walls and the consequent effects on the physical properties of the stems. The architecture of the walls in the mutant was drastically altered compared to the wild type with the cell walls having a massively expanded and diffuse appearance. The *irx4* cell walls often occupied a large proportion of the total cell volume. Furthermore, the walls stained unevenly, compared to wild-type cell walls, indicative of modifications in their composition. The diffuse nature of *irx4* cell walls provides clear evidence for the role of lignin in maintaining the structural integrity of the wall and anchoring the components of the wall together. As a consequence of these ultrastructural modifications in *irx4* cell walls, stems from the mutant display

significantly altered physical properties with both the strength and stiffness of the stems severely reduced (Jones *et al.*, 2001). Phenotypic observations and mapping data suggested that the cinnamoyl-CoA reductase (CCR) gene, an enzyme involved in the penultimate step in the lignin biosynthetic pathway, was a good candidate for the identity of the *IRX4* gene. Further analysis indicated the presence of a mutation within this gene in *irx4* plants and definitive evidence of the identity of IRX4 was achieved by complementation with the wild-type CCR gene (Jones *et al.*, 2001).

The *irx4* mutant confirms that CCR is an ideal candidate for modifying the total lignin content of secondary cell walls. This mutant has also provided valuable insight into the functional role of lignin within the cell wall and its influence on the mechanical properties of plant tissue.

Regulatory mutants

The mutants described above have proved important in providing insight into the structure and biosynthesis of lignin and cellulose. To date, however, they have only provided information on individual steps in the pathway. At least 17 genes are required to synthesise lignin and how the co-ordinate regulation of these individual steps in this pathway is achieved in a cell-specific manner is an area of intense interest.

In addition, the pathways for lignin, cellulose and non-cellulosic glycan biosynthesis must also be regulated in a highly co-ordinated manner to achieve the proper deposition of these polymers during secondary wall formation. How this co-ordination of different pathways is achieved also remains unclear. Recent work on several different *Arabidopsis* mutants, however, offers an opportunity to address these questions.

Ectopic lignification mutants

Two mutants, *ectopic lignin deposition* (*elp1*) (Zhong *et al.*, 2000) and *ectopic lignification* (*eli1*) (Caño-Delgado *et al.*, 2000), appear to alter the normal pattern of lignin deposition and therefore offer an opportunity to examine the spatial control of lignin deposition. The *elp1* mutant in *Arabidopsis* demonstrates altered lignin deposition patterns in the stem. Ectopic deposition of lignin occurs in the walls of pith parenchyma cells in addition to the normal deposition of lignin in the walls of xylem and fibre cells. Lignin

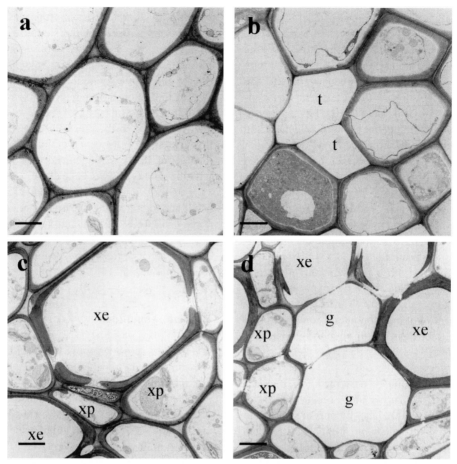

Figure 3. Phenotype of *gpx* plants. Transmission electron micrographs showing interfascicular cells (a, b) and xylem (c, d) from wild-type (a, c) and *gpx* (b, d) mutant plants. Note the thin-walled cells (t) present in the interfascicular region and gaps (g) in the xylem in sections from *gpx* plants, but not the wild-type plants. Xylem elements (xe) and xylem parenchyma (xp) are also indicated. Scale bar represents 7 μm.

appeared to be deposited in regions of parenchyma cells in the pith of both young and mature stems. Furthermore, stems from *elp1* plants had ca. 20% more lignin than wild-type stems and this increase in lignin content was accompanied by an increase in the activities of enzymes involved in the lignin biosynthetic pathway, such as PAL, CCoAOMT and CCR. Ectopic expression of the enzyme CCoAOMT was also identified in pith cells by immunolocalisation. Interestingly, however, this increase in lignification in *elp1* stems was not accompanied by secondary cell wall thickening, suggesting that other secondary cell wall components were unaffected by this mutation. The most likely role of the *ELP1* gene product is as a negative regulator of the lignin biosynthetic pathway that under normal circumstances suppresses lignin deposition in the stem (Zhong *et al.*, 2000).

In contrast to *elp1*, the primary roots of *eli1* plants demonstrate abnormal lignification patterns with lignin present in cell types not normally lignified, as well as an absence of lignin in cells that are generally lignified, such as the xylem. This ectopic lignification pattern is associated with reduced cell elongation and shorter, thicker primary roots (Caño-Delgado *et al.*, 2000). The connection between abnormal cell shape and lignification was further verified by examining previously described mutations with altered cell elongation. Mutants with cell elongation defects, such as *lit*, *rsw1* and *kor*, all show ectopic lignification, suggesting that the ectopic lignification of *eli1* may be a consequence of a cell elongation defect (Caño-Delgado *et al.*, 2000). The link between altered secondary cell wall deposition and cell elongation is also suggested by work on the *Arabidopsis* mutant *elongation defective1* (*eld1*). *eld1* plants exhibit decreased cell elongation, altered xylem development and ectopic deposition of suberin (Cheng *et al.*, 2000).

Clearly both the *ELI1* and *ELD1* genes will provide valuable information on the mechanisms involved in regulating the process of lignification and as such are good candidates for genetically engineering and specifically increasing the lignin composition of cell walls from particular cell types.

Mutants exhibiting decreased secondary cell wall thickening

The *Arabidopsis* mutant *interfascicular fiberless*1 (*ifl1*) was initially described as a mutation affecting the regulation of secondary cell wall deposition (Zhong *et al.*, 1997). Whilst the secondary cell wall appears to form relatively normally within the xylem of these plants, the fibres of the interfascicular region appear to undergo little or no secondary cell wall thickening (Zhong *et al.*, 1997). Map-based cloning of the *IFL* gene has revealed it to be a homeodomain-leucine zipper protein (Zhong and Ye, 1999). Independent work that led to the cloning of the *revoluta* (*rev*) gene (Ratcliffe *et al.*, 2000) has demonstrated that *rev* and *ifl* are allelic. The *rev* mutant was initially described as having altered patterns of cell division in the apical meristem (Talbert *et al.*, 1995). These observations suggest that the phenotype of the mutation is in fact highly pleiotropic and the reduction in cell wall formation in the interfascicular region may be the consequence of reduced auxin flow from an abnormal apical meristem (Ratcliffe *et al.*, 2000). Whatever the initial mode of action of the *REV* gene, understanding how it affects cell wall formation in the interfascicular region is likely to reveal interesting insights into how this process is controlled.

In contrast to *ifl/rev*, which appear to disrupt secondary cell wall formation in the interfascicular region, the *gapped xylem* (*gpx*) mutant appears to affect secondary cell wall formation in both the xylem and the interfascicular region. In the xylem, gaps are present in positions where the water-conducting xylem elements normally exist (Figure 3). Since the procambial cells of wild-type and *gpx* plants are indistinguishable, these gaps apparently arise due to a defect in normal xylem element development. In the interfascicular regions, very thick-walled cells are adjacent to cells that contain little or no secondary cell wall (Figure 3). The interpretation of the gapped xylem phenotype stems from considering the difference between the development of xylem elements and the development of interfascicular cells. In xylem elements, the formation of a patterned cell wall is followed by programmed cell death, which is responsible for removing the contents of the cells. During this process of cell death, the only parts of the primary cell wall that remain are those that have been overlaid by a secondary cell wall. Consequently, if a cell fails to initiate a secondary cell wall the entire cell would be removed leaving a gap, such as those observed in the *gpx* mutant (Figure 3). In contrast, interfascicular cells form a secondary cell wall that is not patterned and do not undergo programmed cell death. In this case, if some cells failed to form a secondary cell wall, these cells would not be removed but would remain as cells bounded only by a thin primary cell wall, a phenotype also observed in *gpx* plants (Figure 3) (Turner and Hall, 2000).

Some differences, such as lignin subunit composition (Chapple *et al.*, 1992), exist between secondary cell wall deposition in the xylem and the interfascicular region. The *gpx* mutant, however, suggests that at least some aspects of the regulation of secondary cell wall formation are regulated in a common manner. In addition, the *gpx* mutant suggests that cell death and secondary cell wall formation in xylem elements are regulated independently, since cell death appears to occur in the absence of a secondary cell wall. This separation of cell death from cell wall formation does not support the idea that the signal for cell death accumulates as an integral part of secondary cell wall formation (Groover and Jones, 1999).

Whilst some cells in *gpx* fail to form a secondary cell wall, those that do form a secondary wall often form a wall that is considerably thicker than that normally observed in the wild type (Turner and Hall, 2000). How these thick-walled cells arise and why only a proportion of the cells form a secondary cell wall remains unclear. Progress on addressing these questions should be forthcoming once the *GPX* gene has been cloned.

Conclusion

It is clear that secondary cell wall mutants provide a complementary approach to antisense/sense strategies as a means of probing the structure, synthesis and assembly of the secondary cell wall. In some areas, such as the regulation of secondary cell wall deposition, analysis of mutants is likely to be the most productive avenue of research. To date, however,

218

most secondary cell wall mutants have been caused by mutations in rather obvious targets, such as the biosynthetic pathway of cellulose and lignin biosynthesis. One challenge for the future is to use genetic analysis to reveal completely novel components essential for secondary cell wall biosynthesis.

References

Arioli, T., Peng, L., Betzner, A.S., Burn, J., Wittke, W., Herth, W., Camilleri, C., Hofte, H, Plazinski, J., Birch, R., Cork, A., Glover, J., Redmond, J. and Williamson, R.E. 1998. Molecular analysis of cellulose biosynthesis in *Arabidopsis*. Science 279: 717–720.

Bacic, A., Harris, P.J. and Stone, B.A. 1988. Structure and function of plant cell walls. In: P.K. Stumpf (Ed.) The Biochemistry of Plants, Academic Press, New York, pp. 297–371.

Barrière, Y. and Argillier, O. 1993. Brown-midrib genes of maize: a review. Agronomie 13: 865–876.

Baskin, T.I., Betzner, A.S., Hoggart, R., Cork, A. and Williamson, R.E. 1992. Root morphology mutants in *Arabidopsis thaliana*. Aust. J. Plant Physiol. 19: 427–437.

Baucher, M., Monties, B., Van Montagu, M. and Boerjan, W. 1998. Biosynthesis and genetic engineering of lignin. Crit. Rev. Plant Sci. 17: 125–197.

Boudet, A.-M. 1998. A new view of lignification. Trends Plant Sci. 3: 67–71.

Campbell, M. and Rogers, L. 2001. Spatial and temporal regulation of lignin biosynthesis. Plant Mol. Biol., this issue.

Campbell, M.M. and Sederoff, R.R. 1996. Variation in lignin content and composition. Plant Physiol. 110: 3–13.

Caño-Delgado, A., Metzlaff, K. and Bevan, M. 2000. The *eli1* mutation reveals a link between cell expansion and secondary cell wall formation in *Arabidopsis thaliana*. Development 127: 3395–3405

Chapple, C.C.S., Vogt, T., Ellis, B.E. and Somerville, C.R. 1992. An *Arabidopsis* mutant defective in the general phenylpropanoid pathway. Plant Cell 4: 1413–1424.

Cheng, J.-C, Lertpiriyapong, K., Wang, S., and Sung, Z.R. 2000. The role of the *Arabidopsis ELD1* gene in cell development and photomorphogenisis in darkness. Plant Physiol. 123: 509–520.

Cherney, J.H., Cherney, D.J.R., Akin, D.E. and Axtell, J.D. 1991. Potential of brown-midrib, low-lignin mutants for improving forage quality. Adv. Agron. 46: 157–198.

Cutler, S. and Somerville, C.R. 1997. Cellulose synthesis: cloning in silico. Curr. Biol. 7: R108–R111.

Englehardt, J. 1995. Sources, industrial derivatives and commercial applications of cellulose. Carbohydrate Res. 12: 5–14.

Franke, R., McMichael, C.M., Meyer, K, Shirley, A.M., Cusumano, J.C and Chapple, C. 2000. Modified lignin in tobacco and popular plants overexpressing the *Arabidopsis* gene encoding ferulate 5-hydroxylase. Plant J. 22: 223–234.

Gorshkova, T.A., Wyatt, S.E., Salnikov, V.V. Gibeaut, D.M. Ibragimov, M.R. Lozovaya, V.V. and Carpita, N.C. 1996. Cell-wall polysaccharides of developing flax plants. Plant Physiol. 110: 721–729.

Grand, C., Parmentier, P., Boudet, A. and Boudet, A.M. 1985. Comparison of lignins and of enzymes involved in lignification in normal and brown midrib (*bm3*) mutant corn seedlings. Physiol. Vég. 23: 905–911.

Groover, A. and Jones, A.M. 1999. Tracheary element differentiation uses a novel mechanism co-ordinating programmed cell death and secondary cell wall synthesis. Plant Physiol. 119: 375–384.

Halpin, C., Holt, K., Chojecki, J., Oliver, D., Chabbert, B., Monties, B., Edwards, K., Barakate, A. and Foxon, G.A. 1998. *Brown-midrib* maize (*bm1*): a mutation affecting the cinnamyl alcohol dehydrogenase gene. Plant J. 14: 545–553.

Halpin, C., Barakate, A., Askari, B. Abbott, J. and Ryan, M. 2001. Enabling technologies for manipulating multiple genes on complex pathways. Plant Mol. Biol., this issue.

Hori, H. and Elbein, A.D 1985. The biosynthesis of plant cell wall polysaccharides. In: T. Higuchi (Ed.) Biosynthesis and Biodegradation of Wood Components, Academic Press, Orlando, FL, pp. 109–139.

Humphreys, J.M., Hemm, M.R. and Chapple, C. 1999. New routes for lignin biosynthesis defined by biochemical characterization of recombinant ferulate 5-hydroxylase, a multifunctional cytochrome P450-dependent monooxygenase. Proc. Natl. Acad. Sci. USA 96: 10045–10050.

Jones, L. Ennos, A.R. and Turner S.r. 2001. Cloning and characterisation of irregular xylem4 (irx4) a severely lignin deficient mutant of Arabidopsis. Plant J 26: 205–216.

Kimura, S., Sakurai, N. and Itoh, T. 1999. Different distribution of cellulose synthesizing complexes in brittle and non-brittle strains of barley. Plant Cell Physiol. 40: 335–338.

Kohel, R.J., Benedict, C.R. and Jividen, G.M. 1993. Incorporation of [^{14}C]glucose into crystalline cellulose in aberrant fibers of a cotton mutant. Crop Sci 33: 1036-1040.

Kokubo, A., Kuraishi, S. and Sakurai, N. 1989. Culm strength of barley. Plant Physiol. 91: 876–882.

Kokubo, A., Sakurai, N., Kuraishi, S. and Takeda, K. 1991. Culm brittleness of barley (*Hordeum vulgare* L.) mutants is caused by smaller number of cellulose molecules in cell wall. Plant Physiol. 97: 509–514.

Lapierre, C., Pollet, B., Mackay, J.J. and Sederoff, R.R. 2000. Lignin structure in a mutant pine deficient in cinnamyl alcohol dehydrogenase. J. Agric. Food Chem. 48: 2326–2331.

Lewis, N.G. and Yamamoto, E. 1990. Lignin: occurrence, biogenesis and biodegradation. Annu. Rev. Plant Physiol. Plant Mol. Biol. 41: 455–496.

Mackay, J.J., O'Malley, D.M., Presnell, T., Booker, F.L., Campbell, M.M., Whetten, R.W. and Sederoff, R.R. 1997. Inheritance, gene expression, and lignin characterization in a mutant pine deficient in cinnamyl alcohol dehydrogenase. Proc. Natl. Acad. Sci. USA 94: 8255–8260.

Marita, J.M., Ralph, J., Hatfield, R.D. and Chapple, C. 1999. NMR characterisation of lignins in *Arabidopsis* altered in the activity of ferulate 5-hydroxylase. Proc. Natl. Acad. Sci. USA 96: 12328–12332.

Mellerowicz, E.J., Baucher, M., Sundberg, B. and Boerjan, W. 2001. Unravelling cell wall formation in the woody dicot stem. Plant Mol. Biol., this issue

Meyer, K., Cusumano, J.C., Somerville, C. and Chapple, C.C.S. 1996. Ferulate 5-hydroxylase from *Arabidopsis thaliana* defines a new family of cytochrome P450-dependent monooxygenases. Proc. Natl. Acad. Sci. USA 93: 6869–6874.

Meyer, K., Shirley, A.M., Cusumano, J.C., Bell-Lelong, D.A. and Chapple, C. 1998. Lignin monomer composition is determined by the expression of a cytochrome P450-dependent monoxygenase in *Arabidopsis*. Proc. Natl. Acad. Sci. USA 95: 6619–6623.

Morrow, S.L., Mascia, P., Self, K.A. and Altschuler, M. 1997. Molecular characterization of a *brown midrib3* deletion mutation in maize. Mol. Breed. 3: 351–357.

Pear, J.P., Kawagoe, Y., Schreckengost, W.E., Delmer. D.P. and Stalker, D.M. 1996. Higher plants contain homologs of the bacterial *cel*A genes encoding the catalytic subunit of cellulose synthase. Proc. Natl. Acad. Sci. USA 93: 12637–12642.

Potikha, T. and Delmer, D.P. 1995. A mutant of *Arabidopsis thaliana* displaying altered patterns of cellulose deposition. Plant J. 7: 453–460.

Ralph, J., Mackay, J.J., Hatfield, R.D., O'Malley, D.M., Whetten, R.W. and Sederoff, R.R. 1997. Abnormal lignin in a loblolly pine mutant. Science 277: 235–239.

Ratcliffe, O.J., Riechmann, J.L. and Zhang, J. 2000. *Interfascicular fiberless1* is the same gene as *revoluta*. Plant Cell 12: 315–317.

Talbert, P.B., Adler, H.T. and Parks, B. 1995. The *revoluta* gene is necessary for apical meristem development and for limiting cell division in the leaves and stems of *Arabidopsis thaliana*. Development 121: 2723–2735.

Taylor, N.G., Scheible, W.-R., Cutler, S., Somerville, C.R. and Turner, S.R. 1999. The irregular *xylem3* locus of *Arabidopsis* encodes a cellulose synthase required for secondary cell wall synthesis. Plant Cell 11: 769–779.

Taylor, N.G., Laurie, S. and Turner S.R. 2000. Multiple cellulose synthase catalytic subunits are required for cellulose synthesis in *Arabidopsis*. Plant Cell 12: 2529–2540.

Turner S.R. and Hall, M. 2000. The *gapped xylem* mutant identifies a common regulatory step in secondary cell wall deposition. Plant J. 24: 477–488.

Turner, S.R. and Somerville, C.R. 1997. Collapsed xylem phenotype of *Arabidopsis* identifies mutants deficient in cellulose deposition in the secondary cell wall. Plant Cell 9: 689–701.

Vignols, F., Rigau, J., Torres, M.A., Capellades, M. and Puigdomènech, P. 1995. The *brown midrib3* (*bm3*) mutation in maize occurs in the gene encoding caffeic acid *O*-methyltransferase. Plant Cell 7: 407–416.

Wu, R.L., Remington, D.L., Mackay, J.J., Mckeand, S.E. and O'Malley, D.M. 1999. Average effect of a mutation in lignin biosynthesis in loblolly pine. Theor. Appl. Genet. 99: 705–710.

Yeo, U.-D., Soh, W.-Y., Tasaka, H., Sakurai, N., Kuraishi, S. and Takeda, K. 1995. Cell wall polysaccharides of callus and suspension-cultured cells from three cellulose-less mutants of barley (*Hordeum vulgare* L.). Plant Cell Physiol. 36: 931–936.

Zhong, R.Q. and Ye, Z.H. 1999. IFL1, a gene regulating interfascicular fiber differentiation in *Arabidopsis*, encodes a homeodomain-leucine zipper protein. Plant Cell 11: 2139–2152.

Zhong, R., Taylor, J.J. and Ye, Z.-H. 1997. Disruption of interfascicular fiber differentiation in an *Arabidopsis* mutant. Plant Cell 9: 2159–2170.

Zhong, R., Ripperger, A. and Ye, Z.-H. 2000. Ectopic deposition of lignin in the pith of stems of two *Arabidopsis* mutants. Plant Physiol. 123: 59–69.

Plant Molecular Biology **47**: 221–238, 2001.
© 2001 *Kluwer Academic Publishers. Printed in the Netherlands.*

Differential expression of cell-wall-related genes during the formation of tracheary elements in the *Zinnia* mesophyll cell system

Dimitra Milioni[+], Pierre-Etienne Sado[+], Nicola J. Stacey, Concha Domingo[1], Keith Roberts and Maureen C. McCann*
*Department of Cell and Developmental Biology, John Innes Centre, Norwich Research Park, Colney, Norwich NR4 7UH, UK (*author for correspondence; e-mail maureen.mccann@bbsrc.ac.uk); [1]Current address: IBMCP, Universidad Politecnica, Camino de Vera s/n, Valencia 46022, Spain; [+]these authors contributed equally to this work*

Key words: gene expression, lignification, secondary wall formation, tracheary element, trans-differentiation, *Zinnia*

Abstract

Plants, animals and some fungi undergo processes of cell specialization such that specific groups of cells are adapted to carry out particular functions. One of the more remarkable examples of cellular development in higher plants is the formation of water-conducting cells that are capable of supporting a column of water from the roots to tens of metres in the air for some trees. The *Zinnia* mesophyll cell system is a remarkable tool with which to study this entire developmental pathway *in vitro*. We have recently applied an RNA fingerprinting technology, to allow the detection of DNA fragments derived from RNA using cDNA synthesis and subsequent PCR-amplified fragment length polymorphisms (cDNA-AFLP), to systematically characterize hundreds of the genes involved in the process of tracheary element formation. Building hoops of secondary wall material is the key structural event in forming functional tracheary elements and we have identified over 50 partial sequences related to cell walls out of 600 differentially expressed cDNA fragments. The *Zinnia* system is an engine of gene discovery which is allowing us to identify and characterize candidate genes involved in cell wall biosynthesis and assembly.

Abbreviations: AGP, arabinogalactan-protein; BAP, benzylaminopurine; (cDNA)-AFLP, (complementary DNA)-amplified fragment length polymorphism; EST, expressed sequence tag; GH, glycosyl hydrolase; GRP, glycine-rich protein; GT, glycosyltransferase; HRGP, hydroxyproline-rich protein; NAA, naphthaleneacetic acid; PGase, polygalacturonase; PME, pectin methylesterase; PRP, proline-rich protein, TE, tracheary element; XET, xyloglucan endotransglycosylase

Introduction

For many types of plant cell, the process of differentiation is accompanied by the assembly of a distinct secondary wall on the plasma-membrane. Regardless of chemical composition, the primary wall is always defined as the structure that participates in irreversible expansion of the cell. The deposition of the secondary wall begins as cells stop growing. Not all wall secondary thickenings represent distinct secondary walls. Some thickened walls have a composition typical of a primary wall but simply containing many more

lamellae. Epidermal cells thicken the wall facing the environment to a much greater degree than side-walls or the inward-facing wall. Pairs of stomatal guard cells contain thickenings of cellulose microfibrils needed to create a curved wall that defines a stomatal aperture. Epidermal cells form specialized exterior layers of cutin and suberin to prevent the loss of water vapour, and the endodermal cells suberize their contiguous side walls to force the water to move symplastically into the stele (Esau, 1977).

Secondary walls often exhibit elaborate specializations. The cotton fibre, for example, consists of nearly 98% cellulose at maturity (Meinert and Delmer, 1997). In some cells, like sclereids, and vascular fibres, and the stone cells of pear, the secondary wall becomes uniformly thick, composed largely of cellulose microfibrils that almost fill the entire lumen of the cell. The secondary wall may, however, contain additional non-cellulosic polysaccharides, proteins and aromatic substances. The most obvious distinguishing feature of secondary walls is lignin, complex networks of phenylpropanoids. In tracheary elements (TEs), secondary walls can display annular or helical coils or reticulate and pitted patterns (Esau, 1977). These walls typically contain glucuronoxylans or 4-O-methylglucuronoxylans in addition to cellulose (Bacic et al., 1988). The production of a thick secondary wall of carbohydrate, structural protein, and lignin is essential for TE function, as the wall must be reinforced to resist compressive forces from the other cells that arise as a consequence of the extreme negative pressures that may develop in actively transpiring plants.

The formation of xylem elements or TEs involves several processes fundamental to plant development, including cell division, local cell signalling, cell elongation, cell specification, cell wall synthesis and deposition, lignification and programmed cell death, in total involving the activity of many hundreds of genes (McCann, 1997). Many of these genes have been identified in two large-scale screens. These involved cDNA sequencing using material derived from young xylem tissue from loblolly pine (Allona et al., 1998; Whetten et al., 2001, this issue) and from poplar trees (Sterky et al., 1998; Mellerowicz et al., 2001, this issue). The large number of genes identified in this way is impressive, but it remains to be seen how many of them are really involved in xylogenesis as these systems represent a mixture of cell types and developmental stages. Recent progress in identifying cell-wall-related genes and their functions from the loblolly pine (Whetten et al., 2001) and poplar (Mellerowicz et al., 2001) EST databases is described in these two companion articles. Two alternative generic strategies have been used to identify genes involved in the various stages of xylem formation and to investigate their function: the use of Arabidopsis mutants (Turner et al., 2001, this issue) and the use of an in vitro cell system, the Zinnia mesophyll cell system. In combination, these approaches will provide a clear description of the genes that encode the biosynthetic and hydrolytic enzymes involved in secondary wall formation and lignification. In addition, the regulatory genes that are responsible for inducing and co-ordinating secondary

wall formation can also be obtained. In this review, we focus on the use of the Zinnia model system as an engine of gene discovery to identify cell-wall-related genes involved specifically in xylogenesis.

The *Zinnia* mesophyll cell system

One of the more remarkable examples of cell development in vitro is the formation of TEs, including the deposition of secondary walls, by isolated mesophyll cells from the leaves of Zinnia elegans. Intact, single cells are obtained aseptically by gently mashing young leaves of Zinnia in a mortar and pestle and incubating them in a medium containing cytokinin and auxin (Fukuda and Komamine, 1980). Over a time-course of several days, about one-half of the cells undergo xylogenesis synchronously (McCann et al., 2000). Thus, one cell type can be reproducibly and synchronously switched, by known external signals, into a different cell type in culture.

Several laboratories have characterized a handful of the genes involved in different stages of the developmental pathway to TE fate using the Zinnia system (Demura and Fukuda, 1993; Ye and Varner, 1993; McCann et al., 2000). As molecular markers, these genes have proved to be extremely useful, and Igarashi et al. (1998) were able to demonstrate that the promoter element from one such gene, TED3, was effective in promoting xylem-specific expression of a GUS reporter gene in immature xylem cells of Arabidopsis. By differential screening methods, cDNA clones of transcripts that are up-regulated in inductive medium, at the time when cellulosic thickenings are beginning to be made, have been isolated (Ye and Varner, 1993; Demura and Fukuda, 1993). We have used a similar approach to identify transcripts associated with cell commitment, rather than the later differentiation events (McCann, 1997). Subtractive hybridization of cDNA libraries, made at time-points before and after the time at which the cells become re-specified to their new fate (Stacey et al., 1995), has shown that several transcripts are up-regulated between these times. One such gene encoded a pectate lyase (Domingo et al., 1998). Expression of the pectate lyase gene in Escherichia coli confirmed that the protein had the correct enzyme activity, and in situ hybridization showed that the gene was transcribed in cells associated with vascularization in the Zinnia plant. This approach has resulted in the identification of about 30 differentiation-related gene sequences (Fukuda, 1996;

McCann, 1997). However, only a few of these were seemingly related to secondary wall formation.

The time-courses for differentiation are somewhat variable in the various laboratories, perhaps as a consequence of the different cultivars and cell isolation procedures adopted. Therefore, it is important to define the sequential cellular events with respect to markers related to developmental state rather than simply the time in culture (Stacey *et al.*, 1995). Also, although differentiation is semi-synchronous in the *Zinnia* system, other processes, such as wound response, cell division and cell elongation are occurring, and the timing of TE formation may vary considerably among different cell preparations. Given that the expectation is for a large number of cell-wall-related genes to be involved in the process of TE formation, the synchrony of the system is critical.

The timing of growth factor requirements in the Zinnia mesophyll cell system

Church and Galston (1988) reported that auxin was required for 56 h while cytokinin was required for only 24 h. Fukuda and Komamine (1985) reported that the growth factors are not required in the first 12 h of culture. We investigated the precise times in the differentiation time-course at which the hormones are required in the culture medium. We observed that there is a window in the time-course at which the cells respond maximally to added auxin and cytokinin, resulting in increased numbers of TEs, and this is generally between 46 and 50 h of culture. Within this window, the growth factors need only be present for 10 min.

Cells were isolated from *Zinnia* leaves and cultured at 10^6 cells/ml in maintenance medium (medium without growth factors) for 1, 21, 26 or 47 h before transfer into inductive medium (medium with 1.0 mg/l of both auxin and cytokinin). The cells then remained in inductive medium until 48 h when they were transferred into fresh maintenance medium (Figure 1). Before transfer, the cells were washed twice in 0.2 M mannitol to minimize carry-over of exogenous hormone. The density of cells was maintained at 10^6 cells/ml after transfer into fresh medium. Progressively increasing numbers of TEs were observed at 120 h of culture as the time of exposure to inductive medium decreased (Figure 1). As just 1 h of exposure to exogenous growth factors at 48 h was sufficient to induce TE formation by 96 h, we continued to narrow this window, finding that just 5 to 10 min exposure to inductive

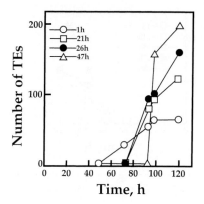

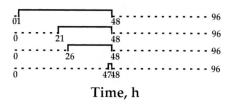

Figure 1. Graph showing the number of TEs formed when cells are transferred from maintenance medium to inductive medium at 1, 21, 26 or 47 h, and then transferred to fresh maintenance medium at 48 h, as illustrated in the schematic below (dashed line, maintenance medium; solid line, inductive medium). The number of TEs per mm^2 was averaged from three fields of view in each of four independent experiments for each time-point. Error limits for each point are less than ± 25 TEs.

medium at 48 h is sufficient to induce differentiation by 96 h. Further, the brief exposure to growth factors at around 48 h resulted in an enhanced level of differentiation (about 80% of living cells), with the culture apparently more synchronous than if the cells are cultured continuously in inductive medium (about 50% of living cells). To eliminate the possibility that transfer into fresh medium was itself enhancing TE formation, auxin and cytokinin were added directly to cells kept in maintenance medium for 48 h and the culture dish re-sealed. Enhanced levels of differentiation by 96 h were still observed.

To determine whether the enhanced level of TE differentiation was a consequence of a brief exposure to inductive medium specifically at 48 h, we transferred cells to inductive medium for 1 h at 1, 21, 28, 47 and 48 h (Figure 2). Almost double the number of TEs are formed when the window of induction is at 48 h than at earlier time-points (Figure 2). The cells still remain competent for TE formation at 66, 72 or even 90 h of culture but the number of TEs finally formed is reduced (data not shown). Transferring cells at 48 h from inductive medium into fresh inductive

224

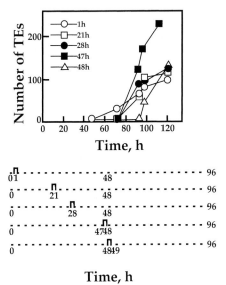

Figure 2. Graph showing the number of TEs formed when cells are transferred to inductive medium for 1 h at 1, 21, 28, 47 or 48 h, and then transferred to fresh maintenance medium, as illustrated in the schematic below (dashed line, maintenance medium; solid line, inductive medium). The number of TEs per mm^2 was averaged from three fields of view in each of four independent experiments for each time-point. Error limits for each point are less than 25 TEs.

medium improves the percentage of TE differentiation only slightly.

The position of molecular markers on the time-course of TE differentiation

Genes such as *TED2, 3* and *4* (Demura and Fukuda, 1994) and *ZePel1* (Domingo *et al.,* 1998) are useful molecular markers for early stages of the developmental pathway. We have mapped the time of expression of these marker genes, both in continuous inductive culture and when differentiation is induced by transfer of cells from maintenance medium into inductive medium at 48 h. The molecular marker *ZePel1* is expressed at 48 h in inductive medium (Domingo *et al.,* 1998), and the *TED2* and *3* genes are expressed at 36 and 60 h of culture respectively (Demura and Fukuda 1994). We isolated mRNA at various times during the time-course when cells were transferred from maintenance medium into inductive medium after 48 h and used this in RNA gel blot analyses. Both *ZePel1* and *TED3* are expressed at 57 h, just 9 h after transfer, whereas *TED2* is expressed later (Figure 3). These results demonstrate the compressed time-course and the improved synchrony of TE formation under these culture conditions.

Changes in pH during the culture period

Roberts and Haigler (1994) suggested that, although an alkaline pH in the culture medium did not prevent differentiation, there was a correlation between acidic pH and optimal levels of TE differentiation. Their experiments were done by buffering the pH of the medium into which the mesophyll cells were cultured. We have followed pH changes at intervals throughout each time-course when the cells are cultured continuously in maintenance, inductive, and auxin-only or cytokinin-only media. In all culture conditions, the pH of the medium rises by about one-half of a pH unit in the first 24 h, which is then succeeded by an acidification event (Figure 4a). In maintenance medium or with high levels of cytokinin only (1.0 mg/l BAP), acidification occurs roughly 24 h later than in inductive medium. However, the presence of high levels of auxin (1.0 mg/l NAA) also induces comparatively rapid acidification (Figure 4a).

If the cells are first incubated in maintenance medium for 48 h and then transferred into fresh maintenance medium at pH 5.6, then there is a second alkalinization phase followed by acidification (Figure 4b). If the cells are transferred into inductive medium or auxin-only medium (both also at pH 5.6), then the medium rapidly acidifies further. Transfer into cytokinin-only medium (pH 5.6) at 48 h does not result in rapid acidification. The pH minimum in inductive medium is lower by 0.2 to 0.3 pH units if the cells have been pre-incubated in maintenance medium for 48 h (Figure 4). As this was correlated with increased numbers of TEs by 96 h, we wondered if acidification itself was enhancing the level of differentiation. However, adjusting the pH of continuous inductive culture to be more acidic at 48 h did not affect the proportion of TE differentiation.

In all culture conditions, we observe a transient alkalinization of the culture medium which may be a wound response on subculture, such as has been observed in *Lycopersicon peruvianum* suspension culture (Felix and Boller, 1995). Roberts and Haigler (1994) reported an acidification of the culture medium by one pH unit prior to TE differentiation, and we observe the acidification both in continuous culture in inductive medium and when the cells have been transferred from maintenance medium into inductive medium at 48 h. The presence of auxin in the medium is sufficient to induce acidification, but not for differentiation, and Roberts and Haigler (1994) noted that TEs eventually form even when the medium is

225

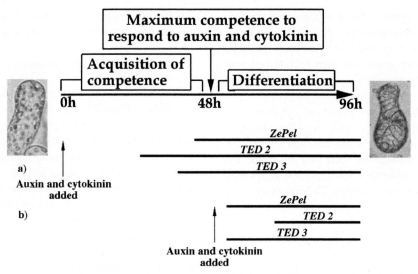

Figure 3. Schematic of the time-courses of *ZePel*, *TED2* and *TED3* expression mapped with respect to: the early phase in which the cells are not yet determined; the point at which the cells are determined with respect to auxin and cytokinin; and the late phase of differentiation in (a) a standard time-course in which auxin and cytokinin are added at time 0, or (b) with a pre-incubation in non-inductive medium in which auxin and cytokinin are added at 48 h (after McCann *et al.*, 2000).

buffered at high pH. A pH optimum between 5 and 6 has been noted for TE differentiation in citrus vesicle cultures (Khan *et al.*, 1986), and we observe a general correlation between relative acidification of the culture medium and numbers of TEs. However, we conclude that shifts in extracellular pH are neither sufficient nor necessary for cell commitment or differentiation.

A molecular approach to understanding TE formation

The improved synchrony of the time-course of xylogenesis in the *Zinnia* system raises the possibility of identifying very early events in the process, the signals that initiate the process of secondary wall deposition as well as all of the biosynthetic and hydrolytic enzymes involved. In short, we require a broad-based screen. We have recently applied an RNA fingerprinting technology, cDNA-AFLP (Bachem *et al.*, 1996, Durrant *et al.*, 2000), which allows us to detect differentially regulated genes across the time-course of xylogenesis (Milioni *et al.*, unpublished). cDNAs are synthesized from mRNA populations isolated from the *Zinnia* cultures at five time-points, digested with a pair of restriction enzymes, adaptor-ligated and amplified by PCR to produce the primary template. A subset of this population of fragments is selectively amplified using degenerate primers with two selective nucleotides, and then analysed on polyacrylamide gels. We selected over 600 genes, whose transcription show overt changes in abundance over time, and ob-

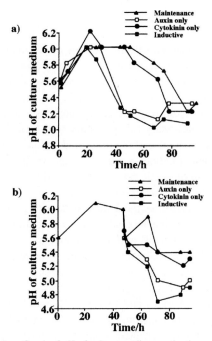

Figure 4. a. Graph of pH of culture media over the time-course of 96 h for *Zinnia* cells cultured in maintenance, inductive, auxin-only or cytokinin-only medium. There is a transient alkalinisation of all of the media but acidification occurs more rapidly in inductive and auxin-only conditions. b. Graph of pH of culture media over the time-course of 96 h when *Zinnia* cells are transferred from maintenance medium at 48 h into fresh maintenance, or inductive medium, or auxin-only or cytokinin-only media, as illustrated in Figure 1. The transfer into inductive medium results in the most rapid acidification.

tained partial sequences. These sequences were then compared with public databases that allowed us to assign an identity to about one-half of the predicted gene products, including about 10% that encode cell wall biosynthetic enzymes, hydrolytic enzymes or structural proteins. It remains to be established by *in situ* hybridization whether these represent xylem-specific members of their respective gene families. As these fragments may represent about one-half of the differentially expressed genes (one pair of restriction enzymes), we estimate that 1000 to 1500 genes may be differentially regulated during TE formation.

Construction of secondary walls and functional TEs

The final stage in the generation of a functional TE is autolysis of the protoplasmic contents to leave only the surrounding cell wall. Prior to this autolysis, all modifications must occur to the structures of both the original primary wall surrounding the mesophyll cell and the secondary wall deposited after commitment to the new cell fate. As observed in the light microscope, the deposition of hoops of secondary wall material occurs between 42 and 48 h after addition of auxin and cytokinin. However, other changes in cell wall architecture are occurring at much earlier times. In addition to increased cellulose and lignin synthesis necessary for secondary wall formation, an increase in total carbohydrate and xylose in alkali-extractable fractions, and a change in the relative proportion of pectic polysaccharides that are EDTA-extractable, are also correlated with the time course of TE formation in the *Zinnia* system (Ingold *et al.*, 1988). Sugar and linkage analysis of culture media show that a relatively unbranched rhamnogalacturonan is enriched in inductive medium from about halfway through the time-course and increases rapidly in concentration (Stacey *et al.*, 1995). Secretion of pectic polysaccharides and fucosylated xyloglucans into inductive culture medium has been detected with monoclonal antibodies (Stacey *et al.*, 1995) and the activity of a xylan synthase increases (Suzuki *et al.*, 1991). Changes occur in the ability to generate protoplasts shortly after cell isolation and these continue throughout the time-course (Stacey and McCann, unpublished results). Genes of interest that are likely to be developmentally regulated in this system include cell-wall biosynthetic enzymes, structural proteins and enzymes involved in re-modelling of the wall architecture for

TE function, and many hydrolytic enzymes, perhaps involved in the perforation of the wall.

Biosynthesis of cell wall polymers

Polysaccharides are not primary gene products, and it has proven difficult to genetically analyse the dynamic role that each component plays in the overall mechanical and functional properties of the cell wall, or of the tissues that contain them. Over 1000 gene products are probably involved in cell wall biosynthesis, assembly and turnover. *Arabidopsis* mutants mapping to 11 different loci in which one or several specific sugars are over- or under-represented were compared with the sugar composition in wild-type plants (Reiter *et al.*, 1997). Of these, the *mur1* defect has been traced to a GDP-mannose 4,6-dehydratase, and *mur4* to a C-4 epimerase (see Reiter and Vanzin, 2001, this issue). However, many cell wall mutants have also been selected on the basis of a growth or developmental phenotype. A temperature-sensitive mutant in primary wall cellulose synthase, *rsw1*, was selected by a root radial swelling phenotype at restrictive temperatures (Arioli *et al.*, 1998), while a secondary wall cellulose synthase mutant, *irx3*, was selected by a collapsed xylem phenotype (Taylor *et al.*, 1999). Two glycosyl transferases, a galactosyl transferase (Edwards *et al.*, 1999) and a fucosyl transferase (Perrin *et al.*, 1999), were cloned after isolation of the biochemical activities and obtaining protein sequence. However, these examples hardly begin to approach the complexity of structures required to build wall polysaccharides. It has been estimated that as many as 46 glycosyl-transferases are likely to be required to build pectic polysaccharides alone (Mohnen, 1999). Figure 5 shows the timing of expression of six genes encoding cell wall-related biosynthetic enzymes during TE formation in the *Zinnia* system.

Cellulose

Bands of cellulose microfibrils form one of the major elements of the TE wall thickenings. In the *Zinnia* system, secondary wall deposition is a hierarchical process in which the deposition of cellulose reflects the patterning of cortical microtubules. Disassembly of microtubules by depolymerizing agents results in random deposition of secondary wall material over the entire surface of the inner wall. Plasma-membrane sucrose synthase, thought to channel the immediate substrate for cellulose synthase (UDP-Glc), has been immuno-localized to sites of secondary thickenings

(Harrison *et al.*, 1997), indicating that the cellulose synthase complexes may be restricted to regions of the plasma membrane by the presence of cortical microtubules. Deposition of lignin in ordered patterns depends on the prior deposition of cellulose (Suzuki *et al.*, 1992), and inhibition of cellulose synthesis by the herbicide 2,6-dichlorobenzonitrile disrupts lignin patterning and also causes the loss of xylans from the cellulose-depleted thickenings (Taylor *et al.*, 1992; Taylor and Haigler, 1993). In contrast, inhibitors of lignification do not affect the patterning of thickenings or subsequent cell autolysis (Ingold *et al.*, 1990).

In *Arabidopsis*, the cellulose synthase (*CesA*) gene family comprises 12 members, none of which has been fully characterized (see Richmond and Somerville, 2001, this issue; and http://cellwall.stanford.edu/). It is not known whether the structure of the cellulose microfibrils in secondary walls is different from that of primary walls, or whether the ancillary proteins required to produce a secondary wall microfibril are different necessitating a different cellulose synthase. However, it appears that a subset of the *CesA* family may be associated specifically with secondary wall formation. The *IRREGULAR XYLEM* 3 locus encodes a cellulose synthase that is required in developing xylem vessel elements (Taylor *et al.,* 1999). Another xylem-specific cellulose synthase from poplar is up-regulated in tension wood and during normal growth but down-regulated in compression wood, in which the relative proportion of crystalline cellulose to lignin is known to decrease (Wu *et al.*, 2000). This observation provides the beginnings of a mechanism whereby wall composition can be altered in response to biomechanical signals (Wu *et al.*, 2000). We have isolated two *Zinnia* ESTs from the cDNA-AFLP screen that share very high identity with the cellulose synthase catalytic subunit (Figure 5). Both share the highest similarity with the cotton cellulose synthase, *GhCesA*-2, which is expressed in differentiating cotton fibres actively engaged in secondary wall synthesis (Pear *et al.*, 1996; Holland *et al.*, 2000), with *Arabidopsis CesA*-4 (Richmond and Somerville, 2000) and with a xylem-specific cellulose synthase gene from aspen (Wu *et al.*, 2000). The identities with the cotton and poplar genes suggest a specific role in the formation of secondary walls.

Matrix polymers

We also isolated cDNA-AFLP fragments encoding two glycosyltransferases, and a cellulose-synthase-like gene (Figure 5). The cellulose synthase-like (*Csl*) family members are divided into 6 families (A–F),

our fragment showing highest identity with an A family member, *AtCsl*A-2 (Richmond and Somerville, 2001; and http://cellwall.stanford.edu). It is speculated that these family members are sufficiently like *CesA* family members (containing the D,D,D,QxxRW motifs) to function as processive glycosyltransferases, but sufficiently different to be involved in binding different substrates. *Csl* family members are therefore candidates to encode biosynthetic enzymes making non-cellulosic polymer backbones.

There are two categories of glycosyltransferases (GTs): (1) GTs containing a PSPG consensus motif (thought to represent the nucleotide-diphosphate-sugar binding site) that catalyses reactions in secondary metabolism, and (2) GTs involved in plant cell wall biosynthesis (Vogt and Jones, 2000). The majority of β-linked plant polysaccharides are synthesized by glycosyl transferase family 2 (GT-2) enzymes, with $(1\rightarrow3)\beta$-D-glucan synthases being the exception in family GT-48 (Henrissat and Davies, 2000). The *Zinnia* galactosyl transferase (Figure 5) may transfer galactose residues onto a polymer backbone such as xyloglucan. However, glucose is commonly used by plant enzymes in a wide variety of transfer reactions from UDP-glucose to aglycones, including plant hormones, secondary metabolites and xenobiotics (Li *et al.*, 2000). The UDP-glucose-dependent glycosyltransferase gene family in *Arabidopsis* consists of at least 88 members, with the activity of some subgroups of the UGT family highly conserved between different plant species while others change their substrate specificity (Li *et al.*, 2000). Lignols may be glycosylated in reactions associated with the ER and Golgi apparatus, and this glycosylation may be necessary for membrane transport and targeting. Five of the *Arabidopsis* GT genes have been functionally characterized recently and shown to encode enzymes that can glycosylate sinapic acid, sinapyl alcohol and their related phenylpropanoids (Lim *et al.*, 2000). One of our clones shares significant identity with probable UDP-glucosyl-dependent glycosyltransferases from garden pea (T06371) and *Arabidopsis* (AB016819), but it remains to be confirmed whether this is involved in glucosylation of cell wall polymers. Another clone shares similarity with a UDP-glucose flavonoid 3-*O*-glucosyltransferase 1 gene from cassava (Q40284) that has been proposed to be involved in secondary plant metabolism (Hughes and Hughes, 1994) and also to an unknown flavonol 3-*O*-glucosyltransferase from *Arabidopsis* (AC005496). A third is similar to an *Arabidopsis* glucosyltransferase (AC002396) but it

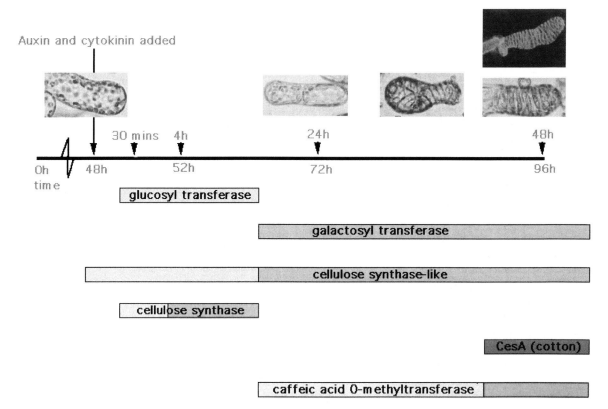

Figure 5. Time course of the trans-differentiation of isolated *Zinnia* mesophyll cells to TEs and the expression patterns of six cell wall biosynthetic genes identified so far by the cDNA-AFLP screen. Micrographs, left to right. 1. A palisade parenchyma cell appears unchanged after 48 h in liquid culture without TE-inducing growth factors. Auxin and cytokinin are added at 48 h, and RNA is extracted at this time and then 30 min, 4 h, 24 h and 48 h later. 2. At the 24 h point after induction, the chloroplasts move to the cell edges, and most of the cells have divided. 3. By 42 h after induction, the secondary wall thickenings have been deposited. 4. By 48 h after induction, the thickenings become lignified, and can be stained with phloroglucinol (above). Shading intensities of the bars indicate relative abundance of the cDNA-AFLP fragment in the gel (after McCann *et al.*, 2001).

also shares similarity with the *Arabidopsis* sequence AC007153. The latter sequence is homologous to maize indole-3-acetate β-glucosyltransferase. Seven genes encoding hormone glucosyltransferases have only recently been identified: the *IAGLU* gene from maize (Szerszen *et al.*, 1994), two zeatin glucosyltransferases (Martin *et al.*, 1999a, b) and four from *Arabidopsis* (Jackson *et al.*, 2001).

Lignin

At late times in the *Zinnia* culture system, genes relevant to lignification (cinnamyl alcohol dehydrogenase *ZCAD1*, phenylalanine ammonia-lyases *ZePAL1*, 2 and 3, cinnamic acid-4-hydroxylase *ZC4H*, and peroxidase *ZPO*-C (Fukuda, 1997) are up-regulated. The synthesis of monolignols is well documented in plants, and all synthetic reactions appear to occur in the cytosol. However, the extent to which monolignols begin

to condense and form associations with carbohydrates or other materials during secretion is not known. Once in the wall, monolignols and their initial condensation products are polymerized to form lignin. Laccase, a member of the 'blue copper oxidase' family of enzymes (O'Malley *et al.*, 1993), plays an important role in lignin biosynthesis in the formation of dilignols, and two cDNA-AFLP fragments encoding laccases are expressed only at the very late stage of TE formation.

In addition to laccases, the peroxidases are also good candidates for the polymerization of monolignols (Ostergaard *et al.*, 2000). One *Zinnia* EST shows high similarities with plant peroxidases from *Arabidopsis* (accession number AB010692 and Medline 98344145) and with a lignin-forming anionic peroxidase expressed in mesophyll cells of *Nicotiana silvestris* (Criqui *et al.*, 1992, accession number Q02200 and Medline 93041285). However the pattern of early expression of the *Zinnia* gene obtained from the cDNA-AFLP does not support a role in lignification. The peroxidase gene family is very large, with more than 20 hits (with high score) within the *Arabidopsis* genome and no published function for any of them. Careful study will be needed to elucidate the function of this particular enzyme amongst the other family members that might also be expressed.

According to the traditional pathway of lignin biosynthesis, caffeate *O*-methyltransferase (COMT) acts to methylate caffeic acid to produce ferulic acid (Whetten *et al.*, 1998). It has also been proposed that caffeoyl-CoA *O*-methyltransferase (CCoAOMT) is involved in an alternative methylation pathway of lignin biosynthesis in *Zinnia* (Ye *et al.*, 1994). Using an antisense approach, Zhong *et al.* (1998) demonstrated that methylation reactions in lignin biosynthesis are catalysed by both CCoAOMT and COMT in transgenic tobacco plants. Recently, Jouanin *et al.* (2000) showed that lignin levels in transgenic trees were substantially reduced in CAOMT down-regulated poplar trees. Several genes encoding CAOMT have been cloned from various species including *Zinnia* (Ye and Varner, 1995), poplar (Tsai *et al.*, 1998), tobacco (Martz *et al.*, 1998) and *Stylosanthes* (McIntyre *et al.*, 1995). A cDNA-AFLP fragment encoding COMT was isolated and found to share a high degree of identity with the COMT from *Stylosanthes humilis* (1582580) and with a COMT clone isolated from basil (AF154917) (Figure 5) but is unrelated to other *Zinnia* COMT clones already available in the data bank.

A mutant screen has exploited the simple pattern of protoxylem elements in the seedling root of *Arabidopsis* to uncover a wide spectrum of mutant phenotypes including the timing of protoxylem differentiation, the number of protoxylem strands and ectopic lignified cells (Caño-Delgado *et al.*, 2000). One of these (*eli1*) shows disrupted protoxylem, and lignification in the stem pith cells that appears to be related to expansion of these cells. This is an intriguing connection, particularly as the authors have found that some other cell expansion mutants, including *rsw1*, also display ectopic lignification (Caño-Delgado *et al.*, 2000). Similar ectopic deposition of lignin, also in the pith of the stem of *Arabidopsis*, has been found in the *elp1* mutant (Zhong *et al.*, 2000).

Remodelling of wall architecture

Structural proteins

Although the structural framework of the cell wall is largely carbohydrate, structural proteins may also form networks in the wall. There are four major classes of structural proteins; three of them are named for their uniquely enriched amino acid: the hydroxyproline-rich proteins (HRGPs), the proline-rich proteins (PRPs) and the glycine-rich proteins (GRPs). All of them are developmentally regulated, with relative amounts varying among tissues and species (Keller and Lamb, 1989; Ye *et al.*, 1991; Wyatt *et al.*, 1992; Santino *et al.*, 1997; Bernhardt and Tierney, 2000; Dubreucq *et al.*, 2000; Merkouropoulos *et al.*, 2000). Many structural cell-wall proteins are specifically associated with secondary thickenings. In bean, GPRs are synthesized in the xylem parenchyma cells and exported into the walls of protoxylem vessels (Ryser and Keller, 1992). Arabinogalactan-proteins (AGPs) (Schindler *et al.*, 1995), an extensin-like protein (Bao *et al.*, 1992) and a tyrosine- and lysine-rich protein (Domingo *et al.*, 1994) have been found in maize, loblolly pine and tomato xylem, respectively. Some PRPs concentrate in the secondary walls of protoxylem elements of bean. We have found cDNA-AFLP fragments encoding both a glycine-rich protein and an extensin-like protein (Figure 6).

Extensin, encoded by a multigene family, is one of the best-studied HRGPs of plants. Extensin consists of repeating Ser-(Hyp)4 and Tyr-Lys-Tyr sequences that are important for secondary and tertiary structure: the repeating Hyp units predict a 'polyproline II' rod-like molecule. HRGPs are generally found at low levels in the primary walls of all tissues, although they are particularly abundant in phloem. A Thr-rich, extensin-like protein of maize is more abundant in the secondary walls of the firm pericarp of popcorn. It has been suggested that synthesis, deposition and cross-linking of extensins helps to increase the mechanical strength of the cell wall. Both HRGPs and PRPs are also considered to be involved in the responses of plants to environmental factors, such as wounding and infection (Sheng *et al.*, 1991; Ebener *et al.*, 1993). Recently, it has been demonstrated that specific extensins are expressed in elongating cells and may have an important role in cell wall structure (Dubreucq *et al.*, 2000). It has also been suggested that the activation of genes encoding specific structural proteins would provide a mechanism for morphogenetic control of cell

230

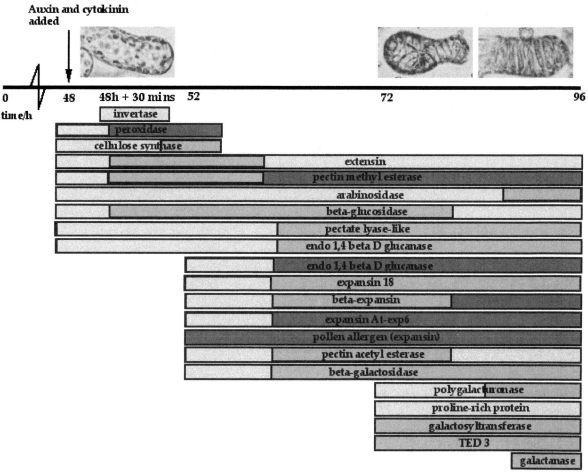

Figure 6. Time course of the trans-differentiation of isolated *Zinnia* mesophyll cells to TEs and the expression patterns of genes encoding cell wall enzymes identified so far by the cDNA-AFLP screen. Micrographs as detailed in Figure 4. Shading intensities of the bars indicate the relative abundance of the cDNA-AFLP fragment in the gels.

wall architecture during cellular differentiation (Keller and Lamb, 1989).

The AGPs are more aptly named proteoglycans, as they can consist of more than 95% carbohydrate (Du *et al.*, 1996). AGPs constitute a broad class of molecules that are located in Golgi-derived vesicles, the plasma membrane, and the cell wall. Of the few core proteins that have been characterized, they are enriched in Pro(Hyp), Ala, and Ser/Thr. They possess no distinguishing common motifs, but contain domains with similarity to some PRPs, extensins, and the solanaceous lectins (Gaspar *et al.*, 2001, this issue). No clear-cut function has been described for AGPs, or indeed for any of the cell-wall structural proteins. In the *Zinnia* system, an AGP is detected using a monoclonal antibody in the primary walls of a sub-population of cells 24 h before secondary thickenings

are visible, and, in addition, AGPs are secreted into the culture medium. This AGP epitope is present in the secondary thickenings of mature TEs but not in the surrounding primary walls (Stacey *et al.*, 1995). *TED3* (Demura and Fukuda, 1993) encodes a hydrophilic protein whose sequence contains an Asn-Gly-Tyr motif repeated 15 times and 3 repeats of 23 amino acids. With the correction of a sequencing error (van Hengel, personal communication), it shares a high degree of similarity with 2 cDNAs encoding the protein backbone of 2 AGPs, one from pear suspension cultures (*PcAGP2*) and the other from tobacco suspension culture (*NaAGP-2*) (Mau *et al.*, 1995). *TED3* transcripts accumulate 12 to 24 h before the beginning of secondary wall thickening in the *in vitro Zinnia* system. Expression analysis of the *TED3* promoter revealed that the promoter is not significantly wound-inducible in mesophyll and epidermal cells and confirmed the TE preferential expression of *TED3* (Igarashi *et al.*, 1998). One cDNA-AFLP clone was identified through our screen and this is homologous to the TED3 sequence. The expression pattern coincides with that predicted by northern analysis (Figure 6).

Expansins

Zinnia posseses a xyloglucan- and pectin-rich 'Type I' cell wall typical of all dicots and several monocots, which is distinct from the arabinoxylan-rich, pectin-poor 'Type II' wall of grasses and related monocots (Carpita and Gibeaut, 1993). Because grass expansins induce extension of tissues with Type I walls, it is attractive to think that expansins are ubiquitous enzymes involved in the rapid growth responses of both TypeI and Type II walls (Cosgrove, 2000). However, a second multigene family of β-expansins, predominantly found in the grasses, have no appreciable activity on Type I cell walls (Cosgrove *et al.*, 1997). Expansins catalyse wall extension *in vitro* without any detectable hydrolytic or transglycosylation events. These proteins probably catalyse breakage of hydrogen bonds between cellulose and the load-bearing cross-linking glycans (McQueen-Mason and Rochange, 1998; Cosgrove, 2000). During the past few years, molecular studies of the expansins as well as the genomic sequencing projects have led to isolation of several related genes in various higher-plant species. The analysis of the complete *Arabidopsis* genome suggests that there might be more than 24 members of the α-expansin family in *Arabidopsis* (http://www.bio.psu.edu/expansins; Cosgrove, 2000). On the basis of sequence similarities, expansins are classified into two multigene families, the α- and β-expansins. A notable difference between the two groups is the extensive glycosylation of β-expansins which appears to be absent in the α-expansins. The α-expansin genes have also been identified in tomato meristems (Fleming *et al.*, 1997), in rice seedlings (Huang *et al.*, 2000), in *Zinnia* xylem cells (Im *et al.*, 2000), in pine (Hutchinson *et al.*, 1999), in *Marsilea quadrifolia* (fern) (Kim *et al.*, 1999) and other species. Three *Zinnia* expansin genes have recently been described (Im *et al.*, 2000). The mRNA corresponding to each is localized to stem tissue cells, two being localized to the apical end and one to the basal end of putative cambial cells. Two α-expansin sequences were isolated through our cDNA-AFLP screen that show similarity to *ZeEXP1* from *Zinnia* (Im *et al.*, 2000). They also show similarity to a puta-

tive expansin from *Arabidopsis* as well as to *LeExp*18 from tomato, whose tissue-specific expression suggests its involvement in distinct processes involving differentiation (Reinhardt *et al.*, 1998).

The β-expansin family members are expressed in grass vegetative tissues and share homology with pollen allergen genes (Cosgrove *et al.*, 1997). At least 19 maize and 10 rice β-expansin genes have been described to date as well as four β-expansins identified in the *Arabidopis* genome and some in tomato EST databank (Cosgrove, 2000). Three putative members of the β-expansin family have been detected in *Zinnia* in our cDNA-AFLP screen. The first clone shares significant similarity with an *Arabidopsis* β-expansin while the second shows similarity to an *Arabidopsis* clone (AC001229), a *Holcus major* pollen allergen gene, and to a cytokinin-induced mRNA (*CIM1*) from soybean, that is postulated to be part of the mechanism by which cytokinin induces cell proliferation (Downes and Crowell, 1998). The third clone shares a high degree of homology with a β-expansin from *Zea mays* (2315515A) and a putative *Arabidopsis* β-expansin.

Hydrolytic activities

The conducting component of xylem tissue consists of cells formed into continuous tubes by perforation of the ends of contiguous TE or xylem elements. Many hydrolytic enzyme activities have been detected in the *Zinnia* culture, including polygalacturonase, pectin methyl esterase, fucosidase, and xyloglucan endo-transglycosylase (XET) (Stacey *et al.*, 1995). These enzymes, which appear to increase in activity at a late stage, may have a role in perforating the end-walls of the TEs, as recently observed in the scanning electron microscope (McCann *et al.*, 2000; Nakashima *et al.*, 2000).

Pectin-degrading enzymes

Pectate lyases (EC 4.2.2.2) have previously been described as microbial extracellular enzymes that assist pathogenesis by cleaving polygalacturonate blocks in the plant host cell wall (Davis *et al.*, 1984; Collmer and Keen, 1986). Enzymatic cleavage of $(1\rightarrow4)\alpha$-linked galacturonosyl residues occurs at a pH optimum of 8–11 through a β-elimination mechanism, resulting in an unsaturated C-4/C-5 bond in the galacturonosyl moiety at the non-reducing end of the polysaccharide produced at the cleavage site (Rombouts and Pilnik, 1980). This mechanism is common to another class of pectin-degrading enzyme, pectin lyase. However, pectate lyase is distinguished by its preference for a glycosidic linkage next to a free carboxyl group rather than to an esterified carboxyl group, and by its pH optimum (Pilnik, 1990). All pectate lyases show calcium dependence. Given the complex structure of pectic polysaccharides, their enzymic degradation by pathogens is usually accomplished by the synergistic action of enzymes such as pectin methyl esterase, pectin lyase, endo- and exo-polygalacturonase and pectate lyase. Pectate lyase is particularly effective, and soft-rot symptoms in various plant species can be induced by applying recombinant pectate lyases from *E. coli* (Keen and Tamaki, 1986; Bartling *et al.*, 1995). In plants, pectate lyase genes have been described that are expressed primarily in pollen and in the transmitting tissue of plants, where the enzyme may facilitate pollen tube growth (Ori *et al.*, 1990).

The *ZePel*1 gene encoding a *Zinnia* pectate lyase (Domingo *et al.*, 1998) is auxin-inducible, and its expression is associated with vascular bundles and shoot primordia. With a pH optimum of 10, and only residual enzymatic activity below pH 7.5, the pectate lyase is likely to be sub-optimally active *in vivo*, and this may be a necessary regulatory control for an enzyme that can produce soft-rot symptoms in plants. The natural substrates of pectate lyases are not clearly defined. Three *Erwinia* pectate lyase isoforms obtained by expression of three *Pel* genes in *E. coli* act synergistically to extend the range of pectin substrates which the bacterium can degrade (Bartling *et al.*, 1995). *ZePel* may assist in the removal and modification of an existing pectin matrix to allow the deposition of newly-synthesized wall polymers for a specialized function. A further possibility is that pectic fragments released by pectate lyase may act as oligosaccharins – polysaccharide fragments that act as cell-signalling molecules to elicit a range of cellular responses (Mohnen and Hahn, 1993). The enzymes from *Erwinia* liberate oligosaccharides from purified pectic substances and higher plant cell walls that function as elicitors of active plant defence reactions (Gardner and Kado, 1976; Davis *et al.*, 1984). The cDNA-AFLP screen has identified a second pectate-lyase-like protein associated with TE formation.

Other pectin-degrading enzymes also show differential gene expression patterns across the time-course. Two *Zinnia* ESTs have similarities with a pectin acetyl esterase gene from *Arabidopsis* (accession number AC01293, unpublished), called putative pectin acetyl esterase from its high similarity (44% identities and 60% similarities) with a pectin acetyl esterase from

mung bean (Breton *et al.*, 1996, accession number S68805). The search generated more than 10 hits in the *Arabidopsis* genome database suggesting the presence of a large gene family. The patterns of expression of the two ESTs favour a possible role in the final stage of TE differentiation. Another *Zinnia* EST has similarities with pectin methyl esterase. The search generated only three significant hits in the *Arabidopsis* genome, suggesting a sub-family of the large family of PMEs (about 50 members).

Multiple genes encoding polygalacturonases (PGases) have been described in a number of species, including tomato (Kalaitzis *et al.*, 1997), melon (Hadfield *et al.*, 1998), *Brassica napus* (Jenkins *et al.*, 1996), maize (Allen and Lonsdale, 1992), peach (Lester *et al.*, 1994), *Arabidopsis* (Torki *et al.*, 2000) and soybean (Mahalingam *et al.*, 1999). PGases are involved in the disassembly of pectin that accompanies many developmental processes including fruit ripening, abscission/dehiscence, pathogenesis, and cell expansion. PGase mRNA and protein are localized to the developing vascular system in several young, growing tissues which suggests that PGase may be involved in xylogenesis and disassembly of the xylem vessel primary cell wall (Dubald *et al.*, 1993; Sitrit *et al.*, 1996). In *Zinnia*, PGases are represented by three sequences pulled from the AFLP screen. The first is similar to a polygalacturonase β-subunit from tomato (U63374 and U64790), which has been shown to play a significant role in regulating pectin metabolism in tomato fruit by limiting the extent of pectin solubilization and depolymerization that can occur during ripening (Watson *et al.*, 1994). The second cDNA-AFLP clone shares a high degree of homology with two PGases from soybean, which show up-regulated expression in the roots after infection with soybean cyst nematode, supporting the hypothesis that precocious expression of developmentally programmed cell wall dissolution may be responsible for syncytium formation (Bird, 1996; Mahalingam *et al.*, 1999). The last sequence shares similarity with a putative PGase from *Arabidopsis* (AC002292) and with a PGase-like protein from *Arabidopsis* (T48638)

O-glycosyl hydrolases
O-glycosyl hydrolases are a widespread group of enzymes which hydrolyse the glycosidic bond between two or more carbohydrates or between a carbohydrate and a non-carbohydrate moiety. A classification system for glycosyl hydrolases (GH), based on sequence similarity, has led to the definition of up to 68 different

families that are divided into 8 clans (GH-A, -B, -C, -D, -E, -F, -G and -H) on the basis of structural and mechanistic information (Henrissat *et al.*, 2001, this issue). Some GHs are multifunctional enzymes that contain catalytic domains that belong to different GH families or to other protein activities, such as kinases (Henrissat and Davies, 2000).

β-galactosidases from mammals, fungi, plants and the bacterium *Xanthomonas manihotis* are evolutionarily related (Henrissat and Bairoch, 1993, 1996). They belong to family 35 (EC 3.2.1.23) in the classification of GHs with a putative active site: G-G-P-[LIVM](2)-x(2)-Q-x-E-N-E-[FY]. A cDNA-AFLP clone shares homology with β-galactosidase from *Cicer arientium* (AJ006771) and with an exo-$(1\rightarrow4)\beta$-galactanase that is involved in cell wall degradation during tomato ripening (Carey *et al.*, 1995). A second cDNA-AFLP sequence is similar to exo-galactanase from *Lupinus angustifolius* (AJ011047) and to a β-galactosidase cloned from harvested asparagus tips (King and Davies, 1995). Arabinofuranosidases are classified to GH families 43 (EC 3.2.1.37) and 51 (EC 3.2.1.55). Family 43 contains 26 different enzymes, including β-xylosidase, bifunctional β-xylosidase/α-L-arabinofuranosidase, β-xylanase, endo-arabinase and α-L-arabinofuranosidase. One *Zinnia* putative α-L-arabinofuranosidase clone was isolated, sharing similarities with two *Arabidopsis* clones from the genome sequencing project (AAF19575.1 and AAD40132). Two ESTs have been identified from the screen that show high similarity with plant endoglucanases. The first gives highest similarity with a strawberry endoglucanase (Trainotti *et al.*, 1999, accession number AJ006349 and Medline 99339255). The second gives the highest score with a gene from *Arabidopsis* (AC066689_2) that is annotated as an endo-$(1\rightarrow4)\beta$-glucanase, based on its similarity to other characterized genes. Endoglucanase activities have been reported in most plant tissues and endoglucanase gene expression can be auxin-induced (Wu *et al.*, 1996).

Invertase

Invertases catalyse the hydrolysis of terminal non-reducing β-D-fructofuranoside residues in β-D-fructofuranosides. There are two categories of invertases according to pH optimum: (1) the acid invertases which are either ionically bound to the cell wall (cell-wall invertases) or accumulate as soluble proteins in the vacuole (vacuolar invertases) and (2) the neutral and alkaline invertases with neutral or slightly alkaline pH optimum which are cytoplasmic proteins. Invertase sequences have been described in a variety of species such as *Arabidopsis* (Tymowske-Lalanne and Kreis, 1998), oat (Mercier *et al.*, 1993), tomato (Ohyama *et al.*, 1998), potato (Zrenner *et al.*, 1996), maize (Xu *et al.*, 1995), onion (Vijn *et al.*, 1998) and three sequences with very good hits exist in the loblolly pine EST database. Two cDNA-AFLP *Zinnia* clones show similarity to invertases. Based on amino acid comparisons, one of them shares homology with a β-fructofuranosidase precursor from *Daucus carota* (S23217) and also with an acid invertase from *Vigna radiata*, which was shown to appear after germination and maintained at high levels in rapidly growing tissues (Arai *et al.*, 1992).

Prospects

The plant cell wall contains structural proteins as well as hundreds of different cell-wall enzymes. The cell wall polysaccharides that dominate the wall architecture are the products of biosynthetic enzymes that are located in the Golgi, or, in the case of cellulose and callose synthases, at the plasma membrane. It appears that the genes that encode these classes of cell-wall-related proteins generally belong to large multi-gene families. Much work remains to be done to elucidate the functions of the different family members, and to identify cell-wall-related genes that currently lack homologues in the databases. By looking at one specific developmental pathway, the formation of a TE, we hope to uncover those genes involved specifically in secondary wall formation in this particular cell type. About 40% of our sequences are unknowns – either hypothetical proteins in the *Arabidopsis* database or not represented in any of the plant or animal databases. Since known genes encoding cell wall-related proteins comprise about 10% of our 600 sequences, we anticipate that a further 30 sequences may represent cell-wall-related proteins of totally unknown function.

The collections of ESTs sequenced from xylem-enriched tissues of loblolly pine (Whetten *et al.*, 2001) and poplar (Mellerowicz *et al.*, 2001) are key resources of candidate genes involved in xylogenesis. Mining the information represented in these collections using microarray technologies will provide a comprehensive approach to defining the complete range of genes involved in xylogenesis and fibre formation.

Several mutant screens in *Arabidopsis* have successfully identified plants that are affected in normal xylem development, some of which display a cell-wall phenotype. Stems of *irx3* plants, with a collapsed xylem phenotype, have less than 20% of the cellulose content of the wild-type (Taylor *et al.*, 1999). A cellulose synthase gene was identified at the *irx3* map location, and the wild type gene complemented the mutation by restoring cellulose deposition in the secondary walls of xylem vessels. Other irregular xylem (*irx*) genes are currently being cloned. Another screen for vascular mutants, but this time taking advantage of the simplicity of the seedling root xylem, has identified many mutants, of which one displaying ectopic lignification (*eli1*) has been described in detail (Caño-Delgado *et al.*, 2000). The *eli1* mutant shows disorganized xylem with discontinuous elements, as well as inappropriate cells, possibly connected with altered expansion growth, developing secondary walls that become lignified. Another screen for altered lignification patterns has identified two mutant alleles of the *ELP*1 gene on chromosome 1, whose wild-type function may be to repress lignin deposition in pith tissues (Zhong *et al.*, 2000). Further screens for altered patterns of venation (Deyholos *et al.*, 2000; Koizumi *et al.*, 2000) and for vascular development (Zhong *et al.*, 1999) are ongoing, and may lead to identification of key transcription factors that regulate secondary wall formation and lignification. Candidate genes such as *MYB* family members have already been implicated in the regulation of lignification (Newman and Campbell, 2000).

Genes and mutants are key resources, but polysaccharides are secondary gene products and further subject to a wide range of post-synthetic modifications that may be critical for their function. Methods such as the use of antibody probes that directly detect the presence of a particular epitope on a molecule are also important in identifying the final products used in secondary wall construction. Exploiting the antibody-phage display method, Shinohara *et al.* (2000) identified an epitope in the cell wall cross-linking glycan fraction that is distributed in a polarized way in immature tracheids both *in planta* and in the *Zinnia* cell system. Development of further sensitive methods of detection for polysaccharide structure will be critical.

With the completion of the *Arabidopsis* genome (Arabidopsis Genome Initiative, 2000), the availability of sequence information and the tools of reverse genetics will make it possible to move rapidly between an excellent system for molecular genetics and other systems in which the biochemistry and cell biology of the process can be defined more easily. This powerful combined approach can be expected to contribute to the rapid elucidation of function.

Acknowledgements

Many thanks to Dr Taku Demura and Professor Hiroo Fukuda for the *TED2* and *TED3* clones, and to Sue Bunnewell for photographic assistance. K.R., N.J.S. and M.C.M. gratefully acknowledge the financial support of the BBSRC and the Leverhulme Trust. P.S. is funded by a BBSRC special studentship, D.M. is funded by an EU Marie Curie fellowship and M.C.M. is funded by a Royal Society University Research Fellowship.

References

Allen, R.L. and Lonsdale, D.M. 1992. Sequence analysis of three members of the maize polygalacturonase gene family expressed during pollen development. Plant Mol. Biol. 20: 343–345.

Allona, I., Quinn, M., Shoop, E., Swope, K., St. Cyr, S., Carlis, J. *et al.* 1998. Analysis of xylem formation in pine by cDNA sequencing. Proc. Natl. Acad. Sci. USA 95: 9693–9698.

Arabidopsis Genome Initiative. 2000. Analysis of the genome sequence of the flowering plant *Arabidopsis thaliana*. Nature 408: 796–815.

Arai, M., Mori, H. and Imaseki, H. 1992. Cloning and sequence of cDNAs for an intracellular acid invertase from etiolated hypocotyls of mung bean and expression of the gene during growth of seedlings. Plant Cell Physiol. 33: 245–252.

Arioli, T., Peng, L., Betzner, A.S., Burn, J., Wittke, W., Herth, W. et al. 1998. Molecular analysis of cellulose biosynthesis in *Arabidopsis*. Science 279: 717–720.

Bachem, C.W.B., van der Hoeven, R.S., de Bruijn, S.M., Vreugdenhil, D., Zabeau, M. and Visser, R.G.F. 1996. Visualization of differential gene expression using a novel method of RNA fingerprinting based on AFLP: analysis of gene expression during potato tuber development. Plant J. 9: 745–753.

Bacic, A., Harris, P.J. and Stone, B.A. 1988. Structure and function of plant cell walls. In: J. Preiss (Ed.) The Biochemistry of Plants, vol. 14, Academic Press, New York, pp. 297–371.

Bao, W.L., O'Malley, D.M. and Sederoff, R.R. 1992. Wood contains a cell-wall structural protein. Proc. Natl. Acad. Sci. USA 89: 6604–6608.

Bartling, S., Wegener, C. and Olsen, O. 1995. Synergism between *Erwinia* pectate lyase isoenzymes that depolymerize both pectate and pectin. Microbiology 141: 873–881.

Bird, D.M. 1996. Manipulation of host gene expression by root-knot nematodes. J. Parasitol. 82: 881–888.

Bernhardt, C. and Tierney, M.L. 2000. Expression of AtPRP3, a proline-rich structural cell wall protein from arabidopsis, is regulated by cell-type-specific developmental pathways involved in root hair formation. Plant Physiol. 122: 705–714.

Breton, C., Bordenave, M., Richard, L., Pernollet, J.C., Huet, J.C., Perez, S. and Goldberg, R. 1996. PCR cloning and expression analysis of a cDNA encoding a pectinacetylesterase from *Vigna radiata* L. FEBS Lett. 388: 139–142.

Caño-Delgado, A., Metzlaff, K. and Bevan, M.W. 2000. The *eli1* mutation reveals a link between cell expansion and secondary cell wall formation in *Arabidopsis thaliana*. Development 127: 3395–3405.

Carey, A.T., Holt, K., Picard, S., Wilde, R., Tucker, G.A., Bird, C.R. *et al.* 1995. Tomato exo-(1→4)-β-D-galactanase. Isolation, changes during ripening in normal and mutant tomato fruit, and characterization of a related cDNA clone. J. Plant Physiol. 108: 1099–1107.

Carpita, N.C. and Gibeaut, D.M. 1993. Structural models of primary cell walls in flowering plants: consistency of molecular structure with the physical properties of the walls during growth. Plant J. 3: 1–30.

Church, D.L. and Galston, A.W. 1988. Kinetics of determination in the differentiation of isolated mesophyll cells of *Zinnia elegans* to tracheary elements. Plant Physiol. 88: 92–96.

Collmer, A. and Keen, N.T. 1986. The role of pectic enzyme in plant pathogenesis. Annu. Rev. Phytopath. 24: 383–409.

Cosgrove, D.J. 2000. New genes and new biological roles for expansins. Curr. Opin. Plant Biol. 3: 73–78.

Cosgrove D.J., Bedinger, P. and Durachko, D.M. 1997. Group I allergens of grass pollen as cell wall-loosening agents. Proc. Natl. Acad. Sci. USA 94: 6559–6564.

Criqui, M.C., Plesse, B., Durr, A., Marbach, J., Parmentier, Y., Jamet, E. and Fleck, J. 1992. Characterization of genes expressed in mesophyll protoplasts of *Nicotiana sylvestris* before the re-initiation of the DNA replicational activity. Mech. Dev. 38: 121–132.

Davis, K.R., Lyon, G.D., Darvill, A.G. and Albersheim, P. 1984. Host-pathogen interactions. XXV. Endopolygalacturonic acid lyase from *Erwinia carotovora* elicits phytoalexin accumulation by releasing plant cell wall fragments. Plant Physiol. 74: 52–60.

Demura, T. and Fukuda, H. 1993. Molecular cloning and characterization of cDNAs associated with tracheary element differentiation in cultured *Zinnia* cells. Plant Physiol 103: 815–821.

Demura, T. and Fukuda, H. 1994. Novel vascular cell-specific genes whose expression is regulated temporally and spatially during vascular system development. Plant Cell 6: 967–981.

Deyholos, M.K., Cordner, G., Beebe, D. and Sieburth, L.E. 2000. The *SCARFACE* gene is required for cotyledon and leaf vein patterning. Development 127: 3205–3213.

Domingo, C., Gomez, M.D., Canas, L., Hernandez-Yago, J., Conejero, V. and Vera, P. 1994. A novel extracellular matrix protein from tomato associated with lignified secondary cell walls. Plant Cell 6: 1035–1047.

Domingo, C., Roberts, K., Stacey, N.J., Connerton, I., Ruíz-Teran, F. and McCann, M.C. 1998. A pectate lyase from *Zinnia elegans* is auxin inducible. Plant J. 13: 17–28.

Downes, B.P. and Crowell, D.N. 1998. Cytokinin regulates the expression of a soybean β-expansin gene by a post-transcriptional mechanism. Plant Mol. Biol. 37: 437–444.

Du, H., Clarke, A.E. and Bacic, A. 1996. Arabinogalactan-proteins: A class of extracellular matrix proteoglycans involved in plant growth and development. Trends Cell Biol. 6: 411–414.

Dubald, M., Barakate, A., Mandaron, P. and Mache, R. 1993. The ubiquitous presence of exopolygalacturonase in maize suggests a fundamental cellular function for this enzyme. Plant J. 4: 781–791.

Dubreucq, B., Berger, N., Vincent, E., Boisson, M., Pelletier, G., Caboche, M. and Lepiniec, L. 2000. The *Arabidopsis AtEPR1* extensin-like gene is specifically expressed in endosperm during seed germination. Plant J. 23: 643–652

Durrant, W.E., Rowland, O., Piedras, P., Hammond-Kosack, K.E. and Jones, J.D.G. 2000. cDNA-AFLP reveals a striking overlap in race-specific resistance and wound response gene expression profiles. Plant Cell 12: 163.

Ebener, W., Fowler, T.J., Suzuki, H., Shaver, J. and Tierney, M.L. 1993. Expression of *DcPRP*1 is linked to carrot storage root formation and is induced by wounding and auxin treatment. Plant Physiol. 101: 259–265.

Edwards, M.E., Dickson, C.A., Chengappa, S., Sidebottom, C., Gidley, M.J. and Reid, J.S.G. 1999. Molecular characterisation of a membrane-bound galactosyltransferase of plant cell wall matrix polysaccharide biosynthesis. Plant J. 19: 691–697.

Esau, K. 1977. Anatomy of Seed Plants. John Wiley, New York.

Felix, G. and Boller, T. 1995. Systemin induces rapid ion fluxes and ethylene biosynthesis in *Lycopersicon peruvianum* cells. Plant J. 7: 381–389.

Fleming, A.J., McQueen-Mason, S.J., Mandel, T. and Kuhlemeier, C. 1997. Induction of leaf primordia by the cell wall protein expansin. Science 276: 1415–1418.

Fukuda, H. and Komamine, A. 1980. Establishment of an experimental system for the tracheary element differentiation from single cells isolated from the mesophyll of *Zinnia elegans*. Plant Physiol. 5: 57–60.

Fukuda, H. 1996. Xylogenesis: initiation, progression, and cell death. Annu. Rev. Plant Physiol. Plant Mol. Biol. 47: 299–325.

Fukuda, H. 1997. Tracheary element differentiation. Plant Cell 9: 1147–1156.

Fukuda, H. and Komamine, A. 1985. Cytodifferentiation. In: I.K. Vasil (Ed.) Cell Culture and Somatic Cell Genetics of Plants, vol. 2, Academic Press, New York, pp. 149–212.

Gardner, J.M. and Kado, C.I. 1976. Polygalacturonic acid *trans*-eliminase in the osmotic shock fluid of *Erwinia rubrifaciens*: characterisation of the purified enzyme and its effect on plant cells. J. Bact. 127: 451–460.

Gaspar, Y.M., Johnson, K.L., McKenna, J.A., Bacic, A. and Schultz, C.J. 2001. The complex structures of arabinogalactan proteins and the journey towards understanding function. Plant Mol. Biol., this issue.

Hadfield, K.A., Rose, J.K.C., Yaver, D.S., Berka, R.M. and Bennett, A.B. 1998. Polygalacturonase gene expression in ripe melon fruit supports a role for polygalacturonase in ripening-associated pectin disassembly. Plant Physiol. 117: 363–373.

Harrison, M.J., Delmer, D.P., Amor, Y., Grimson, M.J., Johnson, S. and Haigler, C.H. 1997. Localization of sucrose synthase in differentiating tracheary elements of *Zinnia elegans*. Plant Physiol. 114 (Suppl. 3): 349.

Henrissat, B. and Bairoch, A. 1993. New families in the classification of glycosyl hydrolases based on amino acid sequence similarities. Biochem. J. 293: 781–788.

Henrissat, B. and Bairoch, A. 1996. Updating the sequence-based classification of glycosyl hydrolases. Biochem. J. 316: 695–696.

Henrissat, B. and Davies, G.J. 2000. Glycoside hydrolases and glycosyltransferases. Families, modules, and implications for genomics. Plant Physiol. 124: 1515–1519.

Henrissat, B., Coutinho, P. M. and Davies, G. J. 2001. A census of carbohydrate-active enzymes in the genome of *Arabidopsis thaliana*. Plant Mol. Biol., this issue.

Holland, N., Holland, D., Helentjaris, T., Dhugga, K., Xoconostle-Cazares, B. and Delmer, D.P. 2000. A comparative analysis of the plant cellulose synthase (*CesA*) gene family. Plant Physiol. 123: 1313.

Huang, J., Takano, T. and Akita, S. 2000. Expression of α-expansin genes in young seedlings of rice. Planta 211: 467–473.

236

Hughes, J. and Hughes, M.A. 1994. Multiple secondary plant product UDP-glucose glycosyltransferase genes expressed in cassava (*Manihot esculenta* Crantz) cotyledons. DNA Sequences 5: 41–49.

Hutchinson, K.W., Singer, P.B., Diaz-Sala, C. and Greenwood, M.S. 1999. Expansins are conserved in conifers and expressed in response to exogenous auxin. Plant Physiol. 120: 827–832.

Igarashi, M., Demura, T. and Fukuda, H. 1998. Expression of the *Zinnia TED3* promoter in developing tracheary elements of transgenic *Arabidopsis*. Plant Mol. Biol. 36: 917–927.

Im, K.H., Cosgrove, D.T. and Jones, A.M. 2000. Subcellular localization of expansin mRNA in xylem cells. Plant Physiol. 123: 463–470.

Ingold, E., Sugiyama, M. and Komamine, A. 1988. Secondary cell wall formation: changes in cell wall constituents during the differentiation of isolated mesophyll cells of *Zinnia elegans* to tracheary elements. Plant Cell Physiol. 29: 295–303.

Ingold, E., Sugiyama, M. and Komamine, A. 1990. L-α-aminooxy-β-phenylpropionic acid inhibits lignification but not the differentiation to tracheary elements of isolated mesophyll cells of *Zinnia elegans*. Physiol. Plant. 78: 67–74.

Jackson, R.G., Lim, E.-K., Kowalczyk, M., Sandberg, G., Hoggett, J., Ashford, D.A. and Bowles, D.J. 2001. Identification and biochemical characterisation of an *Arabidopsis* indole-3-acetic acid glucosyltransferase. J. Biol. Chem. 276: 4350–4356.

Jenkins, E.S., Paul, W., Coupe, S.A., Bell, S.J., Davies, E.C. and Roberts, J.A. 1996. Characterization of an mRNA encoding a polygalacturonase expressed during pod development in oilseed rape (*Brassica napus* L.). J. Exp. Bot. 47: 111–115.

Jouanin, L., Goujon, T., de Nada, V., Martin, M.-T., Mila, I., Vallet, C. *et al.* 2000. Lignification in transgenic poplars with extremely reduced caffeic acid *O*-methyltransferase activity. Plant Physiol. 123: 1363–1374.

Kalaitzis, P., Solomon, T. and Tucker, M.L. 1997. Three different polygalacturonases are expressed in tomato leaf and flower abscission, each with a different temporal expression pattern. Plant Physiol. 113: 1303–1308.

Keen, N.T. and Tamaki, S. 1986. Structure of two pectate lyase genes from *Erwinia chrysanthemi* EC16 and their high-level expression in *Escherichia coli*. J. Bact. 168: 595–606.

Keller, B. and Lamb, C.J. 1989. Specific expression of a novel cell wall hydroxyproline-rich glycoprotein gene in lateral root initiation. Genes Dev. 3.1639–1646.

Khan, A., Chauhan, Y.S. and Roberts, L.W. 1986. In vitro studies on xylogenesis in citrus fruit vesicles. 2. Effect of pH of the nutrient medium on the induction of cytodifferentiation. Plant Sci. 4: 213–216.

Kim, J.-H., Cho, H.T. and Kende, H. 1999. Presence and expression of an expansin gene in fern *Marsilea quadrifolia*. Abstract 994 of ASPP meeting.

King, G.A. and Davies, K.M. 1995. Cloning of a harvest-induced β-galactosidase from tips of harvested asparagus spears. Plant Physiol. 108: 419–420.

Koizumi, K., Sugiyama, M. and Fukuda, H. 2000. A series of novel mutants of *Arabidopsis thaliana* that are defective in the formation of continuous vascular network: calling the auxin signal flow canalization hypothesis into question. Development 127: 3197–3204.

Lester, D.R., Speirs, J., Orr, G. and Brady, C.J. 1994. Peach (*Prunus persica*) endopolygalacturonase cDNA isolation and mRNA analysis in melting and non-melting peach cultivars. Plant Physiol. 105: 225–231.

Li, Y., Baldauf, S., Lim, E.-K. and Bowles, D.J. 2001. Phylogenetic analysis of the UDP-glucosyltransferase multigene family of *Arabidopsis thaliana*. J. Biol. Chem. 276: 4338–4343.

Lim, E.-K., Li, Y., Parr, A., Jackson, R., Ashford, D.A. and Bowles, D.J. 2001. Identification of glucosyltransferase genes involved in sinapate metabolism and lignin synthesis in *Arabidopsis*. J. Biol. Chem. 276: 4344–4349.

Mahalingam, R., Wang, G. and Knap, H.T. 1999. Polygalacturonase and polygalacturonase inhibitor protein: gene isolation and transcription in *Glycine max-Heterodera glycines* interactions. Mol. Plant-Microbe Interact. 12: 490–498.

Martin, R.C., Mok, M.C. and Mok, D.W. 1999a. Isolation of a cytokinin gene, *ZOG1*, encoding zeatin *O*-glucosyltransferase from *Phaseolus lunatus*. Proc. Natl. Acad. Sci. USA 96: 284–289.

Martin, R.C., Mok, M.C. and Mok, D.W. 1999b. A gene encoding the cytokinin enzyme zeatin *O*-xylosyltransferase of *Phaseolus vulgaris*. Plant Physiol. 12: 553–558.

Martz, F., Maury, S., Pincon, G. and Legrand, M. 1998. cDNA cloning, substrate specificity and expression study of tobacco caffeoyl-CoA 3-*O*-methyltransferase, a lignin biosynthetic enzyme. Plant Mol. Biol. 36: 427–443.

Mau, S.L., Chen, C.G., Pu, Z.Y., Moritz, R.L., Simpson, R.J., Bacic, A. and Clarke, A.E. 1995. Molecular cloning of cDNAs encoding the protein backbones of arabinogalactan-proteins from the filtrate of suspension-cultured cells of *Pyrus communis* and *Nicotiana alata*. Plant J. 8: 269–281.

McCann, M.C. 1997. Tracheary element formation: building up to a dead end. Trends Plant Sci. 2: 333–338.

McCann, M.C., Domingo, C., Stacey, N.J., Milioni, D. and Roberts, K. 2000. Tracheary element formation in an *in vitro* system. In: R. Savidge, J. Barnett and R. Napier (Eds.) Cambium: The Biology of Wood Formation, BIOS Scientific Publishers, Oxford, UK, Chap. 37, pp. 457–470.

McIntyre, C.L., Rae, A.L., Curtis, M.D. and Manners, J.M. 1995. Sequence and expression of a caffeic acid *O*-methyltransferase cDNA homologue in the tropical forage legume *Stylosanthes humilis*. Aust. J. Plant Physiol. 22: 471–478.

McQueen-Mason, S.J. and Rochange, F. 1998. Expansins in plant growth and development: an update on an emerging topic. Plant Biol. 1: 19–25.

Mellerowicz, E.J., Baucher, M., Sundberg, B. and Boerjan, W. 2001. Unravelling cell wall formation in the woody dicot stem. Plant Mol. Biol., this issue.

Mercier, R.W., Chaivisuthangkurar, P. and Gogarten, J.P. 1993. Invertase encoding cDNA from oat. Plant Mol. Biol. 23: 229–230.

Merkouropoulos, G., Barnett, D.C. and Shirsat, A.H. 1999. The *Arabidopsis* extensin gene is developmentally regulated, is induced by wounding, methyl jasmonate, abscisic and salicylic acid, and codes for a protein with unusual motifs. Planta 208: 212–219.

Mohnen, D. 1999. Biosynthesis of pectins and galactomannans. In: D. Barton, K. Nakanishi and O. Meth-Cohn (Eds.) Comprehensive Natural Products Chemistry, vol. 3, Elsevier Science, Amsterdam, pp. 497–527.

Mohnen, D. and Hahn, M.G. 1993. Cell wall carbohydrates as signals in plants. Semin. Cell Biol. 4: 93–102.

Nakashima, J., Takabe, K., Fujita, M. and Fukuda H. 2000. Autolysis during *in vitro* tracheary element differentiation: formation and location of the perforation. Plant Cell Physiol. 41: 1267–1271.

Newman, L.J. and Campbell, M.M. 2000. MYB proteins and xylem differentiation. In: R. Savidge, J. Barnett and R. Napier (Eds.)

Cell & Molecular Biology of Wood Formation, BIOS Scientific Publishers, Oxford, UK, pp. 437–444.

Ohyama, A., Nishimura, S. and Hirai, M. 1998. Cloning of cDNA for a cell wall-bound acid invertase from tomato (*Lycopersicon esculentum*) and expression of soluble and cell wall-bound invertases in plants and wounded leaves of *L. esculentum* and *L. peruvianum*. Genes Genet Syst. 73: 149–157.

O'Malley, D.M., Whetten, R., Bao, W.L., Chen, C.L. and Sederoff, R.R. 1993. The role of laccase in lignification. Plant J. 4: 751–757.

Ori, N., Sessa, G., Lotan, T., Himmelhoch, S., and Fluhr, R. 1990. A major stylar matrix polypeptide (sp41) is a member of the pathogenesis-related proteins superclass. EMBO J. 9: 3429–3436.

Ostergaard, L., Teilum, K., Mirza, O., Mattsson, O., Petersen, M., Welinder, K.G., Mundy, J., Gajhede, M. and Henriksen, A. 2000. *Arabidopsis* ATP A2 peroxidase. Expression and high-resolution structure of a plant peroxidase with implications for lignification. Plant Mol. Biol. 44: 231–243.

Pear, J.R., Kawagoe, Y., Schreckengost, W.E., Delmer, D.P. and Stalker, D.M. 1996. Higher plants contain homologs of the bacterial *celA* genes encoding the catalytic subunit of cellulose synthase. Proc. Natl. Acad. Sci. USA 93: 12637–12642.

Perrin, R.M., DeRocher, A.E., Bar-Peled, M., Zeng, W.Q., Norambuena, L., Orellana, A., Raikhel, N.V. and Keegstra, K. 1999. Xyloglucan fucosyltransferase, an enzyme involved in plant cell wall biosynthesis. Science 284: 1976–1979.

Pilnik, W. 1990. Pectin: a many splendoured thing. In: G.O. Phillips, D.J. Wedlock and P.A. Williams (Eds.) Gums and Stabilizers in the Food Industry, vol. 5, Oxford University Press, Oxford, UK, pp. 209–221.

Reinhardt, D., Witter, F., Mandel, T. and Kuhlemeier, C. 1998. Localized up-regulation of a new expansin gene predicts the site of leaf formation in the tomato meristem. Plant Cell 10: 1427–1437.

Reiter, W.D., Chapple, C. and Somerville, C.R. 1997. Mutants of *Arabidopsis thaliana* with altered cell wall polysaccharide composition. Plant J. 12: 335–345.

Reiter, W.-D. and Vanzin, G. 2001. Molecular genetics of nucleotide sugar interconversion pathways. Plant Mol. Biol., this issue.

Richmond, T.A. and Somerville, C.R. 2000. The cellulose synthase superfamily. Plant Physiol. 124: 495–498.

Richmond, T. A. and Somerville, C.R. 2001. Integrative approaches to determining *Csl* function. Plant Mol. Biol., this issue.

Roberts, A.W. and Haigler, C.H. 1994. Cell expansion and tracheary element differentiation are regulated by extracellular pH in mesophyll cultures of *Zinnia elegans* L. Plant Physiol. 105: 699–706.

Rombouts, F.M. and Pilnik, W. 1980. Pectic enzymes. In: A.H. Rose (Ed.) Economic Microbiology, vol. 5: Microbial Enzymes and Bioconversions, Academic Press, New York, pp. 228–282.

Ryser, U. and Keller, B. 1992. Ultrastructural localisation of a bean glycine-rich protein in unlignified primary walls of protoxylem cells. Plant Cell 4: 773–783.

Santino, C.G., Stanford, G.L. and Conner, T.W. 1997. Developmental and transgenic analysis of two tomato fruit enhanced genes. Plant Mol. Biol. 33: 405–416.

Schindler, T., Bergfeld, R. and Schopfer, P. 1995. Arabinogalactan proteins in maize coleoptiles: developmental relationship to cell death during xylem differentiation but not to extension growth. Plant J. 7: 25–36.

Sheng, J., D'Ovidio, R. and Mehdy, M.C. 1991. Negative and positive regulation of a novel proline-rich protein mRNA by fungal elicitor and wounding. Plant J. 1: 345–354.

Shinohara, N., Demura, T. and Fukuda, H. 2000. Isolation of a vascular cell wall-specific monoclonal antibody recognizing a cell polarity by using a phage display subtraction method. Proc. Natl. Acad. Sci. USA 97: 2585–2590.

Shirzadegan, M., Christie, P. and Seemann, J.R. 1991. An efficient method for isolation of RNA from tissue cultured plant cells. Nucl. Acids Res. 19: 6055.

Sitrit, Y., Downie, B., Bennett, A.B. and Bradford, K.J. 1996. A novel exo-polygalacturonase is associated with radicle protrusion in tomato (*Lycopersicon esculentum*) seeds (abstract 752). Plant Physiol. 111: 161.

Stacey, N.J., Roberts, K., Carpita, N.C., Wells, B. and McCann, M.C. 1995. Dynamic changes in cell surface molecules are very early events in the differentiation of mesophyll cells from *Zinnia elegans* into tracheary elements. Plant J. 8: 891–906.

Sterky, F., Regan, S., Karlsson, J., Hertzberg, M., Rohde, A., Holmberg, A. *et al.* 1998. Gene discovery in the wood-forming tissues of poplar: analysis of 5,692 expressed sequence tags. Proc. Natl. Acad. Sci. USA 95: 13330–13335.

Suzuki, K., Ingold, E., Sugiyama, M. and Komamine, A. 1991. Xylan synthase activity in isolated mesophyll cells of *Zinnia elegans* during differentiation to tracheary elements. Plant Cell Physiol. 32: 303–306.

Suzuki, K., Ingold, E., Sugiyama, M., Fukuda, H. and Komamine, A. 1992. Effects of 2,6-dichorobenzonitrile on differentiation to tracheary elements of isolated mesophyll cells of *Zinnia elegans* and formation of secondary cell walls. Physiol. Plant. 86: 43–48.

Szersen, J.B., Szczyglowski, K. and Bandurski, R.S. 1994. *iaglu*, a gene from *Zea mays* involved in conjugation of growth hormone indole-3-acetic acid. Science 265: 1699–1701.

Taylor, J.G., Owen, T.P., Koonce, L. and Haigler, C.H. 1992. Dispersed lignin in tracheary elements treated with cellulose synthesis inhibitors provides evidence that molecules of the secondary cell wall mediate wall patterning. Plant J. 2: 959–970.

Taylor, J.G. and Haigler, C.H. 1993. Patterned secondary cell-wall assembly in tracheary elements occurs in a self-perpetuating cascade. Acta Bot. Neerland. 42: 153–163.

Taylor, N.G., Scheible, W.R., Cutler, S., Somerville, C.R. and Turner, S.R. 1999. The *irregular xylem*3 locus of *Arabidopsis* encodes a cellulose synthase required for secondary cell wall synthesis. Plant Cell 11: 769–779.

Torki, M., Mandaron, P., Mache, R. and Falconet, D. 2000. Characterization of a ubiquitous expressed gene family encoding polygalacturonase in *Arabidopsis thaliana*. Gene 242: 427–436.

Trainotti, L., Spolaore, S., Pavanello, A., Baldan, B. and Casadoro, G. 1999. A novel E-type endo-β-1,4-glucanase with a putative cellulose-binding domain is highly expressed in ripening strawberry fruits. Plant Mol. Biol. 40: 323–332.

Tsai, C.-J., Popko, J.L., Mielke, M.R., Hu, W.-J., Podila, G.K. and Chiang, V.L. 1998. Suppression of *O*-methyltransferase gene by homologous sense transgene in quaking aspen causes red-brown wood phenotypes. Plant Physiol. 117: 101–112.

Turner, S., Taylor, N. and Jones, L. 2001. Mutations of the secondary wall. Plant Mol. Biol., this issue.

Tymowska-Lalanne, Z. and Kreis, M. 1998. Expression of the *Arabidopsis thaliana* invertase gene family. Planta 207: 259–265.

Vijn, I., van Dijken, A., Luscher, M., Bos, A., Smeets, E., Weisbeek, P. *et al.* 1998. Cloning of sucrose:sucrose 1-fructosyltransferase from onion and synthesis of structurally defined fructan molecules from sucrose. Plant Physiol. 117: 1507–1513.

Vogt, T. and Jones, P. 2000. Glycosyltransferases in plant natural product synthesis: characterisation of a supergene family. Trends Plant Sci. 5: 380–386.

238

Watson, C.F., Zheng, L. and DellaPenna, D. 1994. Reduction of tomato polygalacturonase β subunit expression affects pectin solubilization and degradation during fruit ripening. Plant Cell 6: 1623–1634.

Whetten, R.W., Mackay, J.J. and Sederoff, R.R. 1998. Recent advances in understanding lignin biosynthesis. Annu. Rev. Plant Physiol. Plant Mol. Biol. 49: 585–609.

Whetten, R., Sun, Y.-H., Zhang, Y. and Sederoff, R. 2001. Functional genomics and cell wall biosynthesis in loblolly pine. Plant Mol. Biol., this issue.

Wu, S.C., Blumer, J.M. Darvill, A.G. et al., 1996. Characterization of an endo-β-1,4-glucanase gene induced by auxin in elongating pea epicotyls. Plant Physiol. 110: 163–170.

Wu, L.G., Joshi, S.P. and Chiang, V.L. 2000. A xylem-specific cellulose synthase gene from aspen (Populus tremuloides) is responsive to mechanical stress. Plant J. 22: 495–502.

Wyatt, R.E., Nagao, R.T. and Key, J.L. 1992. Patterns of soybean proline-rich protein gene expression. Plant Cell 4: 99–110.

Xu, J., Pemberton, G.H., Almira, E.C., McCarty, D.R. and Koch, K.E. 1995. The Ivr1 gene for invertase in maize. Plant Physiol. 108: 1293–1294.

Ye, Z.-H, and Varner, J.E. 1993. Gene expression patterns associated with in vitro tracheary element formation in isolated single mesophyll cells of Zinnia elegans. Plant Physiol. 103: 805–813.

Ye, Z.-H. and Varner, J.E. 1995. Differential expression of two O-methyltransferases in lignin biosynthesis in Zinnia. Plant Physiol. 108: 459–467.

Ye, Z.-H., Song, Y.-R., Marcus, A. and Varner, J.E. 1991. Comparative localization of three classes of cell wall proteins. Plant J. 1: 175–183.

Ye, Z.-H., Kneusel, R.E., Matern, U. and Varner, J.E. 1994. An alternative methylation pathway in lignin biosynthesis in Zinnia. Plant Cell 6: 1427–1443.

Zhong, R. and Ye, Z.-H. 1999. IFL1, a gene regulating interfascicular fiber differentiation in Arabidopsis, encodes a homeodomain-leucine zipper protein. Plant Cell 11: 2139–2152.

Zhong, R., Morrison, W.H., Negrelc, J. and Ye, Z.-H. 1998. Dual methylation pathways in lignin biosynthesis. Plant Cell 10: 2033–2046.

Zhong, R., Ripperger, A. and Ye, Z.-H. 2000. Ectopic deposition of lignin in the pith of stems of two Arabidopsis mutants. Plant Physiol. 123: 59–69.

Zrenner, R., Schuler, K. and Sonnewald, U. 1996. Soluble acid invertase determines the hexose-to-sucrose ratio in cold-stored potato tubers. Planta 198: 246–252.

Plant Molecular Biology **47:** 239–274, 2001.
© 2001 *Kluwer Academic Publishers. Printed in the Netherlands.*

239

Unravelling cell wall formation in the woody dicot stem

Ewa J. Mellerowicz[1], Marie Baucher[2], Björn Sundberg[1,*] and Wout Boerjan[3]

[1]*Department of Forest Genetics and Plant Physiology, Swedish University of Agricultural Sciences, 90183 Umeå, Sweden (*author for correspondence; e-mail bjorn.sundberg@genfys.slu.se);* [2]*Laboratory of Plant Biotechnology, Free University of Brussels (ULB), 1160 Brussels, Belgium;* [3]*Department of Plant Genetics, Flanders Interuniversity Institute for Biotechnology (VIB), Ghent University, 9000 Gent, Belgium*

Key words: expressed sequence tag, hybrid aspen, *Populus*, vascular cambium, wood formation, xylem cell wall

Abstract

Populus is presented as a model system for the study of wood formation (xylogenesis). The formation of wood (secondary xylem) is an ordered developmental process involving cell division, cell expansion, secondary wall deposition, lignification and programmed cell death. Because wood is formed in a variable environment and subject to developmental control, xylem cells are produced that differ in size, shape, cell wall structure, texture and composition. Hormones mediate some of the variability observed and control the process of xylogenesis. High-resolution analysis of auxin distribution across cambial region tissues, combined with the analysis of transgenic plants with modified auxin distribution, suggests that auxin provides positional information for the exit of cells from the meristem and probably also for the duration of cell expansion. Poplar sequencing projects have provided access to genes involved in cell wall formation. Genes involved in the biosynthesis of the carbohydrate skeleton of the cell wall are briefly reviewed. Most progress has been made in characterizing pectin methyl esterases that modify pectins in the cambial region. Specific expression patterns have also been found for expansins, xyloglucan endotransglycosylases and cellulose synthases, pointing to their role in wood cell wall formation and modification. Finally, by studying transgenic plants modified in various steps of the monolignol biosynthetic pathway and by localizing the expression of various enzymes, new insight into the lignin biosynthesis *in planta* has been gained.

Abbreviations: 4CL, 4-coumarate:coenzyme A ligase; ACC, 1-aminocyclopropane-1-carboxylate; C3H, coumarate 3-hydroxylase; C4H, cinnamate 4-hydroxylase; CAD, cinnamyl alcohol dehydrogenase; CCoA3H, coumaroyl-coenzyme A 3-hydroxylase; CCoAOMT, caffeoyl-coenzyme A *O*-methyltransferase; CCR, cinnamoyl-coenzyme A reductase; COMT/AldOMT, caffeate/5-hydroxyconiferaldehyde *O*-methyltransferase; DDC, dehydrodiconiferyl alcohol; DMSO, dimethyl sulfoxide; EDTA, ethylenediaminetetraacetic acid; EST, expressed sequence tag; F5H/CAld5H, ferulic acid/coniferaldehyde 5-hydroxylase; FCC, fusiform cambial cell; G, guaiacyl; GA, gibberellin; GCA, O^3-β-D-glucosyl-caffeic acid; GSA, O^4-β-D-glucosyl-sinapic acid; GVA, O^4-β-D-glucosyl-vanillic acid; H, *p*-hydroxyphenyl; IAA, indole-3-acetic acid ; IDDDC, isodihydrodehydrodiconiferyl alcohol; MIOP, *myo*-inositol oxidation pathway; NMR, nuclear magnetic resonance; PAL, phenylalanine ammonia-lyase; PCBER, phenylcoumaran benzylic ether reductase; PCD, programmed cell death; PME, pectin methyl esterase; QTL, quantitative trait locus; RCC, ray cambial cell; RE, radial expansion; S, syringyl; SAM, *S*-adenosyl-L-methionine; SHMT, serine hydroxymethyltransferase; TE, tracheary element; XET, xyloglucan endotransglycosylase

Introduction

Wood is a product of the vascular cambium (Larson, 1994), a lateral meristem that develops in conifer and most dicot land plants but that contributes substantially to plant biomass only in perennial tree species.

Lignified xylem cells, which are functionally competent in water transport and mechanical support, are formed by cambial derivatives that undergo terminal differentiation ended by the autolysis of the cell protoplast. In this review, we will focus on the organismal approach to study wood formation or xylogenesis

Table 1. Most common poplar species and their systematic classification (http://willow.ncfes.umn.edu/silvics_manual/volume_2/populus/populus.htm).

Section	Species
Leuce (aspen type)	*P. grandidentata* (Michx.)
	P. alba (L.)
	P. tremula (L.)
	P. tremuloides (Michx.)
Aigeiros (cottonwood or poplar type)	*P. deltoides* (Bartr. ex. Marsh.)
	P. sargentii (Dode)
	P. fremontii (Wats.)
	P. nigra (L.)
Tacamahaca (balsam poplar type)	*P. balsamifera* (L.)
	P. maximowiczii (Henry)
	P. trichocarpa (Torr. & Gray)
	P. angustifolia (James)
Leucoides (swamp poplar type)	*P. heterophylla* (L.)
Turanga	*P. euphratica* (Olivier)

using functional genomics in a dicot tree. First, the cellular process of wood formation and the variability in wood properties brought about by developmental and environmental stimuli are described. Second, some aspects of hormonal control of wood formation are highlighted. Finally, the current progress in understanding the biosynthesis of the carbohydrate skeleton and lignification of the xylem cell wall is presented. The aim is to put detailed knowledge obtained from less complex systems, such as the xylogenic *Zinnia* cell culture (Fukuda, 1992, 1997; Milioni *et al.*, 2001, this issue), *Arabidopsis* (Baima *et al.*, 1995: Turner and Somerville, 1997; Taylor *et al.*, 1999; Zhong and Ye, 1999), and suspensions of primary walled cells (Takeda *et al.*, 1996; Kakegawa *et al.*, 2000; Thompson and Fry, 2000), into an organismal perspective. This aspect is important to understand fibre biogenesis in a tree and is a basis for the use of genetic engineering and marker-assisted selection to modify wood properties according to human needs.

It is not our intention to cover the extensive older literature on wood development. For this, the reader is referred to the appropriate reviews. Instead, our aim is to highlight current developments and concepts. The literature on poplar has been preferentially reviewed because poplar has emerged as a model tree for research on wood formation.

Populus as a model

A woody perennial model species is needed if we are to understand such phenomena as shoot dormancy, adaptations to deep frost including the deciduous habit, the presence of juvenile and mature phases, and the extensive secondary growth. Poplar (*Populus* spp.) combines the advantages of being a suitable model for experimental research with its economic importance as a forestry species (Klopfenstein *et al.*, 1997). The genus *Populus* belongs to the Salicaceae family and comprises about 30 species that are native to the Northern hemisphere and are classified into five sections (Table 1). *Populus* offers several advantages as a model tree: it grows fast, is easy to propagate and can be transformed (Klopfenstein *et al.*, 1997). Transformation protocols based on *Agrobacterium*-mediated gene transfer have been successfully applied in the sections *Leuce, Tacamahaca* and *Aigeiros* (Kim *et al.*, 1997; Han *et al.*, 1997, 2000). In addition, *Populus* species have a small genome of 1.1 pg/2C, i.e. 550 Mb (Dhillon, 1987; Bradshaw and Settler, 1993; Wang and Hall, 1995), which is only ca. 5-fold larger than that of *Arabidopsis thaliana*, facilitating map-based approaches. Genetic maps have been made for several *Populus* species (Bradshaw *et al.*, 1994; Cervera *et al.*, 2001; Yin *et al.*, 1999) and can be used for quantitative trait locus (QTL) analysis and map-based cloning (Wu and Stettler, 1994, 1997; Wu *et al.*, 1997, 1998; Wu, 1998; Frewen *et al.*, 2000). The detection of QTLs for wood properties is of particular interest.

High-throughput screening methods for wood properties such as computer tomography X-ray densiometry and pyrolysis molecular beam mass spectrometry will increase our knowledge on the variability and genetics of wood characters (Tuskan *et al.*, 1999).

These advantages have given us and others the incentive to initiate functional genomics programs in *Populus*. Our approach is based on large-scale expressed sequence tag (EST) sequencing (Sterky *et al.*, 1998a; http://www.biochem.kth.se/PopulusDB/). Initially, ESTs were obtained from a cDNA library of the cambial region of hybrid aspen (*P. tremula* L. × *P. tremuloides* Michx.), here denoted the cambium library, and a cDNA library of differentiating xylem of black cottonwood (*P. trichocarpa* Torr. & Gray), denoted the xylem library. More recent sequencing from various tissues and organs within the Swedish initiative has currently increased the number of ESTs to 30 000. These libraries are mines for finding genes based on homology searches. About half of the ESTs from the cambium and xylem libraries correspond to genes for which no function has been described in any other system (Sterky *et al.*, 1998a). Several of these genes are apparently expressed at a high level and may represent novel enzymes (Table 2). Genes related to cell wall formation are represented by 4% and 7% of the ESTs from the cambium and xylem library, respectively. The functional analysis of genes involved in wood formation requires knowledge of the cellular and developmental context in which they operate, which is described briefly below.

Wood cell wall formation: a morphological perspective

The meristematic stage: the cambium

Xylogenesis is initiated in the vascular cambium. The term cambium is defined here as a tissue comprising meristematic cells organized in radial files, which give rise to the secondary xylem and phloem (Larson, 1994; Figure 1). Conceptually, each file contains one initial cell that remains in the meristem, cells that are destined to become phloem, called phloem mother cells, and cells destined to become xylem, called xylem mother cells. The initial cells set the pattern of meristem organization by regulating the number of radial files through anticlinal divisions and by establishing the direction of intrusive tip growth resulting in the spiral fibre orientation that is frequently seen in

wood (Wloch and Polap, 1994). It has been suggested that the initials function as a reservoir of genetically sound cells (stem cells) by keeping the frequency of periclinal cell division low and that most cell divisions occur in the mother cells (Gahan, 1988); however, experimental evidence supporting this idea is scarce. The initials and the mother cells are cytologically identical (except for a small difference in length) and there is no evidence for their determination (reviewed by Larson, 1994; Savidge, 2000). Because most published data do not distinguish between the initials and the mother cells of the cambium, the term cambial cells will be used hereafter to denote both the initials and the mother cells.

Whereas the main function of the cambium is cell division and setting out patterns for differentiation similar to other meristems, several aspects are unique to the vascular cambium. Unlike apical meristems, the cambium is a complex tissue containing two morphologically distinct cell types: axially elongated fusiform cambial cells (FCC) and somewhat isodiametrical ray cambial cells (RCC) (Figure 1). These cells give rise to the axial and horizontal cell systems in the secondary xylem and phloem. The identity of cambial cells is determined by positional cues rather than by cell lineage, because the interconversion between a FCC and a RCC is a common phenomenon (reviewed by Iqbal and Ghouse, 1990; Larson, 1994). Most divisions of FCC (90% in *Acer pseudoplatanus* (L.); Catesson, 1964) are periclinal, i.e. new cells are added within a radial file towards either the secondary xylem or the secondary phloem (Figure 1B). The wall that needs to be formed at each periclinal division of a FCC is the largest possible partition within the cell unlike in other cell types where it is usually the smallest possible partition. To cope with such a task, the rate of biosynthesis of the cell plate, the middle lamella, the plasma membrane and the primary cell wall must be exceptionally high in the rapidly growing cambium. This tissue may therefore be the richest source of mRNA and proteins involved in these biosynthetic processes. Indeed, ultramicroscopy of FCC has revealed all features characteristic for high rates of protein biosynthesis and secretory activity (Catesson, 1990).

The speed of new periclinal wall formation in the FCC of *Pinus strobus* (L.) varies from 47 to 105 μm/h, as estimated by the progress of the phragmoplast. This makes the time of completing the phragmoplast movement ca. 20–40 h in this species (Wilson, 1964). Shorter FCCs of dicot trees would probably require slightly shorter times. Nevertheless, the formation of

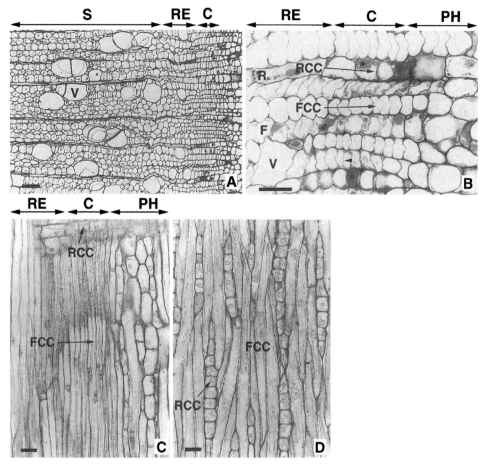

Figure 1. Vascular cambium of hybrid aspen. A. Overview of the cambial region in a transverse section. B–D. Transverse (B), radial (C) and tangential (D) views of the cambium. FCC, fusiform cambial cell; RCC, ray cambial cell; C, cambium; R, ray; RE, the zone of radial cell expansion; S, the zone of secondary wall deposition; PH, phloem; V, vessel element; F, fibre. The arrowhead in B points to a recent periclinal division. Sectioned material provided by Laurence Puech and Siegfried Fink. Bars: A = 50 μm, B, C and D = 20 μm.

the periclinal cell plate and later on the primary cell wall probably limits the frequency of cell divisions. The shortest average duration of a cell cycle for a population of FCCs across the cambium has been estimated to be 7–11 days by counting the number of newly formed cells within a defined period in various conifer species (Mellerowicz *et al.*, 1992; Larson, 1994). Therefore, when high rates of cell formation are required, trees must increase the size of the cambial cell population. Indeed, a good correlation has been observed between the number of cambial cells and the rate of xylem cell formation (Gregory, 1971; Uggla *et al.*, 1998).

Periclinal divisions create the additional difficulty of guiding the phragmoplast and the newly formed cell plate towards the upper and lower ends of a FCC. Furthermore, the new cell plate must traverse

the large central vacuole present in FCCs. Sinnott and Bloch (1940) were the first to observe a rim of the cytoplasm in the plane of the future cell plate in vacuolated cells, which they called a phragmosome. The phragmosome has been also found in FCCs and it contained longitudinally oriented microtubules that probably function as guides for the movement of the phragmoplast in the FCCs (Goosen-de Rao *et al.*, 1984). Vesicles containing cell wall material are guided along perpendicularly oriented microtubules of the phragmoplast and coalesce at the equatorial plane to release their contents, which form the cell plate. The cell plate is probably made of callose initially and of cellulose at a later developmental stage, as observed in apical meristems and in tobacco BY2 cells (Northcote *et al.*, 1989; Samuels *et al.*, 1995; Vaughn *et al.*, 1996; Sonobe *et al.*, 2000). Xyloglucan,

Table 2. Assembled clusters that correspond to the highest expressed genes from the cambium and the xylem libraries. Total number of ESTs in each cluster and their frequency in each library is given. Reprinted from Sterky *et al.* (1998a) with permission.

Putative gene identification	Number of ESTs	%
Cambial-region ESTs		
Cyclophilin	30	0.62
Unknown (I)	28	0.58
Translationally controlled tumour protein	26	0.54
Unknown (II)	23	0.48
Blue copper protein	21	0.44
ADP ribosylation factor (I)	21	0.44
HMG protein 1	17	0.35
ADP ribosylation factor (II)	16	0.33
Developing-xylem ESTs		
Nodulin	17	1.93
Laccase	14	1.59
Unknown	10	1.13
S-adenosyl-L-methionine synthase	8	0.91
Elongation factor 1-α	7	0.79
14-3-3-like protein	7	0.79

rhamnogalacturonans, and arabinogalactans have also been immunolocalized to the developing cell plates in these systems. The nascent periclinal walls of FFCs are very resistant to ethylenediaminetetraacetic acid (EDTA) and dimethylsulfoxide (DMSO) extraction, indicating that they do not contain acidic pectins or cross-linking glycans but are rather of cellulosic nature with a high content of methylated pectin (Catesson, 1989, 1990; Catesson et al., 1994; Chaffey et al., 1997b). The merge of the nascent periclinal wall with the pre-existing radial wall is accompanied by local digestion of the radial cell wall until the middle lamella is reached (Catesson and Roland, 1981). The zone containing acidic pectins in the middle lamella of the radial wall then becomes continuous with the middle lamella of the newly formed tangential walls.

The production of xylem cells displaces the cambium centrifugally (outwards) necessitating its extension in girth. This process is accomplished by anticlinal divisions of cambial initials, where the newly formed cell wall is placed radially, thus initiating new radial cell files. The number of newly formed radial files always exceeds the number of new files needed. Therefore, most new files are lost from the meristematic population, which is accomplished by differentiation of their initial cells to xylem elements

(reviewed by Larson, 1994). This mechanism may be important for the elimination of somatic mutations from the meristematic population, a feature probably important for long-lived organisms such as trees (Gahan, 1988; Klekowski and Godfrey, 1989). Anticlinal divisions play an important role in regulating the length of fusiform initials, and indirectly also fiber length (Larson, 1994). This results from the oblique (pseudotransverse) orientation of the newly formed radial wall, which leads to the formation of daughter cells that are shorter than their mother cell. The length of these fusiform initials then gradually increases over several periclinal divisions via intrusive tip growth until it is reduced again by the next anticlinal division. The average frequency of anticlinal divisions for individual initials may be as low as one every two years up to several divisions per year (reviewed by Larson, 1994).

Meristematic activity is often regarded as an important determinant of growth rate. Considering the functioning of the vascular cambium, two aspects appear to be important for the rate of wood production: (1) the number of xylem mother cells, which depends on whether the initials form xylem or phloem mother cells and how fast the xylem mother cells exit from the meristematic zone, and (2) the duration of the cell cycle in xylem mother cells. Each of these aspects may be individually targeted to maximize the rate of wood production.

Stages of wood differentiation: cell expansion and secondary wall deposition

In *Populus*, identical FCCs give rise to three cell types: vessel elements, fibres, and axial parenchyma, whereas identical RCCs produce two cell types: contact and isolation ray cells. How and when cell fate becomes established is one of the most intriguing issues of xylogenesis. In several dicot species, including poplar, cell wall properties vary among cambial cells and their immediate derivatives (i.e. xylem or phloem, vessel elements or fibres), as determined by their differential susceptibility to extraction with solvents that solubilize pectin and matrix glycans. This suggests that cell fate has already been determined at this early stage (Catesson and Roland, 1981; Catesson, 1989; Catesson et al., 1994). However, the exact molecular nature of these differences is not yet fully understood and molecular markers for the various cell types are not yet available. The absence of plasmodesmatal connections between vessel elements and other cell types during the early stages of differentiation has been suggested as a mechanism of determination of these cells (reviewed by Barnett, 1995). Careful investigations in hybrid aspen, however, have now revealed plasmodesmatal connections between vessel elements and other cell types, but their frequency was very low (K. Pickering and J. Barnett, personal communication). After xylem mother cells have left the meristem, they enlarge while in the primary walled stage. This phase corresponds to the radial expansion (RE) zone (Figure 1). It is in this zone where vessel elements and fibres clearly display different morphological characteristics.

The zone of radial cell expansion

All cells in the RE zone undergo enlargement, with cell types differing in the extent, polarity, and type of enlargement. Fibres and axial parenchyma cells expand primarily in the radial direction. In this case, their radial walls expand uniaxially. Vessel elements, in addition to radial expansion, may undergo substantial tangential growth that is accomplished by the uniaxial extension of radial walls and the lateral displacement of adjacent cells. No growth of the tangential wall is observed except for the tangential walls of contacting vessel elements (Catesson, 1989; Barnett, 1992; Catesson et al., 1994). In contrast, the end walls of vessel elements and axial parenchyma cells undergo multiaxial expansion. Fibre elongation is achieved by intrusive tip growth and requires local wall biogenesis and dissolution of the middle lamellae between neighbouring cells. In P. deltoides (Bartr. ex Marsh.), fibre elongation by ca. 1.5-fold was observed during their differentiation from FCCs (Kaeiser, 1964), but in many dicot species the length of the xylem fibres exceeds several-fold that of FCCs. Tip growth is already evident in the cambium, but is most intense in the RE zone (Wenham and Cusick, 1975). It has been suggested that the formation of Ca^{2+}-bound pectins in the middle lamella of neighbouring cells may limit the penetration of the intrusively growing fibre tip (Catesson et al., 1994; Guglielmino et al., 1997a). Ray cells elongate radially by uniaxial wall expansion. Remarkably, the basic pattern of cell arrangement between the different cell types of the xylem is largely conserved despite their different patterns of expansion.

Because the degree of enlargement varies among cell types, mechanisms must exist that differentially regulate cell turgor pressure and/or cell wall plasticity. Numerous observations in dicot trees, including poplar, indicate that changes in the amount and composition of pectins play an important role in radial cell expansion (Roland, 1978; Catesson and Roland, 1981; Catesson, 1989; Barnett, 1992).

Secondary wall deposition

All xylem cells in Populus form a secondary cell wall that is deposited when the radial expansion is completed. In both conifers and dicots, the orientation of cellulose microfibrils in the primary wall is usually random or longitudinal (reviewed by Funada, 2000). The deposition of the secondary wall is marked by the formation of a dense array of helical, almost transverse cellulose microfibrils, which limit the further radial expansion, as demonstrated in Abies sachalinensis (Masters) (Abe et al., 1997). In tracheids of conifers and in fibres of dicot trees, successive layers of the secondary wall have orderly arranged cellulose microfibrils that form a helicoidal or semi-helicoidal structure (Abe et al., 1995; Prodhan et al., 1995; Awano et al., 2000; Funada, 2000). The first formed S1 layer has a flat, almost transverse microfibril angle that gradually changes clockwise, as seen from the cell lumen, to the longitudinal arrangement that characterizes the thickest S2 layer. Finally, the S3 layer is formed after an abrupt reorientation of the microfibrils back to the transverse helix. The changes in microfibril orientation during the formation of successive cell wall layers are always paralleled by the reorientation of cortical microtubules (Abe et al., 1995; Prodhan et al., 1995; Chaffey et al., 1997a, 1999; Funada, 2000). Drugs that disrupt the microtubules, such as colchicine, abolish the normal pattern of secondary wall thickening (Torrey et al., 1971). These observations support the hypothesis that microtubules control the orientation of cellulose microfibrils in xylem cells. Successive changes in microtubule density and orientation have also been observed in developing fibres of hybrid aspen (Figure 2; N. Chaffey, P. Barlow and B. Sundberg, unpublished).

In Populus, the formation of the secondary cell wall starts first in vessel elements and their contact cells (Murakami et al., 1999). The vessel elements have a three-layered secondary cell wall similar to that of the fibres, but the S2 layer is proportionally thinner (Harada and Côté, 1985). The ray cells also develop secondary walls with three S layers (Fuji et al., 1979; Harada and Côté, 1985; Murakami et al., 1999). Contact ray cells develop a tertiary wall, called the protective layer, over the secondary lignified wall after autolysis of the contacting vessel element (Benayoun, 1983). The protective layer remains non-lignified

Figure 2. Reorientation of microtubules during fibre differentiation in hybrid aspen. The radial section through the cambial region shows fibres at successive stages of secondary wall deposition. Microtubules are stained by indirect immunofluorescence. The vascular cambium is to the left. Illustration provided by Nigel Chaffey. Bar = 20 μm.

(Murakami *et al.*, 1999), and is rich in cross-linking glycans (Fujii *et al.*, 1981). It forms the tyloses in non-functional vessels.

Longitudinal co-ordination of vessel differentiation
The formation of cell contacts and perforations as well as the differentiation of the vessel members must occur in a coordinated fashion to ensure root-to-shoot water transport. Hundreds of vessel elements may be joined end-to-end to form a functional vessel in wood (reviewed by Butterfield, 1995). Pits and perforations become evident in expanding vessel elements as microtubule-free areas (Chaffey *et al.*, 1999; Funada, 2000). Secondary walls develop around pits and perforations and subsequently lignify, but no secondary wall deposition or lignification is observed in the pit/perforation area (Butterfield 1995). Whereas simple pits develop over pre-existing plasmodesmata in primary pit fields, the bordered pits develop without any apparent relation to the plasmodesmatal connections (Barnett and Harris, 1975; Yang, 1978). Before the protoplast of a vessel element is autolysed, the primary wall that closes the perforation appears swollen, presumably because of local wall restructuring and enzymatic hydrolysis (Butterfield, 1995). However, no specific enzymatic activity responsible for the formation of the perforation has been identified. In species that have simple perforation plates, including *Populus*, the nature of vessel end walls is very different from that of the lateral walls (Benayoun *et al.*, 1981). Almost all end wall material, except for the pectins

in the middle lamella, is DMSO-extractable, indicating that the end walls contain mostly cross-linking glycans and very little cellulose. The non-cellulosic material is removed first during the formation of the perforation plate, leaving webs of naked cellulose microfibrils that can sometimes be observed by microscopy (Butterfield, 1995). The subsequent steps are not well documented and probably occur quickly. The fibrilar rim of the primary wall may remain after autolysis of the vessel element and its final removal is possibly accomplished by the surge of the transpiration stream. Within one vessel, the primary wall forming the perforation plates might still be present in some elements while it is already removed in others (Butterfield, 1995).

Lignification and programmed cell death
Lignification starts in vessel elements and contact ray cells that have secondary cell walls, whereas fibres and isolation cells lignify later (Murakami *et al.*, 1999). Lignin is first detected in the middle lamella, particularly at cell corners, when cells have completed the deposition of the S1 layer (Bailey, 1954; Terashima *et al.*, 1993). Lignification progresses inwards during the S2 layer formation, concomitantly with cellulose, mannan and xylan deposition. Lignin deposition is most intense when the S3 layer is formed and it progresses towards the cell lumen until all the wall layers are lignified.

Lignin composition changes during cell differentiation in relation to the distance from the cambium; the closer to the cambium, the more *p*-hydroxyphenyl (H) and guaiacyl (G) units are present in lignin, whereas the further away from the cambium, the more syringyl (S) units are incorporated (Terashima *et al.*, 1979, 1993; Yoshinaga *et al.*, 1997b). Thus, cells that complete lignification closer to the cambium would have more H and G lignin. The composition of lignin affects the interaction between lignin and the carbohydrate components of the cell walls. The H and G lignin interact with pectins, while the S lignin form links with cross-linking glycans (Fukuda and Terashima, 1988). In lignified cells, pectin is difficult to extract, stain and label with pectin-specific antibodies (Guglielmino *et al.*, 1997a; Hafren *et al.*, 2000), but it is still present as demonstrated by the presence of radiolabelled galacturonan upon feeding with radiolabelled pectin precursors (Imai and Terashima, 1992). When lignification is completed, vessel elements undergo programmed cell death (PCD), which involves the hydrolysis of the protoplast. Contact ray cells, which produce secondary cell walls and lignify concomitantly with the vessel elements, do not initiate PCD, indicating that PCD is regulated independently from secondary cell wall formation and lignification.

Cell types: wall properties and chemical composition

Within the wood-forming tissues, cell wall composition differs between the primary- and secondary-walled stage as determined by the analysis of bulked tissue samples. Cell walls derived from the cambium, and the primary-walled developing xylem and phloem tissues of *P. tremuloides* contain 47% pectin, 22% cellulose, 18% matrix glycans, 10% protein, and 3% other material (Simson and Timell, 1978a). In contrast, the secondary-walled wood from *P. nigra* (L.) is composed of 48% cellulose, 23% matrix glycans, 19% lignin, and 10% other material (McDougall *et al.*, 1993). Some changes in dry mass composition associated with the transition from primary- to secondary-walled stage are listed in Table 3. The most noticeable change, however, is the decrease in cell wall water content. This is related to a decrease in wall porosity as a result of the tight packing of cellulose microfibrils and the heavy cross-linking of lignin that fills all available space and concomitantly displaces water (Fujino and Itoh, 1998).

In terms of cell types, 33% v/v of poplar wood is composed of vessel elements, 53–55% of fibres, 11–14% of ray parenchyma and less than 1% of axial parenchyma (Panshin and de Zeeuw, 1980). These cell types differ in wall structure and chemical composition as observed in other dicot trees. For example, vessel elements in oak contain more mannose, xylose, rhamnose, and arabinose but less glucose and galactose than do fibres (Yoshinaga *et al.*, 1993). In several other species, vessel elements contain more polyphenols (Watanabe *et al.*, 1997) and have a lower S/G ratio of lignin monomers when compared to fibers and ray parenchyma cells (Fergus and Goring, 1970; Saka and Goring, 1985; Yoshinaga *et al.*, 1993, 1997a, b). Within a complex tissue such as wood, techniques such as immunolabelling with specific antibodies and lectins (Wojtaszek and Bolwell, 1995; Knox, 1997), UV microspectrometry (Goto *et al.*, 1998) or Fourier transformed infrared microspectroscopy (McCann and Roberts, 1994) are required to determine the precise chemical composition of individual cells.

Variation in wood cells and their wall properties: positional and environmental controls

Because wood is formed in a variable environment and subject to developmental control, xylem cells are produced that differ in size, shape, cell wall structure, texture, and chemical composition. These aspects have to be taken into account when genetically modified trees are compared or the effects of various treatments on wood formation are studied. Below, some important sources of within-tree variation are described.

Juvenile/mature wood and longitudinal variability

During the first years of cambial growth, relatively rapid cellular changes take place within the cambium, which are reflected in its derivatives. For example, the length of the FCCs increases from one year to the next, as deduced from the progressive increase in the length of xylem cells derived from these cells (Hejnowicz and Hejnowicz, 1958). Wood produced during this period is called juvenile. In addition to the shorter cell length, juvenile wood is characterized by a lower crystallinity of the fibers, a larger microfibril angle, thinner secondary walls, a higher density of vessels (number per mm^2), and a lower proportion of latewood than the mature wood produced in the subsequent years (reviewed by Zobel and van Buijtenen, 1989; Kroll *et al.*, 1992; Parresol and Cao, 1998). In poplar, the juvenile wood is produced during the first 5 to 10 years of cambial growth (Liese and Ammer, 1958), when the cambium is situated close to foliated branches.

As a consequence of the gradual transition from juvenile to mature wood, a longitudinal variation is observed in fiber and vessel length along the tree trunk in *Populus* (Kaeiser and Stewart, 1955; Hejnowicz and Hejnowicz, 1958; Boyce and Kaeiser, 1961), similar to other tree species. In a thorough analysis of xylem cell length in a single trunk of *P. tremula*, Hejnowicz and Hejnowicz (1958) found that there was no basipetal increase in the degree of intrusive tip growth of the fibers. Therefore, the observed basipetal increase in fibre length can be explained by a comparable increase in the length of the FCCs. They also observed that the FCC length was primarily determined by cambial age rather than by its distance from the ground (Hejnowicz and Hejnowicz, 1958). Similarly, the radial diameter of vessel elements increases basipetally along the trunk and centrifugally from pith to bark (Kroll *et al.*, 1992). Thus, the xylem cell volume increases concomitantly with cambial age.

Table 3. Chemical composition of cell walls in poplar xylem.

Stage of xylem development	Component (% dry weight)	Composition	Reference
Primary-walled stage	Pectins (47%)	galacturonic acid, galactose, arabinose and rhamnose in a molar ratio 5:2:2:1 Main chain: rhamnogalacturonan [$(1\rightarrow4)$-D-galacturonosyl-$(1\rightarrow2)\alpha$-L-rhamnosyl-]. Side-chains branching from rhamnose units at carbon 4 contain: $(1\rightarrow4)\beta$-D-galactan and $(1\rightarrow5)$-L-arabinan, with occasional terminal arabinose, or a fucose	Simson and Timell, 1978a,c
	Cellulose (22%)	$(1\rightarrow4)\beta$-D-glucan, degree of polymerization 4200	Simson and Timell, 1978a,d
	Xylan (11%)	$(1\rightarrow4)\beta$-D-xylan	Simson and Timell, 1978a
	Xyloglucan (6%)	3 species giving an average molecular mass of 62 kDa Main chain: $(1\rightarrow4)\beta$-D-glucan Side-chains (70% of the glucose residues are substituted): xylose residues, some of which are further linked to a terminal galactose, galactose–galactose or galactose–fucose	Simson and Timell, 1978a;b
	Glucomannan (1%)	$(1\rightarrow4)\beta$-D-glucan, $(1\rightarrow4)$-β-D mannan oligomeric units	Simson and Timell, 1978a
	Proteins (10%)	Various structural proteins and cell wall bound enzymes	Simson and Timell, 1978a
Mature wood (secondary wall plus the middle lamella and the primary wall)	Cellulose (43–48%)	$(1\rightarrow4)\beta$-D-glucan, degree of polymerization 9300	Goring and Timell, 1962; Panshin and de Zeeuw, 1980; McDougall *et al.*, 1993
	Xylan (18%–28%)	Main chain: $(1\rightarrow4)\beta$-D-xylan Side-chains: $(1\rightarrow2)\beta$-4-*O*-methyl-α-D-glucuronic acid, acetyl and arabinosyl residues	Jones *et al.*, 1961, Bolwell, 1993; Panshin and de Zeeuw, 1980; McDougall *et al.*, 1993
	Glucomannan (5%)	$(1\rightarrow4)\beta$-D-glucan, $(1\rightarrow4)\beta$-D-mannan oligomer units in a molar ratio 1:2	Sultze, 1957; Northcote, 1972; Panshin and de Zeeuw, 1980
	Pectin and xyloglucan (3%)	As in primary-walled stage	Panshin and de Zeeuw, 1980
	Lignin (19–21%)	H, G and S units	Panshin and de Zeeuw, 1980; McDougall *et al.*, 1993

248

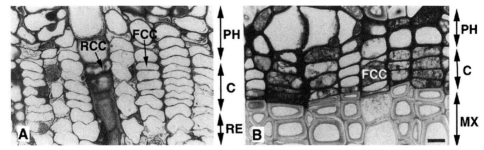

Figure 3. Active (A) and dormant (B) cambium of *P. tremula*. Note the thickened cell walls and the absence of the RE zone in the dormant cambium. MX, mature xylem; other labels as in Figure 1. Bar = 10 μm.

Seasonal variability

Both in conifers and dicot trees, cambial cell walls are thick during the dormant period and thin during the period of active growth (Figure 3) (Catesson, 1964; Riding and Little, 1984; Funada and Catesson, 1991; Chaffey *et al.*, 1998). The question arises whether this thickening represents storage of material that could be metabolized in spring. Storage walls are typically observed in seeds and contain cross-linking glycans and pectins that serve as metabolic reserves during seed germination (Brett and Waldron, 1996). In contrast to seed storage walls, cambial walls of *Aesculus hippocastanum* (L.) display a multi-layered 'herring bone' structure during dormancy, as visualized by transmission electron microscopy, indicating the deposition of layers of helicoidal microfibrils (Chaffey *et al.*, 1998). This is clearly different from the random orientation of microfibrils in the cell walls of the active cambium. Moreover, the cortical microtubules were found to be axially oriented during dormancy and randomly during the period of active growth, conforming to the pattern observed for the microfibrils (Chaffey *et al.*, 1998). Thus, the cellulose microfibrils appear to be one of the components of the thickened cell walls. In poplar, dormant cambial cell walls are enriched in glucan- and xylan-containing hot-water-extractable polysaccharides (Baïer *et al.*, 1994; Vietor *et al.*, 1995; Ermel *et al.*, 2000).

The seasonal cycle of cambial activity and dormancy correlates also with changes in the nature of pectins. Dormant cambium contains less hot-water-extractable pectin than active cambia (Baïer *et al.*, 1994; Ermel *et al.*, 2000). Studies on outdoor-grown *P. × euramericana* (Dode Guinier) and greenhouse-grown hybrid aspen (under natural photoperiod and temperature) indicate an increase in pectin methylation during cambial growth and an up-regulation of the demethylating enzyme, pectin methyl esterase

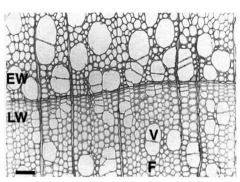

Figure 4. Variability in the secondary xylem of *P. tremula* at the latewood/earlywood boundary. Note the change in size and density of vessel elements. EW, earlywood; LW, latewood; other labels as in Figure 1. Bar = 50 μm.

(PME), during dormancy (Baier *et al.*, 1994; Ermel *et al.*, 2000; Follet-Gueye *et al.*, 2000). Interestingly, a basic isoform of PME, which was active at pH 7.0 but not at pH 6.1, was specifically found in the dormant cambium (Micheli *et al.*, 2000a). A similar basic PME isoform, active at pH 7 to 11, was shown to be specifically induced by the dormancy-breaking chilling treatment in cedar seeds (Ren and Kermode, 2000). It is possible that the B4 form of hybrid aspen PME is involved in the chilling response required to break cambial dormancy (Olsson and Little, 2000).

The time during the growing season at which cambial derivatives are formed affects their properties, thus creating variability within each annual ring. This variability is referred to as the earlywood/latewood continuum. The earlywood and the latewood are formed during the first part and at the end of the growing season, respectively. In poplar, diffuse porous wood is formed that is characterized by a gradual decrease in vessel size and density from earlywood to latewood within an annual ring and by an abrupt increase at the annual ring boundary (Figure 4). Several other features distinguish earlywood and latewood in

Populus. Xylem cell length increases from a minimum in earlywood to a maximum in latewood and sharply decreases at the annual ring boundary (Bissett and Dadswell, 1950; Liese and Ammer, 1958; Hejnowicz and Hejnowicz, 1958). This pattern is explained both by the progressive elongation of FCCs during the growing season before the onset of anticlinal cell divisions in late summer and by the more extensive intrusive fibre tip growth during latewood differentiation. Cell wall crystallinity increases from earlywood to latewood (Parresol and Cao, 1998). In *Fagus crenata* (Bl.), the lignin composition follows a seasonal pattern in vessel elements different from that in fibres (Takabe *et al.* 1992). Within each growing season, the S/G ratio increases in vessel elements whereas it decreases in fibres.

Gravity effects: tension wood

Leaning stems and branches of most dicot trees, including poplar, develop tension wood at their upper side. The extent of tension wood formation correlates with the degree of leaning (Ohta, 1979). Upright-growing stems usually form scattered bands or even isolated tension wood fibers, probably because of bending by wind or growth adjustment (Kaeiser, 1955). Kroll *et al.* (1992) found that 22–63% of the fibres in straight trunks of *P. balsamifera* (L.) had characteristics of tension wood fibres and these characteristics were more often seen in earlywood than in the latewood and in upper parts of the trunk than in the lower parts.

Tension wood contracts longitudinally and pulls the stem back to the vertical position. It is characterized by a high rate of cell production on the tension wood side, which may lead to stem eccentricity (Ohta, 1979). At the same time, wood formation on the opposite side of the stem is inhibited. Other anatomical features include the reduction in vessel size and density and the formation of gelatinous instead of ordinary fibres (Araki *et al.*, 1982). These fibres form a gelatinous cell wall layer (G layer) over the secondary wall inside the cell. In *Populus*, the G layer is formed over the partially developed S2 layer (Figure 5.; Nobushi and Fujitta, 1972; Araki *et al.*, 1982). It is composed of almost pure, high-crystalinity cellulose and has a low lignin content (Norberg and Meier, 1966; Bentum *et al.*, 1969). The microfibril orientation is longitudinal and parallel to the microtubules (Nobushi and Fujitta, 1972; Fujita *et al.*, 1974). The length of gelati-

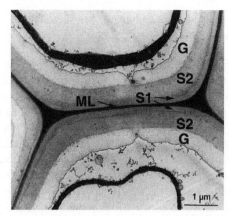

Figure 5. Ultrastructure of developing gelatinous fibres of hybrid aspen. Note the G-layer inside the S2 wall layer. ML, compound middle lamella. Illustration provided by Kathryn Pickering and John Barnett.

nous fibres does not differ significantly from normal fibres (Kaeiser and Stewart, 1955).

Hormones as molecular transducers of positional and environmental information for xylem cell formation in poplar

Plant hormones are key regulators in development and pattern formation, and environmental stimuli often act on plant growth by modulating the hormonal balance. When applied exogenously, plant hormones have been observed to affect most aspects of cambial growth, such as cell division, cell expansion, final cell morphology, the induction of differentiation into different cell types, and cell wall chemistry (Little and Savidge, 1987; Aloni, 1991; Little and Pharis, 1995; Sundberg *et al.*, 2000). This demonstrates the importance of hormones as endogenous regulators of xylem formation and implies that they are involved in controlling at least some aspects of the variation in wood properties described above. However, in most experiments with exogenous hormones, the resulting internal hormone balance in the experimental tissue is not known. This complicates the interpretation of the observed effects of hormonal treatments, and has resulted in conflicting views about the role of hormones in xylem formation. Here we will highlight recent progress on the role of hormones in intact tissues by using modern analytical techniques and molecular tools.

The auxin gradient and its role in cambial cell division and expansion

Auxin is a key signal in xylogenesis. It is the only plant hormone that on its own is sufficient to induce differentiation of vascular elements when applied to plant tissues (Roberts, 1988). Auxin can induce tracheary element (TE) differentiation without prior cell division when applied to *Zinnia* cultures (Fukuda and Komamine, 1980). In contrast, the differentiation of auxin-induced TEs *in planta* is often preceded by cell division (Sussex *et al.*, 1972; Phillips and Arnott, 1983), similar to normal development of vascular bundles. During secondary growth, auxin depletion results in a dedifferentiation of FCC to axial parenchyma cells by transverse divisions (Evert and Kozlowski, 1967; Evert *et al.*, 1972; Savidge, 1983). Thus, a continuous supply of polarly transported auxin is required to maintain the identity of FCC. Numerous experiments have also demonstrated that the application of auxin to cambial tissues stimulates xylem production, i.e. cambial cell division (Sundberg *et al.*, 2000). In a few cases, the indole-3-acetic acid (IAA) level that is induced in the cambial region after exogenous feeding through the polar transport system has been quantified. Both in Scots pine (Sundberg and Little, 1990; Wang *et al.*, 1997) and hybrid aspen (B. Sundberg, unpublished data), feeding with different IAA concentrations resulted in a range of internal IAA levels that was within the variation found in intact plants and showed a positive correlation with the extent of xylem formation. It was further demonstrated that the endogenous supply of IAA in intact conifer shoots was sub-optimal for xylem production, indicating that it plays a regulatory role (Sundberg and Little, 1990).

High-resolution analysis of endogenous IAA in Scots pine and hybrid aspen has demonstrated steep radial concentration gradients across developing xylem and phloem (Uggla *et al.*, 1996; Tuominen *et al.*, 1997). The IAA concentration peaked in the cambium and its most recent derivatives, and reached low levels in maturing xylem and phloem (Figure 6). The visualization of an auxin gradient across developing tissues supports the idea that auxin has a function in positional signalling; cambial derivatives would develop according to their position along the auxin gradient and neighbouring cell files will receive the same information and develop in synchrony. According to this idea, auxin would modify cambial growth by influencing developmental patterns (Sundberg *et al.*, 2000). The cambium would be positioned within a

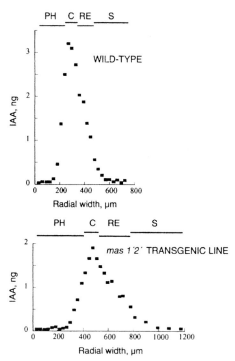

Figure 6. Radial distribution of IAA across the cambial region of hybrid aspen. Each data point represents the IAA content in 1 cm^2 of a 30 μm thick, tangential section. Horizontal bars indicate the approximate position of developmental zones. The IAA distribution is modified by the expression of IAA biosynthetic genes from *Agrobacterium tumefaciens* under the control of the mannopine 1'2' promoter in transgenic poplar. The altered distribution correlates with an altered developmental pattern in the xylem (see text for discussion). Labels are as in Figure 1. Figure redrawn from Tuominen *et al.* (1997).

window of high auxin concentration and the radial population of meristematic cells would be determined by the width of the appropriate IAA concentration window. Accordingly, in Scots pine, xylem cell production is related to the meristem size and the auxin distribution pattern (Uggla *et al.*, 1998). It is clear, however, that auxin is not the only player in providing positional signalling to the cambial meristem. Cells at the phloem side of the meristem remain meristematic at a lower auxin concentration than cells at the xylem side. Thus, the auxin gradient by itself does not provide enough information to position either xylem and phloem mother cells or the cambial initial. The recent demonstration of steep concentration gradients of soluble carbohydrates across the cambium (Uggla *et al.*, 2001), together with accumulating evidence in favour of sugar sensing in plants (Sheen *et al.*, 1999), provide substantial evidence to the concept that auxin/sucrose ratios determine the positioning of the cambium (Warren Wilson and Warren Wilson, 1984). Moreover, the idea that auxin distribution sets the radial limits of the meristem does not exclude the possibility that the amplitude of auxin concentration within the critical window is important for the rate of cell cycling of the meristematic cells.

In a series of experiments with hybrid aspen, the auxin biosynthetic genes from *Agrobacterium tumefaciens* were expressed under the control of different promoters, with the aim of altering the auxin balance in the cambial region (Tuominen *et al.*, 2000a). Surprisingly, neither the strong 35S promoter nor any other promoter used induced an increase in the IAA level comparable to that previously obtained in tobacco and petunia (Klee *et al.*, 1987; Sitbon *et al.*, 1992). The most conspicuous phenotype resulted from the expression of the IAA biosynthetic genes under the mannopine synthase 1′2′ promoter (Tuominen *et al.*, 1995). These trees were smaller than wild-type trees and had an altered development of xylem tissue. Increased IAA levels were only found in the root tips and mature leaves, whereas IAA levels in the apex and stem internodes were similar or lower than those in wild-type. The ectopic expression of the IAA biosynthetic genes in the stem was however demonstrated by the fact that the apical dominance was maintained after decapitation. A detailed examination of the auxin balance in stem tissues of these trees showed a wider radial distribution but with lower concentrations across the cambial region, compared to wild-type plants (Figure 6) (Tuominen *et al.*, 1997). The lower concentration was explained by a decrease in the supply of IAA originating from endogenous genes because of the smaller size of the transgenic trees, whereas ectopic production of IAA induced its wider distribution. The altered IAA balance was interpreted to cause the wider developmental zones of division and expansion observed in the transgenic trees, and to support the idea that IAA has a function in positional signalling. In spite of the wider cambium, the number of cells in the meristem remained the same and the cell production rate was lower, possibly as a result of the lower IAA concentration. When the IAA biosynthetic genes were expressed under the *rolC* promoter from *Agrobacterium rhizogenes*, a 35–40% increase in IAA concentration in the cambial meristem was obtained, but the pattern of radial IAA distribution was unaffected (Tuominen *et al.*, 2000). In this case, no difference in xylem development or anatomy was observed, and the rate of xylem production was similar in transgenic and wild-type trees. This result strengthens the importance of auxin distribution in controlling cambial growth patterns. The small increase in auxin concentration that was induced did not affect cell cycling rate in this experiment.

Cambial derivatives will cease to divide, but continue to expand during their exposure to decreasing IAA concentrations when developing along the auxin gradient. Conceivably, the width of the gradient influences the width of the RE zone, and therefore the duration of expansion, whereas the IAA concentration within this zone can be expected to influence the rate of cell expansion. In conifer trees, both the rate and the duration of expansion have been shown to be important factors that determine the final radial size of the tracheids (Dodd and Fox, 1990). Exogenous feeding experiments in both conifers and angiosperm trees have demonstrated that auxin stimulates xylem cell expansion up to an optimal concentration, while a further increase in the exogenous auxin concentration will reduce the radial size of the derivatives (Sheriff, 1983; Zakrzewski, 1991). Whether the endogenous concentration to which the developing xylem cells are exposed, is sub- or supra-optimal for the rate of radial expansion is unknown. The transgenic hybrid aspen plants that express the IAA biosynthetic genes under the mannopine synthase 1′2′ promoter had radially wider fibres. These fibres had developed in a lower IAA concentration present across a wider radial zone that corresponded to the RE zone (Figure 6). Thus, the increase in radial fibre size was interpreted to be a consequence of a longer duration of fibre expansion. In this context, it should be emphasized that fibres and vessel elements do not develop synchronously: vessels mature earlier than fibres (Murakami *et al.*, 1999). This observation again highlights the fact that the auxin gradient only provides some of the positional information required for xylem development.

Other hormones involved in cambial cell division and expansion

The interpretation of the role for endogenous IAA in cambial growth is complicated by the fact that other hormones often act in synergy with auxin. In particular, gibberellins (GAs) will stimulate meristematic activity and xylem fibre elongation when applied together with auxin (Digby and Wareing, 1966). The role for GAs in wood formation and their potential in tree biotechnology has recently been demonstrated by over-expression of a GA-20 oxidase in hybrid aspen (Eriksson *et al.*, 2000), resulting in a 20-fold increase

252

in the biologically active GA1 and GA4. In line with data from exogenous applications, the transgenic trees exhibited an increased longitudinal and radial growth as well as an increased xylem fibre length.

Although cytokinins have a well-established function in cell division, their role in cambial growth is far from clear (Little and Savidge, 1987; Little and Pharis, 1995). Accurate information on endogenous cytokinin levels in the cambial region of dicot trees is not available. Mass spectrometry quantification in Scots pine revealed nanomolar concentrations in samples comprising the combined tissues of the cambial region from both dormant and actively growing trees (Moritz and Sundberg, 1996). Interestingly, zeatin, which is supposedly a biologically active cytokinin, was only putatively identified and present only in trace amounts. With an auxin concentration in the micromolar range in the same tissues (Uggla et al., 1996), the endogenous auxin/cytokinin ratios would differ strongly from, for example, the ratio used for TE induction in the Zinnia system.

Exogenous ethylene stimulates cambial cell division, possibly by increasing auxin levels through interaction with auxin transport (Eklund and Little, 1996). The concentration of endogenous ethylene in cambial tissues of intact plants cannot be easily assayed because of its gaseous nature. Measurements of ethylene emanating from isolated cambial regions may just reflect the availability of the ethylene precursor 1-aminocyclopropane-1-carboxylate (ACC) or wound-induced ethylene, and should be interpreted with caution. Several recent observations from aspen provide additional evidence for ethylene signalling in cambial growth processes. ACC was identified by mass spectrometry in developing xylem tissues (J. Hellgren, T. Moritz and B. Sundberg, unpublished). ESTs for ACC synthase, ACC oxidase and putative ethylene receptors are present in the cambium library. Only one ACC oxidase gene was found to be expressed in the cambial region, and its mRNA was mainly present in developing xylem at the stage of secondary wall formation (S. Andersson, S. Regan, and B. Sundberg, unpublished). The cellular localization of ACC and ACC oxidase has not yet been established, but it is of particular interest to investigate whether ethylene production takes place in the symplast or in the apoplast, because this will be an important factor determining the ethylene balance in the cambial region.

Induction of xylem cell differentiation

The xylogenic nature of plant hormones has been extensively investigated in cell cultures, callus, and explants (Roberts, 1988). The absolute requirement for auxin in TE induction is without question. Other plant hormones often act in synergy with auxin when applied exogenously, but frequently the observed effects vary between different experimental systems. This can be explained by differences in the resulting hormonal balance in the experimental tissues, which is unknown in most cases. In many systems, including Zinnia cultures, cytokinin is needed for TE induction. When cytokinin is not required, it seems likely that the experimental tissue already contains a sufficient amount of cytokinins. Early studies showed that the number of auxin-induced TEs decreased when ethylene biosynthesis and response inhibitors were applied (Miller and Roberts, 1984), pointing to a role for ethylene in xylogenesis. However, the addition of ethylene has never been demonstrated to be required for TE induction, as is the case for auxin and cytokinin. Possibly, IAA induces the necessary levels of ethylene through the well-established induction of ACC synthase (Abel et al., 1995), but conclusive evidence for the role of ethylene in xylogenesis is still lacking. GAs have been observed to stimulate the production of both phloem and xylem fibres when applied together with auxin to intact tissues (Digby and Wareing, 1966; Aloni, 1979), but evidence for a requirement for GAs in xylem cell differentiation is lacking.

Formation of secondary walls

Most data on the role of plant hormones in secondary wall formation concern the involvement of auxin and ethylene in lignification. Auxin has been shown to enhance lignification in primary phloem fibres of Coleus (Aloni et al., 1990), and in secondary xylem of tobacco (Sitbon et al., 1999). However, at least in the latter case, ethylene production was also increased, and ethylene is well known to induce several enzymes involved in lignin biosynthesis (Sitbon et al., 1999). The endogenous auxin concentration was found to be low in lignifying xylem of Scots pine and hybrid aspen (Uggla et al., 1996; Tuominen et al., 1997) but this observation does not preclude its role in secondary wall formation. Other effects of exogenous hormones on cell wall biosynthesis, such as cytoskeleton organization, induction of genes and enzyme activities, may be direct, particularly in the case of ethylene. However, many of the observed effects of applied hormones are likely to be inductive and result from their primary role in determining developmental fate, cell division, and cell expansion. For example, the induction of TE in *Zinnia* cell cultures requires the presence of IAA and cytokinin only for a very short inductive period prior to any visible sign of differentiation (McCann *et al.*, 2000).

Hormones and variation in wood properties

Little is known about the molecular mechanisms that determine the formation of the different wood types, and about the role of hormones in this process. Many aspects of the spatial and seasonal variability in wood formation have been suggested to be under the control of auxin concentration. However, information on endogenous IAA levels to support this idea is scarce and, when available, does not support traditional views (Sundberg *et al.*, 2000). For instance, the lack of radial expansion during latewood formation has been suggested to result from decreased auxin levels (Larson, 1969). However, high-resolution analysis revealed an increase in auxin concentration during the transition from earlywood to latewood in Scots pine (Uggla *et al.*, 2001). The radial distribution of auxin was narrowed in accordance with the decrease in the width of the RE zone, which is typical of latewood development (Whitmore and Zahner, 1966; Dodd and Fox, 1990). Whether the altered auxin distribution is a cause or a consequence of the altered pattern of development is not known, but it supports the view that auxin is a developmental signal in cambial growth.

The formation of tension wood is an interesting case of a developmental switch that results in an increased formation of xylem having gelatinous fibres with an altered secondary wall composition and structure (see above). From experiments with exogenous auxin and auxin transport inhibitors, it has been proposed that tension wood is induced by auxin deficiency (Little and Savidge, 1987). However, high-resolution analysis of endogenous auxin revealed that the auxin concentration in cambial derivatives developing into tension wood is not significantly affected compared to that in normal xylem development (J. Hellgren and B. Sundberg, unpublished). Instead, the IAA amount and, hence, the width of the radial IAA distribution across the developing xylem tissues was increased. This is in accordance with IAA being a positional signal that stimulates growth by acting on meristem size. Ethylene production is stimulated during tension wood formation and has been suggested to

play a role in this process (Little and Savidge, 1987). In support of this observation, the expression of ACC oxidase is strongly induced in hybrid aspen during tension wood formation (S. Andersson, S. Regan and B. Sundberg, unpublished). However, exogenous ethylene, while stimulating radial growth, has not been observed to induce gelatinous fibres. Taken together, both ethylene and IAA may be involved in growth stimulation accompanying tension wood formation rather than in the induction of gelatinous fibres. Finally, it should be mentioned that exogenous GAs have been observed to induce tension wood characteristics in Japanese cherry (Baba *et al.*, 1995). The endogenous GA balance during tension wood formation is unknown.

Biosynthesis of wall carbohydrates in poplar xylem cells

Precursors for carbohydrate components of the cell wall

UDP-D-glucose is a precursor of all carbohydrates in the cell wall (Gibeaut, 2000; Reiter and Vanzin, 2001, this issue). It is provided either from the reaction (sucrose + UDP → UDP-D-glucose + fructose), catalysed by sucrose synthase (SuSy), or by the reaction catalysed by UDP-glucose pyrophosphorylase (glucose-1-P + UTP → UDP-D-glucose + PP). cDNAs coding for both enzymes are present in the cambium library (Table 4). Correlative evidence indicates that SuSy plays an important role in xylem cell wall formation. The gene is highly expressed in the cambial region as judged from the high frequency of the corresponding ESTs in the cambium library. In addition, the enzymatic activity of SuSy in the developing xylem is correlated with the intensity of xylem wall formation in poplar (Sauter, 2000) and *Robinia pseudoacacia* L. (Hauch and Magel, 1998). In developing xylem of Scots pine, the enzymatic activity of SuSy exhibits a minor peak in the cambium, whereas the highest activity is found in the zone of secondary wall formation (Uggla *et al.*, 2001). This pattern of activity supports a function for the membrane-associated isoforms of SuSy in providing UDP-D-glucose directly to cellulose synthase (Winter and Huber, 2000; Haigler *et al.*, 2001, this issue).

UDP-glucose dehydrogenase, which converts UDP-D-glucose to UDP-D-glucuronic acid, is a key enzyme that regulates the flux of UDP-glucose into

Table 4. Enzymes related to cell wall polysaccharide precursor biosynthesis corresponding to ESTs from the cambial region and developing xylem (Sterky *et al.*, 1998a, 1998b).

Enzyme	Library	Accession		
α,α-trehalose-phosphate synthase (UDP-forming) (UDP-glucose-glucosephosphate glucosyltransferase (EC 2.4.1.15)	Cambium	AI163996		
Diphosphate–fructose-6-phosphate 1-phosphotransferase (EC 2.7.1.90)	Xylem	AI162647	AI163266	
dTDP-glucose 4-6-dehydratase (EC 4.2.1.46)	Cambium	AI161839	AI162550	AI162558
		AI162651	AI163474	AI163546
		AI163581	AI164673	AI165131
		AI165272	AI165288	AI166047
	Xylem	AI166408	AI166808	
Fructokinase (EC 2.7.1.4)	Cambium	AI165611	AI165646	AI166765
		AI162548		
	Xylem	AI162846		
GDP-mannose pyrophosphorylase (EC 2.7.7.13)	Cambium	AI161856	AI162882	AI163264
		AI165115	AI165373	AI166098
myo-inositol 2-dehydrogenase (EC 1.1.1.18)	Cambium	AI163525	AI164586	
myo-inositol-1(or 4)-monophosphatase (EC 3.1.3.25)	Cambium	AI162400	AI165680	
Phosphoglucomutase (EC 5.4.2.2)	Cambium	AI162727 (AF097938)	AI165714	
6-phosphogluconate dehydrogenase (EC 1.1.1.44)	Cambium	AI162532	AI165699	
Phosphomannomutase (EC 5.4.2.8)	Cambium	AI163799	AI162548	AI162846
		AI163832		
Putative sucrose phosphatase (EC 3.1.3.24)	Cambium	AI165881		
Putative UDP-galactose transporter	Cambium	AI165948	AI166124	AI166211
Reversibly glycosylated polypeptide-1	Cambium	AI161671	AI163874	
Sucrose synthase (EC 2.4.1.13)	Cambium	AI162073	AI162587	AI163208
		AI163479	AI163591	AI163699
		AI163955	AI164134	AI164295
		AI164415	AI164510	AI164858
		AI165555	AI165740	AI165766
		AI166066		
	Xylem	AI166709		
UDP-glucose 6-dehydrogenase (EC 1.1.1.22)	Cambium	AI164180	AI162135 (AAF04455)	AI163328
		AI166238	AI166122	
	Xylem	AI166760		
UDP-glucose pyrophosphorylase (EC 2.7.7.9)	Cambium	AI161980		

several other nucleotide-sugars that are substrates for cross-linking glycans and pectin precursors synthesized in the Golgi apparatus (Figure 7; Tenhaken and Thulke, 1996; Seitz *et al.*, 2000). The high abundance of UDP-glucose dehydrogenase transcripts in the cambial region tissues (Table 4) indicates that UDP-D-glucuronic acid may be largely formed directly from UDP-D-glucose. The sugar interconversion reactions also include dehydrogenation, decarboxylation and epimerization reactions that are discussed in detail by Reiter and Vanzin (2001, this issue). Transcripts corresponding to enzymes involved in some of these reactions have been found in the wood-forming tissues of hybrid aspen (Table 4), whereas enzymatic activities for UDP-D-galactose 4-epimerase, UDP-L-arabinose epimerase and UDP-D-glucuronate decarboxylase have been detected in the cambial region of *P. × euramericana* cv. Robusta (Dalessandro and

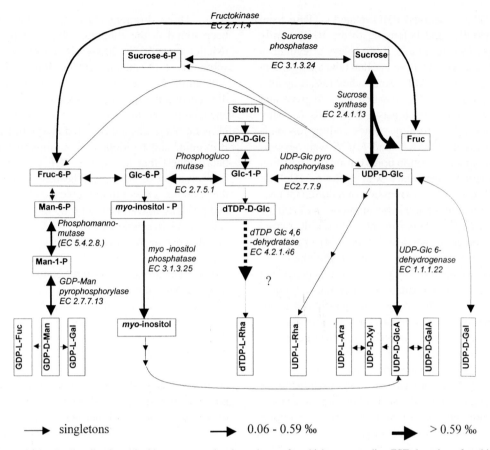

Figure 7. Biosynthesis of cell wall polysaccharide precursors showing enzymes for which corresponding ESTs have been found in the xylem and cambium EST libraries. The thickness of the arrows relates to the abundance of each EST. The dTDP-D-glucose pathway does not have a precedent in plants, and thus the identity of dTDP Gluc 4,6-dehydratase should be considered as tentative.

Northcote, 1977). The composition of glucomannan, synthesized presumably by a single glucomannan synthase, is dependent on the ratio of GDP-D-glucose to GDP-D-mannose in pea (Piro *et al.*, 1993). These results suggest that it will be possible to genetically engineer wall carbohydrate polymers in trees by modifying sugar pools similarly as monolignol pools have been altered to affect lignin composition (see below).

The reversibly glycosylated polypeptide 1 has been proposed to be involved in sugar transport into the Golgi lumen and in biosynthesis of non-cellulosic polysaccharides (Dhugga *et al.*, 1997; Saxena and Brown, 1999). ESTs for the nucleotide sugar transporter-like protein and the reversibly glycosylated polypeptide 1 have been found in the cambium library (Table 4). Several Golgi membrane-bound glycosyltransferases have been identified and are thought to synthesize cell wall matrix components (see Per-

rin *et al.*, 2001, this issue). Some candidates for glycosyltransferases from poplar are listed in Table 4.

Another pathway that can be involved in sugar interconversion is the *myo*-inositol oxidation pathway, providing UDP-activated sugars from the *myo*-inositol (Figure 7; Loewus and Murthy, 2000). The presence in the cambium library of ESTs for *myo*-inositol-1-phosphatase and phosphoglucomutase, the enzyme that provides glucose-6-phosphate to *myo*-inositol, indicates that this pathway operates in the cambial region tissues.

Secretion: the role for RHOs

Pre-formed matrix carbohydrates and probably also lignin precursors are secreted to the wall via exocytosis. Because numerous vesicles are usually observed near to the wall, the process of fusion itself has been suggested to be a limiting step that regulates the rate of wall biosynthesis (Northcote, 1989). In pollen tubes, a small GTPase, denoted RHO-related GTPase from Plants 1 (ROP1) and belonging to the RHO family of GTPases (Valster *et al.*, 2000), plays a regulatory role in vesicle exocytosis and pollen tube enlargement (Zheng and Yang, 2000). Antibodies raised against a related protein, RAC1, labelled the cell plate and Golgi compartments in tobacco BY2 cells (Couchy *et al.*, 1998). Interestingly, genes coding for similar proteins, called *RAC13* and *RAC9,* were specifically up-regulated in cotton fibres during the transition from primary to secondary wall secretion (Delmer *et al.*, 1995). Transient expression of a dominant negative *rac13* mutant or antisense suppression of *RAC13* in soybean and *Arabidopsis* cell cultures decreased the H_2O_2 secretion, whereas ectopic over-expression caused an increase in H_2O_2 secretion (Potikha *et al.*, 1999). The presence of H_2O_2, in turn, was correlated with secondary wall formation and cellulose biosynthesis in cotton. This raises the question whether RHO proteins play a role in cell wall formation in the secondary xylem. Transcripts of *RHO* genes are exceptionally abundant in the cambium EST library. Over 58 (1%) ESTs that encode small GTPases could be found and 25 of them had a BLAST similarity score of 200 or more to RAC13. Such a high similarity indicates that homologues of RAC13 are present in poplar and that they are particularly highly expressed in the cambial region. Analysis of the expression of these proteins will tell us whether they are candidates to be regulators of the primary to secondary wall transition in poplar.

Expanding walls in developing xylem

Cellulose microfibrils coated with xyloglucan and linked by xyloglucan bridges form the main load-bearing component of the primary wall in dicot plants (Hayashi, 1989; Carpita and Gibeaut, 1993; Nishitani, 1998; Whitney *et al.*, 1999; Cosgrove, 2000). Thus, the expansion of the wall depends heavily on the ability to cut the xyloglucan bridges and/or break the hydrogen bonding in the xyloglucan-cellulose interaction. Xyloglucan endotransglycosylases (XETs), endoglucanases and expansins are the most important proteins involved in these processes (McQueen-Mason, 1997; Cosgrove, 2000; Darley *et al.*, 2001, this issue). Only preliminary evidence exists for a role for these proteins in secondary xylem development.

Among the ESTs from differentiating xylem of pine, Allona *et al.* (1998) have found two *XET* genes that were differentially expressed in upright and bent trees. Three different cDNAs encoding XET have been found in the cambium library of hybrid aspen (Mellerowicz *et al.*, 2000), all belonging to the *XET* subfamily I (Nishitani, 1997). Immunolocalization with antibodies against one of these XET proteins indicated that this particular XET was present in cells engaged in wall biosynthesis of secondary xylem and phloem, but it was not detected in xylem RE zone (V. Bourguin, and E. Mellerowicz, unpublished). These data suggest that the poplar XET is involved in the incorporation of newly synthesized xyloglucan into the existing cross-linking glycan-cellulose network and possibly in wall restructuring in long-lived xylem cells. It is still unknown which *XET* genes play a role in radial expansion of cambial derivatives.

The hydrolysis of xyloglucan by xyloglucan-specific glucanases also might contribute to cell expansion (Matsumoto *et al.*, 1997). Endoglucanases are enzymes that catalyse the endo-hydrolysis of $(1\rightarrow4)$-β-D-glucan. They can also be involved in the degradation of non-crystalline cellulose (Ohmiya *et al.*, 2000) or in the biosynthesis of cellulose during growth (del Campillo, 1999). ESTs encoding two different endoglucanases have been found in the cambium library of hybrid aspen, all belonging to the glycosyl hydrolase family 9 (Mellerowicz *et al.*, 2000). Both ESTs were singletons and preliminary sequence comparisons suggest that one corresponds to a soluble, and the other to a membrane-bound enzyme. The role of endoglucanases in wood formation is hypothetical. They may be involved in the regulation of fibre length by acting on cell elongation during primary growth and thus affecting the length of the procambial cells. Longer procambial cells would give rise to longer FCCs and, hence, longer fibres and vessel elements in the secondary xylem. This is supported by the analysis of transgenic *P. tremula* plants over-expressing the *Arabidopsis* endoglucanase *CEL1* gene under the control of the CaMV 35S promoter, which have significantly longer internodes as well as longer fibers (Shani *et al.*, 1999).

Expansins are cell-wall-bound proteins that mediate acid growth and are considered the main agent regulating cell wall rheological properties (McQueen-Mason, 1997; Cosgrove, 1997, 2000). They are proposed to intercalate between the cellulose microfibrils and the xyloglucan polymers in the primary cell wall and thus to disrupt the hydrogen bonding between them. However, an enzymatic activity of expansins has not yet been documented. Interest in the role of expansins in xylem formation has been stimulated by the discovery that transcripts of three xylem-specific expansins were localized either to the apical or to the basal tip of axial xylem parenchyma cells in the primary xylem of *Zinnia* (Im *et al.*, 2000). The meaning of this peculiar localization needs to be further investigated. ESTs for two expansin genes have been identified in the hybrid aspen cambium library, both corresponding to α-expansins (Mellerowicz *et al.*, 2000). One of these ESTs is abundantly present in the library and is specifically expressed in the xylem mother cells and expanding xylem cells, whereas the other is expressed at a low level in the cambium closer to the phloem (M. Grey-Mitsumune and E.J. Mellerowicz, unpublished). Thus, these expansins may be involved in the radial expansion and/or tip growth of FCCs and developing xylem fibre cells.

The stiffening of the wall during cell differentiation may be related to the cross-linking of extensin (and other hydroxyproline-rich proteins) with the cellulose-xyloglucan-pectin network, which locks the microfibrils in place (Carpita and Gibeaut, 1993; Brett and Waldron, 1996). In addition, the incorporation of Ca^{2+} into pectins causes their gelation and makes the wall unstretchable. Parallel neighbouring chains of demethylated (acidic) pectins can ionically bind Ca^{2+} to form an 'egg-box' structure that stiffens the wall. Acidic pectins, on the other hand, while not engaged in Ca^{2+} bonding, lower the pH of the cell wall and therefore may stimulate the activity of wall-bound hydrolases and expansins that are important for wall plasticity (McQueen-Mason, 1997). Thus, the status of pectin methylation, regulated by PME, is an important factor in controlling wall plasticity. Pectin methylation and the composition of the side-chains are developmentally regulated in the cambial region of *Populus*. Labelling with the monoclonal antibodies JIM5 and JIM7, which distinguish homogalacturans with low and high levels of methyl esterification, respectively (Knox *et al.*, 1990), indicates that radial walls of FCC and expanding xylem cells are rich in pectins of both kinds and that the degree of methylation decreases during xylem development (Guglielmino *et al.*, 1997a). This decrease in pectin methylation is accompanied by a shift in the localization of PME from the Golgi (in the cambium) to the cell wall (in the RE zone), as judged from immuno-labeling with polyclonal antibodies against flax PME (Guglielmino *et al.*, 1997b). This observation led the authors to propose that the cambium can synthesize but not secrete PME. Micheli *et al.* (2000a) found wall-bound and soluble PME activities at pH 6.1 in the cambium and a soluble PME activity in the RE zone. Ten PME isoforms have been detected in the cambial region of hybrid aspen by isoelectric focusing and activity assays at pH 7 (Micheli *et al.*, 2000a, b). The expression of several isoforms seemed to be developmentally regulated. Five putative *PME* cDNAs have been cloned from the hybrid aspen cambium library (Micheli *et al.*, 2000b). Their deduced molecular mass was 34 kDa and the deduced pI was between 9 and 10. It will be important to find out which gene corresponds to which isoform and to establish whether the different isoforms are transcriptionally or post-transcriptionally regulated and differentially secreted to the cell wall.

Monoclonal antibodies have been made that recognize specific side-chains of rhamnogalacturonan I, namely LM5 and LM6 that recognize $(1\rightarrow4)$-L- or D-galactosyl residues and 5 to 6 units of $(1\rightarrow5)\alpha$-L-arabinan, respectively (Jones *et al.*, 1997; Willats *et al.*, 1998, 1999, and 2001, this issue). In hybrid aspen, the arabinan epitope was found in phloem cells, particularly in the sieve tubes, in the cambium and in the RE zone, whereas the galactan epitope was restricted to the cambium and the RE zone (Ermel *et al.*, 2000). Whereas the antibodies clearly differentiate between cell populations with different destinies, i.e. differentiating to either xylem or phloem tissues, they do not differentially mark meristematic and expanding cells as was found for carrot suspension cultures (Willats *et al.*, 1999).

Pectins determine porosity and thus appoplastic exclusion limits of primary walls. Therefore, a degradation of the pectin network may make walls more accessible to other modifying enzymes (Carpita and Gibeaut, 1993). Pectin metabolism appears to play an important role in xylem cell development. Pectate lyase hydrolyses demethylated pectin (pectate). In *Zinnia*, a pectate lyase (*ZePel*) mRNA was found to be associated with vascular tissue by *in situ* hybridization technique and the gene was up-regulated in the xylogenic cell culture at the early stage of induction, before the formation of secondary cell walls (Domingo *et al.*, 1998). Poplar EST clones similar to *ZePel* are very abundant in the cambium and xylem EST libraries (Sterky *et al.*, 1998a). Another enzyme involved in pectin degradation is endopolygalacturonase (pectinase). It is encoded by a multigene family of at least 19 members in *Arabidopsis* and is ubiquitously expressed (Torki *et al.*, 2000). Four ESTs corresponding to pectinase have been found in the xylem and cambium libraries.

Cellulose biosynthesis

Cellulose is synthesized by plasma membrane-bound enzyme complexes that can be seen as rosettes in freeze fracture preparations of plant cells (Mueller and Brown, 1980; Brett, 2000). Rosettes are also found in developing primary and secondary xylem (Haigler and Brown, 1986; Fujino and Itoh, 1998). Cellulose synthase has been recently immunolocalized to the freeze-fractured rosette complex of *Vigna angularis* (Kimura *et al.*, 1999). At least 10 cellulose synthase genes are present in *Arabidopsis*, all of them carrying the type A catalytic domain and therefore called *CesA* genes (Holland *et al.*, 2000; Richmond and Somerville, 2001, this issue). Some of them, such as *RSW*1 and *PRC*1, are involved in primary wall biosynthesis (Arioli *et al.*, 1998; Fagard *et al.*, 2000), whereas others are active in secondary wall biosynthesis (*IRX3* and *IRX1*) (Taylor *et al.*, 1999; Turner *et al.*, 2001, this issue). Recently, the *CesA* genes have been cloned from the wood-forming tissues of tree species, including *Populus* (Sterky *et al.*, 1998a; Wu *et al.*, 2000) and pine (Allona *et al.*, 1998).

Two *Populus* genes, *PtCesA*1 and *PtCesA*2 (also called *PtCesA*) from *P. alba* (L.) × *P. tremula* and *P. tremuloides*, respectively, are similar to the cotton *CesA* genes involved in secondary wall formation (Holland *et al.*, 2000; Wu *et al.*, 2000). *In situ* hybridization showed that *PtCesA2* is expressed specifically during secondary wall biosynthesis in developing protoxylem, metaxylem, and secondary xylem, but that the expression is not detectable in the phloem fibers that also develop a secondary cell wall (Wu *et al.*, 2000). Thus, the *CesA* genes have cell-specific expression patterns. Curiously, the *PtCesA2* promoter drives GUS expression in tobacco in developing primary and secondary xylem cells in upright-growing stems, but upon bending of the stem, the expression is induced in the phloem fibres of the convex (upper) and down-regulated in the xylem of the concave (lower) side of the stem. This indicates that the *PtCesA2* promoter responds to mechanical and/or gravitational stimuli. It would be of interest to investigate whether this gene is involved in tension wood formation in gravistimulated stems of trees. Four different cDNAs encoding CesA have been found in the cambium and xylem libraries (Mellerowicz *et al.*, 2000). One of these ESTs was present in many copies and probably represents the major cellulose synthase involved in xylem cell wall formation.

Other carbohydrates present in the secondary wall

Pectin biosynthetic enzymes are down-regulated during secondary wall formation in the xylem (Bolwell and Northcote, 1981; Bolwell *et al.*, 1985; Fukuda, 1992; Gregory *et al.*, 1998), but pectins are still present in the lignified walls of mature xylem cells (Imai and Terashima, 1992). The strong reaction with the JIM7 antibody indicates the prominent presence of methylated pectin in the middle lamellae and primary cell wall of Scots pine tracheids and ray cells, as well as in the membranes of bordered pits (Hafren *et al.*, 2000). Reaction with the JIM5 antibody was weaker but showed a similar pattern, indicating the presence of acidic pectins in the same areas.

The occurrence of xyloglucan in the secondary wall has not been well established. In cotton fibres, the amount of 24% KOH-extractable xyloglucan was positively correlated with cell elongation and decreased to 20% of its maximum when the secondary wall was formed (Shimizu *et al.*, 1997). Surprisingly, *XET* mRNA has been detected at all stages of fibre development in cotton, including the stage of secondary wall formation. The monoclonal antibody CCRC-M1 that recognizes fucosyl side-chains that are typically present in the xyloglucan (Puhlmann *et al.*, 1994), has been used to immunolocalize xyloglucan in *Zinnia* TEs (Stacey *et al.*, 1995). Xyloglucan was found in the cell wall layer corresponding to the primary wall, but not in secondary wall thickening. However, large amounts of xyloglucan were secreted into the culture medium and were present in the protoplasts of tracheary elements undergoing secondary wall deposition.

Xylan synthesis seems to be specifically up-regulated during secondary wall formation. A membrane-bound UDP-D-xylose-xylan transferase activity was shown to increase in *Zinnia* cultures and differentiating secondary xylem of several dicot species during secondary cell wall formation (Suzuki *et al.*, 1991; Bolwell and Northcote, 1981, 1983; Gregory *et al.*, 1998). Correspondingly, xylan deposition increases in the secondary walls of tracheary elements when compared to the primary walls (Ingold *et al.*, 1988; Northcote *et al.*, 1989). In *Zinnia*, xylan deposition depends on the undisturbed synthesis of cellulose (Taylor and Haigler, 1993). This observation, together with the spontaneous binding of xylan to cellulose microfibrils (Reis *et al.*, 1994), indicate that xylan coats cellulose microfibrils in secondary walls, similarly to the xyloglucan coating of microfibrils in primary walls. By using immuno-field emission scanning electron microscopy, Awano *et al.* (2000) have detected an association of xylan with the thick cellulose microfibrils in the S1 and S2 layers, but not with the thin (5 nm) microfibrils present in the S1 layer of fibres of beech wood. Immuno- and enzymatic labelling experiments have demonstrated that xylan is not uniformly distributed within the wall in wood cells of dicot plants (Vian *et al.*, 1992; Awano *et al.*, 1998, 2000). Labelling was particularly strong in the outer S2 layer and it appeared to increase in the S1 layer during S2 layer deposition (Awano *et al.*, 1998, 2000). It was suggested that the newly synthesized xylan might migrate to the deeper wall layers and associate there with uncoated cellulose microfibrils.

Xylanases are $(1\rightarrow4)\beta$-D-xylan endohydrolases that are able to attack intact polymers of xylan. A xylanase cDNA has been found in the cambium library of hybrid aspen (Sterky *et al.*, 1998a). The corresponding protein belongs to the glucan hydrolase family 10 (Mellerowicz *et al.*, 2000), and is similar to that found in *Arabidopsis,* in which the xylanase-catalytic domain is preceded by an additional domain of unknown function. However, the function of this protein in *Arabidopsis* is still unknown. The plant xylanases studied so far appear to be involved in wall degradation in connection with seed and pollen germination, aerenchyma formation, or fruit ripening (Bragina *et al.*, 1999; Benjavongkulchai and Spencer, 1986; Bih *et al.*, 1999). Wall turnover (Gorshkova *et al.*, 1997) or restructuring may also require a xylanase activity and the hybrid aspen xylanase might be involved in any of these activities.

Lignification of xylem cell walls in poplar

Lignin is a heterogeneous phenolic polymer that is mainly present in the secondary thickened cell wall. It is essential for mechanical support and plays a role in defence towards pathogens. Because of its hydrophobic nature, it provides impermeability to tracheary elements and therefore allows the transport of water and solutes through the vascular system. Numerous efforts have been made over the past decade to understand the lignification process because of its economical relevance; lignin is considered as a negative factor in paper-making and limits the digestibility of fodder crops.

Lignin is mainly derived from the dehydrogenative polymerization of three different hydroxycin-

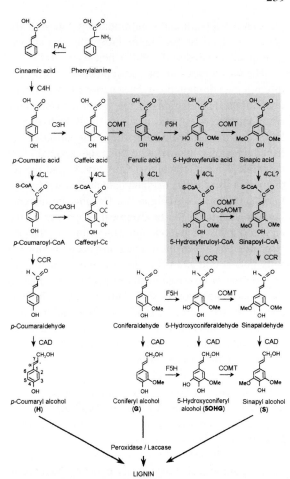

Figure 8. Monolignol biosynthesis pathways. The enzymatic steps shown in the grey box probably do not play a major role *in vivo.* CAD, cinnamyl alcohol dehydrogenase; C3H, coumarate 3-hydroxylase; C4H, cinnamate 4-hydroxylase; CCoA3H, coumaroyl-coenzyme A 3-hydroxylase; CCoAOMT, caffeoyl-coenzyme A *O*-methyltransferase; CCR, cinnamoyl-coenzyme A reductase; COMT, caffeate/5-hydroxyconiferaldehyde *O*-methyltransferase; F5H, ferulic acid/coniferaldehyde 5-hydroxylase; 4CL, 4-coumarate:coenzyme A ligase; PAL, phenylalanine ammonia-lyase.

namyl alcohols (or monolignols), *p*-coumaryl alcohol, coniferyl alcohol, and sinapyl alcohol (Figure 8). These monolignols give rise to the H, G, and S units of the lignin polymer, respectively, and differ from each other only by their degree of methoxylation. The content and composition of lignin are known to vary among taxa, tissues, cell types, and cell wall layers and to depend on the developmental stage of the plant and the environmental conditions (Côté, 1977; Campbell and Sederoff, 1996; He and Terashima, 1991; Joseleau and Ruel, 1997; Monties, 1998). The overall lignin heterogeneity is thought to result from the spatio-temporal and conditional regulation of the genes involved in the lignin biosynthesis pathway (Campbell and Sederoff, 1996; Chen *et al.*, 2000).

The biosynthesis of the lignin precursors proceeds through the common phenylpropanoid pathway, starting with the deamination of phenylalanine to cinnamic acid. Further enzymatic reactions include hydroxylations of the aromatic ring, the methylation of selected phenolic hydroxyl groups, the activation of cinnamic acids to cinnamoyl-CoA esters, and the reduction of these esters to cinnamaldehydes and further to cinnamyl alcohols (Figure 8). Despite the increasing knowledge on the biochemical properties of the enzymes involved in monolignol biosynthesis, the precise order in which these reactions occur is not yet fully understood and many uncertainties remain concerning the *in vivo* role of the respective enzymes in the monolignol biosynthesis pathway. Also little is known on the storage form, the intracellular transport through the cell wall as well as the mechanisms of polymerization of lignin precursors. Recently, it has been demonstrated by *in vitro* enzymatic assays that the hydroxylation and methylation reactions occur preferentially at the cinnamaldehyde and the cinnamyl alcohol level (Figure 8) (Chen *et al.*, 1999; Humphreys *et al.*, 1999; Osakabe *et al.*, 1999; Li *et al.*, 2000; Matsui *et al.*, 2000).

Genes and cDNAs for most of the known enzymes of the monolignol biosynthesis pathway have been cloned (reviewed by Baucher *et al.*, 1998; Christensen *et al.*, 2000) and ESTs for all of these enzymes are present in the xylem and cambium libraries of hybrid aspen (Sterky *et al.*, 1998a). As reviewed below, it is now possible to affect lignin content and composition by genetic modification of the expression of these genes (Table 5). The results obtained have made it possible to redraw the lignin biosynthesis pathway that had been described for many years in biochemistry textbooks (for recent reviews see Baucher *et al.*, 1998; Whetten *et al.*, 1998; Grima-Pettenati and Goffner, 1999; Boudet, 2000). Furthermore, it has been demonstrated that the lignin polymer can be built with different types of phenylpropanoid units (Sederoff *et al.*, 1999; Jouanin *et al.*, 2000; Kim *et al.*, 2000; Ralph *et al.*, 2001). Consequently, for the same overall amount of lignin, different chemical properties of the polymer can be obtained.

Genetic modification of lignin biosynthesis in poplar

Down-regulation of phenylalanine ammonia-lyase (PAL) or cinnamate 4-hydroxylase (C4H) in transgenic tobacco has been shown to dramatically reduce lignin content (Sewalt *et al.*, 1997), but because these enzymes are involved in early steps of the phenylpropanoid pathway, pleiotropic effects could be expected. Indeed, alterations in development were observed; the PAL-down-regulated plants were stunted and had curled leaves. The walls of secondary xylem cells were thinner and contained markedly less lignin than those of wild-type, as monitored by either histochemical staining of cross-sections with toluidine blue or by UV-fluorescence (Elkind *et al.*, 1990; Bate *et al.*, 1994). Interestingly, PAL activity is decreased in plants down-regulated for C4H, indicating a feedback regulation by the product of PAL, cinnamic acid (Blount *et al.*, 2000).

Transgenic poplars with a 95% (Van Doorsselaere *et al.*, 1995), a 78% (Tsai *et al.*, 1998), and a more than 97% (Jouanin *et al.*, 2000) reduction in caffeate *O*-methyltransferase (COMT/AldOMT) activity in the developing xylem have been generated. This enzyme has recently been demonstrated to use preferentially 5-hydroxyconiferaldehyde as substrate *in vitro* (Figure 8; Li *et al.*, 2000). A reduction in Klason lignin content (by 17%) has been detected only in the transgenic lines described by Jouanin *et al.* (2000), showing that COMT activity has to be decreased strongly to affect lignin content. Consistent with the substrate specificity of COMT described by Li *et al.* (2000), the lignin composition of the COMT-down-regulated poplars is characterized by a drastic decrease in the amount of S units and by the incorporation of an unusual lignin monomer, 5-hydroxyguaiacyl (5OHG), as demonstrated by thioacidolysis and nuclear magnetic resonance (NMR) experiments (Van Doorsselaere *et al.*, 1995; Lapierre *et al.*, 1999; Jouanin *et al.*, 2000; Ralph *et al.*, 2001). Modifications in the frequency of the linkages between monomers (lower frequency of β-O-4 ether linkages and a higher proportion of biphenyl (5-5) and phenylcoumaran (β-5) carbon-carbon linkages) indicated an increase in the degree of condensation of the lignin in the transgenic plants (Lapierre *et al.*, 1999). Moreover, a novel dimer corresponding to an α-β-diether structure involving a G unit and a 5-OHG unit has been identified by thioacidolysis (Jouanin *et al.*, 2000). Furthermore, a lower amount of free phenolic groups in β-O-4-linked S and G units has been detected

Table 5. Poplar lignin mutants obtained by genetic engineering.

Gene	Residual enzyme activity, %	Lignin content	Lignin composition	References
P. tremula × *P. alba*				
COMT	5	no changes	S/G decreased G increased, 5OHG	Van Doorsselaere *et al.* (1995); Lapierre *et al.* (1999)
	<3	reduced	S/G decreased 5OHG	Jouanin *et al.* (2000)
CAD2	30	slight decrease	more aldehydes	Baucher *et al.* (1996); Lapierre *et al.* (1999)
CCR	–	reduced	S/G increased	J.-C. Leplé, C. Lapierre and W. Boerjan, unpublished
CcoAOMT	10% protein amount	reduced	S/G increased	Meyermans *et al.* (2000)
	30	reduced	–	Zhong *et al.* (2000)
LAC	–	no changes	no changes	Ranocha *et al.* (2000)
F5H	–	–	S/G increased	Franke *et al.* (2000)
POX	800	no changes	no changes	J.H. Christensen, C. Lapierre, and W. Boerjan, unpublished
P. tremuloides				
COMT	28	no changes	S/G decreased, 5OHG more coniferaldehyde	Tsai *et al.* (1998)
4CL	10	reduced	no changes	Hu *et al.* (1999)

S/G, syringyl/guaiacyl; POX, peroxidase; LAC, laccase; other abbreviations as in Figure 8; – , not determined.

in the lignin of COMT-down-regulated poplars, resulting in a lower chemical reactivity of the lignin (Lapierre *et al.*, 1999; Jouanin *et al.*, 2000). In accordance with these structural changes in lignin, the wood of the transgenic trees is more resistant to Kraft delignification (Lapierre *et al.*, 1999; Jouanin *et al.*, 2000).

Caffeoyl-CoA *O*-methyltransferase (CCoAOMT) preferentially catalyses the methylation of caffeoyl-CoA over 5-hydroxyferuloyl-CoA *in vitro* (Inoue *et al.*, 1998; Martz *et al.*, 1998; Meng and Campbell, 1998; Grimmig *et al.*, 1999; Li *et al.*, 1999; Maury *et al.*, 1999). Transgenic poplars with 10% residual CCoAOMT protein amount have a 12% reduced Klason lignin content and an increased incorporation of *p*-hydroxybenzoic acid into the lignin, as shown by NMR (Meyermans *et al.*, 2000). Following a similar approach, Zhong *et al.* (2000) obtained transgenic poplars with a 40% reduction in Klason lignin content. Both in transgenic tobacco (Zhong *et al.*, 1998) and poplar (Meyermans *et al.*, 2000; Zhong *et al.*, 2000) down-regulated for CCoAOMT, the decreased

lignin content was due to a decrease in both S and G lignin units, confirming the role of CCoAOMT in the biosynthesis of both S and G lignin. Also, the lignin was characterized by a higher S/G ratio. By diffuse reflectance infrared Fourier transform spectroscopy (DRIFTS), the lignin of CCoAOMT down-regulated poplars was shown to be less cross-linked (Zhong *et al.*, 2000). This decrease in lignin content did not affect the morphology and the growth of the transgenic poplars. Furthermore, down-regulation of CCoAOMT causes an accumulation of methanol-extractable O^3-β-D-glucosyl-caffeic acid (GCA), O^4-β-D-glucosyl-vanillic acid (GVA), and O^4-β-D-glucosyl-sinapic acid (GSA) (Meyermans *et al.*, 2000). Feeding experiments have shown that GCA and GSA are storage or detoxification products of caffeic acid and sinapic acid, respectively. These results provide *in vivo* evidence that the pathway for the methylation/hydroxylation reactions of the lignin precursors does not occur at the cinnamic acid level and that sinapic acid is not a precursor for S lignin.

In vitro studies have revealed that ferulic acid 5-hydroxylase (F5H/CAld5H) catalyses preferentially the hydroxylation of coniferaldehyde over ferulic acid (Humphreys *et al.*, 1999; Osakabe *et al.*, 1999). This enzyme has been shown to be a major regulatory step in the determination of lignin monomer composition. The lignin of *Arabidopsis f5h* mutants is devoid of S units, whereas transgenic *Arabidopsis* overexpressing *F5H* have a lignin almost entirely built of S units (Chapple *et al.*, 1992; Meyer *et al.*, 1998; Jung *et al.*, 1999; Marita *et al.*, 1999). Similarly, transgenic poplars over-expressing the *Arabidopsis F5H* under the control of the cinnamate 4-hydroxylase (*C4H*) promoter have a lignin enriched in S units (Franke *et al.*, 2000).

4-coumarate:CoA ligase (4CL) catalyses the formation of CoA thioesters of hydroxycinnamic acids. No activity towards sinapic acid has been found (Allina *et al.*, 1998; Hu *et al.*, 1998), indicating, in addition to the results of Humphreys *et al.* (1999), Osakabe *et al.* (1999), Matsui *et al.* (2000) and Meyermans *et al.* (2000), that sinapic acid is probably not an important precursor for sinapyl alcohol. A more than 90% reduction in 4CL activity has been obtained in transgenic poplar by the antisense strategy (Hu *et al.*, 1999). This modification has led to a 45% reduction in lignin amount in the xylem. On the other hand, a 15% increase in cellulose was detected. The monomeric composition of lignin as determined by thioacidolysis, as well as the frequency of the main dimeric linkages in lignin (β-O-4, β-5, β-β, and α-keto-β-aryl ethers) as determined by NMR, were similar to those of control plants. An increase in non-lignin cell wall constituents, such as *p*-coumaric, ferulic, and sinapic acid, was evidenced by gas chromatography-mass spectrometry analysis (Hu *et al.*, 1999). In agreement with the data of Zhong *et al.* (2000), these results show that it is possible to largely reduce lignin amount in trees without affecting overall plant viability.

Cinnamoyl-CoA reductase (CCR) catalyses the reduction of the hydroxycinnamoyl-CoA esters to their corresponding aldehydes. Transgenic tobacco down-regulated for CCR have a decreased lignin content, an increased S/G ratio as well as an orange-brown coloration of the xylem (Piquemal *et al.*, 1998). Several chimeric *CCR* constructs under the control of the double *CaMV 35S* promoter have been used to transform poplar. Some transgenic lines showed an orange coloration, a decreased lignin content, and a higher S/G ratio in the lignin, similar to the results obtained in transgenic tobacco plants down-regulated

for CCR (J.-C. Leplé, C. Lapierre and W. Boerjan, unpublished).

Cinnamyl alcohol dehydrogenase (CAD) catalyses the last step in the monolignol biosynthesis pathway, which is the reduction of cinnamaldehydes to cinnamyl alcohols. A reduction of CAD activity has been described in tobacco (Halpin *et al.*, 1994; Hibino *et al.*, 1995, Yahiaoui *et al.*, 1998), alfalfa (Baucher *et al.*, 1999), poplar (Baucher *et al.*, 1996) and in a pine mutant (MacKay *et al.*, 1997) and has been associated with the presence of an increased amount of aldehydes in the lignin. The incorporation of cinnamaldehydes in the lignin of tobacco lines down-regulated for CAD has been demonstrated by Kim *et al.* (2000). In poplar, a 10% lower Klason lignin content has been detected in the wood of the transgenic lines (Lapierre *et al.*, 1999). In addition, the lignin has more free phenolic groups in G and S units, an important parameter that contributes to the solubility of lignin (Lapierre *et al.*, 1999). The wood of down-regulated CAD poplars is more easily delignified than control wood, as shown by the lower kappa number after chemical Kraft pulping (Baucher *et al.*, 1996; Lapierre *et al.*, 1999).

Peroxidases are widely believed to be responsible for the final condensation of cinnamyl alcohols to form lignin. However, the high redundancy in genes (more than 60 peroxidase genes have been identified from the *Arabidopsis* genome sequence; Tognolli *et al.*, 2000) and in functions (low substrate specificity) are the main obstacles to identify a peroxidase isoform that is specifically involved in lignification (reviewed by Christensen *et al.*, 2000). Two anionic peroxidase isoenzymes of poplar, shown to be preferentially expressed in developing xylem and to oxidize syringaldazine (a lignin monomer analogue; Goldberg *et al.*, 1983) have been purified and characterized (Christensen *et al.*, 1998). A cDNA corresponding to one of the syringaldazine-oxidizing peroxidases (PXP 3-4) has been cloned. Transgenic poplars that over produce this peroxidase under the control of the *CaMV 35S* promoter have up to 800-fold higher peroxidase activities in the developing xylem. However, no differences in lignin content or S/G composition were detected in this tissue (J.H. Christensen, C. Lapierre and W. Boerjan, unpublished). In contrast, tobacco lines with 10-fold higher peroxidase activities are characterized by a higher lignin content in the leaves, but also by a reduced growth and a browning of wounded tissues (Lagrimini *et al.*, 1990; Lagrimini, 1991).

Other enzymes believed to polymerize monolignols are the laccases, but the precise role played by these enzymes is unclear. Transgenic yellow poplar (Dean *et al.*, 1998) and poplar (Ranocha *et al.*, 2000) down-regulated for laccase did not have alterations in their lignin content. However, an increase in soluble phenolics has been detected in poplar, suggesting a possible role for laccase in the oxidation of simple phenolics leading to cross-linking in the cell wall (Ranocha *et al.*, 2000).

The altered expression of genes coding for MYB-related (Tamagnone *et al.*, 1998) and LIM-related (Kawaoka *et al.*, 2000) transcription factors has been shown to affect the expression of genes involved in the phenylpropanoid biosynthesis pathway and to affect lignin content and the accumulation of phenolics in transgenic tobacco plants. The *Arabidopsis* mutant *ectopic lignification in the pith (elpi)* shows ectopic deposition of lignin in the pith (Zhong *et al.*, 2000). ELPI is thought to act as a repressor of lignin biosynthesis in the pith because the expression of *PAL*, *CCoAOMT* and *CCR* are induced in the *elpi* mutant.

Temporal and spatial expression of lignin biosynthesis genes

The temporal and spatial expression of a number of genes from the lignin biosynthesis pathway has been studied. A strong activity of the *PAL* promoter has been detected in the developing xylem, in the endodermis as well as in flower tissues of tobacco (Kawamata *et al.*, 1997). *In situ* localization and protein gel blot analysis have shown that *PAL* and *C4H* are expressed in the vascular bundles, the epidermis and the ovules and that the expression of both genes is closely co-regulated in parsley (Koopmann *et al.*, 1999). The *C4H* promoter confers expression in all tissues in *Arabidopsis*, but predominantly in the vascular bundles (Bell-Lelong *et al.*, 1997). Genes coding for two isoforms of 4CL, isolated from poplar, are differentially expressed, one being specifically expressed in developing xylem and probably involved in lignin biosynthesis whereas the other, in the epidermis, is most probably associated with the biosynthesis of non-lignin phenylpropanoids (Hu *et al.*, 1998). Promoter-GUS and *in situ* hybridization studies in poplar have enabled the localization of *CAD* gene expression in the vascular tissues, the periderm and the cambium (Feuillet *et al.*, 1995; Hawkins *et al.*, 1997; Regan *et al.*, 1999). Immunolocalization studies have shown

that COMT is strictly expressed in lignifying cells of the phloem and the xylem. In the developing xylem, COMT is expressed in all cell types (C. Chen and W. Boerjan, unpublished).

Zhong *et al.* (2000) have immunolocalized CCoAOMT in all cell types of the developing xylem. In contrast, Chen *et al.* (2000) have shown by promoter-*GUS* analyses and immunolocalization studies that in poplar xylem, CCoAOMT is preferentially expressed in developing vessels and in contact ray cells, whereas it is barely detectable in isolation ray cells and developing xylem fibres (Chen *et al.*, 2000). As a consequence, down-regulation of CCoAOMT in transgenic poplar is expected to affect lignin biosynthesis in a cell-specific manner. This cell-specific effect could indeed be visualized by UV microscopy that showed an increased fluorescence specifically in the vessel cell walls (Meyermans *et al.*, 2000). Upon bending of the stem, however, the cell-specific expression is lost, and CCoAOMT becomes expressed in all cell types of developing xylem, i.e. fibres, vessel elements, contact and isolation rays cells (Chen *et al.*, 2000).

Subcellular localization of enzymes involved in lignin biosynthesis

Little is known about the mechanisms by which lignifying cells regulate the flux of phenylpropanoid metabolites into the different end products. Metabolic labelling experiments have suggested the channelling of lignin precursors by multi-enzyme complexes (Stafford, 1981; Hrazdina, 1992; Rasmussen and Dixon, 1999). PAL has been localized to the cytoplasm, to Golgi-derived vesicles and to secondary wall thickenings (Smith *et al.*, 1994; Nakashima *et al.*, 1997) and C4H has been shown to be associated with the endoplasmic reticulum membrane and Golgi vesicles (Smith *et al.*, 1994). CAD has been localised in the cytoplasm, on Golgi-derived vesicles, and on secondary wall thickenings (Nakashima *et al.*, 1997; Samaj *et al.*, 1998). CCoAOMT and COMT have been localized in the cytoplasm, not associated with any membranes (Ye, 1997; Chen *et al.*, 2000; C. Chen and W. Boerjan, unpublished).

New tools to study cell wall formation in wood

The vascular cambium is an inaccessible tissue and therefore not easily studied. Large organisms, such as trees, however, have the advantage that significant amounts of developing xylem can be obtained for molecular and chemical characterization. In addition, samples from defined developmental stages of xylogenesis can be obtained by tangential cryosectioning. Consecutive sectioning, in combination with microanalytical techniques, has been used to visualize patterns of gene expression, enzyme activities, amino acids, carbohydrates and plant hormones across developing wood tissues (Uggla and Sundberg, 2001). The exciting perspective is now to combine microanalysis with the emerging high-throughput techniques for large-scale analysis of gene and protein expression in wood forming tissues.

Large-scale EST sequencing of hybrid aspen has provided the necessary tools for global expression analysis of genes during wood formation. Within the Swedish consortium (http://www.biochem.kth.se/PopulusDB), a poplar microarray containing some 3000 unique cambial region ESTs has been produced. To take full advantage of this technology, a 3′ cDNA target amplification protocol has been established and validated that enables transcript profiling from sub-milligram amounts of plant tissue (Hertzberg et al., 2001). This technique has now been used in combination with tangential cryosectioning to visualize global gene expression patterns during xylogenesis from cambium to maturing xylem. The next generation of poplar microarrays currently under production will consist of about 10 000 unique EST sequences selected from a range of poplar cDNA libraries.

However, the expression of genes is not necessarily proportionally related to the abundance of the corresponding proteins. Therefore, the integration of both mRNA and protein data will give a more comprehensive view on xylogenesis. The identification and quantification of the proteins present in a particular tissue or in a particular developmental or environmental condition is possible through the developments in mass spectrometry combined with two-dimensional protein gel electrophoresis. Matrix-assisted laser desorption-ionization/time-of-flight (MALDI-TOF) mass spectrometry and tandem mass spectrometry are powerful high-throughput methods for protein identification, provided the DNA or protein sequences already exist in databases (Chalmers and Gaskell, 2000). In addition, the recent development of protein arrays will definitively allow to link genomics with proteomics (Ge, 2000). In this system, fully active proteins are spotted onto membranes that can be used for several purposes, such as the study of interactions with nucleic acids, proteins, and other ligands.

The identification of proteins that are highly abundant in developing xylem tissue, through two-dimensional protein gel electrophoresis, has been initiated in poplar (Vander Mijnsbrugge et al., 2000) and pine (Plomion et al., 2000). The most abundant soluble protein on two-dimensional protein gels of poplar xylem has been shown to have phenylcoumaran benzylic ether reductase (PCBER) activity in vitro (Gang et al., 1999). PCBER reduces dehydrodiconiferyl alcohol (DDC) to isodihydrodehydrodiconiferyl alcohol (IDDDC) and is thus involved in lignan biosynthesis. The function of another abundant protein found in developing poplar xylem, a serine hydroxymethyltransferase (SHMT) homologue, is still unknown. Interestingly, ESTs for PCBER and SHMT are highly abundant in both the poplar and pine xylem EST libraries. Our sparse knowledge on the biological role of these abundant proteins indicates how little we know about important processes in wood formation. Other highly abundant spots correspond to five S-adenosyl-L-methionine (SAM) synthetases, one COMT and two CCoAOMT, all playing important roles in lignin biosynthesis. For all of these proteins, the corresponding ESTs are abundant in the xylem EST library too. It is conceivable that poplar genomics and proteomics, combined with the Zinnia xylogenic system and with reverse and forward genetics approaches in Arabidopsis, will result in the identification of marker genes for wood formation, in the discovery of new genes and their function, and in novel insight into biological processes of wood formation and of growth and development of woody species.

Acknowledgements

This work was supported by the Swedish Wood Ultrastructure Research Centre (WURC), the Foundation for Strategic Research (SSF) and the Belgian Fonds National de la Recherche Scientifique (FNRS). We thank N. Chaffey and L. Kleczkowski for critical reading of parts of the manuscript, K. Pickering, J. R. Barnett, S. Fink, L. Puech, P. Barlow and N. Chaffey for providing original figures and materials for illustrations, and K. Olofsson for technical assistance with photography.

References

Abe, H., Funada, R., Imaizumi, H., Ohtani, J. and Fukazawa, K. 1995. Dynamic changes in the arrangement of cortical microtubules in conifer tracheids during differentiation. Planta 197: 418–421.

Abe, H., Funada, R., Ohtani, J. and Fukazawa, K. 1997. Changes in the arrangement of cellulose microfibrils associated with the cessation of cell expansion in tracheids. Trees 11: 328–332.

Abel, S., Nguyen, M.D., Chow, W. and Theologis, A. 1995. ASC4, a primary indoleacetic acid-responsive gene encoding 1-aminocyclopropane-1-carboxylate synthase in Arabidopsis thaliana. Structural characterization, expression in Escherichia coli, and expression characteristics in response to auxin. J. Biol. Chem. 270: 19093–19099.

Allina, S.M., Pri-Hadash, A., Theilmann, D.A., Ellis, B.E. and Douglas, C.J. 1998. 4-Coumarate:coenzyme A ligase in hybrid poplar. Plant Physiol. 116: 743–754.

Allona, I., Quinn, M., Shoop, E., Swope, K., St. Cyr, S., Carlis, J., Riedl, J., Retzel, E., Campbell, M.M., Sederoff, R. and Whetten, R.W. 1998. Analysis of xylem formation in pine by cDNA sequencing. Proc. Natl. Acad. Sci. USA 95: 9693–9698.

Aloni, R. 1979. Role of auxin and gibberellin in differentiation of primary phloem fibers. Plant Physiol. 63: 609–614.

Aloni, R. 1991. Wood formation in deciduous hardwood trees. In: A.S. Raghavendra (Ed.) Physiology of Trees, John Wiley, New York, pp. 175–197.

Aloni, R., Tollier, M.T. and Monties, B. 1990. The role of auxin and gibberellin in controlling lignin formation in primary phloem fibers and in xylem of Coleus blumei stems. Plant Physiol. 94: 1743–1747.

Araki, N., Fujita, M., Saiki, H. and Harada, H. 1982. Transition of the fiber wall structure from normal wood to tension wood in Robinia pseudoacacia L. and Populus euroamericana Guinier. Mokuzai Gakkaishi 28: 267–273.

Arioli, T., Peng, L.C., Betzner, A.S., Burn, J., Wittke, W., Herth, W., Camilleri, C., Höfte, H., Plazinski, J., Birch, R., Cork, A., Glover, J., Redmond, J. and Williamson, R.E. 1998. Molecular analysis of cellulose biosynthesis in Arabidopsis. Science 279: 717–720.

Awano, T., Takabe, K. and Fujita, M. 1998. Localization of glucuronoxylans in Japanese beech visualized by immunogold labelling. Protoplasma 202: 213–222.

Awano, T., Takabe, K., Fujita, M. and Daniel, G. 2000. Deposition of glucuronoxylans on the secondary cell wall of Japanese beech as observed by immune-scanning electron microscopy. Protoplasma 212: 72–79.

Baba, K.-i., Adachi, K., Take, T., Yokoyama, T., Itho, T., and Nakamura, T. 1995. Induction of tension wood in GA$_3$-treated branches of the weeping type of Japanese cherry, Prunus spachiana. Plant Cell Physiol. 36: 983–988.

Baïer, M., Goldberg, R., Catesson, A.-M., Liberman, M., Bouchemal, N., Michon, V. and Hervé du Penhoat, C. 1994. Pectin changes in samples containing poplar cambium and inner bark in relation to the seasonal cycle. Planta 193: 446–454.

Bailey, I.W. 1954. Contributions to plant anatomy. Chronanica Botanica, Waltham, MA.

Baima, S., Nobili, F., Sessa, G., Lucchetti, S., Ruberti, I. and Morelli, G. 1995. The expression of the Athb-8 homeobox gene is restricted to provascular cells in Arabidopsis thaliana. Development 121: 4171–4182.

Barnett, J.R. 1992. Reactivation of the cambium in Aesculus hippocastanum L.: a transmission electron microscope study. Ann. Bot. 70: 169–177.

Barnett, J.R. 1995. Ultrastructural factors affecting xylem differentiation. In: M. Iqbal (Ed.) The Cambial Derivatives (Handbuch der Pflanzenanatomie, Band IX, Teil 4: Spezieller Teil), Borntraeger, Berlin, pp. 107–130.

Barnett, J.R. and Harris, J.M. 1975. Early stages of bordered pit formation in radiata pine. Wood Sci. Technol. 9: 233–241.

Bate, N.J., Orr, J., Ni, W., Meromi, A., Nadler-Hassar, T., Doerner, P.W., Dixon, R.A., Lamb, C.J. and Elkind, Y. 1994. Quantitative relationship between phenylalanine ammonia-lyase levels and phenylpropanoid accumulation in transgenic tobacco identifies a rate-determining step in natural product synthesis. Proc. Natl. Acad. Sci. USA 91: 7608–7612.

Baucher, M., Chabbert, B., Pilate, G., Van Doorsselaere, J., Tollier, M.-T., Petit-Conil, M., Cornu, D., Monties, B., Van Montagu, M., Inzé, D., Jouanin, L. and Boerjan, W. 1996. Red xylem and higher lignin extractability by down-regulating a cinnamyl alcohol dehydrogenase in poplar). Plant Physiol. 112: 1479–1490.

Baucher, M., Monties, B., Van Montagu, M. and Boerjan, W. 1998. Biosynthesis and genetic engineering of lignin. Crit. Rev. Plant Sci. 17: 125–197.

Baucher, M., Bernard-Vailhé, M.A., Chabbert, B., Besle, J.-M., Opsomer, C., Van Montagu, M. and Botterman, J. 1999. Down-regulation of cinnamyl alcohol dehydrogenase in transgenic alfalfa (Medicago sativa L.) and the impact on lignin composition and digestibility. Plant Mol. Biol. 39: 437–447.

Bell-Lelong, D.A., Cusumano, J.C., Meyer, K. and Chapple, C. 1997. Cinnamate-4-hydroxylase expression in Arabidopsis. Plant Physiol. 113: 729–738.

Benayoun, J. 1983. A cytochemical study of cell wall hydrolysis in the secondary xylem of poplar (Populus italica Moench). Ann. Bot. 52: 189–200.

Benayoun, J., Catesson, A.M. and Czaninski, Y. 1981. A cytochemical study of differentiation and breakdown of vessel end walls. Ann. Bot. 47: 687–698.

Benjavongkulchai, E. and Spencer, M.S. 1986. Purification and characterization of barley aleurone xylanase. Planta 169: 415–419.

Bentum, A.L.K., Côté, W.A. Jr., Day, A.C. and Timell, T.E. 1969. Distibution of lignin in normal and tension wood. Wood Sci. Technol. 3: 218–231.

Bih, F.Y., Wu, S.S.H., Ratnayake, C., Walling, L.L., Nothnagel, E.A. and Huang, A.H.C. 1999. The predominant protein on the surface of maize pollen is an endoxylanase synthesized by a tapetum mRNA with a long 5′ leader. J. Biol. Chem. 274: 22884–22894.

Bisset, I.J.W. and Dadswell, H.E. 1950. The variation in cell length within one growth ring of certain angiosperms and gymnosperms. Aust. For. 14: 17–29.

Blount, J.W., Korth, K.L., Masoud, S.A., Rasmussen, S., Lamb, C. and Dixon, R.A. 2000. Altering expression of cinnamic acid 4-hydroxylase in transgenic plants provides evidence for a feedback loop at the entry point into the phenylpropanoid pathway. Plant Physiol. 122: 107–116.

Bolwell, G.P. 1993. Dynamic aspects of the plant extracellular matrix. Int. Rev. Cytol. 146: 261–324.

Bolwell, G.P. and Northcote, D.H. 1981. Control of hemicellulose and pectin synthesis during differentiation of vascular tissue in bean (Phaseolus vulgaris) callus and in bean hypocotyl. Planta 152: 225–233.

Bolwell, G.P. and Northcote, D.H. 1983. Induction by growth factors of polysaccharide synthases in bean cell suspension cultures. Biochem. J. 210: 509–515.

Bolwell, G.P., Dalessandro, G. and Northcote, D.H. 1985. Decrease of polygalacturonic acid synthase during xylem differentiation in sycamore. Phytochemistry 24: 699–702.

Bonin, C.P., Potter, I., Vanzin, G.F. and Reiter, W.-D. 1997. The *MUR1* gene of *Arabidopsis thaliana* encodes an isoform of GDP-D-mannose-4,6-dehydratase, catalyzing the first step in the *de novo* synthesis of GDP-L-fucose. Proc. Natl. Acad. Sci. USA 94: 2085–2090.

Boudet, A.-M. 2000. Lignin and lignification: selected issues. Plant Physiol. Biochem. 38: 81–96.

Boyce, S.G. and Kaeiser, M. 1961. Environmental and genetic variability in the length of fibers of eastern cottonwood. TAPPI 44: 363–366.

Bradshaw, H.D. Jr. and Stettler, R.F. 1993. Molecular genetics of growth and development in *Populus*. I. Triploidy in hybrid poplars. Theor. Appl. Genet. 86: 301–307.

Bradshaw, H.D. Jr., Villar, M., Watson, B.D., Otto, K.G., Stewart, S. and Stettler, R.F. 1994. Molecular genetics of growth and development in *Populus*. III. A genetic linkage map of a hybrid poplar composed of RFLP, STS, and RAPD markers. Theor. Appl. Genet. 89: 167–178.

Bragina, T.V., Martinovitch, L.I., Rodionova, N.A., Bezborodov, A.M., and Grineva, G.M. 1999. Effects of stress induced by total submergence on cellulase and xylanase activities in adventitious roots of maize. Appl. Biochem. Microbiol. 35: 407–410.

Brett, C.T. 2000. Cellulose microfibrils in plants: biosynthesis, deposition, and integration into the cell wall. Int. Rev. Cytol. 199: 161–199.

Brett, C.T. and Waldron, K.W. 1996. Physiology and Biochemistry of Plant Cell Walls. Chapman and Hall, London.

Butterfield, B.G. 1995. Vessel element differentiation. In: M. Iqbal (Ed.) The Cambial Derivatives (Handbuch der Pflanzenanatomie, Band IX, Teil 4: Spezieller Teil), Borntraeger, Berlin, pp. 93–106.

Campbell, M.M. and Sederoff, R.R. 1996. Variation in lignin content and composition. Mechanisms of control and implications for the genetic improvement of plants. Plant Physiol. 110: 3–13.

Carpita, N.C. and Gibeaut, D.M. 1993. Structural models of primary cell walls in flowering plants: consistency of molecular structure with the physical properties of the walls during growth. Plant J. 3: 1–30.

Catesson, A.-M. 1964. Origine, fonctionnement et variations cytologiques saisonnières du cambium de l'*Acer pseudoplatanus* L. (Acéracées). Ann. Sci. Nat. Bot. (12ème série) 5: 229–498.

Catesson, A.-M. 1989. Specific characters of vessel primary walls during the early stages of wood differentiation. Biol. Cell 67: 221–226.

Catesson, A.-M. 1990. Cambial cytology and biochemistry. In: M. Iqbal (Ed.) The Vascular Cambium (Research Studies in Botany and Related Applied Fields, vol. 7), Research Studies Press, Taunton, USA, pp. 63–112.

Catesson, A.M. and Roland, J.C. 1981. Sequential changes associated with cell wall formation and fusion in the vascular cambium. IAWA Bull. 2: 151–162.

Catesson, A.M., Funada, R., Robertbaby, D., Quinetszely, M., Chuba, J. and Goldberg R. 1994. Biochemical and cytochemical cell-wall changes across the cambial zone. IAWA J. 15: 91–101.

Cervera, M.-T., Storme, V., Ivens, B., Gusmão, J., Liu, B.H., Hostyn, V., Van Slycken, J., Van Montagu, M., Boerjan, W. 2001. Dense genetic linkage maps of three Populus species (*Populus deltoides*, *P. nigra* and *P. trichocarpa*) based on Amplified Fragment Length Polymorphism and Microsatellite Markers. Genetics 158: 787–809.

Chaffey, N., Barnett, J. and Barlow, P. 1997a. Endomembranes, cytoskeleton, and cell walls: aspects of the ultrastructure of the vascular cambium of taproots of *Aesculus hippocastanum* L. (Hippocastanaceae). Int. J. Plant Sci. 158: 97–109.

Chaffey, N., Barlow, P. and Barnett, J. 1997b. Cortical microtubules rearrange during differentiation of vascular cambial derivatives, microfilaments do not. Trees Struct. Funct. 11: 333–341.

Chaffey, N.J., Barlow, P.W. and Barnett, J.R. 1998. A seasonal cycle of cell wall structure is accompanied by a cyclical rearrangement of cortical microtubules in fusiform cambial cells within taproots of *Aesculus hippocastanum* (Hippocastanaceae). New Phytol. 139: 623–635.

Chaffey, N., Barnett, J. and Barlow, P. 1999. A cytoskeletal basis for wood formation in angiosperm trees: the involvement of cortical microtubules. Planta 208: 19–30.

Chalmers, M.J. and Gaskell, S.J. 2000. Advances in mass spectrometry for proteome analysis. Curr. Opin. Biotechnol. 11: 384–390.

Chapple, C.C.S., Vogt, T., Ellis, B.E. and Somerville, C.R. 1992. An *Arabidopsis* mutant defective in the general phenylpropanoid pathway. Plant Cell 4: 1413–1424.

Chen, F., Yasuda, S. and Fukushima, K. 1999. Evidence for a novel biosynthetic pathway that regulates the ratio of syringyl to guaiacyl residues in lignin in the differentiating xylem of *Magnolia kobus* DC. Planta 207: 597–603.

Chen, C., Meyermans, H., Burggraeve, B., De Rycke, R.M., Inoue, K., De Vleesschauwer, V., Steenackers, M., Van Montagu, M.C., Engler, G.J. and Boerjan, W.A. 2000. Cell-specific and conditional expression of caffeoyl-CoA O-methyltransferase in poplar. Plant Physiol. 123: 853–867.

Christensen, J.H., Bauw, G., Welinder, K.G., Van Montagu, M. and Boerjan, W. 1998. Purification and characterization of peroxidases correlated with lignification in poplar xylem. Plant Physiol. 118: 125–135.

Christensen, J.C., Baucher, M., O'Connell, A.P., Van Montagu, M. and Boerjan, W. 2000. Control of lignin biosynthesis. In: S.M. Jain and S.C. Minocha (Eds.) Molecular Biology of Woody Plants, vol. 1 (Forestry Sciences, vol. 64), Kluwer Academic Publishers, Dordrecht, Netherlands, pp. 227–267.

Cosgrove, D.J. 1997. Assembly and enlargement of the primary cell wall in plants. Annu. Rev. Cell Dev. Biol. 13: 171–201.

Cosgrove, D.J. 2000. Expansive growth of plant cell walls. Plant Physiol. Biochem. 38: 109–124.

Côté, W.A. 1977. Wood ultrastructure in relation to chemical composition. In: F.A. Loewus and V.C. Runeckles (Eds.) The Structure, Biosynthesis, and Degradation of Wood (Recent Advances in Phytochemistry, vol. 11), Plenum, New York, pp. 1–44.

Couchy, I., Minic, Z., Laporte, J., Brown, S. and Satiat-Jeunemaitre, B. 1998. Immunodetection of Rho-like plant proteins with Rac1 and Cdc42Hs antibodies. J. Exp. Bot. 49: 1647–1659.

Dalessandro, G. and Northcote, D.H. 1977. Changes in enzymatic activities of nucleoside diphosphate sugar interconversions during differentiation of cambium to xylem in sycamore and poplar. Biochem. J. 162: 267–279.

Darley, C.P., Forrester, A.M. and McQueen-Mason, S.J. 2001. The molecular basis of plant cell wall extension. Plant Mol. Biol. 47: 171–187.

Dean, J.F.D., LaFayette, P.R., Rugh, C., Tristram, A.H., Hoopes, J.T., Eriksson, K.-E.L. and Merckle, S.A. 1998. Laccases associate with lignifying vascular tissues. In: N.G. Lewis and S. Sarkanen (Eds) Lignin and Lignan Biosynthesis (ACS Symposium Series, vol. 697), American Chemical Society, Washington, DC, pp. 96–108.

del Campillo, E. 1999. Multiple endo-1,4–D-glucanase (cellulase) genes in *Arabidopsis*. Curr. Top. Dev. Biol. 46: 39–61.

Delmer, D.P., Pear, J.R., Andrawis, A. and Stalker, D.M. 1995. Genes encoding small GTP-binding proteins analogous to mammalian rac are preferentially expressed in developing cotton fibers. Mol. Gen. Genet. 248: 43–51.

Dhillon, S.S. 1987. DNA in tree species. In: J.M. Bonga and D.J. Durzan (Eds.) Cell and Tissue Culture in Forestry, vol. 1, Martinus Nijhoff, Dordrecht, Netherlands, pp. 298–313.

Dhugga, K.S., Tiwari, S.C. and Ray, P.M. 1997. A reversibly glycosylated polypeptide (RGP1) possibly involved in plant cell wall synthesis: purification, gene cloning, and trans-Golgi localization. Proc. Natl. Acad. Sci. USA 94: 7679–7684.

Digby, J. and Wareing, P.F. 1966. The effect of applied growth hormones on cambial division and the differentiation of the cambial derivatives. Ann. Bot. 30: 539–548.

Dodd, R.S. and Fox, P. 1990. Kinetics of tracheid differentiation in Douglas-fir. Ann. Bot. 65: 649–657.

Domingo, C., Roberts, K., Stacey, N.J., Connerton, I., Ruíz-Teran, F. and McCann, M.C. 1998. A pectate lyase from *Zinnia elegans* is auxin inducible. Plant J. 13: 17–28.

Eklund, L. and Little, C.H.A. 1996. Laterally applied Etherel causes local increases in radial growth and indole-3-acetic acid concentration in *Abies balsamea* shoots. Tree Physiol. 16: 509–513.

Elkind, Y., Edwards, R., Mavandad, M., Hedrick, S.A., Ribak, O., Dixon, R.A. and Lamb, C.J. 1990. Abnormal plant development and down-regulation of phenylpropanoid biosynthesis in transgenic tobacco containing a heterologous phenylalanine ammonia-lyase gene. Proc. Natl. Acad. Sci. USA 87: 9057–9061.

Eriksson, M.E., Israelsson, M., Olsson, O. and Moritz, T. 2000. Increased gibberellin biosynthesis in transgenic trees promotes growth, biomass production and xylem fiber length. Nature Biotechnol. 18: 784–788.

Ermel, F.F., Follet-Gueye, M.-L., Cibert, C., Vian, B., Morvan, C., Catesson, A.-M. and Goldberg, R. 2000. Differential localization of arabinan and galactan side chains of rhamnogalacturonan 1 in cambial derivatives. Planta 210: 732–740.

Evert, R.F. and Kozlowski, T.T. 1967. Effect of isolation of bark on cambial activity and development of xylem and phloem in trembling aspen. Am. J. Bot. 54: 1045–1054.

Evert, R.F., Kozlowski, T.T. and Davis, J.D. 1972. Influence of phloem blockage on cambial growth of sugar maple. Am. J. Bot. 59: 632–641.

Faik, A., Bar-Peled, M., DeRocher, A.E., Zeng, W., Perrin, R.M., Wilkerson, C., Raikhel, N.V. and Keegstra, K. 2000. Biochemical characterization and molecular cloning of an α-1,2-fucosyltransferase that catalyzes the last step of cell wall xyloglucan biosynthesis in pea. J. Biol. Chem. 275: 15082–15089.

Fagard, M., Desnos, T., Desprez, T., Goubet, F., Refregier, G., Mouille, G., McCann, M., Rayon, C., Vernhettes, S. and Höfte, H. 2001. *PROCUSTE1* encodes a cellulose synthase required for normal cell elongation specifically in roots and dark-grown hypocotyls of arabidopsis. Plant Cell 12: 2409–2423.

Fergus, B.J. and Goring, D.A.I. 1970. The location of guaiacyl and syringyl lignins in birch xylem tissue. Holzforschung 24: 113–117.

Feuillet, C., Lauvergeat, V., Deswarte, C., Pilate, G., Boudet, A. and Grima-Pettenati, J. 1995. Tissue- and cell-specific expression of a cinnamyl alcohol dehydrogenase promoter in transgenic poplar plants. Plant Mol. Biol. 27: 651–667.

Follet-Gueye, M.L., Ermel, F.F., Vian, B., Catesson, A.M. and Goldberg, R. 2000. In: R. Savidge, J. Barnett and R. Napier (Eds)

Cell and Molecular Biology of Wood Formation, BIOS Scientific Publications, Oxford, pp. 289–294.

Franke, R., McMichael, C.M., Meyer, K., Shirley, A.M., Cusumano, J.C. and Chapple, C. 2000. Modified lignin in tobacco and poplar plants over-expressing the *Arabidopsis* gene encoding ferulate 5-hydroxylase. Plant J. 22: 223–234.

Frewen, B.E., Chen, T.H.H., Howe, G.T., Davis, J., Rohde, A., Boerjan, W. and Bradshaw, H.D. Jr. 2000. Quantitative trait loci and candidate gene mapping of bud set and bud flush in *Populus*. Genetics 154: 837–845.

Fujii, T., Harada, H. and Saiki H. 1979. The layered structure of ray parenchyma secondary wall in the wood of 49 Jananese angiosperm species. Mokuzai Gakkaishi 25: 251–257.

Fujii, T., Harada, H. and Saiki, H. 1981. Ultrastructure of 'amorphous layer' in xylem parenchyma cell wall of angiosperm species. Mokuzai Gakkaishi 27: 149–156.

Fujino, T. and Itoh, T. 1998. Changes in the three dimensional architecture of the cell wall during lignification of xylem cells in *Eucalyptus tereticornis*. Holzforschung 52: 111–116.

Fujita, M., Sakai, H. and Harada, H. 1974. Electron microscopy of microtubules and cellulose microfibrils in secondary wall formation of poplar tension wood. Mokuzai Gakkaishi 20: 147–156.

Fukuda, H. 1992. Tracheary element formation as a model system of cell differentiation. Int. Rev. Cytol. 136: 289–332.

Fukuda, H. 1997. Tracheary element differentiation. Plant Cell 9: 1147–1156.

Fukuda, H. and Komamine, A. 1980. Direct evidence for cytodifferentiation to tracheary elements without intervening mitosis in a culture of single cells isolated from the mesophyll of *Zinnia elegans*. Plant Physiol. 65: 61–64.

Fukuda, T. and Terashima, N. 1988. Heterogeneity in formation of lignin XII. Deposition of chemical components during the formation of cell walls of black pine and poplar. Makuzai Gakkaishi 34: 604–608.

Funada, R. 2000. Control of wood structure. In: P. Nick (Ed.) Plant Microtubules, Springer-Verlag, Berlin, pp. 51–82.

Funada, R. and Catesson, A.M. 1991. Partial cell-wall lysis and the resumption of meristematic activity in *Fraxinus excelsior* cambium. IAWA Bull. 12: 439–444.

Gahan, P.B. 1988. Xylem and phloem differentiation in perspective. In: L.W. Roberts, P.B. Gahan and R. Aloni (Eds.) Vascular Differentiation and Plant Growth Regulators, Springer-Verlag, Berlin, pp. 1–21.

Gang, D.R., Kasahara, H., Xia, Z.-Q., Vander Mijnsbrugge, K., Bauw, G., Boerjan, W., Van Montagu, M., Davin, L.B. and Lewis, N.G. 1999. Evolution of plant defense mechanisms. Relationships of phenylcoumaran benzylic ether reductases to pinoresinol-lariciresinol and isoflavone reductases. J. Biol. Chem. 274: 7516–7527.

Ge, H. 2000. UPA, a universal protein array system for quantitative detection of protein-protein, protein-DNA, protein-RNA and protein-ligand interactions. Nucl. Acids Res. 28: e3.

Gibeaut, D.M. 2000. Nucleotide sugars and glycosyltransferases for synthesis of cell wall matrix polysaccharides. Plant Physiol. Biochem. 38: 69–80.

Goldberg, R., Catesson, A.-M. and Czaninski, Y. 1983. Some properties of syringaldazine oxidase, a peroxidase specifically involved in the lignification processes. Z. Pflanzenphysiol. 110: 267–279.

Goosen-de Rao, L., Bakhuizen, R., van Spronsen, P. and Libbenga, K.R. 1984. The presence of extended phragmosomes containing cytoskeletal elements in fusiform cambial cells of *Fraxinus excelsior* L. Protoplasma 122: 145–152.

Goring, D.A.I. and Timell, T.E. 1962. Molecular weight of native celluloses. TAPPI 45: 454–460.

Gorshkova, T.A., Chemikosova, S.B., Lozovaya, V.V. and Carpita, N.C. 1997. Turnover of galactans and other cell wall polysaccharides during development of flax plants. Plant Physiol. 114: 723–729.

Goto, M., Takabe, K. and Abe, I. 1998. Histochemistry and UV-microspectrometry of cell walls of untreated and ammonia-treated barley straw. Can. J. Plant Sci. 78: 437–443.

Gregory, A.C.E., O'Connell, A.P. and Bolwell, G.P. 1998. Xylans. Biotechnol. Genet. Eng. Rev. 15: 439–455.

Gregory, R.A. 1971. Cambial activity in Alaskan white spruce. Am. J. Bot. 58: 160–171.

Grima-Pettenati, J. and Goffner, D. 1999. Lignin genetic engineering revisited. FEBS Lett. 145: 51–65.

Grimmig, B., Kneusel, R.E., Junghanns, K.T. and Matern, U. 1999. Expression of bifunctional caffeoyl-CoA 3-O-methyltransferase in stress compensation and lignification. Plant Biol. 1: 299–310.

Guglielmino, N., Liberman, M., Jauneau, A., Vian, B., Catesson, A.M. and Goldberg, R. 1997a. Pectin immunolocalization and calcium visualization in differentiating derivatives from poplar cambium. Protoplasma 199: 151–160.

Guglielmino, N., Liberman, M., Catesson, A.M., Mareck, A., Prat, R., Mutaftschiev, S. and Goldberg, R. 1997b. Pectin methylesterases from poplar cambium and inner bark: localization, properties and seasonal changes. Planta 202: 70–75.

Hafren, J., Daniel, G. and Westmark, U. 2000. The distribution of acidic and esterified pectin in cambium, developing xylem and mature xylem of Pinus sylvestris. IAWA J. 21: 157–168.

Haigler, C.H. and Brown, R.M. Jr. 1986. Transport of rosettes from the Golgi apparatus to the plasma membrane in isolated mesophyll cells of Zinnia elegans during differentiation to tracheary elements in suspension culture. Protoplasma 134: 111–120.

Haigler, C.H., Ivanova-Datcheva, M., Hogan, P.S., Salnikov, V.V., Hwang, S., Martin, L.K. and Delmer, D.P. 2001. Carbon partitioning to cellulose synthesis. Plant Mol. Biol., this issue.

Halpin, C., Knight, M.E., Foxon, G.A., Campbell, M.M., Boudet, A.M., Boon, J.J., Chabbert, B., Tollier, M.-T. and Schuch, W. 1994. Manipulation of lignin quality by downregulation of cinnamyl alcohol dehydrogenase. Plant J. 6: 339–350.

Han, K.-H., Gordon, M.P. and Strauss, S.H. 1997. High-frequency transformation of cottonwoods (genus Populus) by Agrobacterium rhizogenes. Can. J. For. Res. 27: 464–470.

Han, K.-H., Meilan, R., Ma, C. and Strauss, S.H. 2000. An Agrobacterium tumefaciens transformation protocol effective on a variety of cottonwood hybrids (genus Populus). Plant Cell Rep. 19: 315–320.

Harada, H. and Côté, W.A. Jr. 1985. Structure of wood. In: T. Higuchi (Ed.) Biosynthesis and Biodegradation of Wood Components, Academic Press, Orlando, FL, pp. 1–44.

Hauch, S. and Magel, E. 1998. Extractable activities and protein content of sucrose-phosphate synthase, sucrose synthase and neutral invertase in trunk tissues of Robinia pseudoacacia L. are related to cambial wood production and heartwood formation. Planta 207: 266–274.

Hawkins, S., Samaj, J., Lauvergeat, V., Boudet, A. and Grima-Pettenati, J. 1997. Cinnamyl alcohol dehydrogenase: identification of new sites of promoter activity in transgenic poplar. Plant Physiol. 113: 321–325.

Hayashi, T. 1989. Xyloglucans in the primary cell wall. Annu. Rev. Plant Physiol. Plant Mol. Biol. 40: 139–168.

He, L. and Terashima, N. 1991. Formation and structure of lignin in monocotyledons. IV. Deposition process and structural diversity of the lignin in the cell wall of sugarcane and rice plants studied by ultraviolet microscopic spectroscopy. Holzforschung 45: 191–198.

Hejnowicz, A. and Hejnowicz, Z. 1958. Variation of length of vessel members and fibres in the trunk of Populus tremula L. Acta Soc. Bot. Pol. 27: 131–159.

Hertzberg, M., Sievertzon, M., Aspeborg, H., Nilsson, P., Sandberg, G. and Lundeberg, J. 2001. cDNA microarray analysis of small plant tissue samples using a cDNA tag target amplification protocol. Plant J., in press.

Hibino, T., Takabe, K., Kawazu, T., Shibata, D. and Higuchi, T. 1995. Increase of cinnamaldehyde groups in lignin of transgenic tobacco plants carrying an antisense gene for cinnamyl alcohol dehydrogenase. Biosci. Biotech. Biochem. 59: 929–931.

Holland, N., Holland, D., Helentjaris, T., Dhugga, K.S., Xoconostle-Cazares, B. and Delmer, D.P. 2000. A comparative analysis of the plant cellulose synthase (CesA) gene family. Plant Physiol. 123: 1313–1323.

Hrazdina, G. 1992. Compartmentation in aromatic metabolism. In: H.A. Stafford and R.K. Ibrahim (Eds.) Phenolic Metabolism in Plants (Recent Advances in Phytochemistry, vol. 26), Plenum, New York, pp. 1–23.

Hu, W.-J., Kawaoka, A., Tsai, C.-J., Lung, J., Osakabe, K., Ebinuma, H. and Chiang, V.L. 1998. Compartmentalized expression of two structurally and functionally distinct 4-coumarate:CoA ligase genes in aspen (Populus tremuloides). Proc. Natl. Acad. Sci. USA 95: 5407–5412.

Hu, W.-J., Harding, S.A., Lung, J., Popko, J.L., Ralph, J., Stokke, D.D., Tsai, C.-J. and Chiang, V.L. 1999. Repression of lignin biosynthesis promotes cellulose accumulation and growth in transgenic trees. Nature Biotechnol. 17: 808–812.

Humphreys, J.M., Hemm, M.R. and Chapple, C. 1999. New routes for lignin biosynthesis defined by biochemical characterization of recombinant ferulate 5-hydroxylase, a multifunctional cytochrome P450-dependent monooxygenase. Proc. Natl. Acad. Sci. USA 96: 10045–10050.

Im, K.-H., Cosgrove, D.J. and Jones, A.M. 2000. Subcellular localization of expansin mRNA in xylem cells. Plant Physiol. 123: 463–470.

Imai, T. and Terashima, N. 1992. Determination of the distribution and reaction of polysaccharides in wood cell-walls by the isotope tracer technique III. Visualisation of the deposition and distribution of galacturanan in the cell walls of magnolia (Magnolia kobus DC.) xylem by microautoradiography. Mokuzai Gakkaishi 38: 475–481.

Ingold, E., Sugiyama, M. and Komamine, A. 1988. Secondary cell wall formation: changes in cell wall constituents during the differentiation of isolated mesophyll cells of Zinnia elegans to tracheary elements. Plant Cell Physiol. 29: 295–303.

Inoue, K., Sewalt, V.J.H., Ballance, G.M., Ni, W., Stürzer, C. and Dixon, R.A. 1998. Developmental expression and substrate specificities of alfalfa caffeic acid 3-O-methyltransferase and caffeoyl coenzyme A 3-O-methyltransferase in relation to lignification. Plant Physiol. 117: 761–770.

Iqbal, M. and Ghouse, A.H.M. 1990. Cambial concept and organisation. In: M. Iqbal (Ed.) The Vascular Cambium (Research Studies in Botany and Related Applied Fields, vol. 7), Research Studies Press, Taunton, USA, pp. 1–36.

Jones, J.K.N., Purves, C.B. and Timell, T.E. 1961. Constitution of a 4-O-methylglucuronoxylan from the wood of trembling aspen (Populus tremuloides Michx.). Can. J. Chem. 39: 1059–1066.

Jones, L., Seymour, G.B. and Knox, J.P. 1997. Localization of pectic galactan in tomato cell walls using a monoclonal antibody specific to (14)-D-galactan. Plant Physiol. 113: 1405–1412.

Joseleau, J.-P. and Ruel, K. 1997. Study of lignification by non-invasive techniques in growing maize internodes. An investigation by Fourier transform infrared cross-polarization-magic angle spinning ^{13}C-nuclear magnetic resonance spectroscopy and immunocytochemical transmission electron microscopy. Plant Physiol. 114: 1123–1133.

Jouanin, L., Goujon, T., de Nadaï, V., Martin, M.-T., Mila, I., Vallet, C., Pollet, B., Yoshinaga, A., Chabbert, B., Petit-Conil, M. and Lapierre, C. 2000. Lignification in transgenic poplars with extremely reduced caffeic acid O-methyltransferase activity. Plant Physiol. 123: 1363–1373.

Jung, H.-J.G., Ni, W., Chapple, C.C.S. and Meyer, K. 1999. Impact of lignin composition on cell-wall degradability in an *Arabidopsis* mutant. J. Sci. Food Agric. 79: 922–928.

Kaeiser, M. 1955. Frequency and distribution of gelatinous fibers in eastern cottonwood. Am. J. Bot. 42: 331–334.

Kaeiser M. 1964. Vascular cambial initials in Eastern cottonwood in relation to mature wood cells derived from them. Trans. Ill. Acad. Sci. 57: 182–184.

Kaeiser, M. and Stewart, K.D. 1955. Fiber size in *Populus deltoides* Marsh in relation to lean of trunk and position in trunk. Bull. Torrey Bot. Club 82: 57–61.

Kakegawa, K., Edashige, Y. and Ishii, T. 2000. Metabolism of cell wall polysaccharides in cell suspension cultures of *Populus alba* in relation to cell growth. Physiol. Plant. 108: 420–425.

Kawamata, S., Shimoharai, K., Imura, Y., Ozaki, M., Ichinose, Y., Shiraishi, T., Kunoh, H. and Yamada, T. 1997. Temporal and spatial pattern of expression of the pea phenylalanine ammonia-lyase gene1 promoter in transgenic tobacco. Plant Cell Physiol. 38: 792–803.

Kawaoka, A., Kaothien, P., Yoshida, K., Endo, S., Yamada, K. and Ebinuma, H. 2000. Functional analysis of tobacco LIM protein Ntlim1 involved in lignin biosynthesis. Plant J. 22: 289–301.

Kim, M.-S., Klopfenstein, N.B. and Chun, Y.W. 1997. *Agrobacterium*-mediated transformation of *Populus* species. In: N.B. Klopfenstein, Y.W. Chun, M.-S. Kim and M.R. Ahuja (Eds.) Micropropagation, Genetic Engineering, and Molecular Biology of *Populus* (General Technical Report RM-GTR-297), Rocky Mountain Forest and Range Experiment Station, Fort Collins, USA, pp. 51–59.

Kim, H., Ralph, J., Yahiaoui, N., Pean, M. and Boudet, A.-M. 2000. Cross-coupling of hydroxycinnamyl aldehydes into lignins. Org. Lett. 2: 2197–2200.

Kimura, S., Laosinchai, W., Itoh, T., Cui, X., Linder, C.R. and Brown, R.M. Jr. 1999. Immunogold labeling of rosette terminal cellulose-synthesizing complexes in the vascular plant *Vigna angularis*. Plant Cell 11: 2075–2085.

Klee, H.J., Horsch, R.B., Hinchee, M.A., Hein, M.B., and Hoffmann, N.L. 1987. The effects of overproduction of two *Agrobacterium tumefaciens* T-DNA auxin biosynthetic gene products in transgenic petunia plants. Genes Dev. 1: 86–96.

Klekowski, E.J. and Godfrey, P.J. 1989. Ageing and mutation in plants. Nature 340: 389–391.

Klopfenstein, N.B., Chun, Y.W., Kim, M.-S. and Ahuja, M.R. 1997. Micropropagation, Genetic Engineering, and Molecular Biology of *Populus* (General Technical Report RM-GTR-297), Rocky Mountain Forest and Range Experiment Station, Fort Collins, USA.

Knox, J.P. 1997. The use of antibodies to study the architecture and developmental regulation of plant cell walls. Int. Rev. Cytol. 171: 79–120.

Knox, J.P., Linstead, P.J., King, J., Cooper, C. and Roberts, K. 1990. Pectin esterification is spatially regulated both within cell walls and between developing tissues of root apices. Planta 181: 512–521.

Koopmann, E., Logemann, E. and Hahlbrock, K. 1999. Regulation and functional expression of cinnamate 4-hydroxylase from parsley. Plant Physiol. 119: 49–55.

Kroll, R.E., Ritter, D.C., Gertjejansen, R.O. and Au, K.C. 1992. Anatomical and physical properties of balsam poplar (*Populus balsamifera* L.) in Minnesota. Wood Fiber Sci. 24: 13–24.

Lagrimini, L.M. 1991. Wound-induced deposition of polyphenols in transgenic plants overexpressing peroxidase. Plant Physiol. 96: 577–583.

Lagrimini, L.M., Bradford, S. and Rothstein, S. 1990. Peroxidase-induced wilting in transgenic tobacco plants. Plant Cell 2: 7–18.

Lapierre, C., Pollet, B., Petit-Conil, M., Toval, G., Romero, J., Pilate, G., Leplé, J.-C., Boerjan, W., Ferret, V., De Nadai, V. and Jouanin, L. 1999. Structural alterations of lignins in transgenic poplars with depressed cinnamyl alcohol dehydrogenase or caffeic acid O-methyltransferase activity have opposite impact on the efficiency of industrial Kraft pulping. Plant Physiol. 119: 153–163.

Larson, P.R. 1969. Wood formation and the concept of wood quality. Yale Univ. Sch. Forestry Bull. 74: 1–54.

Larson, P.R. 1994. The Vascular Cambium. Springer-Verlag, Berlin.

Li, L., Osakabe, Y., Joshi, C.P. and Chiang, V.L. 1999. Secondary xylem-specific expression of caffeoyl-coenzyme A 3-O-methyltransferase plays an important role in the methylation pathway associated with lignin biosynthesis in loblolly pine. Plant Mol. Biol. 40: 555–565.

Li, L., Popko, J.L., Umezawa, T. and Chiang, V.L. 2000. 5-Hydroxyconiferyl aldehyde modulates enzymatic methylation for syringyl monolignol formation, a new view of monolignol biosynthesis in angiosperms. J. Biol. Chem. 275: 6537–6545.

Liese, W. and Ammer, U. 1958. Investigation on the length of wood fibers in poplars. Holzforschung 11: 69–174.

Little, C.H.A. and Pharis, R.P. 1995. Hormonal control of radial and longitudinal growth in the tree stem. In: B.L. Gartner (Ed.) Plant Stems: Physiology and Functional Morphology (Physiological Ecology Series), Academic Press, San Diego, CA, pp. 281–319.

Little, C.H.A. and Savidge, R.A. 1987. The role of plant growth regulators in forest tree cambial growth. Plant Growth Regul. 6: 137–169.

Loewus, F.A. and Murthy, P.P.N. 2000. *myo*-Inositol metabolism in plants. Plant Sci. 150: 1–19.

MacKay, J.J., O Malley, D.M., Presnell, T., Booker, F.L., Campbell, M.M., Whetten, R.W. and Sederoff, R.R. 1997. Inheritance, gene expression, and lignin characterization in a mutant pine deficient in cinnamyl alcohol dehydrogenase. Proc. Natl. Acad. Sci. USA 94: 8255–8260.

Marita, J.M., Ralph, J., Hatfield, R.D. and Chapple, C. 1999. NMR characterization of lignins in *Arabidopsis* altered in the activity of ferulate 5-hydroxylase. Proc. Natl. Acad. Sci. USA 96: 12328–12332.

Martz, F., Maury, S., Pinçon, G. and Legrand, M. 1998. cDNA cloning, substrate specificity and expression study of tobacco caffeoyl-CoA 3-O-methyltransferase, a lignin biosynthetic enzyme. Plant Mol. Biol. 36: 427–437.

Matsui, N., Chen, F., Yasuda, S. and Fukushima, K. 2000. Conversion of guaiacyl to syringyl moieties on the cinnamyl alcohol pathway during the biosynthesis of lignin in angiosperms. Planta 210: 831–835.

Matsumoto, T., Sakai, F. and Hayashi, T. 1997. A xyloglucan-specific endo-1,4-β-glucanase isolated from auxin-treated pea stems. Plant Physiol. 114: 661–667.

270

Maury, S., Geoffroy, P. and Legrand, M. 1999. Tobacco *O*-methyltransferases involved in phenylpropanoid metabolism. The different caffeoyl-coenzyme A/5-hydroxyferuloyl-coenzyme A 3/5-*O*-methyltransferase and caffeic acid/5-hydroxyferulic acid 3/5-*O*-methyltransferase classes have distinct substrate specificities and expression patterns. Plant Physiol. 121: 215–223.

McCann, M.C. and Roberts, K. 1994. Changes in cell-wall architecture during cell elongation. J. Exp. Bot. 45: 1683–1691.

McDougall, G.J., Morrison, I.M., Stewart, D., Weyers, J.D.B. and Hillman, J.R. 1993. Plant fibres: botany, chemistry and processing for industrial use. J. Sci. Food Agric. 62: 1–20.

McQueen-Mason, S. 1997. Plant cell walls and the control of growth. Biochem. Soc. Trans. 25: 204–214.

Mellerowicz, E.J., Coleman, W.K., Riding, R.T. and Little, C.H.A. 1992. Periodicity of cambial activity in *Abies balsamea*. I. Effects of temperature and photoperiod on cambial dormancy and frost hardiness. Physiol. Plant. 85: 515–525.

Mellerowicz, E.J., Blomqvist, K., Bourquin, V., Brumer, H., Christiernin, M., Denman, S., Djerbi, S., Eklund, M., Gray-Mitsumine, M., Kallas, Å., Lehtiö, J., Raza, S., Regan, S., Rudsander, U., Sundberg, B. and Teeri, T.T. 2000. Cell wall enzyme discovery using high throughput sequencing and in-depth expression analysis in poplar wood forming tissues. Proceedings of the Symposium on Friendly and Emerging Technologies for a Sustainable Pulp and Paper Industry, Taiwan Research Institute, Taipei, Taiwan, 25–27 April 2000.

Meng, H. and Campbell, W.H. 1998. Substrate profiles and expression of caffeoyl coenzyme A and caffeic acid *O*-methyltransferases in secondary xylem of aspen during seasonal development. Plant Mol. Biol. 38: 513–520.

Meyer, K., Shirley, A.M., Cusumano, J.C., Bell-Lelong, D.A. and Chapple, C. 1998. Lignin monomer composition is determined by the expression of a cytochrome P450-dependent monooxygenase in *Arabidopsis*. Proc. Natl. Acad. Sci. USA 95: 6619–6623.

Meyermans, H., Morreel, K., Lapierre, C., Pollet, B., De Bruyn, A., Busson, R., Herdewijn, P., Devreese, B., Van Beeumen, J., Marita, J.M., Ralph, J., Chen, C., Burggraeve, B., Van Montagu, M., Messens, E. and Boerjan, W. 2000. Modification in lignin and accumulation of phenolic glucosides in poplar xylem upon down-regulation of caffeoyl-coenzyme A *O*-methyltransferase, an enzyme involved in lignin biosynthesis. J. Biol. Chem. 275: 36899–36909.

Micheli, F., Sundberg, B., Goldberg, R. and Richard, L. 2000a. Radial distribution pattern of pectin methylesterases across the cambial region of hybrid aspen at activity and dormancy. Plant Physiol. 124: 191–199.

Micheli, F., Bordenave, M. and Richard, L. 2000b. Pectin methylesterases: possible marker for cambial derivative differentiation. In: R. Savidge, J. Barnett and R. Napier (Eds) Cell and Molecular Biology of Wood Formation, BIOS Scientific Publications, Oxford, pp. 295–304.

Milioni, D., Sado, P-E., Stacey, N. J., Domingo, C., Roberts, K. and McCann, M.C. 2001 Differential expression of cell-wall-related genes during formation of tracheary elements in the *Zinnia* mesophyll cell system. Plant Mol. Biol., this issue.

Miller, A.R. and Roberts, L.W. 1984. Ethylene biosynthesis and xylogenesis in *Lactuca* explants cultured *in vitro* in the presence of auxin and cytokinin: the effect of ethylene precursors and inhibitors. J. Exp. Bot. 35: 691–698.

Monties, B. 1998. Novel structures and properties of lignins in relation to their natural and induced variability in ecotypes, mutants and transgenic plants. Polymer Degrad. Stabil. 59: 53–64.

Moritz, T. and Sundberg, B. 1996. Endogenous cytokinins in the vascular cambial regions of *Pinus sylvestris* during activity and dormancy. Physiol. Plant. 98: 693–698.

Mueller, S.C. and Brown, R.M. Jr. 1980. Evidence for an intramembrane component associated with a cellulose microfibril-synthesizing complex in higher plants. J. Cell. Biol. 84: 315–326.

Murakami, Y., Funada, R., Sano, Y. and Ohtani, J. 1999. The differentiation of contact cells and isolation cells in the xylem ray parenchyma of *Populus maximowiczii*. Ann. Bot. 84: 429–435.

Nakashima, J., Awano, T., Takabe, K., Fujita, M. and Saiki, H. 1997. Immunocytochemical localization of phenylalanine ammonia-lyase and cinnamyl alcohol dehydrogenase in differentiating tracheary elements derived from *Zinnia* mesophyll cells. Plant Cell Physiol. 38: 113–123.

Nishitani, K. 1997. The role of endoxyloglucan transferase in the organization of plant cell walls. Int. Rev. Cytol. 173: 157–206.

Nishitani, K. 1998. Construction and restructuring of the cellulose-xyloglucan framework in the apoplast as mediated by the xyloglucan-related protein family: a hypothetical scheme. J. Plant Res. 111: 159–166.

Nobushi, T. and Fujitta, M. 1972. Cytological structure of differentiating tension wood fibres of *Populus euroamericana*. Mokuzai Gakkaishi 18: 137–144.

Norberg, P.H. and Meier, H. 1966. Physical and chemical properties of the gelatinous layer in tension wood fibers of aspen (*Populus tremula* L.). Holzforschung 20: 174–178.

Northcote, D.H. 1972. Chemistry of the plant cell wall. Annu. Rev. Plant Physiol. 23: 113–132.

Northcote, D.H. 1989. Control of plant cell wall biosynthesis: an overview. In: N.G. Lewis, and M.G. Paice (Eds.) Plant Cell Wall Polymers (ACS Symposium Series vol. 399), American Chemical Society, Washington, DC, pp. 1–15.

Northcote, D.H., Davey, R. and Lay, J. 1989. Use of antisera to localize callose, xylan and arabinogalactan in the cell-plate, primary and secondary walls of plant cells. Planta 178: 353–366.

Ohmiya, Y., Samejima, M., Shiroishi, M., Amano, Y., Kanda, T., Sakai, F. and Hayashi, T. 2000. Evidence that endo-1,4-β-glucanases act on cellulose in suspension-cultured poplar cells. Plant J. 24: 147–158.

Ohta, S. 1979. Tension wood from stems of poplar (*Populus euroamericana*) with various degree of leaning. Mokuzai Gakkaishi 25: 610–614.

Olsson, O. and Little, C.H.A. 2000. Molecular control of the development and function of the vascular cambium. In: S.M. Jain, and S.C. Minocha (Eds.) Molecular Biology of Woody Plants, vol. 1 (Forestry Sciences vol. 64), Kluwer Academic Publishers, Dordrecht, Netherlands, pp. 155–180.

Osakabe, K., Tsao, C.C., Li, L., Popko, J.L., Umezawa, T., Carraway, D.T., Smeltzer, R.H., Joshi, C.P. and Chiang, V.L. 1999. Coniferyl aldehyde 5-hydroxylation and methylation direct syringyl lignin biosynthesis in angiosperms. Proc. Natl. Acad. Sci. USA 96: 8955–8960.

Panshin, A.J. and de Zeeuw, C. 1980. Textbook of Wood Technology. McGraw-Hill, New York.

Parresol, B.R. and Cao, F. 1998. An investigation of crystalline intensity of the wood of poplar clones grown in Jiangsu Province, China. Research Paper, Southern Research Station, USDA Forest Service, No. SRS-11, 7 pp.

Perrin, R.M., DeRocher, A.E., Bar-Peled, M., Zeng, W., Norambuena, L., Orellana, A., Raikhel, N.V. and Keegstra, K. 1999. Xyloglucan fucosyltransferase, an enzyme involved in plant cell wall biosynthesis. Science 284: 1976–1979.

Perri, R., Wilkerson, C. and Keegstra, K. 2001. Golgi enzymes that synthesize plant cell wall polysaccharides: finding and evaluating candidates in the genomic era. Plant Mol. Biol. 47: 109–124.

Phillips, R. and Arnott, S.M. 1983. Studies on induced tracheary element differentiation in cultured tissues of tubers of the Jerusalem artichoke, *Helianthus tuberosus*. Histochem. J. 15: 427–436.

Piquemal, J., Lapierre, C., Myton, K., O'Connell, A., Schuch, W., Grima-Pettenati, J. and Boudet, A.-M. 1998. Down-regulation in cinnamoyl-CoA reductase induces significant changes of lignin profiles in transgenic tobacco plants. Plant J. 13: 71–83.

Piro, G., Zuppa, A., Dalessandro, G. and Northcote, D.H. 1993. Glucomannan synthesis in pea epicotyls: the mannose and glucose transferases. Planta 190: 206–220.

Plomion, C., Pionneau, C., Brach, J., Costa, P. and Baillères, H. 2000. Compression wood-responsive proteins in developing xylem of maritime pine (*Pinus pinaster* Ait.). Plant Physiol. 123: 959–969.

Potikha, T.S., Collins, C.C., Johnson, D.I., Delmer, D.P. and Levine, A. 1999. The involvement of hydrogen peroxide in the differentiation of secondary walls in cotton fibers. Plant Physiol. 119: 849–858.

Prodhan, A.K.M.A., Funada, R., Ohtani, J., Abe, H. and Fukazawa, K. 1995. Orientation of microfibrils and microtubules in developing tension-wood fibres of Japanese ash (*Fraxinus mandshurica* var. *japonica*). Planta 196: 577–585.

Puhlmann, J., Bucheli, E., Swain, M.J., Dunning, N., Albersheim, P., Darvill, A.G. and Hahn, M.G. 1994. Generation of monoclonal antibodies against plant cell-wall polysaccharides. I. Characterization of a monoclonal antibody to a terminal α-(1→2)-linked fucosyl-containing epitope. Plant Physiol. 104: 699–710.

Rajagopal, J., Das, S., Khurana, D.K., Srivastava, P.S. and Lakshmikumaran, M. 1999. Molecular characterization and distribution of a 145-bp tandem repeat family in the genus *Populus*. Genome 42: 909–918.

Ralph, J., Lapierre, C., Lu, F., Marita, J.M., Van Doorsselaere, J., Pilate, G., Boerjan, W. and Jouanin, L. 2001. NMR evidence for benzodioxane structures resulting from incorporation of 5-hydroxyconiferyl alcohol into lignins of *O*-methyltransferase-deficient poplars. J. Agric. Food Chem. 49: 86–91.

Ranocha, P., Goffner, D. and Boudet, A.M. 2000. Plant laccases: are they involved in lignification. In: R. Savidge, J. Barnett, and R. Napier (Eds.) Cell and Molecular Biology of Wood Formation, BIOS Scientific Publications, Oxford, pp. 397–410.

Rasmussen, S. and Dixon, R.A. 1999. Transgene-mediated and elicitor-induced perturbation of metabolic channeling at the entry point into the phenylpropanoid pathway. Plant Cell 11: 1537–1551.

Regan, S., Bourquin, V., Tuominen, H., Sundberg, B. 1999. Accurate and high resolution *in situ* hybridization analysis of gene expression in secondary stem tissues. Plant J. 19: 363–369.

Reis, D., Vian, B. and Roland, J.C. 1994. Cellulose-glucuronoxylans and plant-cell wall structure. Micron 25: 171–187.

Reiter, W.-D. and Vanzin, G. 2001. Molecular genetics of nucleotide sugar interconversion pathways. Plant Mol. Biol., this issue.

Ren, C. and Kermode, A.R. 2000. An increase in pectin methyl esterase activity accompanies dormancy breakage and germination of yellow cedar seeds. Plant Physiol. 124: 231–242.

Richmond, T.A. and Somerville, C.R. 2001. Integrative approaches to determining *Csl* function. Plant Mol. Biol., this issue.

Riding, R.T. and Little, C.H.A. 1984. Anatomy and histochemistry of *Abies balsamea* cambial zone cells during the onset and breaking of dormancy. Can. J. Bot. 62: 2571–2579.

Roberts, L.W. 1988. Hormonal aspects of vascular differentiation In: Vascular Differentiation and Plant Growth Regulators, Springer-Verlag, Berlin, pp. 22–38.

Roland, J.C. 1978. Early differences between radial walls and tangential walls of actively growing cambial zone. IAWA Bull. 1978: 7–10.

Saka, S. and Goring, D.A.I. 1985. Localization of lignins in wood cell walls. In: T. Higuchi (Ed.) Biosynthesis and Biodegradation of Wood Components, Academic Press, Orlando, FL, pp. 51–62.

Samaj, J., Hawkins, S., Lauvergeat, V., Grima-Pettenati, J. and Boudet, A. 1998. Immunolocalization of cinnamyl alcohol dehydrogenase 2 (CAD 2) indicates a good correlation with cell-specific activity of CAD 2 promoter in transgenic poplar shoots. Planta 204: 437–443.

Samuels, A.L., Giddings, T.H. Jr and Staehelin, L.A. 1995. Cytokinesis in tobacco BY-2 and root tip cells: a new model of cell plate formation in higher plants. J. Cell Biol. 130: 1345–1357.

Sauter, J.J. 2000. Photosynthate allocation to the vascular cambium: facts and problems. In: R. Savidge, J. Barnett and R. Napier (Eds) Cell and Molecular Biology of Wood Formation, BIOS Scientific Publications, Oxford, pp. 71–83.

Savidge, R.A. 1983. The role of plant hormones in higher plant cellular differentiation. II. Experiments with the vascular cambium, and sclereid and tracheid differentiation in the pine, *Pinus contorta*. Histochem. J. 15: 447–466.

Savidge, R.A. 2000. Biochemistry of seasonal cambial growth and wood formation: an overview of the challenges. In: R. Savidge, J. Barnett and R. Napier (Eds) Cell and Molecular Biology of Wood Formation, BIOS Scientific Publications, Oxford, pp. 1–30.

Saxena, I.M. and Brown, R.M. Jr. 1999. Are the reversibly glycosylated polypeptides implicated in plant cell wall biosynthesis non-processive β-glycosyltransferases? Trends Plant Sci. 4: 6–7.

Sederoff, R.R., MacKay, J.J., Ralph, J. and Hatfield, R.D. 1999. Unexpected variation in lignin. Curr. Opin. Plant Biol. 2: 145–152.

Seitz, B., Klos, C., Wurm, M. and Tenhaken, R. 2000. Matrix polysaccharide precursors in *Arabidopsis* cell walls are synthesized by alternate pathways with organ-specific expression patterns. Plant J. 21: 537–546.

Sewalt, V.J.H., Ni, W., Blount, J.W., Jung, H.G., Masoud, S.A., Howles, P.A., Lamb, C. and Dixon, R.A. 1997. Reduced lignin content and altered lignin composition in transgenic tobacco down-regulated in expression of L-phenylalanine ammonia-lyase or cinnamate 4-hydroxylase. Plant Physiol. 115: 41–50.

Shani, Z., Dekel, M., Tsabary, G., Jensen, C.S., Tzfira, T., Goren, R., Altman, A. and Shoseyov, O. 1999. Expression of *Arabidopsis thaliana*, endo-1,4-β-glucanase (*cel*1) in transgenic poplar plants. In: A. Altman, M. Ziv, and S. Izhar (Eds.) Plant Biotechnology and In Vitro Biology in the 21st Century (Current Plant Science and Biotechnology in Agriculture vol. 36), Kluwer Academic Publishers, Dordrecht, Netherlands, pp. 209–212.

Sheen, J., Zhou, L. and Jang, J.-C. 1999. Sugars as signaling molecules. Curr. Opin. Plant Biol. 2: 410–418.

Sheriff, D.W. 1983. Control by indole-3-acetic acid of wood production in *Pinus radiata* D. Don segments in culture. Aust. J. Plant Physiol. 10: 131–135.

Shimizu, Y., Aotsuka, S., Hasegawa, O., Kawada, T., Sakuno, T., Sakai, F. and Hayashi T. 1997. Changes in levels of mRNAs for cell wall-related enzymes in growing cotton fiber cells. Plant Cell Physiol. 38: 375–378.

Simson, B.W. and Timell, T.E. 1978a. Polysaccharides in cambial tissues of *Populus tremuloides* and *Tilia americana*. 1. Isolation, fractionation, and chemical composition of the cambial tissues. Cell. Chem. Technol. 12: 39–50.

272

Simson, B.W. and Timell, T.E. 1978b. Polysaccharides in cambial tissues of *Populus tremuloides* and *Tilia americana*. II. Isolation and structure of a xyloglucan. Cell. Chem. Technol. 12: 51–62.

Simson, B.W. and Timell, T.E. 1978c. Polysaccharides in cambial tissues of *Populus tremuloides* and *Tilia americana*. IV. 4-*O*-methylglucuronoxylan and pectin. Cell. Chem. Technol. 12: 79–84.

Simson, B.W. and Timell, T.E. 1978d. Polysaccharides in cambial tissues of *Populus tremuloides* and *Tilia americana*. V. Cellulose. Cell. Chem. Technol. 12: 137–141.

Sinnott, E.W. and Bloch, R. 1940. Cytoplasmic behaviour during division of vacuolate plant cells. Proc. Natl. Acad. Sci. USA 26: 223–227.

Sitbon, F., Hennion, S., Sundberg, B., Little, C.H.A., Olsson, O. and Sandberg, G. 1992. Transgenic tobacco plants coexpressing the *Agrobacterium tumefaciens iaaM* and *iaaH* genes display altered growth and indoleacetic acid metabolism. Plant Physiol. 99: 1062–1069.

Sitbon, F., Hennion, S., Little, C.H.A. and Sundberg, B. 1999. Enhanced ethylene production and peroxidase activity in IAA-overproducing transgenic tobacco plants is associated with increased lignin content and altered lignin composition. Plant Sci. 141: 165–173.

Smith, C.G., Rodgers, M.W., Zimmerlin, A., Ferdinando, D. and Bolwell, G.P. 1994. Tissue and subcellular immunolocalisation of enzymes of lignin synthesis in differentiating and wounded hypocotyl tissue of French beans (*Phaseolus vulgaris* L.). Planta 192: 155–164.

Sonobe, S., Nakayama, N., Shimmen, T. and Sone, Y. 2000. Intracellular distribution of subcellular organelles revealed by antibody against xyloglucan during cell cycle in tobacco BY-2 cells. Protoplasma 213: 218–227.

Stacey, N.J., Roberts, K., Carpita, N.C., Wells, B. and McCann, M.C. 1995. Dynamic changes in cell surface molecules are very early events in the differentiation of mesophyll cells from *Zinnia elegans* into tracheary elements. Plant J. 8: 891–906.

Stafford, H.A. 1981. Phenylalanine ammonia-lyase. In: E.E. Conn (Ed.) Secondary Plant Products (The Biochemistry of Plants vol. 7), Academic Press, New York, pp. 117–137.

Sterky, F., Regan, S., Karlsson, J., Hertzberg, M., Rohde, A., Holmberg, A., Amini, B., Bhalerao, R., Larsson, M., Villarroel, R., Van Montagu, M., Sandberg, G., Olsson, O., Teeri, T.T., Boerjan, W., Gustafsson, P., Uhlén, M., Sundberg, B. and Lundeberg, J. 1998a. Gene discovery in the wood-forming tissues of poplar: analysis of 5692 expressed sequence tags. Proc. Natl. Acad. Sci. USA 95: 13330–13335.

Sterky, F., Sievertzon, M. and Kleczkowski, L.A. 1998b. Molecular cloning of a cDNA encoding a cytosolic form of phosphoglucomutase (Accession No. AF097938) from cambium of poplar (PGR 98-205). Plant Physiol. 118: 1535.

Sultze, R.F. 1957. A study of the developing tissues of aspen wood. TAPPI 40: 985–994.

Sundberg, B. and Little, C.H.A. 1990. Tracheid production in response to changes in the internal level of indole-3-acetic acid in 1-year-old shoots of Scots pine. Plant Physiol. 94: 1721–1727.

Sundberg, B., Uggla, C. and Tuominen H. 2000. Auxin gradients and cambial growth. In: R. Savidge, J. Barnett and R. Napier (Eds.) Cell and Molecular Biology of Wood Formation (SEB Experimental Biology Reviews), BIOS, Oxford, pp. 169–188.

Sussex, I.M., Clutter, M.E. and Goldsmith, M.H.M. 1972. Wound recovery by pith cell redifferentiation: structural changes. Am. J. Bot. 59: 797–804.

Suzuki, K., Ingold, E., Sugiyama, M. and Komamine, A. 1991. Xylan synthase activity in isolated mesophyll cells of *Zinnia elegans*

during differentiation to tracheary elements. Plant Cell Physiol. 32: 303–306.

Takabe, K., Miyauchi, S., Tsunoda, R. and Fukazawa, K. 1992. Distribution of guaiacyl and syringyl lignins in Japanese beech (*Fagus crenata*): variation within an annual ring. IAWA Bull. 13: 105–112.

Takeda, T., Mitsuishi, Y., Sakai, F. and Hayashi, T. 1996. Xyloglucan endotransglycosylation in suspension-cultured poplar cells. Biosci. Biotech. Biochem. 60: 1950–1955.

Tamagnone, L., Merida, A., Parr, A., Mackay, S., Culianez-Macia, F.A., Roberts, K. and Martin, C. 1998. The AmMYB308 and AmMYB330 transcription factors from *Antirrhinum* regulate phenylpropanoid and lignin biosynthesis in transgenic tobacco. Plant Cell 10: 135–154.

Taylor, J.G. and Haigler, C.H. 1993. Patterned secondary cell-wall assembly in tracheary elements occurs in a self-perpetuating cascade. Acta Bot. Neerl. 42: 153–163.

Taylor, N.G., Scheible, W.-R., Cutler, S., Somerville, C.R. and Turner, S.R. 1999. The *irregular xylem3* locus of *Arabidopsis* encodes a cellulose synthase required for secondary cell wall synthesis. Plant Cell 11: 769–779.

Tenhaken, R. and Thulke, O. 1996. Cloning of an enzyme that synthesizes a key nucleotide-sugar precursor of hemicellulose biosynthesis from soybean: UDP-glucose dehydrogenase. Plant Physiol. 112: 1127–1134.

Terashima, N., Okada, M. and Tomimura, Y. 1979. Heterogeneity in formation of lignin. I. Heterogeneous incorporation of *p*-hydroxybenzoic acid into poplar lignin. Mokuzai Gakkaishi 25: 422–426.

Terashima, N., Fukushima, K., He, L.F. and Takabe, K. 1993. Comprehensive model of the lignified plant cell wall. In: H.G. Jung (Ed.) Forage Cell Wall Structure and Digestibility, ASA-CSSA-SSSA, Madison, WI, pp. 247–270.

Thompson, J.E. and Fry, S.C. 2000. Evidence for covalent linkage between xyloglucan and acidic pectins in suspension-cultured rose cells. Planta 211: 275–286.

Tognolli, M., Overney, S., Penel, C., Greppin, H. and Simon, P. 2000. A genetic and enzymatic survey of *Arabidopsis thaliana* peroxidases. Plant Peroxidase Newsl. 14: 3–12.

Torki, M., Mandaron, P., Mache, R. and Falconet, D. 2000. Characterization of a ubiquitous expressed gene family encoding polygalacturonase in *Arabidopsis thaliana*. Gene 242: 427–436.

Torrey, J.G., Fosket, D.E. and Hepler, P.K. 1971. Xylem formation: a paradigm of cytodifferentiation in higher plants. Am. Scient. 59: 338–352.

Tsai, C.-J., Popko, J.L., Mielke, M.R., Hu, W.-J., Podila, G.K. and Chiang, V.L. 1998. Suppression of *O*-methyltransferase gene by homologous sense transgene in quaking aspen causes red-brown wood phenotypes. Plant Physiol. 117: 101–112.

Tuominen, H., Sitbon, F., Jacobsson, C., Sandberg, G., Olsson, O. and Sundberg, B. 1995. Altered growth and wood characteristics in transgenic hybrid aspen expressing *Agrobacterium tumefaciens* T-DNA indoleacetic acid-biosynthetic genes. Plant Physiol. 109: 1179–1189.

Tuominen, H., Puech, L., Fink, S. and Sundberg, B. 1997. A radial concentration gradient of indole-3-acetic acid is related to secondary xylem development in hybrid aspen. Plant Physiol. 115: 577–585.

Tuominen, H., Olsson, O. and Sundberg, B. 2000a. Genetic engineering of wood formation. Expression of bacterial IAA-biosynthetic genes in hybrid aspen (*Populus tremula* × *P. tremuloides*). In: S.M. Jain and S.C. Minocha (Eds.) Molecular Biology of Woody Plants, vol. 1 (Forestry Sciences vol. 64), Kluwer Academic Publishers, Dordrecht, Netherlands, pp. 181–203.

Tuominen, H., Puech, L., Regan, S., Fink, S., Olsson, O. and Sundberg, B. 2000b. Cambial-region-specific expression of the *Agrobacterium iaa* genes in transgenic aspen visualized by a linked *uidA* reporter gene. Plant Physiol. 123: 531–541.

Turner, S.R. and Somerville, C.R. 1997. Collapsed xylem phenotype of *Arabidopsis* identifies mutants deficient in cellulose deposition in the secondary cell wall. Plant Cell 9: 689–701.

Turner, S., Taylor, N. and Jones, L. 2001. Mutations of the secondary wall. Plant Mol Biol, this issue.

Tuskan, G., West, D., Bradshaw, H.D., Neale, D., Sewell, M., Wheeler, N., Megraw, B., Jech, K., Wiselogel, A., Evans, R., Elam, C., Davis, M. and Dinus, R. 1999. Two high-throughput techniques for determining wood properties as part of a molecular genetics analysis of hybrid poplar and loblolly pine. Appl. Biochem. Biotech. 77: 55–65.

Uggla, C. and Sundberg, B., 2001. Sampling of cambial region tissues for high resolution analysis. In: N.J. Chaffey (Ed.) Wood formation in Trees: Cell and Molecular Biology Techniques, Harwood Academic Publishers, in press.

Uggla, C., Moritz, T., Sandberg, G. and Sundberg, B. 1996. Auxin as a positional signal in pattern formation in plants. Proc. Natl. Acad. Sci. USA 93: 9282–9286.

Uggla, C., Mellerowicz, E.J. and Sundberg, B. 1998. Indole-3-acetic acid controls cambial growth in Scots pine by positional signaling. Plant Physiol. 117: 113–121.

Uggla, C., Magel, E., Moritz, T. and Sundberg, B. 2001. Function and dynamics of auxin and carbohydrates during early-wood/latewood transition in *Pinus sylvestris*. Plant Physiol., in press.

Valster, A.H., Hepler, P.K. and Chernoff, J. 2000. Plant GTPases: the Rhos in bloom. Trends Cell Biol. 10: 141–146.

Vander Mijnsbrugge, K., Meyermans, H., Van Montagu, M., Bauw, G. and Boerjan, W. 2000. Wood formation in poplar: identification, characterization, and seasonal variation of xylem proteins. Planta 210: 589–598.

Van Doorsselaere, J., Baucher, M., Chognot, E., Chabbert, B., Tollier, M.-T., Petit-Conil, M., Leplé, J.-C., Pilate, G., Cornu, D., Monties, B., Van Montagu, M., Inzé, D., Boerjan, W. and Jouanin, L. 1995. A novel lignin in poplar trees with a reduced caffeic acid/5-hydroxyferulic acid *O*-methyltransferase activity. Plant J. 8: 855–864.

Vaughn, K.C., Hoffman, J.C., Hahn, M.G. and Staehelin, L.A. 1996. The herbicide dichlobenil disrupts cell plate formation: immunogold characterization. Protoplasma 194: 117–132.

Vian, B., Roland, J.C., Reis, D. and Mosiniak, M. 1992. Distribution and possible morphogenetic role of the xylans within the secondary vessel wall of linden wood. IAWA Bull. 13: 269–282.

Vietor, R.J., Renard, C.M.G.C., Goldberg, R. and Catesson, A.M. 1995. Cell-wall polysaccharides in growing poplar bark tissue. Int. J. Biol. Micromol. 17: 341–344.

Wang, J. and Hall, R. 1995. Comparison of DNA content and chromosome numbers in species and clones of poplar. Abstract presented at the International Poplar Symposium (Seattle, WA, 20–25 August 1995), p. 93.

Wang, Q., Little, C.H.A. and Oden, P.C. 1997. Control of longitudinal and cambial growth by gibberellins and indole-3-acetic acid in current-year shoots of *Pinus sylvestris*. Tree Physiol. 17: 715–721.

Warren Wilson, J. and Warren Wilson, P.M. 1984. Control of tissue patterns in normal development and in regeneration. In: P. Barlow and D. Carr (Eds.) Positional Controls in Plant Development. Cambridge University Press, Cambridge, UK, pp. 225–280.

Watanabe, Y., Fukazawa, K., Kojima, Y., Funada, R., Ona, T. and Asada, T. 1997. Histochemical study on heterogeneity of lignin in Eucalyptus species. 1. Effects of polyphenols. Mokuzai Gakkaishi 43: 102–107.

Wenham, M.W. and Cusick, F. 1975. The growth of secondary fibers. New Phytol. 74: 247–261.

Whetten, R.W., MacKay, J.J. and Sederoff, R.R. 1998. Recent advances in understanding lignin biosynthesis. Annu. Rev. Plant Physiol. Plant Mol. Biol. 49: 585–609.

Whitmore, F.W. and Zahner, R. 1966. Development of the xylem ring in stems of young red pine trees. Forestry Sci. 12: 198–210.

Whitney, S.E.C., Gothard, M.G.E., Mitchell, J.T. and Gidley, M.J. 1999. Roles of cellulose and xyloglucan in determining the mechanical properties of primary plant cell walls. Plant Physiol. 121: 657–663.

Willats, W.G.T., Marcus, S.E. and Knox, J.P. 1998. Generation of a monoclonal antibody specific to (15)-α-L-arabinan. Carbohydrate Res. 308: 149–152.

Willats, W.G.T., Steele-King, C.G., Marcus, S.E. and Knox, J.P. 1999. Side chains of pectic polysaccharides are regulated in relation to cell proliferation and cell differentiation. Plant J. 20: 619–628.

Willats, W.G.T., McCartney, L., Mackie, W. and Knox, J.P. 2001. Pectin: Cell biology and prospects for the functional analysis. Plant Mol. Biol., this issue.

Wilson, B.F. 1964. A model for cell production by the cambium in conifers. In: M.H. Zimmerman (Ed.) The Formation of Wood in Forest Trees, Academic Press, New York, pp. 19–36.

Winter, H. and Huber, S.C. 2000. Regulation of sucrose metabolism in higher plants: localization and regulation of activity of key enzymes. Crit. Rev. Plant Sci. 19: 31–67.

Wloch, W. and Polap, E. 1994. The intrusive growth of initial cells in re-arrangement of cells in cambium of *Tilia cordata* Mill. Acta Soc. Bot. Pol. 63: 109–116.

Wojtaszek, P. and Bolwell, G.P. 1995. Secondary cell-wall-specific glycoproteins(s) from French bean hypocotyls. Plant Physiol. 108: 1001–1012.

Wu, L.R. 1998. Genetic mapping of QTLs affecting tree growth and architecture in *Populus*: implication for ideotype breeding. Theor. Appl. Genet. 96: 447–457.

Wu, R. and Stettler, R.F. 1994. Quantitative genetics of growth and development in *Populus*. I. A three-generation comparison of tree architecture during the first 2 years of growth. Theor. Appl. Genet. 89: 1046–1054.

Wu, R. and Stettler, R.F. 1997. Quantitative genetics of growth and development in *Populus*. II. The partitioning of genotype x environment interaction in stem growth. Heredity 78: 124–134.

Wu, R., Bradshaw, H.D. Jr. and Stettler, R.F. 1997. Molecular genetics of growth and development in *Populus* (Salicaceae). V. Mapping quantitative trait loci affecting leaf variation. Am. J. Bot. 84: 143–153.

Wu, R., Bradshaw, H.D. Jr. and Stettler, R.F. 1998. Developmental quantitative genetics of growth in *Populus*. Theor. Appl. Genet. 97: 1110–1119.

Wu, L., Joshi, C.P. and Chiang, V.L. 2000. A xylem-specific cellulose synthase gene from aspen (*Populus tremuloides*) is responsive to mechanical stress. Plant J. 22: 495–502.

Wulff, C., Norambuena, L. and Orellana, A. 2000. GDP-fucose uptake into the Golgi apparatus during xyloglucan biosynthesis requires the activity of a transporter-like protein other than the UDP-glucose transporter. Plant Physiol. 122: 867–877.

Yahiaoui, N., Marque, C., Myton, K.E., Negrel, J. and Boudet, A.M. 1998. Impact of different levels of cinnamyl alcohol dehydrogenase down-regulation on lignins of transgenic tobacco plants. Planta 204: 8–15.

274

Yang, K.C. 1978. The fine structue of pits in yellow birch (*Betula alleghaniensis* Britton). IAWA Bull. 1978: 71–77.

Ye, Z.-H. 1997. Association of caffeoyl coenzyme A 3-*O*-methyltransferase expression with lignifying tissues in several dicot plants. Plant Physiol. 115: 1341–1350.

Yin, T.M., Huang, M.R., Wang, M.X., Zhu, L.H., He, P. and Zhai, W.X. 1999. RAPD linkage mapping in a *Populus adenopoda* × *P. alba* F-1 family. Acta Bot. Sin. 41: 956–961.

Yoshinaga, A., Fujita, M. and Saiki, H. 1993. Compositions of lignin building units and neutral sugars in oak xylem tissue. Mokuzai Gakkaishi 39: 621–627.

Yoshinaga, A., Fujita, M. and Saiki, H. 1997a. Cellular distribution of guaiacyl and syringyl lignins within an annual ring in oak wood. Mokuzai Gakkaishi 43: 384–390.

Yoshinaga,, A., Fujita, M. and Saiki, H. 1997b. Secondary wall thickening and lignification of oak xylem components during latewood formation. Mokuzai Gakkaishi 43: 377–383.

Zablackis, E., York, W.S., Pauly, M., Hantus, S., Reiter, W.-D., Chapple, C.C.S., Albersheim, P. and Darvill, A. 1996. Substitution of L-fucose by L-galactose in cell walls of *Arabidopsis mur1*. Science 272: 1808–1810.

Zakrzewski, J. 1991. Effect of indole-3-acetic acid (IAA) and sucrose on vessel size and density in isolated stem segments of oak (*Quercus robur*). Physiol. Plant. 81: 234–238.

Zhang, G.F. and Staehelin, L.A. 1992. Functional compartmentation of the Golgi apparatus of plant cells. Immunocytochemical analysis of high-pressure frozen- and freeze-substituted sycamore maple suspension culture cells. Plant Physiol. 99: 1070–1083.

Zheng, Z.-L. and Yang, Z. 2000. The Rop GTPase switch turns on polar growth in pollen. Trends Plant Sci. 5: 298–303.

Zhong, R. and Ye, Z.-H. 1999. *IFL1*, a gene regulating interfascicular fiber differentiation in *Arabidopsis*, encodes a homeodomain-leucine zipper protein. Plant Cell 11: 2139–2152.

Zhong, R., Morrison, W.H. III, Negrel, J. and Ye, Z.-H. 1998. Dual methylation pathways in lignin biosynthesis. Plant Cell 10: 2033–2046.

Zhong, R., Ripperger, A. and Ye, Z.-H. 2000. Ectopic deposition of lignin in the pith of stems of two *Arabidopsis* mutants. Plant Physiol. 123: 59–69.

Zobel, B.J. and van Buijtenen, J.P. 1989. Wood Variation: Its Causes and Control. Springer-Verlag, New York.

Plant Molecular Biology **47**: 275–291, 2001.
© 2001 *Kluwer Academic Publishers. Printed in the Netherlands.*

Functional genomics and cell wall biosynthesis in loblolly pine

Ross Whetten*, Ying-Hsuan Sun, Yi Zhang and Ron Sederoff
*Forest Biotechnology Group, 2500 Partners II, Centennial Campus, Campus Box 7247, North Carolina State University, Raleigh, NC 27695, USA (*author for correspondence; e-mail ross_whetten@ncsu.edu)*

Key words: EST sequencing, microarrays, *Pinus taeda*, xylogenesis, wood formation

Abstract

Loblolly pine (*Pinus taeda* L.) is the most widely planted tree species in the USA and an important tree in commercial forestry world-wide. The large genome size and long generation time of this species present obstacles to both breeding and molecular genetic analysis. Gene discovery by partial DNA sequence determination of cDNA clones is an effective means of building a knowledge base for molecular investigations of mechanisms governing aspects of pine growth and development, including the commercially relevant properties of secondary cell walls in wood. Microarray experiments utilizing pine cDNA clones can be used to gain additional information about the potential roles of expressed genes in wood formation. Different methods have been used to analyze data from first-generation pine microarrays, with differing degrees of success. Disparities in predictions of differential gene expression between cDNA sequencing experiments and microarray experiments arise from differences in the nature of the respective analyses, but both approaches provide lists of candidate genes which should be further investigated for potential roles in cell wall formation in differentiating pine secondary xylem. Some of these genes seem to be specific to pine, while others also occur in model plants such as *Arabidopsis*, where they could be more efficiently investigated.

Abbreviations: AGP, arabinogalactan protein; APRP, adhesive proline-rich protein; EST, expressed sequence tags; GRP, glycine-rich protein; OMT, *O*-methyltransferase; PHY, phytocyanin; PRP, proline-rich protein; XET, xyloglucan endotransglycosylase

Introduction

Mankind has long relied on wood from forest trees and woody stems of grasses for structural materials. More recently, pulp and paper production has taken on increased importance in the forest products industry; about 30% of the US wood harvest goes to pulp and paper manufacturing (Saltman *et al.*, 1998). The properties of wood are functions of the size, shape and arrangement of cells in wood, as well as the structure and chemistry of the cell walls. The importance of wood as an industrial raw material has motivated extensive analysis of its structure and chemistry (Lewin and Goldstein, 1991; Biermann, 1993; Higuchi, 1997).

Wood formation involves the specialized biosynthesis of a secondary cell wall in xylogenesis of woody plants. Specialized cell walls are found in many plant organs; however, the formation of secondary thickenings in vascular tissue is particularly important in the evolution of higher land plants. Well-developed vasculature was apparent in fossil trees as early as the Devonian (Meyer-Berthaud *et al.* 1999), and has been the basis for structural support and for water transport, both essential for the large size of woody plants. In gymnosperms such as loblolly pine (*Pinus taeda* L.), only a few cell types are present in secondary xylem (Harada and Côté, 1985). Wood is formed from the terminal differentiation of cells in xylem, a process that typically continues throughout the annual growing season of a woody plant. The specialized nature of the cell walls in woody tissues provides favorable material for the investigation of many aspects of cell wall structure and biogenesis. Wood is almost entirely composed of cell wall material, and differentiating secondary xylem is rich in cell wall biosynthetic enzymes (Sederoff *et al.*, 1994; Allona *et al*, 1998; Sterky *et al.*, 1998).

A relatively small number of forest tree species have been subjected to intensive molecular genetic analysis. Trees in general are difficult experimental organisms, due to their large size and long generation times, and so attention has been focused on those species of greatest commercial importance. In the USA, loblolly pine is the most widely planted species, with over 10^9 seedlings planted per year. Pines have the additional disadvantage, as experimental organisms, of extremely large genomes, ranging from 20 to 40 pg DNA per haploid genome equivalent (Wakiyama *et al.*, 1993) or about 200–400 times larger than the genome of *Arabidopsis thaliana* (Somerville and Somerville, 2000). The pine genome is rich in repetitive DNA (Kriebel, 1985), at least some of which is due to the presence of abundant retrotransposon (Kamm *et al.*, 1996; Kossack and Kinlaw, 1999). The haploid megagametophyte of pines and other conifers does, however, provide unique advantages for genetic analysis of forest trees in natural and managed ecosystems (O'Malley *et al.* 1996).

The properties of wood, and the process of wood formation, are due to the action of the genes and proteins in differentiating secondary xylem. These genes and proteins need to be studied to understand wood formation, and are potential targets for the directed modification of wood properties. The regulation of these genes in response to developmental and environmental cues is likely to determine variation in wood properties. The tracheid length, diameter and wall thickness all affect the strength and density of wood. When lignin is removed, these properties also determine the strength, coarseness, and density of pulp and paper products.

Wood in many trees varies greatly in structure, composition or both, during ontogeny, during the growing season, and under mechanical stress (Megraw, 1985; Zobel and van Buitenen, 1989). During the growing season there are major changes in the structure of the tracheids (springwood versus summerwood), which affect their ability to transport water under wet or dry conditions. In pines, the composition and morphology changes as the vascular cambium ages (the juvenile/mature transition), and in response to mechanical stress (reaction wood). Springwood (or earlywood) is characterized by large lumens and thinner walls and is lower in density than latewood, which

has smaller lumens and thicker walls. Juvenile wood has lower density, shorter fibers and a higher lignin content than mature wood (Zobel and Sprague, 1998).

A change in the orientation of a tree stem with respect to gravity frequently stimulates the formation of a specialized type of wood termed reaction wood. The reaction wood formed in pines and other conifers is called compression wood, and is formed on the underside of a bent stem, serving to reorient the stem or branch to a vertical position. Compression wood differs in morphology and composition from normal wood, and from wood on the lateral and upper sides of the bent stem. Tracheids in compression wood are characterized by a round rather than rectangular cross-sectional profile, a higher ratio of secondary wall thickness to cell diameter, a decrease in cell length, an increase in the angle of cellulose microfibrils in the secondary wall to the long axis of the cell, an increase in the fraction of *p*-hydroxyphenyl subunits in lignin and an increase in lignin content (Timell, 1986). Changes in the relative abundance of specific transcripts between normal and compression wood provide insight into the developmental transitions that underlie the modifications in cell wall structure and composition induced by mechanical stress. Side wood does not show these changes and serves as an internal control.

Xylogenesis in pine begins with cell division of two different types of cambial cells, known as fusiform initials and ray initials, to give rise to mother cells which will eventually become tracheids and ray parenchyma cells, respectively. After cell division of fusiform initials and mother cells, the daughter cells undergo radial expansion, secondary wall deposition, lignification and (in the case of tracheids) programmed cell death (Figure 1). Trees have advantages for the study of xylem differentiation because of the ability to obtain large amounts of differentiating xylem for biochemical studies from field-grown trees, allowing genomic approaches to understanding the molecular events in wood formation. During active stem growth of pine, the bark can be removed leaving the immature xylem attached to the wood. Scraping the surface removes the non-lignified differentiating xylem, while deeper planing removes tracheids undergoing lignification and programmed cell death, as well as some mature tracheids and fully differentiated ray cells (Figure 2).

We have used a genomic approach to identifying genes and proteins involved in cell wall biosynthesis during xylogenesis in loblolly pine (Allona *et al.*,

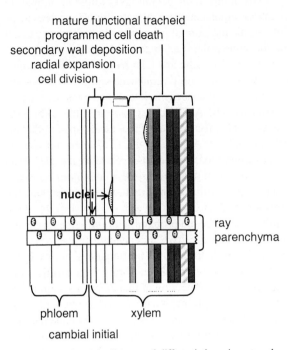

mature functional tracheid
programmed cell death
secondary wall deposition
radial expansion
cell division

nuclei →

ray
parenchyma

phloem | xylem

cambial initial

Figure 1. A schematic diagram of differentiating pine secondary xylem. The cambium is a meristem comprised of a single cell layer. This layer contains both elongated fusiform initials that gives rise to phloem precursors to the outside and xylem precursors to the inside, as well as ray initials that give rise to parenchyma cells that form continuous rays across the xylem and phloem.

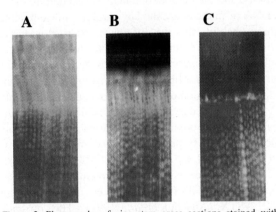

A   **B**   **C**

Figure 2. Photographs of pine stem cross sections stained with phloroglucinol to show lignified cells walls as red. A. A stem section taken before removal of phloem and external tissues. B. A stem section taken after removal of phloem and external tissues, leaving the differentiating secondary xylem on the surface of the wood. C. A stem section taken after harvest of non-lignified differentiating secondary xylem by scraping the surface of the wood with a vegetable peeler. Lignified differentiating xylem remains, but can be harvested with a carpenter's plane.

1998; Zhang *et al.*, 2000). In this report, we present results from identification of large numbers of expressed genes based on cDNA sequencing, and from preliminary microarray analysis of relative expression levels of a subset of these genes. In addition, we have identified and partially characterized several cell wall-associated proteins by this approach.

Materials and methods

The methods used to obtain differentiating pine secondary xylem, prepare cDNA libraries, and carry out partial DNA sequence determination have been described (Allona *et al.*, 1998). The libraries from which pine ESTs have been sampled to date were constructed from RNA obtained from a variety of tissues and organs, including several different types of differentiating xylem. Non-xylem tissues and organs include immature male strobili, or pollen cones, and shoot tips, or the terminal 2 cm of elongating primary growth from branches. Differentiating xylem samples include compression wood, side-wood (morphologically normal tissue from the side of the same stems, serving as a control), and normal vertical xylem. Pollen cones were collected from lower branches of a single individual mature (ca. 30- year old) tree in the early spring after reaching full size, but prior to dehiscence, and represent late stages of pollen differentiation. Shoot tips were collected from lower branches of a different individual tree (ca. 10 years old) in late spring during the period of active shoot elongation. The pollen cone and shoot tip libraries should therefore contain no more than two alleles for any single gene, because they are each derived from a single diploid individual. The alleles present in the two libraries may differ, however. Compression wood and side-wood libraries were made from pooled RNA samples obtained from three trees (6 years old), as previously described (Allona *et al.*, 1998). The normal vertical xylem library was made from RNA from a single individual tree (ca. 35 years old). Differentiating xylem samples collected from an individual tree were all pooled prior to RNA isolation, i.e. no separation of juvenile wood from mature wood was attempted during library construction. The compression wood and side wood libraries may contain up to six different alleles of a single gene, while the normal vertical wood library should contain no more than two alleles. Sampling from these libraries has largely been random, although an early project did construct subtracted libraries and sequence

278

relatively small numbers of clones from them (Allona et al., 1998). The fact that most cDNAs were sampled at random from the different libraries allows statistical analysis of the frequency with which particular sequences appear in the resulting data set.

Annotation of the pine EST sequences obtained at North Carolina State University has been carried out in collaboration with Ernest Retzel and his colleagues at the Computational Biology Center of University of Minnesota, and the results are displayed at http://web.ahc.umn.edu/biodata/doepine/ and http://web.ahc.umn.edu/biodata/nsfpine/. One important analysis carried out on the pine EST data sets was assembly of 'contigs', or clusters of overlapping EST sequences all apparently derived from the same mRNA. Contigs were assembled by PHRAP (Ewing and Green, 1997; see also http://www.phrap.org). The PHRAP parameters used for contig assembly were a minimum match of 40 and a minimum score of 80; as a result, sequences with more than 98% identity are sometimes placed into different contigs. These sequence differences may be allelic variation, or variation between members of gene families. Sequencing errors are unlikely to result in formation of multiple contigs from otherwise identical ESTs, because the PHRAP program uses error probability scores produced by the PHRED base-calling program in evaluating the probability that two ESTs are representatives of the same sequence. The assembly of contigs used in the analysis presented here used 4557 sequences from the compression wood library and 8490 sequences from the normal vertical wood library.

Contig assembly reduces the numbers of sequences to be analyzed, and can also aid in identifying allelic variation and defining members of multigene families. These contigs represent the more abundantly expressed genes in differentiating pine secondary xylem, and have better coverage of the coding sequences than single ESTs. Analysis of abundantly expressed genes is likely to be more informative regarding function in tissues than trying to analyze all ESTs, because the most abundantly expressed genes in a tissue are good candidates to have important functions in that tissue. Using contigs, we are also less likely to misidentify a gene because only a small amount of sequence has been obtained for that gene. We have restricted our analysis to those ESTs that have at least 200 nucleotides of high-quality sequence (PHRED score greater than 20, equivalent to error probabilities of less than 10^{-2}, per nucleotide) and occur in contigs of 4 or more sequences.

The numbers of specific ESTs found in samples of cDNAs sequenced from different libraries represent an electronic measure of gene expression for many genes at the same time, if analyzed with appropriate statistical methods (Audic and Claverie, 1997). The number of loblolly pine EST sequences similar to a specific pine contig sequence was determined by using the pine contig as a BLASTN query to search the 'otherests' section of GenBank, with 'Pinus taeda' specified in the organism field (Altschul et al., 1997). The numbers of ESTs similar to the query at expect (E) values less than 10^{-5} were counted, and these values used to calculate the probability of differential expression using the method of Audic and Claverie (1997). Briefly, these authors describe the probability of the same rare event occurring twice in two independent trials, based on the Poisson distribution. The frequency of any given cDNA in a non-normalized library is generally sufficiently low that detection of that cDNA by sequencing during an EST project is a rare event. Sequencing from two different libraries constitutes independent trials, and identification of the same cDNA by assembly of EST contigs from different libraries is an example of the same rare event occurring twice. The difference in the number of ESTs corresponding to a particular cDNA between the two libraries may be due to a difference in the representation of that cDNA in the libraries, or it could be due to random chance in sampling. The total number of ESTs obtained from each library, and the number obtained corresponding to the cDNA of interest, are the key variables in calculating the probability that the difference observed is due to differential representation rather than to chance fluctuations. The equation used to calculate the probability of y occurrences of ESTs derived from a gene of interest in a population of N_2 total sequences from library 2, given the occurrence of x ESTs derived from the same gene in a population of N_1 total sequences from library 1, and assuming only random fluctuation, is $p(y|x) = (N_2/N_1)^y [(x+y)!/\{x!y!(1+N_2/N_1)^{(x+y+1)}\}]$.

Searches for similarity between pine EST sequences and ESTs from poplar were carried out with TBLASTX (Altschul et al., 1997). This program translates the query sequence in all six reading frames, and searches a six-frame translation of the EST collection for similarity between predicted polypeptides. This provides a greater level of sensitivity for detection of diverged coding regions than do similarity searches at the nucleotide sequence level. Twenty pine contigs lacking significant similarity to sequences in GenBank, ten from the 'vertical' collection and ten from the 'compression wood' collection, were used as queries. These contigs correspond to transcripts present at a frequency of greater than 0.1% in these two cDNA libraries, based on the numbers of sequences comprising each contig. Use of contigs as queries reduces the probability that similarity would fail to be detected simply because the query sequence is too short to overlap with possible similar ESTs in the database, and the use of the TBLASTX search tool provides a sensitive search for similarity at the protein level. The ten contigs from the vertical EST collection were 1460, 1463, 1493, 1496, 1500, 1507, 513, 1538, 1549, and 1553, from the 23 August 2000 contig set presented at http://web.ahc.umn.edu/biodata/nsfpine/contig_dir0. The compression wood contigs were 629, 634, 644, 648, 656, 661, 668, 679, 694, and 723, from the September 19, 2000 contig set presented at http://web.ahc.umn.edu/biodata/nsfpine/contig_dir1.

Comparisons of relative levels of gene expression between pine and *Arabidopsis* were carried out with the full-length GenBank sequences most similar to individual pine EST contigs as TBLASTN queries to search both *Arabidopsis* and pine EST collections in dbEST. The numbers of ESTs from each organism similar to the query sequence at E values of less than 10^{-5} were counted for use in the statistical comparisons. This approach provides a first approximation of the relative abundance of entire classes of mRNAs, but does not discriminate well between different members of gene families. The method does provide an overview of the relative abundance of RNAs encoding particular metabolic functions. Metabolic function is typically an important criterion for interpreting the biological significance of any predicted differential abundance of mRNAs based on relative frequency of ESTs, so the fact that the estimates of differential abundance presented here represent entire gene families rather than individual genes focuses attention on metabolism rather than on specific genes. Complex gene families are relatively common in pine (Kinlaw and Neale, 1997), and high-throughput methods for accurately determining the contributions of individual members of gene families to overall levels of gene expression will be needed to explore questions of differential gene regulation more fully.

Microarray methods

Methods for printing PCR-amplified cDNA inserts from plasmid clones onto glass slides in high-density arrays have been described (Winzeler *et al.*, 1999). The first-generation pine microarrays described here were printed at the Carnegie Institute of Washington, in collaboration with Shauna Somerville's group. These arrays consist of about 3000 pine cDNAs; 1100 of these clones were derived from a suppression subtractive PCR experiment between differentiating juvenile and mature secondary xylem (Sun, unpublished work based on the method of Diatchenko *et al.*, 1996) and the remaining clones were those available from early EST sequencing projects (Kinlaw *et al.*, 1996; Allona *et al.*, 1998). No selection of cDNA clones from the available pool was done; instead, all available cDNAs from differentiating pine xylem were arrayed. Two different pairs of samples were used in these experiments, representing one pair of developmental stages and one environmental stimulus. The developmental stages examined were differentiating juvenile wood, from the apical two meters of three 20-year old trees (of different genotypes), and differentiating mature wood from the basal two meters of the same trees. The environmental stimulus examined was mechanical stress, applied to one ramet of a pair of clonal individuals, 6 years of age, by bending one tree away from a vertical position and tying it to a stake driven into the ground so that the apical portion of the stem was parallel to the ground. The control ramet remained in a vertical orientation for the 6 days of treatment, and both trees were then harvested. The final data set used for most of the analyses presented here includes data from three complete replicates for all three genotypes in the juvenile-mature comparison and four replicates for the compression wood versus normal wood comparison. After removal of data points for which various quality control measures indicated potential problems, a final set of about 2400 elements remained suitable for analysis.

Raw images acquired from the arrays were analyzed with ScanAlyze (M.B. Eisen; http://rana.lbl.gov/), and the two signal channels were balanced to yield equivalent total signal summed across all elements. Background was adjusted by a local regression method to be described in detail elsewhere (Sun, manuscript in preparation). The raw signal intensity data and background-adjusted normalized data are available by anonymous ftp from ftp.ncsu.edu in the directory /pub/unity/lockers/ftp/rosswhet. Testing for differential gene expression was carried out by several approaches. The first compared ratios of the signal intensity from a 'treatment' cDNA preparation to that obtained from a 'control' cDNA preparation for each element on the same array. This method of comparison of signal intensity ratios within the data set derived from each individual array will be referred to as 'within-slide' comparison. This analysis incorporated data from replicated arrays by testing the consistency with which a particular element yielded a ratio different from the mean ratio of all elements on each array. This method identified genes that changed in relative level of expression by a greater amount than the other genes tested on the array, focusing attention on the largest changes in relative expression levels. For the 'within-slide' test, array elements were designated as differentially expressed when the natural logarithm of the ratio (ln ratio) of background-corrected signal intensities on a given slide differed from the mean ln ratio for that slide by a set amount on at least 50% of the replicate slides. The criterion for deviation from the mean was 1.6 times the slide standard deviation for comparison of transcript levels between vertical wood RNA and compression wood RNA, and 1.0 times the slide standard deviation for comparison of transcript levels in juvenile wood RNA and mature wood RNA. This difference in criterion was determined subjectively, based on the variability of replicated slides within each set of treatments.

An alternative approach uses statistical methods to determine the reproducibility of each measurement, both within an individual array and across duplicate arrays. This recognizes that the largest magnitude of change in relative gene expression level is not necessarily the most interesting, and a reproducible change may be interesting regardless of its relative magnitude. One relatively simple version of this statistical approach is to compare the ratio of channel intensities for replicates of each element (across multiple arrays) to that of a set of replicated 'control' elements that do not show significant differences in signal between the RNA samples being compared. This requires some means of normalizing relative intensities between arrays, and so positive control elements on the array and corresponding synthetic mRNAs added to each labeling reaction are essential to this simple approach. It is also possible to use statistical methods to account for experimental variation, for example, in labeling efficiency. Appropriate models can be used to assign components of the observed variation in signal intensities to the various experimental sources, and variation not accounted for by noise is used as an estimate of differences in transcript levels. Several groups are pursuing this approach in slightly different ways, ranging from Bayesian methods (Friedman *et al.*, 2000; New-

ton *et al.*, 2000) to ANOVA methods based on those used in agricultural field trials (Kerr *et al.*, 2000; Kerr and Churchill, 2000; Wolfinger *et al.*, 2000).

The statistical test (Dunnett, 1955) used in this study yields an experiment-wide significance level, so that the chosen significance level (e.g. $P<0.05$) signifies the probability of Type I error for the entire set of genes designated as differentially expressed, rather than for each gene individually. This is particularly important for microarrays with hundreds or thousands of elements, because of the many comparisons that must be made. A probability of 0.01 of Type I error in each of 2000 comparisons allows an average of 20 false-positive errors, which can be a major problem if only 200 genes are determined to be differentially expressed. Dunnett's test was conducted by identifying a set of 68 elements as a 'control set', defined as elements whose ratios varied from the mean by greater than 0.5 standard deviation no more than once in the four slides used for the mechanical stress experiment. None of the control set ratios deviated from the mean by 0.5 standard deviation in any of the ten slides used for juvenile-mature comparisons. The ratios from all replicates of each element were tested for significant difference from the ratios of the control set.

Results

EST sequencing in loblolly pine

In October 2000, the dbEST section of GenBank contained 22 233 loblolly pine ESTs, a substantial increase from the 750 ESTs present in February 1998. About 60% of these were from differentiating xylem, 5624 from a shoot tip library, and 1507 from a pollen cone library. Several libraries have been made from wood forming tissues under different conditions, for comparison to normal vertical wood formed in the spring. Sequencing of pine ESTs is ongoing, and the numbers of pine sequences in GenBank will continue to increase. These EST collections have been used to address questions of unique genes in pines, relative gene expression in different types of differentiating xylem, and relative levels of gene expression in pine versus *Arabidopsis*.

Are there genes unique to gymnosperms or to differentiating secondary xylem?

Only recently have EST projects in pine and poplar begun to provide information on the genes expressed in woody plants in wood-forming tissues, to address the question of whether there may be unique genes expressed during secondary xylem formation (Allona *et al.*, 1998; Sterky *et al.*, 1998; Mellerowicz *et al.*, 2001, this issue). There has also been little information on the possible number of unique genes present in pines and other gymnosperms that are not found in herbaceous or woody angiosperms. One approach to the first question is to ask about the number of expressed genes not found in herbaceous annuals, but common to pines and woody angiosperms. The current EST collections from poplar and pine are not sufficient to allow a comprehensive answer to this question, but do provide some insight into the relationships between abundant transcripts in differentiating secondary xylem of these two tree species (see Mellerowicz *et al.*, 2001, this issue).

About 7% of contigs from loblolly pine show no similarity to DNA or protein sequences in GenBank, as of August 2000. These are candidates for genes that are either specific to gymnosperms, to woody tissues, or to both. Comparison of the most abundant of these pine sequences with poplar ESTs was carried out to test the hypothesis that some of these genes are common to woody tissues of both gymnosperms and woody dicots. Sterky *et al.* (1998) described a collection of EST sequences from differentiating poplar secondary vascular tissues. These authors noted that almost 350 of the set of 5692 ESTs showed no similarity to any sequences in the public DNA and protein sequence databases, and that three of these unknown genes were among the most common transcripts detected. Only one of the twenty pine contigs tested, contig 648 from compression wood, was similar to a poplar EST; the other 19 pine contigs showed no significant similarities to any poplar EST.

Relative levels of gene expression in compression wood and normal wood

Contigs encompassing multiple ESTs can be used to estimate significant differences in relative abundance of cDNAs corresponding to a particular gene in different libraries. It is not possible to attribute statistical significance to the presence or absence of single ESTs (Audic and Claverie, 1997). Comparisons of sequence similarity among the 50 largest contigs of the compression wood and vertical wood libraries (i.e. the 50 contigs from each library comprised of the most sequences) show that a total of 69 different sequences are present, after removing redundancy within and between the contig sets (Table 1). Of these 69 most abundant transcripts, 33 are likely to be more abundant in the compression wood library than in the normal wood library ($P < 0.05$ for difference due to random fluctuation, as calculated by the method of Audic and Claverie, 1997), and only three are likely to be more abundant in the normal wood library than in the compression wood library by the same criterion. The relevance of these putative differentially expressed genes to cell wall formation remains to be proven, but the identities assigned by BLAST searches suggest roles for many of them in cell wall biosynthesis. Among the cDNAs likely to be more abundant in the compression wood library are several related to monolignol biosynthesis (4-coumarate CoA ligase, caffeoyl CoA *O*-methyltransferase, glycine hydroxymethyltransferase, and *S*-adenosyl methionine synthetase) and several putative cell wall proteins (described in more detail in Table 2). Two pine cDNAs with no similarity to any protein or DNA sequence, and three pine cDNAs similar to unknown *Arabidopsis* genes predicted from genomic and EST sequences are also more abundant in the compression wood library. The three cDNAs found to be more abundant in the vertical wood library are predicted to encode a Skp1-like protein, a xyloglucan endotransglycosylase (XET)-like protein, and a protein similar to pollen allergens. Skp1 of yeast is involved in the ubiquitin-proteosome protein turnover pathway (Bai *et al.*, 1996) and the *Arabidopsis* homologue is expressed in meristematic tissues (Porat *et al.*, 1998). *Arabidopsis* Skp1-like proteins have also been implicated in ubiquitin-mediated proteolysis (Schouten *et al.*, 2000), but a strong connection with cell wall formation is not clear. A relationship of an XET to cell wall restructuring seems clear, however, and the change in abundance of the pollen-allergen-like cDNA seems to implicate it also in some aspect of the morphological or chemical differences between compression wood and vertical wood.

Contigs are also useful to identify allelic variation and to identify gene families. The PHRAP program parameters used to assemble pine EST sequences into contigs resulted in creation of different contigs if otherwise identical sequences contained a few single nucleotide differences in DNA sequence. There are several cases of groups of pine EST contigs which show considerable similarity in GenBank hits among the members of the group (Table 1). Such groups may represent allelic variation or duplicate members of a multigene family. Members of multigene families would be expected to show higher levels of sequence

Table 1. Contigs of abundant ESTs analyzed for differential expression. The probabilities shown in column four are calculated by the method of Audic and Claverie (1997). Values shown are the upper limits of the probability of observed numbers of ESTs arising by chance alone if the cDNA is equally represented in both libraries being compared. The table shows the presence of a contig in both libraries only when it was among the top 50 contigs ranked by size (number of ESTs) in both libraries. The contigs are ordered in the table from rank 50 of normal wood to rank 1 of normal wood, then rank 50 of compression wood to rank 1 of compression wood, with appropriate cross-references to other contigs similar to the same GenBank record. Contigs shown as present only in the top 50 contigs of one library are still similar to ESTs from the other library, but the corresponding contig was not large enough to rank in the top 50.

Library, line and contig number	Hit in GenBank	Accession number	Differentially expressed?	Number of ESTs in contig
Normal xylem: 1509	probable aquaporin	O65045	no	11
1510	probable thioredoxin H	O65049	no	11
1511	3-deoxy-D-arabino-heptulosonate 7-phosphate synthase	O24051	no	11
1512, 1559	allergen-like protein	AAF16869	$P < 0.001$, normal	11+48
1513	no hits		no	11
1514	calmodulin 3	CAA09302	no	11
1515, 1541	ADP-ribosylation factor	D17760	no	12+17
1516, comp. wood 748	4-coumarate CoA ligase	T09775	$P < 0.001$, compression	12+17
1517	GA-regulated protein	O82328	no	12
1518	T19F11.6 protein	AC009918	no	12
1519, comp. wood 732, 751	AGP-like protein	S52995	$P < 0.001$, compression	12+9+18
1520, 1525, 1531, 1532	S-adenosylmethionine synthase	P50300	$P < 0.001$, compression	12+13 +14+14
1521, comp. wood 731	Low-MW heat shock protein	S71768	$P < 0.001$, compression	12+8
1522	cellulase	T07612	no	13
1523, comp. wood 718	R40g2 from rice	T03960	no	13+7
1524	ELI-3	O82550	no	13
1526, 1527, comp. wood 719	methionine synthase	Q42699	no	13+13+7
1528, comp. wood 734	putative laccase	Q9SIY8	no	13+9
1529	fructose-bisphosphate aldolase	T12416	no	13
1530	MJK13.14 protein	AAF35414	$P < 0.001$, compression	14
1533	unknown protein	BAA92731	no	14
1534	pine LP6 protein	Q41083	no	14
1535	putative UDPG-glucosyltransferase	Q9SK82	$P < 0.01$, compression	14
1536, comp wood 728	actin-depolymerizing factor	P30175	nNo	15+8
1537	Skp1-like protein	AF135596	$P < 0.02$, normal	15
1538	no hits		$P < 0.001$, compression	16
1539	caffeoyl-CoA OMT	AAD02050	$P < 0.02$, compression	16
1540	aquaporin	O81186	$P < 0.001$, compression	17
1542	dTDP-glucose 4,6-dehydratase	CAB61752	no	18
1543	expansin	Q9XGI6	see Table 2	18
1544, comp wood 755	glycine hydroxymethyltransferase	B71400	$P < 0.001$, compression	19+22
1545, comp wood 754, 756	S-adenosylhomocysteine hydrolase	O23255	$P < 0.001$, compression	20+21+23
1546	actin	AF172094	no	21
1547, comp wood 741	translationally controlled tumor protein	AJ012484	no	22+13

Table 1 continued.

Library, line and contig number	Hit in GenBank	Accession number	Differentially expressed?	Number of ESTs in contig
1548, comp wood 735	phenylcoumaran benzylic ether reductase	AAF64176	no	22+9
1549	no hits		no	22
1550	dicyanin	AAF66242	no	24
1551, 1554, 1555, comp wood 717, 744, 757	α-tubulin	P33629	$P<0.001$, compression	24+28+30 +7+13+28
1552	unknown protein	O04324	no	24
1553	no hits		no	27
1556	XET	S61555	$P<0.01$, normal wood	30
1557, comp wood 713	polyubiquitin	CAB81047	no	31+7
1558, comp wood 745	elongation factor 1-α	AAD56020	$P<0.01$, compression	35+14
Compression wood 708	60S ribosomal protein L12	AB005246	$P<0.002$, compression	7
709	allyl alcohol dehydrogenase	AB036735	$P<0.001$, compression	7
710	histone H2A	P35063	$P<0.004$, compression	7
711	glutamine synthetase	AJ005119	$P<0.03$, compression	7
712	collagen 1	A48295	repetitive sequence[a]	7
714	glycine decarboxylase H-protein	AC004667	$p<0.05$, compression	7
715	histone H3.2	P11105	$P<0.03$, compression	7
716	glyceraldehyde 3-phosphate dehydrogenase	S51836	$P<0.001$, compression	7
720	β-tubulin	U76746	$P<0.04$, compression	7
721	nucleotide translocator	444790	$P<0.03$, compression	8
722	acidic ribosomal protein P2a-2	U62748	$p<0.001$, compression	8
723	no hits		$P<0.001$, compression	8
724	cytosolic malate dehydrogenase	T02935	$p<0.02$, compression	8
726	unknown protein	AC001645	no	8
727	unknown protein	AC007017	$p<0.001$, compression	8
729, 738	AGP4	AF101790	see Table 2	8
730, 747	glycine-rich RNA binding protein	AF109917	$p<0.001$, compression	8+16
733	methionine synthase	P93263	$p<0.04$, compression	9
736	proline-rich protein	AF101789	see Table 2	9
737, 746	peptidyl-prolyl isomerase	S54833	$P<0.001$, compression	9+14
739	unknown protein	AC009918.6	no	10
740	porin Mip1	T14863	$P<0.002$, compression	12
742, 750	AGP6	AF101785	see Table 2	13+17
743	RNA-binding protein 3	T15047	$P<0.001$, compression	13
749	unknown protein	AAF30339	$P<0.001$, compression	13
753	tonoplast intrinsic protein	T10804	$p<0.001$, compression	20

[a]The sequence of this contig was so repetitive as to prevent meaningful comparisons of numbers of similar sequences in the two EST collections.

variation than allelic variants, but this assumption has not yet been extensively tested in gymnosperms. In pines it is possible to distinguish between these alternatives by analysis of the haploid megagametophyte, which is derived from the same meiotic product as the maternal contribution to the zygote. Any individual megagametophyte should contain only one allele from a single locus, but will contain alleles from all members of a gene family (O'Malley et al., 1996). The challenge will be to carry out segregation analysis on thousands of different sequence variants identified in EST collections to determine which represent allelic variation and which are different gene family members. A high-throughput, cost-effective method for parallel identification of thousands of sequence variants in small amounts of genomic DNA from a few dozen individuals will be essential to gaining a more complete understanding of the size and complexity of gene families in pine.

What types of genes are abundantly expressed in wood-forming tissues?

Among the most abundantly expressed genes in wood-forming tissues are many genes expected to be involved in the formation of the wood cell wall. Genes involved in monolignol precursor biosynthesis are found, including genes encoding 4-coumarate CoA ligase (4CL, contig 1516 in Table 1) and caffeoyl-CoA *O*-methyltransferase (CCoAOMT, contig 1539 in Table 1). Other genes of less certain relationship to lignin biosynthesis are also highly expressed, such as a gene (contig 1524) very similar to ELI3 of parsley, which encodes a protein with benzyl alcohol dehydrogenase activity (Somssich et al., 1996). At least one transcript encoding a protein similar to laccase is highly abundant, while none of the abundant transcripts encode peroxidases, consistent with findings of pilot-scale EST studies of pine and poplar (Allona et al., 1998; Sterky et al., 1998; Mellerowicz et al., 2001, this issue). Also consistent with these previous studies is the finding of abundant transcripts related to methyl-transfer reactions involving *S*-adenosyl methionine (SAM), such as methionine synthetase, *S*-adenosylmethionine synthetase, glycine hydroxymethyltransferase, and *S*-adenosylhomocysteine hydrolase. These results suggest that the supply of methyl groups for lignin biosynthesis is a significant factor for the high level of carbon flux into lignin and that one-carbon metabolism could affect the ratio of methylated and unmethylated lignin

precursors. SAM is also a precursor for other biosynthetic pathways, as well as a methyl group donor, and the prevalence of enzymes involved in SAM formation and turnover may be a signal that some of these other pathways are important in xylem formation as well.

It might also be expected that genes involving formation of polysaccharides would be abundant, and several such genes are present, including genes homologous to XET (contig 1556, vertical wood), cellulase (contig 1522, vertical wood), and RGP1 (contig 1153, vertical wood), a gene thought to be involved in xyloglucan biosynthesis (Dhugga et al., 1997). Transcripts encoding expansin (EXP), proline-rich protein (PRP), glycine-rich protein (GRP), adhesive proline-rich protein (APRP) and phytocyanin (PHY) are also abundant in differentiating xylem (Zhang et al., 2000). It is less expected to find that several transcripts predicted to encode arabinogalactan (AGP) proteins are highly expressed.

Loopstra and Sederoff (1995) identified two abundant and specifically expressed AGPs in differentiating xylem. Zhang et al. (2000) and Loopstra et al. (2000) have now identified a total of six different genes encoding AGP-like proteins that are abundantly expressed in differentiating xylem. AGPs are unusual in the repeated motif protein structure, and in the high level of glycosylation. Many AGPs contain a sequence indicating the presence of a glycerophosphatidylinositol (GPI) anchor, for attachment to the cell membrane. All are abundant in the immature xylem cDNA libraries, compared to shoot tips. Some increase in expression significantly in the formation of compression wood compared to normal wood (Table 2). These results suggest an important but as yet undefined role of AGPs in the differentiation of pine secondary xylem, perhaps in formation of the secondary cell wall. In one case, the immunolocalization of AGP6 is restricted to radially expanded cells just preceding the thickening of their secondary walls (Y. Zhang, unpublished results). Sequences encoding phytocyanin-like proteins are of interest because they contain both a domain similar to proteins associated with cell walls, and a laccase-like domain, both associated with activities thought to have a role in wood formation (O'Malley et al., 1993; Zhang et al., 2000).

Comparing gene expression in differentiating pine xylem with gene expression in Arabidopsis

Database searches can contribute ideas about possible gene functions, by allowing researchers to compare

Table 2. Significance testing of digital differential gene expression profiles. The estimates of relative abundance of mRNA for AGPs and some other cell wall-associated proteins are compared based on numbers of corresponding ESTs found in pine cDNA libraries from differentiating normal xylem, differentiating compression xylem, shoot tips, and immature pollen cones. Values in the table are probabilities as in Table 1. Values in bold indicate significantly higher gene expression in the experimental condition on the corresponding row than in the condition in the corresponding column. Values in italics indicate higher gene expression in the experimental condition given in the column relative to the row.

	Compression	Shoot tip	Pllen cone		Compression	Shoot tip	Pollen cone
AGP1				**PRP1**			
vertical	$P<0.3$	**$P<0.002$**	$P<0.2$	vertical	*$P<0.007$*	$P<0.3$	$P<0.7$
compression		$P<0.07$	$P<0.6$	compression		**$P<0.001$**	$P<0.06$
Shoot tip			$P<0.5$	shoot tip			$P<0.8$
AGP2				**GRP2**			
vertical	*$P<0.001$*	$P<0.006$	$P<0.2$	vertical	$P<0.7$	$P>0.9$	$P<0.8$
compression		$P<0.001$	**$P<0.03$**	compression		$P<0.7$	$P<0.6$
Shoot tip			$P<0.09$	shoot tip			$P<0.8$
AGP3				**EXP1**			
vertical	**$P<0.05$**	**$P<0.001$**	**$P<0.03$**	vertical	$P<0.9$	$P<0.3$	$P<0.3$
compression		$P<0.06$	$P<0.4$	compression		$P<0.3$	$P<0.3$
Shoot tip			$P<0.8$	shoot tip			$P<0.7$
AGP4				**APRP1**			
vertical	*$P<0.001$*	**$P<0.006$**	$P<0.6$	vertical	$P<0.5$	$P>0.02$	$P<0.5$
compression		**$P<0.001$**	**$P<0.004$**	compression		$P<0.2$	$P<0.8$
Shoot tip			$P<0.3$	shoot tip			$P<0.5$
AGP5				**PHY2**			
vertical	$P<0.2$	**$P<0.001$**	**$P<0.05$**	vertical	$P<0.2$	**$P>0.009$**	$P<0.08$
compression		**$P<0.001$**	**$P<0.007$**	compression		$P<0.5$	$P<0.5$
Shoot tip			$P<0.9$	shoot tip			$P<0.8$
AGP6							
vertical	*$P<0.001$*	**$P<0.002$**	$P<0.2$				
compression		**$P<0.001$**	**$P<0.001$**				
Shoot tip			$P<0.5$				

apparent expression levels between organisms. There are over 111 000 *Arabidopsis* ESTs in in the dbEST division of GenBank (as of August 2000), derived from a variety of different libraries. TBLASTN searches were conducted using full-length protein sequences to which abundant pine ESTs show similarity (angiosperm protein sequences were used when possible, to avoid biasing search results against *Arabidopsis*). Some genes expressed in both pine and *Arabidopsis* are far more abundant in the pine EST data set, largely derived from wood-forming tissues, than in the *Arabidopsis* EST data set, derived from many different tissues (Table 3). These differences in abundance provide evidence for a potential role of these gene products in secondary cell wall biosynthesis or some other aspect of pine secondary xylem differentiation. Linking the pine ESTs to databases which organize and curate the growing volume of information about

other plant genes will be an important step for the future, to take full advantage of all the information gained in model plant systems.

Microarray results

Comparison of the 'within-slide' and Dunnett's test analyses of pine microarray data shows that the methods agree in large part, but that the Dunnett's test is more conservative in declaring a particular gene to be differentially expressed. In the vertical wood versus compression wood comparison, 156 of the 2300 elements were designated as differentially expressed by the within-slide test, 85 up-regulated and 71 down-regulated in compression wood. Dunnett's test confirmed up-regulation of 36 of the 85 elements called up-regulated by the within-slide test, but called 6 of the 85 down-regulated. Of the 71 elements called

286

Table 3. Comparisons of EST abundance in *Arabidopsis* and in pine. The table shows protein sequences used as search queries, the number of ESTs from *Arabidopsis* and pine that are similar to each query at an E value of less than 1×10^{-5}, the fold-difference in representation of that EST in the pine collection versus the *Arabidopsis* collection, and the probability that the difference in EST abundance is due to chance alone (Audic and Claverie, 1997). These calculations do not take into consideration that 40% of the pine ESTs are from tissues or organs other than differentiating xylem.

Protein	Hits in *Arabidopsis*	Hits in pine	Fold difference	Probability (P)
pirT03962 r40g3 stress-induced protein	4 ESTs (0.0036%)	47 ESTs (0.26%)	72-fold more in pine	<0.001
spQ40854 metallothionein-like protein	74 ESTs (0.067%)	204 ESTs (1.1%)	16-fold more in pine	<0.001
Gi 2765366, similar to pollen allergens	24 ESTs (0.022%)	53 ESTs (0.29%)	13-fold more in pine	<0.001
Gi 1563719 cyclophilin	119 ESTs (0.11%)	158 ESTs (0.88%)	8-fold more in pine	<0.001
pirT07139 cysteine proteinase inhibitor	34 ESTs (0.031%)	33 ESTs (0.18%)	5.8-fold more in pine	<0.001
pirT05667 (former) auxin-independent growth regulator	33 ESTs (0.030%)	30 ESTs (0.17%)	5.5-fold more in pine	<0.001
Gi 1419088 calreticulin	52 ESTs (0.046%)	28 ESTs (0.16%)	3.4-fold more in pine	<0.001
spP28014 translationally controlled tumor protein	169 ESTs (0.15%)	57 ESTs (0.31%)	2.1-fold more in pine	<0.001
pirT05950 lipid transfer protein	293 ESTs (0.26%)	73 ESTs (0.40%)	1.5-fold more in pine	<0.002
Gi 2995990 dormancy-associated protein	144 ESTs (0.13%)	26 ESTs (0.14%)	1.1-fold more in pine	<0.6

down-regulated by the within-slide test, 24 were confirmed by Dunnett's test, and 47 were not detected as down-regulated. Dunnett's test did not designate any element as differentially expressed which was not so designated by the within-slide test (Table 4).

In the juvenile wood versus mature wood comparison, 188 elements were designated as differentially expressed by the within-slide test, 113 as up-regulated in mature wood and 75 as up-regulated in juvenile wood. Of these 113 up-regulated elements, 94 were confirmed by Dunnett's test along with another 12 elements not designated as differentially expressed by the within-slide test. Of the 75 elements called up-regulated in juvenile wood by the within-slide test, 60 were confirmed by Dunnett's test, along with another 5 elements not designated as differentially expressed by the within-slide test. Table 3 shows known genes to which array elements showing differential expression by both methods are similar. The remaining elements showing differential expression either showed no similarity to any sequence in GenBank, or showed similiarity only to uncharacterized genomic or EST sequences.

Conclusions

Only one of the 20 largest pine contigs (of those without similarity to sequences in GenBank) showed significant similarity to one of the abundant poplar ESTs of unknown function. This result suggests that many of the genes highly expressed during secondary xylem formation in poplar are significantly different, either in sequence or in relative abundance, from the genes expressed during secondary xylem formation in pine. A more complete analysis of this interesting question awaits the completion of larger sets of both pine and poplar ESTs, or a separate project specifically focused on addressing the question of how many unique genes may be involved in secondary xylem formation in poplar and in pine. An equally interesting question, which must also await additional information, is whether any genes expressed during secondary xylem formation are truly specific to that process, or if secondary xylem formation uses largely the same genes as formation of primary xylem.

Comparisons of relative EST abundance in different libraries do not always agree with the microarray results about relative levels of gene expression in dif-

Table 4. Genes detected as differentially expressed by microarray analysis of juvenile vs. mature and compression wood vs. normal wood. EST sequences from elements of microarrays showing differences in expression in both 'within-slide' and Dunnett's test analyses were used to search Gen-Bank (BLASTX) for similar sequences of known function, and the resulting GenBank accessions and descriptions are shown. If multiple elements were similar to a single type of sequence, then only a single entry is present in the table.

Tissue specificity	Similar to accession number	GenBank description
Juvenile wood abundant	AAC39360	low-MW heat shock protein
	P36182	heat shock protein 82
	T09248	HSP23.5
	T09253	HSP17
	A49539	XET
	P35694	BRU-1, XET-like protein
	JE0184	chitinase
	P21563	peptidyl-prolyl isomerase
	AAF17645	pectinesterase-like
	BAB01177	lipid transfer protein
	T14889	porin Mip2
	BAB09857	Phi-1-like protein
	AAC67358	acid phosphatase-like
	444790	nucleotide translocator
	CAA05979	adenine nucleotide translocator
	Q42679	*S*-adenosylmethionine decarboxylase
	Q05212	DNA damage/repair protein DRT102
	P35063	histone H2A
	AAC64128	actin
	AAG02215	Class III peroxidase
	BAA86060	senescence-related protein
	P54778	26S protease regulatory subunit 6B
	S31035	retroviral gag protein-like
	S58500	auxin-induced protein IAA9
Mature wood abundant	AAD50628	α-tubulin
	P41636	4-coumarate CoA ligase
	AAD02050	caffeoyl-CoA *O*-methyltransferase
	AAD23378	*trans*-cinnamate 4- hydroxylase
	P52777	phenylalanine ammonia-lyase
	AAG02215	Class III peroxidase
	S52995	arabinogalactan-like protein
	AAF75827	pine AGP5
	AAF75821	pine AGP6
	U09554	pine AGP-like 3H6
	CAB88264	callose synthase catalytic subunit-like protein
	BAB09063	cellulose synthase catalytic subunit-like protein
	AAF76468	contains a peptidase S8 domain
	BAB09397	cysteine protease-like
	AAA34123	hexameric polyubiquitin
	AAD39373	plasma membrane intrinsic protein 1
	P50300	*S*-adenosylmethionine synthase

Table 4 continued.

Tissue specificity	Similar to accession number	GenBank description
Mature wood abundant	AAC39360	low-MW heat shock protein
	P36182	heat shock protein 82
	CAA07232	putative Pi-starvation induced protein
	AAD21718	putative ribose phosphate pyrophosphokinase
	AAC33203	ATP-citrate-lyase-like
	BAA94511	ABC transporter-like
	S71769	low-MW heat shock protein
	T00801	homeobox protein-like
	T01643	DNA-J-protein-like
	U10432	lipid transfer protein
	Z11487	pine globulin-2
Compression wood abundant	U09554	AGP-like protein 3H6
	AAF75826	AGP4
	P52777	phenylalanine ammonia-lyase
	AAD21718	putative ribose phosphate pyrophosphokinase
	CAA94437	ABC transporter-like
	P93263	methionine synthase
	B71400	glycine hydroxymethyltransferase
Normal wood abundant	BAA94511	ABC transporter-like
	T16974	pectinesterase
	BAB09857	Phi-1-like protein
	AAC67358	acid phosphatase-like

ferent tissue types, and several reasons could account for the differences. Different trees were harvested at different times for the two types of studies, so true biological variation in the response is probably partially responsible. The differences could also be due to variation in the nature of the experimental methods. Microarray hybridizations and washes were carried out under relatively low-stringency conditions, so cross-hybridization between members of gene families could have occurred. The extent of cross-hybridization undoubtedly differs between elements on the array, both because the sizes of multi-gene families vary, and because some cDNA clones are full-length while others include only a small portion of the protein coding sequence and the 3′-untranslated region. Comparisons of EST abundance between libraries were carried out by using pine EST contigs as BLASTN queries of the pine EST collection, and counting the numbers of sequences from different libraries that show similarity at expect values less than 10^{-5}. This could lead to inclusion of several members of a gene family in the count for a particular contig, depending on the presence of conserved domains in some families of proteins. Some libraries used for the EST projects could contain up to six different alleles for each gene, and allelic variation within pine is sufficiently high to make it difficult to distinguish between alleles and different members of a gene family without genetic segregation data.

The two methods used for analysis of the microarray data each have strengths and weaknesses. The 'within-slide' approach is relatively easy to apply, and does not require positive control elements and construction of synthetic mRNAs. It does, however, have the disadvantage that the significance of any change in relative transcript levels for each element is dependent on all the other elements on the array. The same cDNA fragment spotted on two different arrays, and hybridized to the same two labeled cDNA pools, could yield data deemed significant from an array with other elements that show little change in expression, and yet the same ratio of differential expression might not be significant on a second array with other elements that show large changes in expression. This use of the other elements on the array as a test for significance limits the ability of researchers to compare results between arrays that contain different but overlapping subsets of the total complement of genes in any genome. Dunnett's test requires some objective means of balancing the relative signal strength from the two channels, but does provide a measure of the experiment-wide false-positive rate.

The within-slide test as used in this study is not a stringent test for differential expression, requiring less than two standard deviations difference from the mean on a majority of arrays, but most elements on the array fail it. This test provides no means of determining the robustness of the conclusion other than permutation testing or bootstrapping, which are relatively computationally intensive methods of deriving an estimate of error rates from the data-set under study. Dunnett's test or ANOVA methods, in contrast, can be conducted at any desired level of experiment-wide error probability, and can also be adjusted to minimize the false-negative error rate or the false-positive error rate. The latter is an important consideration for investigators using microarrays to screen thousands of genes for those deserving more detailed study in a particular biological context – arguably a false-negative result could be more damaging to the outcome of such an experiment than a false-positive result, because of the failure to detect a gene of true significance to the process of interest.

Genes expressed during wood formation

Many of the genes expressed during pine secondary xylem differentiation are those genes expected to be present from previous work: cellulose synthase subunits, sucrose synthase, cell wall proteins, glucosidases and glucosyltransferases, and enzymes of lignin biosynthesis. Some pine cDNAs abundant in differentiating xylem cDNA libraries are similar to known genes which have never previously been linked with xylem differentiation or cell wall formation. One possibility is that those genes serve functions common to many cell types, but unrelated to cell wall formation, in differentiating xylem. Another possibility is that those genes have functions not yet known, and that they do play roles relevant to cell wall formation in differentiating xylem. The last class of genes are those identified by pine cDNAs which show no similarity to any known gene from any source. These may truly be pine-specific or secondary xylem-specific genes, and at least some of these genes are likely to play roles

in determining some of the aspects of secondary cell walls unique to pines.

Testing the roles in cell wall formation of candidate genes identified in pine will be difficult, due to the formidable challenges pine presents as an experimental system. Functional analysis of pine genes similar to *Arabidopsis* genes can readily be carried out in *Arabidopsis*, however, and the results confirmed in other model plants as appropriate. Pine genes not present in *Arabidopsis* clearly must be analyzed in pine, unless they can be found in maize or another model plant which provides the opportunity to carry out both forward and reverse genetics. These seem, however, to be a small fraction of all the interesting candidate genes identified in pine EST projects, so it may not be an overwhelming task to identify naturally occurring genetic variants in those genes in pine populations, then analyze the effects of those variants in controlled crosses. Natural populations of pines have relatively high frequencies of null alleles at isozyme loci, averaging about 0.3% across 22 isozyme loci (Allendorf *et al.*, 1982). This is equivalent to about one plant heterozygous for a deleterious allele per 300 plants in the population, or approximately the same frequency of mutant alleles observed in mutagenized populations of *Arabidopsis*. This approach has been fruitful in analyzing the effects of variation in cinnamyl alcohol dehydrogenase activity on lignin composition in loblolly pine (Ralph *et al.*, 1997).

The cell wall protein genes show a range of different patterns of expression. The transcripts of putative AGPs 2, 4 and 6 show similar probabilities of differential representation between vertical wood and compression wood libraries, vertical wood and shoot tip libraries, and compression wood and pollen cone libraries. Putative AGPs 1, 3, and 5 all have different patterns of differential expression among the libraries examined here, and EXP1 and GRP2 show little indication of differential expression in this analysis.

The most abundant cDNA clones identified by sequencing occur at a frequency of less than 1% of the total number of clones sequenced (50 of ca. 8500), and the frequency distribution drops rapidly from that point. There are 106 contigs that contain eight or more ESTs in a set of about 8500 normal wood ESTs analyzed, so we can estimate that about 100 transcripts occur at a frequency greater than 0.1% in differentiating normal wood. This set of 106 contigs encompasses a total of 1389 ESTs, or about 16% of the total number in the normal wood EST collection. This suggests that the hundred most abundant transcripts in differentiat-

290

ing pine secondary xylem account for only about 16% of the total number of mRNA molecules, and the other 84% constitute transcripts that are less abundant. This suggests that the mRNA pool in differentiating secondary xylem is quite complex, and that further EST sequencing will continue to yield novel genes, albeit with increasing numbers of cDNAs similar to those genes already identified.

Most of the contigs composed of the most abundant ESTs show sequence similarity to proteins of known or presumed function, but a significant minority (20 of 106) are either similar only to proteins of unknown function or show no similarity to public sequences. The latter group of genes may represent functions common to many or all plants, but abundant in differentiating pine xylem because of the specialized nature of the cells in that tissue. Functional genomics in differentiating pine xylem has the potential to contribute not to our understanding of wood formation, but to our understanding of cell wall formation and cellular differentiation in all plants.

Acknowledgements

The research described in this report has been supported by grants from USDA (95-373000-1591), US Department of Energy (DE-FC07-97ID13550) and NSF (9975806). The long-term collaboration of Ernest Retzel and the Computational Biology Center at University of Minnesota is gratefully acknowledged, as are the contributions of past and present members of the Forest Biotechnology Group at North Carolina State University and the Institute of Forest Genetics of the Pacific Southwest Forest and Range Research Station of the USDA-Forest Service. The first-generation pine microarrays described here benefited greatly from the advice and assistance of Shauna Somerville and Per Villand of Carnegie Institute of Washington, Palo Alto, CA, Patrick Hurban of Paradigm Genetics, Research Triangle Park, NC, and Ernest Kawasaki of GSI Lumonics, Watertown, MA.

References

Allona, I., Quinn, M., Shoop, E., Swope, K., St. Cyr, S., Carlis, J., Riedl, J., Retzel, E., Campbell, M.M., Sederoff, R. and Whetten, R. 1998. Analysis of xylem formation in pine by cDNA sequencing. Proc. Natl. Acad. Sci. USA 95: 9693–9698.

Altschul, S.F., Madden, T.L., Schaffer, A.A, Zhang, J., Zhang, Z., Miller, W. and Lipman, D.J. 1997 Gapped BLAST and PSI-BLAST: a new generation of protein database search programs. Nucl. Acids Res. 25: 3389–3402.

Audic, S. and Claverie, J.M. 1997 The significance of digital gene expression profiles. Genome Res. 7: 986–995. Software available through http://igs-server.cnrs-mrs.fr

Bennett, M.D. and Leitch, I.J. 1995. Nuclear DNA amounts in angiosperms. Ann. Bot. 76: 113–176.

Biermann, C.J. 1993. Essentials of pulping and papermaking. Academic Press, San Diego, CA.

Chapple, C. and Carpita, N. 1998. Plant cell walls as targets for biotechnology. Curr. Opin. Plant Biol. 1: 179–185.

Diatchenko, L., Lau, Y.F., Campbell, A.P., Chenchik, A., Moqadam, F., Huang, B., Lukyanov, S., Lukyanov, K., Gurskaya, N., Sverdlov, E.D. and Siebert P.D. 1996. Suppression subtractive hybridization: a method for generating differentially regulated or tissue-specific cDNA probes and libraries. Proc. Natl. Acad. Sci. USA. 93: 6025–6030.

Dhugga, K.S., Tiwari, S.C. and Ray, P.M. 1997. A reversibly glycosylated polypeptide (RGP1) possibly involved in plant cell wall synthesis: purification, gene cloning, and trans-Golgi localization. Proc. Natl. Acad. Sci. USA 94: 7679–7684.

Dunnett, C.W. 1955. A multiple comparison procedure for comparing several treatments with a control. J. Am. Statist. Ass. 50: 1096–1121.

Eisen, M.B., Spellman, P.T., Brown, P.O. and Botstein, D. 1998. Cluster analysis and display of genome-wide expression patterns. Proc. Natl. Acad. Sci. USA 95:14863–14868.

Ewing, B. and Green, P. 1998. Base-calling of automated sequencer traces using phred. II. Error probabilities. Genome Res 8: 186–194.

Friedman, N., Linial, M., Nachman, I. and Pe'er, D. 2000. Using Bayesian networks to analyze expression data. J. Comput. Biol., in press.

Harada, H. and Côté, W.A. 1985. The structure of wood. In: T. Higuchi (Ed.) Biosynthesis and Biodegradation of Wood Components, Academic Press, Orlando, FL, pp. 1–42.

Higuchi, T. 1997. Biochemistry and Molecular Biology of Wood. Springer-Verlag, Berlin.

Kamm, A., Doudrick, R.L., Heslop-Harrison, J.S. and Schmidt, T. 1996. The genomic and physical organization of Ty1-copia-like sequences as a component of large genomes in *Pinus elliottii* var. *elliottii* and other gymnosperms. Proc. Natl. Acad. Sci. USA 93: 2708–2713.

Kerr, M.K. and Churchill, G.A. 2000. Experimental design for gene expression microarrays. Submitted; manuscript available at http://www.jax.org/research/churchill/pubs/index.html.

Kerr, M.K., Martin, M. and Churchill, G.A. 2000. Analysis of variance for gene expression microarray data. Submitted; manuscript available at http://www.jax.org/research/churchill/pubs/index.html.

Kinlaw, C.S., Ho, T., Gerttula, S.M., Gladstone, E. and Harry, D.E. 1996. Gene discovery in loblolly pine through cDNA sequencing. In: Somatic Cell Genetics and Molecular Genetics of Trees (Forestry Sciences vol. 49), Kluwer Academic Publishers, Dordrecht, Netherlands, pp. 175–182.

Kinlaw, C. and Neale, D. 1997. Complex gene families in pine genomes. Trends Plant Sci. 2: 356–359.

Kossack, D. 1989. The IFG copia-like element: characterization of a transposable element present in high copy number in *Pinus* and a history of the pines using IFG as a marker. Ph.D. dissertation, University of California at Davis, CA.

Kossack, D.S. and Kinlaw, C.S. 1999 IFG, a gypsy-like retrotransposon in *Pinus* (Pinaceae), has an extensive history in pines. Plant Mol. Biol. 39: 417–426.

Kriebel, H.B. 1985. DNA Sequence components of *Pinus strobus* nuclear genome. Can. J. For. Res. 15: 1–4.

Lewin, M. and Goldstein, I.S. 1991. Wood Structure and Composition. Marcel Dekker, New York.

Loopstra, C.A. and Sederoff, R.R. 1995. Xylem-specific gene expression in loblolly pine. Plant Mol. Biol. 27: 277–291.

Loopstra, C.A., Puryear, J.D. and No, E.G. 2000. Purification and cloning of an arabinogalactan-protein from xylem of loblolly pine. Planta 210: 686–689.

Megraw, R.A. 1985. Wood Quality Factors in Loblolly Pine: the influence of tree age, position in tree, and cultural practice on wood specific gravity, fiber length, and fibril angle. TAPPI Press, Atlanta, GA.

Mellerowicz, E.J., Baucher, M., Sundberg, B. and Boerjan, W. 2001. Unravelling cell wall formation in the woody dicot stem. Plant Mol. Biol., this issue.

Meyer-Berthaud, B., Scheckler, S.E. and Wendt, J. 1999. *Archaeopteris* is the earliest known modern tree. Nature 398: 700–701.

Murray, B.G. 1998. Nuclear DNA amounts in gymnosperms. Ann. Bot. 82: 3–15.

Newton, M.A., Kendziorski, C.M., Richmond, C.S., Blattner, F.R. and Tsui, K.W. 2000. On differential variability of expression ratios: Improving statistical inference about gene expression changes from microarray data. J. Comput. Biol., in press.

O'Malley, D., Whetten, R., Bao, W., Chen, C.-L. and Sederoff, R.R. 1993. The role of laccase in lignification. Plant J. 4: 751–757.

O'Malley, D.M., Grattapaglia, D., Chaparro, J.X., Wilcox, P.L., Amerson, H.V., Liu, B.-H., Whetten, R., McKeand, S.E., Kuhlman, E.G., McCord , S., Crane, B. and Sederoff, R.R. 1996. Molecular markers, forest genetics and tree breeding. In: J.P Gustafson and R.B. Flavell (Eds.) Genomes of Plants and Animals: Proceedings of the 21st Stadler Symposium (Columbia, MO), Plenum, New York, pp. 87–102.

Ralph, J., MacKay, J.J., Hatfield, R.D., O'Malley, D.M., Whetten, R.W. and Sederoff, R.R. 1997. Abnormal lignin in a loblolly pine mutant. Science 277: 235–239.

Reiter, W.D. 1998. The molecular analysis of cell wall components. Trends Plant Sci. 3: 27–32.

Saltman, D., Thompson, L. and Bennett, K.M. 1998. Pulp and Paper Primer. TAPPI Press, Atlanta, GA.

Schouten, J., de Kam, R.J., Fetter, K. and Hoge, J.H. 2000. Overexpression of *Arabidopsis thaliana* SKP1 homologues in yeast inactivates the Mig1 repressor by destabilising the F-box protein Grr1. Mol. Gen. Genet. 263: 309–319.

Sederoff, R., Campbell, M., O'Malley, D. and Whetten, R. 1994. Genetic regulation of lignin biosynthesis and the potential modification of wood by genetic engineering in loblolly pine. Rec. Adv. Phytochem. 28: 313–355.

Somerville, C. and Somerville, S. 1999. Plant functional genomics. Science 285: 380–383.

Somssich, I.E., Wernert, P., Kiedrowski, S. and Hahlbrock, K. 1996. *Arabidopsis thaliana* defense-related protein ELI3 is an aromatic alcohol:NADP(+) oxidoreductase. Proc. Natl. Acad. Sci. USA 93: 14199–14203.

Sterky, F., Regan, S., Karlsson, J., Hertzberg, M., Rohde, A., Holmberg, A., Amini, B., Bhalerao, R., Larsson, M., Villarroel, R., Van Montagu, M., Sandberg, G., Olsson, O., Teeri, T.T., Boerjan, W., Gustafsson, P., Uhlen, M., Sundberg, B. and Lundeberg, J. 1998. Gene discovery in the wood-forming tissues of poplar: analysis of 5692 expressed sequence tags. Proc. Natl. Acad. Sci. USA 95: 13330–13335.

Timell, T.E. 1986 Compression Wood in Gymnosperms (3 vols.). Springer-Verlag, Berlin.

Wakamiya, I., Newton, R.J., Johnston, J.S. and Price, H.J. 1993. Genome size and environmental factors in the genus *Pinus*. Am. J. Bot. 80: 1235–1241.

Winzeler, E.A., Schena, M. and Davis, R.W. 1999. Fluorescence-based expression monitoring using microarrays. Meth. Enzymol. 306: 3–18.

Wojtaszek, P. 2000. Genes and plant cell walls: a difficult relationship. Biol. Rev. Camb. Phil. Soc. 75: 437–475.

Wolfinger, R.D., Gibson, G., Wolfinger, E.D., Bennett, L., Hamadeh, H., Bushel, P., Afshari, C. and Paules, R.S. 2000. Assessing gene significance from cDNA microarray expression data via mixed models. Manuscript available from http://statgen.ncsu.edu/ggibson/Publications/WGetc.pdf

Zhang, Y., Sederoff, R.R. and Allona, I. 2000. Differential expression of genes encoding cell wall proteins in vascular tissues from vertical and bent pine trees. Tree Physiol 20: 457–466.

Zobel, B.J. and Sprague, J.R. 1998. Juvenile Wood in Forest Trees. Springer-Verlag, Berlin.

Zobel, B.J. and van Buitenen, J. P.1989. Wood Variation: Its Causes and Control. Springer-Verlag, Berlin.

SECTION 6

CELL WALL BIOTECHNOLOGY

Plant Molecular Biology **47**: 295–310, 2001.
© 2001 *Kluwer Academic Publishers. Printed in the Netherlands.*

Enabling technologies for manipulating multiple genes on complex pathways

Claire Halpin[1,*], Abdellah Barakate[1], Barak M. Askari, James C. Abbott[1] and Martin D. Ryan[2]
[1]*Division of Environmental and Applied Biology, School of Life Sciences, University of Dundee, Dundee DD1 4HN, UK (*author for correspondence; e-mail c.halpin@dundee.ac.uk); [2]Division of Biomedical Science, School of Biology, University of St. Andrews, Fife KY16 9ST, UK*

Key words: co-ordinate expression, lignin, multiple transgenes, polyproteins, pyramiding, transgene silencing

Abstract

Many complex biochemical pathways in plants have now been manipulated genetically, usually by suppression or over-expression of single genes. Further exploitation of the potential for plant genetic manipulation, both as a research tool and as a vehicle for plant biotechnology, will require the co-ordinate manipulation of multiple genes on a pathway. This goal is currently very difficult to achieve. A number of approaches have been taken to combine or 'pyramid' transgenes in one plant and have met with varying degrees of success. These approaches include sexual crossing, re-transformation, co-transformation and the use of linked transgenes. Novel, alternative 'enabling' technologies are also being developed that aim to use single transgenes to manipulate the expression of multiple genes. A chimeric transgene with linked partial gene sequences placed under the control of a single promoter can be used to co-ordinately suppress numerous plant endogenous genes. Constructs modelled on viral polyproteins can be used to simultaneously introduce multiple protein-coding genes into plant cells. In the course of our work on the lignin biosynthetic pathway, we have tested both conventional and novel methods for achieving co-ordinate suppression or over-expression of up to three plant lignin genes. In this article we review the literature concerning the manipulation of multiple genes in plants. We also report on our own experiences and results using different methods to perform directed manipulation of lignin biosynthesis in tobacco.

Abbreviations: CAD, cinnamyl alcohol dehydrogenase; CaMV, cauliflower mosaic virus; CAT, chloramphenicol acetyltransferase; CCoAOMT, caffeoyl-CoA O-methyltransferase; CCR, cinnamoyl-CoA reductase; COMT, caffeate/5-hydroxyferulate O-methyltransferase; F5H, ferulate 5-hydroxylase; FMDV, foot-and-mouth disease virus; G, guaiacyl; GFP, green fluorescent protein; GUS, β-glucuronidase; IPT, isopentenyltransferase; MAR, matrix-attachment region; MAT, multi-auto-transformation; NIa protease, nuclear inclusion protease; PE, pectin esterase; PG, polygalacturonase; PHB, polyhydroxybutyrate; PSY, phytoene synthase; PTGS, post-transcriptional gene silencing; S, syringyl; T-DNA, transferred DNA; TEV, tobacco etch virus; TVMV, tobacco vein mottling virus

Introduction

Over the past decade, the ability to manipulate plant genomes by transgenesis has provided both a novel route to crop improvement and a versatile research tool for plant biology. Many different heterologous proteins have been introduced into plants and resident enzymes of both primary and secondary metabolism have been manipulated. However despite rapid and continuing progress in this area, the vast majority of experiments reported to date involve the expression or manipulation of single genes. Full exploitation of the potential for manipulating plant metabolism awaits the development and adoption of methods for the routine introduction or manipulation of multiple genes. This is because important developmental or commercially valuable traits often depend on the interaction between a number of genes; for example, the control of flux through many biochemical pathways is shared by multiple enzymes. Suppression or over-expression of single genes will, at best, only partly reveal the control processes involved, particularly as plants are adept at compensating physiologically for small changes in their genetic or external environment. Similarly, the introduction of new biochemical pathways into plants will require the expression of numerous enzymes. Production of the biodegradable plastic polyhydroxybutyrate (PHB) in plants required the introduction of two to three bacterial genes (Poirier et al., 1992; Nawrath et al., 1994) but biosynthesis of a really durable plastic (a co-polymer of PHB and polyhydroxyvalerate) will require the introduction of four to six genes (Poirier, 1999; Slater et al., 1999). Similarly, the recent exciting announcement of the development of 'golden rice', capable of producing provitamin A, depended on the introduction of one bacterial and two daffodil genes into rice (Ye et al., 2000). However, additionally improving resorbable iron in rice may require introduction of another three genes.

For these reasons, the importance of 'pyramiding' or gene stacking (i.e. combining two or more desirable genes or transgenes in one plant) is receiving increased attention in review articles on the future of plant biotechnology, even though relatively few research papers can yet be cited to illustrate the point. Although conventional methods exist for combining transgenes in a single plant, such methods are, for the most part, relatively crude. They generally suffer the drawbacks of being both time-consuming and labour-intensive and cannot ensure co-ordinate expression of different transgenes – this has to be screened and selected for. There is a real and urgent need to improve the efficiency of conventional methods for pyramiding genes or to develop new methods for achieving the same result. Until this happens, the full potential of plant genetic manipulation, either as a basic research tool or as a commercial enterprise, cannot be realised.

We have been researching the lignin biosynthetic pathway for many years and have encountered a number of situations where it would be desirable to manipulate the expression of multiple genes in a co-ordinate fashion. In order to achieve this we have tested both conventional and more novel methods, some of which we have had to develop ourselves. This review surveys the existing literature relevant to the co-ordinate manipulation of multiple genes in plants, presenting both the problems and advantages associated with conventional and novel methods. It also reports on our own experiences comparing some of these methods in order to manipulate lignin biosynthesis in specific ways in transgenic tobacco.

The problem of variability of transgene expression

The level of expression of transgenes is notoriously variable and influenced by many factors, including the site of integration (An, 1986; Matzke and Matzke, 1998). It is assumed that various structural and functional properties of the chromatin region flanking the T-DNA integration site affect expression levels. This can complicate attempts to combine multiple transgenes and achieve co-ordinate expression. As these 'position' effects clearly vary with integration site, unlinked transgenes (combined by crossing or re-transformation, for example) are unlikely to be co-ordinately expressed. Moreover the insertion of transgenes into a plant genome is an uncertain process which may result in the presence of one, or many, copies of a transgene at one, or many, transgenic loci. Transgene stability also varies between transformants and some plants show a variety of instability phenomena as primary transformants or in subsequent generations. Instability of expression can be due to loss, re-arrangement or silencing of transgenes (Cherdshewasart et al., 1993; Finnegan and McElroy, 1994).

Several attempts have been made to reduce the variability in expression between primary transformants by flanking transgenes with nuclear matrix-attachment regions (MARs). These are thought to influence gene expression by anchoring active chromatin to the proteinaceous nuclear matrix. While some reports indicate significant success with this approach (e.g. Mylnarova et al., 1996), others have reported no difference in the variation in transgene expression between individual transformants when using MARs (Liu and Tabe, 1998). However a few reports indicate that the use of MARs still has real benefits. MARs may, in some cases, improve the stability of expression over a number of generations by protecting against transgene silencing (Ulker et al., 1999; Vain et al., 1999), although they are not able to protect transgene expression from strong silencing loci that act in trans (Vaucheret et al., 1998).

The problem of transgene silencing

A second problem that besets attempts at transgene stacking is that of gene silencing. A huge body of literature describes how endogenous genes can be silenced after the introduction of a homologous transgene and some describe how silencing can similarly affect transgenes themselves. Silencing is a homology-based phenomenon: there must be significant homology between the introduced transgene and an existing endogenous gene or transgene for silencing to occur (Thierry *et al.*, 1996). This homology can exist in either the coding region or in the promoter of the transgene (Park *et al.*, 1996; Thierry *et al.*, 1996; Chareonpornwattana *et al.*, 1999). Silencing is typified by a large variation in transgene expression that is not directly associated with copy number, but with some epigenetic effect which leads to a block in transcription (transcriptional silencing) or an inhibition of mRNA accumulation (post-transcriptional silencing). Gene and transgene silencing can probably occur by more than one mechanism and, although these processes are not fully understood, intensive research into the area in recent years has provided much valuable information (for recent reviews see Fagard and Vaucheret, 1996; Matzke and Matzke, 1998; Sijen and Kooter, 2000).

Transcriptional gene silencing is associated with methylation in the promoter region of the silenced gene or transgene and is both mitotically and meiotically heritable (Neuhuber *et al.*, 1994; Park *et al.*, 1996). Post-transcriptional gene silencing (PTGS) is typified by mRNA failing to accumulate in the cytoplasm, even though transcription is occurring in the nucleus (Niebel *et al.*, 1995). Genes silenced by PTGS tend to be reactivated upon meiosis (Dehio and Schell, 1994; Balandin and Castresana, 1997), then re-silenced later in development. Silencing can affect transgene expression in a number of different ways; for example, multiple transgenes integrated at a single locus can silence each other (*cis*-inactivation) or one transgenic locus can be silenced by another, unlinked, one (*trans*-silencing). *Cis*-inactivation has been observed when multiple copies of foreign transgenes under the control of a strong promoter are inserted into the genome (Elmayan and Vaucheret, 1996). The most severely affected plants tend to be haploids and homozygotes suggesting a potential dose effect whereby mRNA accumulates to a threshold that initiates mRNA degradation. *Trans*-inactivation is also more severe in homozygous plants, again suggesting a dose effect (De Carvalho *et al.*, 1992). Silencing therefore tends to be exacerbated by increasing transgene copy number (Vaucheret *et al.*,1997) and by repeated use of homologous promoter or coding sequences (Flavell, 1994). Greatest stability of transgene expression is likely to be achieved by selecting for plants with single-copy transgenes. However even plants with stable transgene expression may behave unpredictably when crossed with plants expressing a different transgene, especially if homologous promoters have been used in an attempt to attain co-ordinate expression.

Combining transgenes by sexual crossing

Probably the most obvious method for combining transgenes is to cross plants containing different transgenes. If both parents are homozygous for the transgene they harbour, then all progeny of the cross will contain both transgenes. This method has been used to spectacular success in one of the examples cited above (production of PHB) and in the production of complex proteins such as functional antibodies, in plants (Ma *et al.*, 1995). While the strategy may be worth adopting in such high-profile cases, it has limited use as a routine research tool as it is slow and cumbersome. For each gene that is to be combined, homozygous parents are usually produced. This typically takes 2 to 3 plant generations. Transgenes can then be combined pairwise, homozygotes again produced (another 2 to 3 generations) and plants containing different pairs of transgenes crossed to produce progeny hemizygous for four different transgenes. Thus it takes 4 to 6 generations to combine 3 or 4 transgenes in one plant (see Figure 1). Moreover obtaining plants homozygous for all transgenes can be virtually impossible due to the complexity of the segregation of the independent transgenes. Hitz (1999) has reviewed the cost implications of introducing unlinked multiple transgenes into a crop breeding programme. He estimates that each unlinked locus to be introduced will increase the size of the breeding population by two, if the population also has to make the same yield improvements as commodity cultivars. Thus incorporating a three-gene trait such as PHB synthesis potentially requires an eight-fold increase in the breeding programme. He concludes that this increased breeding effort limits the number of independent loci that can realistically be incorporated and incurs a cost that will be reflected in higher seed prices or reduced yield. Even at the laboratory scale, the strategy is labour-intensive as it

298

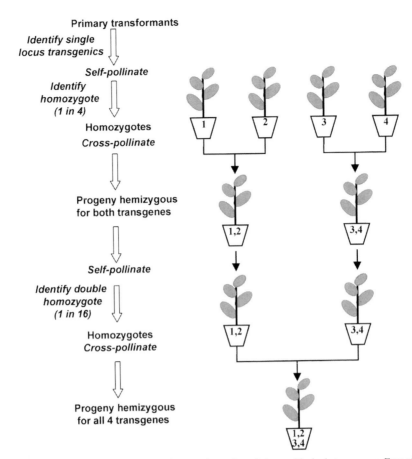

Figure 1. Schematic strategy for combining four transgenes by sexual crossing of plants with single transgenes. Four plants each have a single transgenic locus for one of four different transgenes (denoted 1,2,3,4). These are selfed and 1 in 4 progeny will be homozygous for the transgene. Identification of the homozygotes may be possible in the F1 generation by PCR or Southern-based methods of screening. However it is more likely that homozygous plants will be identified in the F2 generation by the resistance to the appropriate selectable agent (kanamycin etc) of all seedlings germinating from seed of a given F1 plant. Pairs of homozygotes containing different transgenes can then be cross-pollinated to produce progeny hemizygous for both transgenes. After selfing, 1 in 16 progeny will be homozygous for both transgenes. Identification of the double homozygotes will again most likely take place in the next generation. Different double homozygotes can then be cross-pollinated to produce progeny hemizygous for all 4 transgenes. Thus it takes a minimum of 6 generations to combine all four transgenes. Time can be saved by cross-pollinating hemizygous lines but the screening effort is then greatly increased. Taking this strategy further becomes impracticable. If the final hemizygotes are selfed, only 1 in 256 progeny will be homozygous for all 4 transgenes.

involves multiple transformations initially, with rigorous characterisation of each population of plants to identify individuals with stable high expression and simple patterns of single-locus transgene integration. Screening must be maintained in the later stages of the strategy to ensure continued stable expression and to eliminate any crosses where interactions between transgenes induce silencing phenomena and loss or reduction of transgene expression. In our experience, silencing is a particular problem when the same promoter has been used to drive expression of different transgenes in an attempt to co-ordinate expression.

Although most reports in the literature describe the combining of protein-expressing transgenes by crossing, transgenes designed to downregulate gene expression can also be combined in this way. We have performed a number of experiments crossing tobacco plants with different suppressing transgenes (antisense or partial sense gene sequences). Wherever possible we have used well-characterised homozygous parents with a single transgenic locus. In almost all of the crosses we have performed, some degree of transgene silencing has occurred and the required degree of suppression of target gene expression has been lost. For example, a cross was performed between male and female parents respectively harbouring single-locus homozygous antisense transgenes for either caffeate/5-hydroxyferulate O-methyltransferase (COMT) or cinnamyl alcohol dehydrogenase (CAD). Both transgenes had previously shown stable expression patterns and predictable levels of target gene suppression in hemizygous and homozygous plants. Based on this history, we expected that all progeny of the cross to combine the transgenes would display levels of enzyme activity characteristic of hemizygous lines, i.e. 8% of normal COMT activity and 20–30% CAD activity. Surprisingly, we found that, in a population of young progeny of the cross, activity for each enzyme varied from 8% to 100% of wild-type values. In the progeny population, the expected high degree of suppression of target genes became unstable, most likely due to partial silencing of the antisense transgenes.

It is likely that combining transgenes aimed at suppressing the expression of target plant genes is always going to be more difficult than combining protein-producing transgenes. Effective suppression is only ever seen in a small proportion of transformants as target gene activity often has to be severely reduced (by 80–90% or more) before physiological changes consequent on that suppression become evident. Consequently even partial transgene silencing, where only a small proportion of target gene activity is restored, can result in loss of the physiological effect of the transgene.

Combining transgenes by re-transformation

Successive re-transformations have also been used as a method of combining transgenes in the one plant. For many species, the strategy has no particular advantages over crossing. Lengthy screening of transgenics to identify individuals with the desired expression level still has to be performed after each transformation step. As successively introduced transgenes integrate independently they will segregate again in progeny of subsequent sexual crosses. However the strategy can be useful in species where combining transgenes by sexual crossing is impossible or impractical, such as trees. Poplars suppressed in the expression of the two lignin biosynthetic genes CAD and COMT were successfully produced by this method (Lapierre et al., 1999) as were tobacco plants suppressed in the lignin genes COMT and caffeoyl-

CoA O-methyltransferase (CCoAOMT) (Zhong et al., 1998).

Some reports have suggested that re-transformation can induce transgene silencing (Matzke et al., 1989; Fujiwara et al., 1993). A second problem can be the need for a range of selectable marker genes so that a different one can be used with each sequential transformation. In an attempt to get around this problem, Ebinuma and co-workers (Ebinuma et al., 1997; Sugita et al., 2000) have designed a new plant vector system for repeated transformation (called MAT for multi-auto-transformation). A chimeric isopentenyl-transferase (ipt) gene is inserted into the transposable element Ac and used as the selectable marker for transformation. After selection, subsequent excision of the modified Ac produces marker-free transgenic plants, allowing ipt and MAT to be used again for another round of transformation.

Co-transformation with multiple transgenes

Many recent papers describe the successful use of co-transformation as a vehicle for combining multiple transgenes quickly in the one plant. The system appears to work in a wide range of species (Arabidopsis, tobacco, oats, bean, ryegrass, rice, potato, soybean and others) and can be adapted to a variety of transformation methods (biolistics, Agrobacterium co-cultivation/co-infiltration). A surprising conclusion from much of this work is that different T-DNAs introduced together by co-transformation tend to integrate as repeats at single chromosomal positions in a high proportion of transgenics. Estimates of 40–70% of transgenics carrying all co-introduced genes are not unusual (Aragao et al., 1996; De Buck et al., 1998; Maqbool and Christou, 1999). Up to 14 different genes have been introduced in this way (Chen et al., 1998). Many experimenters have found a high copy number of integrated transgenes, especially when using biolistics as the method of transformation (for example, see Hadi et al., 1996; Maqbool and Christou, 1999).

Only a few groups have looked at expression of co-transformed transgenes. When all three genes necessary for PHB synthesis were co-infiltrated into Arabidopsis, 9% of transgenics were found to have incorporated both introduced T-DNAs together at a single locus and to be synthesising detectable amounts of PHB (Poirier et al., 2000). However the majority of plants had more complex integration patterns with 3 or more copies of each T-DNA, and 84% of plants did not produce PHB. Similarly, Tang *et al.* (1999) found that, although 50% of transgenics apparently incorporated and expressed four genes co-bombarded into rice on three separate plasmids, the majority of plants contained multiple copies of one or more transgenes. Only 10% of plants that expressed all four genes had a single transgenic locus. Although two of the genes transferred in this experiment were agronomically relevant, being effective against bacterial blight and sap-sucking insect pests, it was not determined whether the levels of expression achieved were capable of conferring resistance to either pathogen. In experiments where the relative expression levels of co-transformed genes have been investigated in a semi-quantitative fashion, a large variation in the level of expression of different transgenes has been noted (Maqbool and Christou, 1999). Nevertheless, for many applications co-transformation may offer a practical strategy for combining transgenes, particularly when high levels of expression of all transgenes is not needed and the stoichiometry between the products of different transgenes does not have to be too exact. Clearly, with increasing numbers of co-introduced plasmids or T-DNAs, the work involved in identifying plants with simple integration patterns is amplified. Similarly the chances of obtaining plants that can be used to make homozygotes for further breeding (i.e. plants with a single transgenic locus that express all introduced genes) is greatly reduced. The possibility that transgene silencing may also become a problem in subsequent generations, due to the high transgene copy number that frequently characterises co-transformation events, has not yet been sufficiently studied.

Transformation with linked transgenes

In attempts to co-ordinate the expression of co-introduced transgenes, some researchers have linked multiple transgenes on the same T-DNA. Although some have had success with this method (e.g. van Engelen *et al.*, 1994), many have found that even linked transgenes are not necessarily co-ordinately expressed. Expression levels of linked or closely adjacent reporter genes can vary independently, even when controlled by identical promoters (Nagy *et al.*, 1985; An, 1986; Jones *et al.*, 1987). Co-ordinated expression was achieved by maintaining the normal linkage of petunia *Cab22L* and *Cab22R* genes but could be disrupted by changing the position of the transgenes

within the T-DNA or substituting the *Cab* genes with chimeric genes (Gidoni *et al.*, 1988). When CAT and GUS reporter genes were expressed from adjacent and divergent mannopine promoters individual tobacco transgenics showed no correlation between the expression levels of the two genes (Peach and Velten, 1991). Moreover, the lack of convenient restriction sites and the size of the T-DNA quickly become limiting when considering the introduction of three or more transgenes into a plant by this method. Linked transgenes have therefore most successfully been used in combination with co-transformation (for example, Poirier *et al.*, 2000; Ye, 2000), thereby limiting the size of each T-DNA to be introduced.

Alternatives to transgene 'pyramiding'

All of the methods described above for combining or stacking transgenes have been shown to be useful under certain circumstances. However, none offer a simple or rapid method for combining multiple transgenes under conditions where co-ordinate and stable expression of transgenes is favoured. A small number of reports in the literature suggest more novel methods whereby this may be achieved. One approach is to manipulate single regulatory genes such as transcription factors, so as to alter the activity of the set of downstream genes influenced by that transcription factor. In general, this strategy lacks versatility and, in many cases, we do not yet know enough about plant regulatory genes to use it effectively. It will not be discussed further here but interesting results of the use of this strategy to manipulate phenylpropanoid metabolism are discussed elsewhere (Campbell and Rogers, 2001, this issue). Alternative 'single-transgene' methods also exist that have significant advantages over conventional methods as multiple transgenes with associated individual regulatory sequences do not accumulate in plants. In these 'enabling' technologies, a number of protein-coding sequences or gene-suppression cassettes ('effect' sequences) are fused together as a single transcriptional unit and expressed from a single promoter. This immediately ensures co-ordinate expression of the different 'effect' sequences within the transgene. In addition, duplication of regulatory sequences, such as the use of the same promoter on different transgenes, is avoided, so transgene silencing in subsequent generations should be reduced. Subsequent breeding is much faster and cheaper as the 'effect' sequences are absolutely linked and segregate together. In general these methods have not yet been widely adopted despite initial promising results. Here we review the scant published literature concerning these methods and report on our own experiences in testing some of them.

Suppression of multiple genes by single chimeric transgenes

This strategy for suppressing two non-homologous genes using a single chimeric transgene was discovered rather inadvertently by Seymour *et al.* (1993). Polygalacturonase (PG) and pectin esterase (PE) are two pectolytic enzymes found in ripening tomato fruit. Both are made with long N-terminal extensions or pre-sequences that may be involved in regulating the extent of enzyme secretion into the cell wall. In an attempt to test the role of the PG pre-sequence, a chimeric DNA construct was assembled where the pre-sequence for PG was attached to the mature protein sequence of PE (Seymour *et al.*, 1993). The chimeric construct was expressed from the 35S CaMV promoter in tomato plants. Analysis of leaf material showed the expected expression of the PGPE mRNA. In the fruit, however, no transgene transcripts could be detected and both the endogenous PG and PE genes appeared to be down-regulated. Thus a single construct containing sense sequences of two non-homologous genes is capable of triggering silencing of the endogenous copies of both genes by a mechanism presumably akin to co-suppression.

This phenomenon was characterised further by the same group by examining whether two target genes were co-ordinately silenced when chimeric constructs are used (Jones *et al.*, 1998). They used two transgenes that incorporated a 244 bp sense fragment of PG followed by a 414 bp fragment of the PSY (phytoene synthase) gene in either sense or antisense orientation. Phytoene synthase is responsible for the conversion of geranylgeranyl pyrophosphate to phytoene in the carotenoid biosynthesis pathway. Tomato plants expressing antisense RNA for PSY produce yellow fruit. A simple visual screen of fruit can therefore be used to detect areas where PYS has been silenced. Transformation of tomato with either of the PG/PSY chimeric transgenes resulted in the production of three types of fruit: fruit that was all red, fruit that was all yellow and fruit that had both red and yellow sectors. Northern blots indicated that expression of both PG and PSY were co-ordinately suppressed in yellow tissue whereas both were expressed at normal levels in

red tissue. Both constructs gave similar results indicating that partial gene sequences could be incorporated into chimeric constructs in either sense or antisense orientation and still induce silencing. Despite the fact that these chimeric silencing constructs offer great potential for the co-ordinate silencing of multiple genes, the technique has not been widely adopted by other researchers.

Expression of multiple proteins from a single polyprotein transgene

A number of research groups have investigated the possibility of co-ordinating expression of multiple heterologous proteins in plants by expressing them initially as polyproteins. Co-ordinating protein production via polyproteins is a strategy adopted by many viruses, notably the positive-strand RNA viruses. On translation, the polyprotein is cleaved, by proteases encoded within the polyprotein itself, either co-translationally (*in cis*) and/or post-translationally (*in trans*) to yield individual protein products. Many of the proteases involved in these reactions have been well characterised and their specific cleavage recognition sequence is known. For example, the nuclear inclusion (NIa) proteins of plant potyviruses such as tobacco etch virus (TEV) and tobacco vein mottling virus (TVMV) recognise specific heptapeptide sequences and are responsible for several processing events of the large viral polyproteins.

Three groups have independently reported on the use of NIa-containing chimeric polyproteins in plants (Marcos and Beachy, 1994; Beck von Bodman *et al.*, 1995; Dasgupta *et al.*, 1998). In their constructs two 'effect' genes were separated by the coding sequence for the 48 kDa NIa proteinase (from either TEV or TVMV), flanked on both sides by its requisite cleavage sites. When expressed in cell-free systems or transgenic tobacco, the viral proteinase cleaves within its recognition sequence, dissociating the polyprotein and releasing the 'effect' proteins. While polyprotein cleavage in this system appeared to be very efficient, a number of limitations were recognised. Expression levels in plants were low and equimolar amounts of the 'effect' genes were not detected. This might be related to the presence of a nuclear localisation signal within the NIa protease that may target a proportion of the polyprotein to the nucleus while causing premature termination of translation (Marcos and Beachy, 1997). However even after removal of the nuclear localisation signal from NIa, all 'effect' genes in the chimeric polyprotein constructs did not accumulate to equivalent levels. The position of each gene within the polyprotein apparently influenced the level of expression as did the stability of the protein produced (Ceriani *et al.*, 1998).

The utility of the NIa polyprotein expression system was further explored by investigating the possibility of targeting polyprotein proteins to chloroplasts (Dasgupta *et al.*, 1998). Sequences for two reporter genes and the NIa protease were separated by the NIa recognition sequence and organised in a variety of ways in polyprotein constructs. In some constructs a transit peptide from a pea chloroplast gene was incorporated into the polyprotein. The results showed that, in some cases, chloroplast localisation could take place before complete processing of the polyprotein. Processing of NIa acting *in cis* to cleave itself away from the polyprotein occurred more rapidly than processing of other NIa recognition sequences within the polyproteins. Nevertheless when transit peptides, NIa protease and NIa recognition sequences were appropriately organised, targeting to chloroplasts of transit peptide-linked proteins could take place (Dasgupta *et al.*, 1998).

An alternative approach has been taken by Urwin *et al.* (1998) who expressed two proteinase inhibitors as a polyprotein, linking them via a proteinase sensitive sequence taken from the spacer region of a plant metallothionein-like protein. On expression in *Arabidopsis thaliana*, an unidentified plant proteinase cleaved the spacer sequence to release the two protease inhibitors. Again, cleavage was reasonably efficient but as this approach depends upon co-incidence (both temporally and spatially) between the expression of the polyprotein and the endogenous plant proteinase, it may be useful only for certain applications.

Picornaviruses, like potyviruses, encode all of their proteins in a single, long, open reading frame. Full-length translation products are not observed in virus-infected cells due to extremely rapid co-translational, intramolecular (*cis*), primary cleavages. In the aphthovirus (foot-and-mouth disease virus, FMDV) a primary cleavage occurs at the C-terminus of the 2A region. In the entero- and rhinoviruses the analogous primary cleavage is mediated by a well characterised virus-encoded 17 kDa proteinase, 2Apro, but in FMDV the 2A region is only 19 amino acids long (see Figure 2) (Ryan *et al.*, 1991). Despite this, FMDV can mediate a 'cleavage' or polyprotein dissociation at its own C-terminus by an apparently enzyme-independent, novel type of reaction. It is

thought that 2A may function by impairing normal peptide bond formation between glycine and proline residues spanning the 'cleavage' site, while allowing normal translation to continue (Ryan *et al.*, 1999). FMDV 2A can mediate polyprotein dissociation in a heterologous protein context in a range of eukaryotic expression systems including rabbit reticulocyte lysate (Ryan and Drew, 1994; Donnelly *et al.*, 1997) and mammalian (Varnavski *et al.*, 2000), human (Ryan and Drew, 1994), insect (Roosien *et al.*, 1990; Thomas and Maule, 2000) and fungal cells (Suzuki *et al.*, 2000). Most importantly, we have shown that 2A can function in plant extracts and cells (Halpin *et al.*, 1999). A construct in which 2A was inserted between the reporter genes CAT (chloramphenicol acetyltransferase) and GUS (β-glucuronidase) maintaining a single open reading frame, was expressed from the 35S CaMV promoter (see Figure 3A) in transgenic tobacco. Individual plants showed a high degree of correlation between CAT and GUS activities. Individual plants either expressed both enzymes at relatively high levels, expressed both enzymes at low levels, or did not express either enzyme at levels above that of untransformed plants. Western blotting showed that the expressed polyprotein had undergone dissociation, as expected, at the C-terminus of 2A, to yield products of the correct size for CAT2A and GUS (see Figure 3B and C). After dissociation 2A remains attached to the C-terminus of the protein anterior to it in the polyprotein but this has not interfered with enzyme activity in any of the six reporter genes so far tested. A single proline residue remains attached to the protein posterior to 2A in the polyprotein but again, we have not yet encountered a protein where this has interfered with function, and proline will not destabilise proteins according to the N-end rule (Varshavsky, 1992). In tobacco, dissociation of the CAT-2A-GUS polyprotein appeared to be very efficient and little ($< 5\%$) unprocessed precursor could be detected. The length of 2A sequence used is important in determining the efficiency of polyprotein dissociation (Donnelly *et al.*, 1997). We have been working with a 20 amino acid sequence spanning FMDV 2A. Others have sometimes used shorter versions of 16 amino acids that allow significant levels of both intact and dissociated polyprotein to be produced (Santa Cruz *et al.*, 1996; Gopinath *et al.*, 2000; O'Brien *et al.*, 2000).

This work on 2A polyproteins indicates that they could constitute a useful system for ensuring coordinate, stable expression of multiple introduced proteins in plant cells. On translation such polyproteins

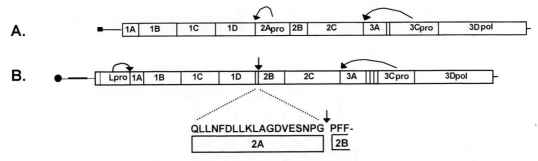

Figure 2. Organisation of picornaviral polyproteins. The organisation of the genome of picornaviruses of the rhino-/enterovirus group (A) and the aphtho- or foot-and-mouth disease virus (FMDV) group (B) are shown. Polyprotein-encoding regions are shown (boxed areas) with positions of 'primary' cleavages mediated by 2A and 3C proteinases indicated (curved arrows). In FMDV the 2A region is just 19 amino acids long. The amino acid sequence spanning this region is shown below the polyprotein in B. The initial proline of FMDV 2B is also essential for 'cleavage' at the carboxy terminus of 2A.

would self-process or self-dissociate at the carboxy terminus of 2A to yield discrete protein products. Very recently, we have further exemplified their usefulness by demonstrating that (1) 2A can be used more than once in a polyprotein enabling its dissociation into 3 or more polypeptide products and (2) each constituent polypeptide can be independently targeted to a variety of sub-cellular locations. We have made a polyprotein where coding sequences for 3 reporter genes (green fluorescent protein (GFP), CAT and GUS) are separated by 2A. On translation in wheat germ lysate the polyprotein construct pGFP-2A-CAT-2A-GUS yields products of GFP-2A, CAT-2A and GUS. Constructs were also prepared to test our assumption that 2A polyproteins are completely compatible with the potential need to target individual constituent proteins to different sub-cellular locations. Clones for two chloroplast proteins were used to make a range of 2A polyprotein constructs. Polypeptide products from *in vitro* transcription/translation reactions were incubated with isolated chloroplasts *in vitro*. Our results show that chloroplast proteins encoded within 2A polyproteins in either anterior or posterior positions can be targeted normally to chloroplasts. Similarly we have constructed a polyprotein encoding a polypeptide destined to remain in the cytoplasm (GUS) linked to an endoplasmic reticulum membrane protein via 2A (i.e. GUS-2A membrane protein). Translation of this construct in the presence of microsome membranes shows unequivocally that GUS-2A remains in the matrix phase while the membrane protein associates and pellets with the microsomes. These results further illustrate the versatility of 2A polyproteins, suggesting they are compatible with the need to co-translationally target proteins to the endomembrane system or post-

translationally target them to chloroplasts or mitochondria in living plant cells. Recently three groups have shown that 2A can be used to express a heterodimeric protein as a self-dissociating polyprotein. When separated by 2A and expressed in *Cos* or insect cells, the two polypeptides of interleukin-12 were correctly dissociated, assembled and secreted in a biologically active form (Chaplin *et al.*, 1999; Kokuho *et al.*, 1999; De Rose *et al.*, 2000). These results indicate that 2A polyproteins should also be useful for the production of complex, multimeric proteins in plant cells.

The 2A polyprotein system, therefore, has a lot of advantages for plant expression systems. It is simple and efficient, as only a short 20 amino acid sequence is needed to link 'effect' proteins. Its mode of action ensures efficient co-translational cleavage and does not depend on the presence of any cellular factors (other than eukaryotic ribosomes). Thus it can be effective in all tissues and cells at all developmental stages. Most of all, it is rapid, allowing multiple proteins to be introduced into plants on one transgene while ensuring they are co-ordinately expressed. However the retention of the 2A peptide at the carboxy terminus of proteins anterior to it in polyproteins may obviously interfere with the activity or sub-cellular targeting of some proteins. Similarly, the difficulty of finding or engineering appropriate restriction sites during cloning may eventually limit the size of 2A polyproteins that can conveniently be made. We have not yet reached those limits, however, and have had no problem in expressing three reasonably large (40–70 kDa) proteins together from a 2A polyprotein.

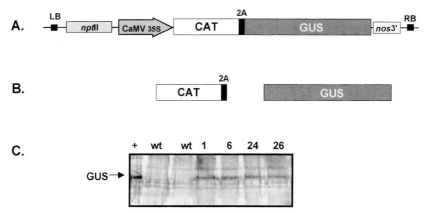

Figure 3. Expression of a CAT-2A-GUS polyprotein in transgenic tobacco. A. Construct designed to express the polyprotein CAT-2A-GUS in transgenic tobacco plants. The reporter genes CAT and GUS, separated by the 20 amino acid sequence spanning FMDV 2A, were encoded as a single open reading frame (CAT-2A-GUS) and placed under control of the 35S CaMV promoter in a Bin 19-based plant transformation vector. The neomycin phosphotransferase (*nptII*) gene was used as a selectable marker. LB and RB denote left and right borders of the T-DNA respectively. This construct was introduced into tobacco by *Agrobacterium*-mediated transformation and transgenic callus and plantlets were recovered on kanamycin-containing medium The protein products expected on translation of the polyprotein are shown in B. After co-translational 'cleavage' at the carboxy terminus of 2A, the protein products are CAT-2A and GUS. 2A 'cleavage' in transgenic callus was confirmed by western blotting of protein extracts run on a 10% SDS-polyacrylamide gel. Blots were probed with anti-GUS serum (1:500 dilution) and detected by chemiluminescence (C). The lane labelled '+' is a positive control comprised of a protein extract from a tobacco plant expressing a GUS transgene. Lanes labelled 'wt' are extracts from untransformed callus. Lanes labelled 1, 6, 24 and 26 are extracts from independent calluses expressing GUS activity. The western blots show that all of the immunoreactive product in these extracts is the expected size for mature 'cleaved' GUS protein.

Genetic manipulation of lignin biosynthesis

A major aim of our research is to understand how the lignin biosynthetic pathway operates *in vivo*. The transgenic approach, producing plants with altered lignification, has provided a powerful route to further elucidating this basic developmental process unique to plants. Lignin also has significant commercial importance, however, and modified-lignin transgenics suggest how lignin might be manipulated to improve plant materials for industrial and agricultural uses (Vailhe *et al.*, 1998; Baucher *et al.*, 1999; Hu *et al.*, 1999; Lapierre *et al.*, 1999). Commercial interest in this area has driven work rapidly forward. Almost all of the genes on this complex pathway have now been cloned and genetically manipulated, sometimes in a variety of species including trees, and field trials underpinning commercial exploitation are underway. Despite this rapid advance towards biotechnological applications, many aspects of the lignin pathway still require investigation and fundamental questions such as the exact sequence and identity of enzymes involved are all areas of intensive current research. The vast majority of work to date has involved the manipulation of single lignin biosynthetic genes. Further elucidation of the pathway and more directed designing of novel lignins will require the co-ordinated manipulation of a num-

ber of genes. Our recent work has concentrated on the production and characterisation of plants where two or three lignin genes have been manipulated simultaneously to either suppress or over-express their activities. We have used a number of methods to achieve this with differing degrees of success. Before describing the results of this work, a brief introduction to the lignin pathway and the relevant existing work on its manipulation is necessary.

Lignin is predominantly a polymer of cinnamyl alcohols or monolignols. These are thought to be synthesised inside the cell then exported to the cell wall where they are polymerised into lignin. The lignin biosynthetic pathway is highly complex involving component steps of both the general phenylpropanoid pathway and reactions more dedicated to lignin and lignan production (for review, see Baucher *et al.*, 1998). The roles of most of the genes contributing directly to the pathway have now been studied in genetically manipulated plants. These include phenylalanine ammonia-lyase and cinnamate 4-hydroxylase (Sewalt *et al.*, 1997), 4-coumarate:CoA ligase (Kajita *et al.*, 1996; Lee *et al.*, 1997), COMT (Atanassova *et al.*, 1995), ferulate 5-hydroxylase (F5H) (Meyer *et al.*, 1998), cinnamoyl-CoA reductase (CCR) (Piquemal *et al.*, 1998) and CAD (Halpin *et al.*, 1994). Although the basic pathway was outlined many years ago, recent data has prompted some revisions and multiple alternative routes have been suggested for the synthesis of certain intermediates.

COMT was originally thought to be the only O-methyltransferase needed for lignin synthesis, methylating both caffeate and 5-hydroxyferulate (Bugos et al., 1992). However, in recent years, circumstantial evidence has accumulated to suggest that another enzyme, caffeoyl-CoA O-methyltransferase (CCoAOMT) might also be involved (Ye et al., 1994; Ye and Varner, 1995; Ye, 1997). Genetic manipulation experiments have recently confirmed this. Zhong et al. (1998) generated transgenic tobacco plants with substantial reductions in CCoAOMT and plants with simultaneous reductions in CCoAOMT and COMT. Reduction of CCoAOMT alone resulted in decreased lignin content but plants with suppression of both CCoAOMT and COMT showed a further reduction in lignin content. This confirmed that both enzymes are indeed involved in methylation reactions in lignin biosynthesis.

Similarly, conversion of guaiacyl (G) lignin precurSors into syringyl (S) lignin precursors has traditionally been believed to occur at the level of the acids, converting ferulic acid into sinapic acid. However two independent groups have recently provided biochemical evidence that suggests that the current model for syringyl lignin biosynthesis is incorrect or incomplete. Work with recombinant *Arabidopsis* suggests that, in addition to ferulic acid, F5H can also use coniferyl aldehyde and coniferyl alcohol as substrates (Humphreys et al., 1999). On the other hand, work with recombinant sweetgum and aspen F5H enzymes, have suggested that the conversion of G lignin precursors to S lignin precursors occurs only at coniferaldehyde (Osakabe et al., 1999; Li et al., 2000). Currently there are therefore three alternative model pathways for S lignin biosynthesis. Clarification of this situation awaits experiments that can determine which pathway really operates *in vivo*.

Co-ordinate manipulation of multiple lignin biosynthetic genes

Manipulating combinations of genes involved in monolignol biosynthesis offers the best potential for gaining additional evidence on the organisation of the pathway *in planta*. For example, determination of lignin structure and composition after suppression of successive genes on the monolignol pathway should

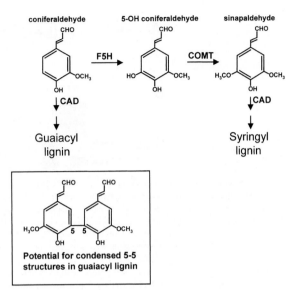

Figure 4. Potential for condensed 5-5 structures in guaiacyl lignin. The lignin pathway is shown in abbreviated form from coniferaldehyde onwards. Although conversion of guaiacyl to syringyl lignin precursors is indicated to occur at the position of the aldehydes, it is currently unclear whether, *in vivo*, this conversion can also take place earlier and later in the pathway. The potential for guaiacyl precursors to form condensed 5-5 inter-unit linkages that resist degradation arises from the fact that the 5 position is free in these precursors. Syringyl precursors are methoxylated at the 5 position and cannot contribute to such structures.

allow the sequence of the enzymes to be confirmed, as the block at the earliest occurring enzyme should exert a dominant effect. In some cases, suppression of particular genes might be combined with over-expression in order to increase flux into specific parts of the pathway so as to more easily study them. Moreover, manipulating multiple genes may give rise to novel lignins with improved pulping characteristics or digestibility. For example, maize plants carrying mutations in the two lignin genes CAD and COMT show greater reduction in lignin content, and better improvement in digestibility than plants carrying either single mutation (Kuc et al., 1968; Lechtenberg et al., 1972).

A particularly valuable manipulation to improve plant materials for pulping would be to increase S lignin content while decreasing G lignin content. This is because condensed bonds in lignin make wood more difficult to pulp as these bonds resist degradation and impede lignin removal. The number of condensed bonds in lignin would be reduced by increasing the proportion of S units. Syringyl monomers are dimethoxylated at the 3′ and 5′ ring positions, inhibiting the formation of condensed 5′-5′ inter-monomer structures (see Figure 4). These structures are com-

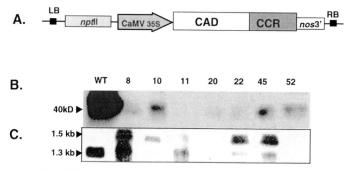

Figure 5. Simultaneous down-regulation of CAD and CCR in tobacco plants. A. Partial sequences of CAD (0.9 kb) and CCR (0.6 kb) cDNAs were linked and inserted in sense orientation between the CaMV 35S promoter and the *nos* terminator in a plant transformation vector. This construct was introduced into leaf discs and plants resistant to kanamycin were regenerated. B and C. Selected plants (denoted 8, 10, 11, 20, 22, 45, 52) and wild-type (WT) plants were grown in tissue culture on MS medium for 6 weeks. Proteins and total RNA were extracted from stems and subjected to western and northern blots to moniter CAD and CCR expression, respectively (Halpin *et al.*, 1998). B. Proteins (20 µg) of WT and different CAD-CCR tobacco lines were separated on 10% gel (SDS-PAGE) then blotted onto nitrocellulose membrane. CAD protein (40 kDa) was detected using anti-CAD antibody (1:10 000 dilution). C. Total RNA (10 µg) of the same lines were separated on a 1% agarose gel then blotted onto nylon membrane and probed with CCR anti-sense DIG-labelled riboprobe. The signals corresponding to the CCR mRNA (1.3 kb) and RNA from the CAD-CCR transgene (1.5 kb) are indicated.

mon in G lignin which is methoxylated only at the 3′ position. Consequently, increasing the S lignin content of wood is a commercially valuable target. One particularly exciting idea is the possibility of engineering gymnosperms to allow them to produce S lignin (for commentary see Baucher *et al.*, 1998; Merkle and Dean, 2000). Although the predominant G lignin in the wood of most gymnosperms makes it difficult to pulp, other characteristics of the wood (e.g. long fibre length) are highly valued. The difference in lignin composition between angiosperms and gymnosperms is often explained in terms of differing substrate specificity of particular lignin biosynthetic enzymes from different species (for review, see Campbell and Sederoff, 1996). For example, CCR and CAD purified from gymnosperms have no or low affinity for S lignin precursors (Luderitz and Grisebach, 1981; Sarni *et al.*, 1984; O'Malley *et al.*, 1992; Galliano *et al.*, 1993). On the other hand, recent work has clearly indicated that F5H plays a critical role in controlling flux into S lignin. No F5H activity has been detected in gymnosperms (Higuchi, 1990), prompting some commentators to speculate that the absence of F5H alone may prevent the synthesis of S lignin by gymnosperms. It is currently unclear how many angiosperm genes would need to be introduced into a gymnosperm to allow it to produce S lignin but it could be as many as four (F5H, COMT, CCR, CAD).

Co-ordinate suppression of lignin biosynthetic genes

Clear reasons exist therefore for attempting to manipulate multiple lignin genes. We have compared two methods for achieving co-ordinate suppression of two lignin biosynthetic genes. We have performed a number of experiments crossing tobacco plants with different transgenes designed to suppress expression of COMT, CAD or CCR. As previously described (see Combining transgenes by sexual crossing), all crosses showed some degree of transgene silencing resulting in reduced suppression of target gene expression. This problem was possibly exacerbated by the repeated use of the 35S CaMV promoter on different transgenes in our experiments. Lignin composition in the progeny of these crosses was less modified than it was in either parent although there were some indications that the effects of moderate down-regulation of both genes had been combined in the lignin.

As an alternative to crossing we have used a single chimeric transgene to down-regulate multiple lignin genes according to the strategy described by Seymour *et al.* (1993). Partial sense sequences for combinations of two, or of all three, of the lignin genes COMT, CCR and CAD, were fused and placed under the control of the 35S CaMV promoter. A total of four constructs were produced and introduced separately into tobacco. In all cases we have been able to identify young tissue culture plants with reduced activity of both, or all three, targeted genes. Figure 5 shows results for a transgene containing partial sense sequences of CCR and CAD. Many independent primary transformants show greatly reduced expression of the 1.3 kb endogenous CCR mRNA (Figure 5C). The same plants also show radically reduced expression of CAD protein on western blots (Figure 5B). Thus the suppression of both genes by the single chimeric transgene is both effective and co-ordinate. In terms of enzyme activity, these plants show greater degrees of down-regulation than corresponding plants produced by crossing parents with single transgenes. A chimeric transgene with partial-sense sequences of COMT, CCR and CAD was also effective in suppressing the activity of all three target genes (data not shown). In our experience, single chimeric transgenes offer a more effective, and far more rapid, strategy for down-regulating multiple genes than does sexual crossing.

Co-ordinate over-expression of lignin biosynthetic genes

We have also tested the single- transgene approach to over-expressing multiple lignin genes by encoding them within self-dissociating 2A polyproteins. Expression of *Arabidopsis* F5H in tobacco results in a significant increase in stem lignin syringyl monomer content (Franke *et al.*, 2000), but it is not known whether co-expression of COMT could have an additional effect. To investigate this we have produced constructs to express two polyproteins, GUS-2A-F5H and GUS-2A-COMT-2A-F5H. These are being introduced into tobacco in order to increase the flux towards S lignin. Translation of both constructs *in vitro* has shown that, on translation, the polyproteins 'cleave' or dissociate as expected to yield products of GUS-2A and F5H or GUS-2A, COMT-2A and F5H respectively (data not shown). Moreover, F5H, a membrane-bound P450 enzyme, appears to insert correctly into added microsome membranes. Primary transformants expressing these polyproteins are currently being screened to select the most appropriate individuals for further study. F5H activity cannot be easily assayed in crude plant extracts. However the presence of GUS in our polyprotein constructs is allowing us to very easily screen primary transformants for transgene expression on the basis of GUS activity. Histochemical staining of GUS will also allow us to monitor the spatial and temporal expression of the polyprotein transgene in tissue sections of transgenic plants taken throughout development. This work further demonstrates the versatility of the 2A-polyprotein

system that has allowed us to rapidly introduce three proteins into tobacco plants and is providing a simple assay to monitor their expression. Monitoring over a number of plant generations will indicate how stable this expression system is as a vehicle for one-step co-ordinate expression of multiple proteins in plants.

Conclusions

A variety of different methods have been used successfully to manipulate multiple genes from complex biochemical pathways in plants. Each method has its own specific problems. In general, conventional approaches can be slow and cumbersome. In addition, methods that promote the accumulation of multiple or complex transgenic loci or allow repeated use of particular regulatory sequences (such as promoters), can suffer from transgene silencing. Novel methods where single transgenes are being used to engineer co-ordinate suppression or over-expression of multiple genes can overcome many of these problems. We have used single transgenes incorporating multiple 'effect' sequences to successfully and simultaneously suppress or over-express up to three genes on the lignin pathway. Analysis of plants manipulated in this way is both yielding useful information on how the pathway operates *in vivo* and producing modified lignins that may be improved for particular commercial uses. Others have also had promising results using similar 'single transgene' approaches. These enabling technologies deserve to be more widely adopted and tested in order to determine just how versatile they can be for routine manipulation of multiple genes on complex biochemical pathways.

Acknowledgements

We would like to thank Zeneca Agrochemicals and the BBSRC for financial support (grants CO5962 and WCP05388) and Jess Searle for technical assistance.

References

An, G. 1986. Development of plant promoter expression vectors and their use for analysis of differential activity of the nopaline synthase promoter in transformed tobacco cells. Plant Physiol. 81: 86–91.

Aragao, F.J.L., Barros, L.M.G., Brasileiro, A.C.M., Ribeiro, S.G., Smith, F.D., Sanford, J.C., Faria, J.C. and Rech, E.L. 1996. Inheritance of foreign genes in transgenic bean (*Phaseolus vulgaris*

308

L) co-transformed via particle bombardment. Theor. Appl. Genet 93: 142–150.

Atanassova, R., Favet, N., Martz, F., Chabbert, B., Tollier, M.T., Monties, B., Fritig, B. and Legrand, M. 1995. Altered lignin composition in transgenic tobacco expressing O-methyltransferase sequences in sense and antisense orientation. Plant J. 8: 465–477.

Balandin, T. and Castresana, C. 1997. Silencing of a β-1,3-glucanase transgene is overcome during seed formation. Plant Mol. Biol. 34: 125–137.

Baucher, M., Vailhe, M.A.B., Chabbert, B., Besle, J.-M., Opsomer, C., Van Montagu, M. and Botterman, J. 1999. Down-regulation of cinnamyl alcohol dehydrogenase in transgenic alfalfa (Medicago sativa L.) and the effect on lignin composition and digestibility. Plant Mol. Biol. 39: 437–447.

Baucher, M., Monties, B., Van Montagu, M. and Boerjan, W. 1998. Biosynthesis and genetic engineering of lignin. Crit. Rev. Plant Sci. 17: 125–197.

Beck von Bodman, S., Domier, L.L. and Farrand, S.K. 1995. Expression of multiple eukaryotic genes from a single promoter in Nicotiana. Bio/technology 13: 587–591.

Bugos, R.C., Chiang, V.L.C. and Campbell, W.H. 1992. Characterization of bispecific caffeic acid/5-hydroxyferulic acid O-methyltransferase from aspen. Phytochemistry 31: 1495–1498.

Campbell, M. and Rogers, L. 2001. Spatial and temporal regulation of lignin biosynthesis. Plant Mol. Biol., this issue.

Campbell, M.M. and Sederoff, R.R. 1996. Variation in lignin content and composition. Plant Physiol. 110: 3–13.

Ceriani, M.F., Marcos, J.F., Esteban Hopp, H. and Beachy, R.N. 1998. Simultaneous accumulation of multiple virus coat proteins from a TEV-Nia based expression vector. Plant Mol. Biol. 36: 239–248.

Chaplin, P.J., Camon, E.B., Villarreal-Ramos, B., Flint, M., Ryan, M.D. and Collins, R.A. 1999. Production of interlukin-12 as a self-processing polypeptide. J. Interferon Cytokine Res. 19: 235–241.

Chareonpornwattana, S., Thara, K.V., Wang, L., Datta, S.K., Panbangred, W. and Muthukrishnan, S. 1999. Inheritance, expression, and silencing of a chitinase transgene in rice. Theor. Appl. Genet. 98: 371–378.

Chen, L.L., Marmey, P., Taylor, N.J., Brizard, J.P., Espinoza, C., D'Cruz, P., Huet, H., Zhang, S.P., de Kochko, A., Beachy, R.N. and Fauquet, C.M. 1998. Expression and inheritance of multiple transgenes in rice plants. Nature Biotechnol. 16: 1060–1064.

Cherdshewasart, W., Gharti-Chhetri, G.B., Saul, M.W., Jacobs, M. and Negrutiu, I. 1993. Expression instability and genetic disorders in transgenic Nicotiana plumbaginifolia L. plants. Transgenic Res. 2: 307–320.

Dasgupta, S., Collins, G.B. and Hunt, A.G. 1998. Co-ordinated expression of multiple enzymes in different subcellular compartments in plants. Plant J. 16: 107–116.

De Buck, S., Jacobs, A., Van Montagu, M. and Depicker, A. 1998. Agrobacterium tumefaciens transformation and cotransformation frequencies of Arabidopsis thaliana root explants and tobacco protoplasts. Mol. Plant-Microbe Interact. 11: 449–457.

De Carvalho, F., Gheysen, G., Kushnir, S., Van Montagu, M., Inzé, D. and Castresana, C. 1992. Suppression of β-1,3-glucanase transgene expression in homozygous plants. EMBO J. 11: 2595–2602.

Dehio, C. and Schell, J. 1994. Identification of plant genetic-loci involved in a posttranscriptional mechanism for meiotically reversible transgene silencing. Proc. Natl. Acad. Sci. USA 91: 5538–5542.

De Rose, R., Scheerlinck, J.P., Casey, G., Wood, P.R., Tennent, J.M. and Chaplin, P.J. 2000. Ovine interleukin-12: analysis of biologic function and species comparison. J. Interferon Cytokine Res. 6: 557–564.

Donnelly, M., Gani, D., Flint, M., Monoghan, S. and Ryan, M.D. 1997. The cleavage activity of aphtho- and cardiovirus 2A proteins. J. Gen. Virol. 78: 13–21.

Ebinuma, H., Sugita, K., Matsunaga, E. and Yamakado, M. 1997. Selection of marker-free transgenic plants using the isopentenyl transferase gene. Proc. Natl. Acad. Sci. USA 94: 2117–2121.

Elmayan, T. and Vaucheret, H. 1996. Expression of single copies of a strongly expressed 35S transgene can be silenced post-transcriptionally. Plant J 9: 787–797.

Fagard, M. and Vaucheret, H. 2000. (Trans)gene silencing in plants: how many mechanisms? Annu. Rev. Plant Physiol. Plant Mol. Biol. 51: 167–194.

Finnegan, J. and McElroy, D. 1994. Transgene inactivation: plants fight back. Bio/technology 12: 883–888.

Flavell, R.B. 1994. Inactivation of gene expression in plants as a consequence of specific sequence duplication. Proc. Natl. Acad. Sci. USA 91: 3409–3496.

Franke, R., McMichael, C.M., Meyer, K., Shirley, A.M., Cusumano, J.C. and Chapple, C. 2000. Modified lignin in tobacco and poplar plants over-expressing the Arabidopsis gene encoding ferulate 5-hydroxylase. Plant J. 22: 223–234.

Fujiwara, T., Lessard, P.A. and Beachy, R.N. 1993. Inactivation of the nopaline synthase gene by double transformation: Reactivation by segregation of the induced DNA. Plant Cell Rep. 12: 133–138.

Galliano, H., Heller, W. and Sandermann, H. 1993. Ozone induction and purification of spruce cinnamyl alcohol-dehydrogenase. Phytochemistry 32: 557–563.

Gidoni, D., Bond-Nutter, D., Brosio, P., Jones, J., Bedbrook, J. and Dunsmuir, P. 1988. Coordinated expression between two photosynthetic petunia genes in transgenic plants. Mol. Gen. Genet. 211: 507–514.

Gopinath, K., Wellink, J., Porta, C., Taylor, K.M., Lomonossoff, G.P. and van Kammen, A. 2000. Engineering cowpea mosaic virus RNA-2 into a vector to express heterologous proteins in plants. Virology 267: 159–173.

Hadi, M.Z., McMullen, M.D. and Finer, J.J. 1996. Transformation of 12 different plasmids into soybean via particle bombardment. Plant Cell Rep. 15: 500–505.

Halpin, C., Cooke, S.E., Barakate, A., El Amrani, A. and Ryan, M.D. 1999. Self-processing 2A-polyproteins: a system for co-ordinate expression of multiple proteins in transgenic plants. Plant J. 17: 453–459.

Halpin, C., Holt, K., Chojecki, J., Oliver, D., Chabbert, B., Monties, B., Edwards, K., Barakate, A. and Foxon, G.A. 1998. Brown-midrib maize (bm1): a mutation affecting the cinnamyl alcohol-dehydrogenase. Plant J. 14: 545–553.

Halpin, C., Knight, M.E., Foxon, G.A., Campbell, M.M., Boudet, A.M., Boon, J.J., Chabbert, B., Tollier, M.-T. and Schuch, W. 1994. Manipulation of lignin quality by downregulation of cinnamyl alcohol dehydrogenase. Plant J. 6: 339–350.

Higuchi, T. 1990. Lignin biochemistry: biosynthesis and biodegradation. Wood Sci. Technol. 24: 23063.

Hitz, B. 1999. Economic aspects of transgenic crops which produce novel products. Curr. Opin. Plant Biol. 2: 135–138.

Hu, W.J., Harding, S.A., Lung, J., Popko, J.L., Ralph, J., Stokke, D.D., Tsai, C.J. and Chiang, V.L. 1999. Repression of lignin biosynthesis promotes cellulose accumulation and growth in transgenic trees. Nature Biotechnol. 17: 808–812.

Humphreys, J.M., Hemm, M.R. and Chapple, C. 1999. New routes for lignin biosynthesis defined by biochemical characterization of recombinant ferulate 5-hydroxylase, a multifunctional cytochrome P450-dependent monooxygenase. Proc. Natl. Acad. Sci. USA 96: 10045–10050.

Jones, C.G., Scothern, G.P., Lycett, G.W. and Tucker, G.A. 1998. The effect of chimeric transgene architecture on co-ordinated gene silencing. Planta 204: 499–505.

Jones, J.D.G., Gilbert, D.E., Grady, K.L. and Jorgensen, R.A. 1987. T-DNA structure and gene expression in petunia cells transformed with *Agrobacterium tumefaciens* C58 derivatives. Mol. Gen. Genet. 207: 478–485.

Kajita, S., Katayama, Y. and Omori, S. 1996. Alterations in the biosynthesis of lignin in transgenic plants with chimeric genes for 4-coumarate:coenzyme A ligase. Plant Cell Physiol. 37: 957–965.

Kokuho, T., Watanabe, S., Yokomizo, Y. and Inumaru, S. 1999. Production of biologically active, heterodimeric porcine interleukin-12 using a monocistronic baculoviral expression system. Jpn. Vet. Immunol. Immunopath. 72: 289–302.

Kuc, J., Nelson, O.E. and Flanagan, P. 1968. Degradation of abnormal lignins in the *brown-midrib* mutants and double mutants of maize. Phytochemistry 7: 1435–1436.

Lapierre, C., Pollet, B., Petit-Conil, M., Toval, G., Romero, J., Pilate, G., Leple, J.-C., Boerjan, W., Ferret, V., De Nadai, V. and Jounain, L. 1999. Structural alterations of lignins in transgenic poplars with depressed cinnamyl alcohol dehydrogenase or caffeic acid *O*-methyltransferase activity have an opposite impact on the effeciency of industrial craft pulping. Plant Physiol. 119: 153–163.

Lechtenberg, V.L., Muller, L.D., Bauman, L.F., Rhykerd, C.L. and Barnes, R.F. 1972. Laboratory and *in vitro* evaluation of inbred and F_2 populations of *brown midrib* mutants of *Zea mays* L. Agron. J. 64: 657–660.

Lee, D., Meyer, K., Chapple, C. and Douglas, C.J. 1997. Antisense suppression of 4-coumarate:coenzyme A ligase activity in *Arabidopsis* leads to altered lignin subunit composition. Plant Cell 9: 1985–1998.

Li, L.G., Popko, J.L., Umezawa, T. and Chiang, V.L. 2000. 5-Hydroxyconiferyl aldehyde modulates enzymatic methylation for syringyl monolignol formation, a new view of monolignol biosynthesis in angiosperms. J. Biol. Chem. 275: 6537–6545.

Liu, J.W. and Tabe, L.M. 1998. The influences of two plant nuclear matrix attachment regions (MARs) on gene expression in transgenic plants. Plant Cell Physiol. 39: 115–123.

Luderitz, T. and Grisebach, H. 1981. Enzymic synthesis of lignin precursors: comparison of cinnamoyl-CoA reductase and cinnamyl alcohol-NADP+ dehydrogenase from spruce (*Picea abies* L.) and soybean (*Glycine max* L.). Eur. J. Biochem. 119: 115–124.

Ma, J.K.-C., Hiatt, A., Hein, M.B., Vine, N.D., Wang, F., Stabila, P., van Dolleweerd, C., Mostov, K. and Lehner, T. 1995. Generation and assembly of secretory antibodies in plants. Science 268: 716–719.

Maqbool, S.B. and Christou, P. 1999. Multiple traits of agronomic importance in transgenic indica rice plants: analysis of transgene integration patterns, expression levels and stability. Mol. Breed. 5: 471–480.

Marcos, J.F. and Beachy, R.N. 1997. Transgenic accumulation of two plant virus coat proteins on a single self-processing polypeptide. J. Gen. Virol. 78: 1771–1778.

Marcos, J.F. and Beachy, R.N. 1994. *In vitro* characterization of a cassette to accumulate multiple proteins through synthesis of a self-processing polypeptide. Plant Mol. Biol. 24: 495–503.

Matzke, A.J.M. and Matzke, M.A. 1998. Position effects and epigenetic silencing of plant transgenes. Curr. Opin. Plant Biol. 1: 142–148.

Matzke, M., Primig, M., Trnovsky, J. and Matzke, A. 1989 Reversible methylation and inactivation of marker genes in sequentially transformed tobacco plants. EMBO J. 8: 643–649.

Merkle, S.A. and Dean, J.F.D. 2000. Forest tree biotechnology. Curr. Opin. Biotechnol. 11: 298–302.

Meyer, K., Shirley, A.M., Cusumano, J.C., Bell-Lelong, D.A. and Chapple, C. 1998. Lignin monomer composition is determined by the expression of a cytochrome P450-dependent monooxygenase in *Arabidopsis*. Proc. Natl. Acad. Sci. USA 95: 6619–6623.

Mylnarova, L., Keizer, L.C.P., Stiekema, W.J. and Nap, J.P. 1996. Approaching the lower limits of transgene variability. Plant Cell 8: 1589–1599.

Nagy, F., Morelli, G., Fraley, R.T., Rogers, S.G. and Chua, N.-H. 1985. Photoregulated expression of a pea *rbcS* gene in leaves of transgenic plants. EMBO J. 12: 3063–3068.

Nawrath, C., Poirier, Y. and Somerville, C. 1994. Targeting of the polyhydroxybutyrate biosynthetic pathway to the plastids of *Arabidopsis thaliana* results in high levels of polymer accumulation. Proc. Natl. Acad. Sci. USA 91: 12760–12764.

Neuhuber, F., Park, Y.D., Matzke, A.J.M. and Matzke, M.A. 1994. Susceptibility of transgene loci to homology-dependent gene silencing. Mol. Gen. Genet. 244: 230–241.

Niebel, F.D., Frendo, P., Van Montagu, M. and Cornelissen, M. 1995. Posttranscriptional cosuppression of β-1,3-glucanase genes does not affect accumulation of transgene nuclear messenger RNA. Plant Cell 7: 347–358.

O'Brien, G.J., Bryant, C.J., Voogd, C., Greenberg, H.B., Gardner, R.C. and Bellamy, A.R. 2000. Rotavirus VP6 expressed by PVX vectors in *Nicotiana benthamiana* coats PVX rods and also assembles into viruslike particles. Virology 270: 444–453.

O'Malley, D.M., Porter, S. and Sederoff, R.R. 1992. Purification, characterization, and cloning of cinnamyl alcohol-dehydrogenase in loblolly-pine (*Pinus taeda* L.). Plant Physiol. 98: 1364–1371.

Osakabe, K., Tsao, C.C., Li, L.G., Popko, J.L., Umezawa, T., Carraway, D.T., Smeltzer, R.H., Joshi, C.P. and Chiang, V.L. 1999. Coniferyl aldehyde 5-hydroxylation and methylation direct syringyl lignin biosynthesis in angiosperms. Proc. Natl. Acad. Sci. USA 96: 8955–8960.

Park, Y.D., Papp, I., Moscone, E.A., Iglesias, V.A., Vaucheret, H., Matzke, A.J.M. and Matzke, M.A. 1996. Gene silencing mediated by promoter homology occurs at the level of transcription and results in meiotically heritable alterations in methylation and gene activity. Plant J. 9: 183–194.

Peach, C. and Velten, J. 1991. Transgene expression variability (position effect) of CAT and GUS reporter genes driven by linked divergent T-DNA promoters. Plant Mol. Biol. 17: 49–60.

Piquemal, J., Lapierre, C., Myton, K., O'Connell, A., Schuch, W., Grima-Pettenati, J. and Boudet, A.M. 1998. Down-regulation of cinnamoyl-CoA reductase induces significant changes of lignin profiles in transgenic tobacco plants. Plant J. 13: 71–83.

Poirier, Y. 1999. Production of a new polymeric compound in plants. Curr. Opin. Biotechnol. 10: 181–185.

Poirier, Y., Denis, D.E., Klomparens, K. and Somerville, C. 1992. Polyhydroxybutyrate, a biodegradable thermoplastic produced in transgenic plants. Science 256: 520–523.

Poirier, Y., Ventre, G. and Nawrath, C. 2000. High-frequency linkage of co-expressing T-DNA in transgenic *Arabidopsis thaliana* transformed by vacuum-infiltration of *Agrobacterium tumefaciens*. Theor. Appl. Genet. 100: 487–493.

310

Roosien, J., Belsham, G.J., Ryan, M.D., King, A.M.Q. and Vlak, J.M. 1990. Synthesis of foot-and-mouth disease virus capsid proteins in insect cells using baculovirus expression vectors. J. Gen. Virol. 71: 1703–1711.

Ryan, M.D. and Drew, J. 1994. Foot-and-mouth disease virus 2A oligopeptide mediated cleavage of an artificial polyprotein. EMBO J. 134: 928–933.

Ryan, M.D., King, A.M.Q. and Thomas, G.P. 1991. Cleavage of foot-and-mouth disease virus polyprotein is mediated by residues located within a 19 amino acid sequence. J. Gen. Virol. 72: 2727–2732.

Ryan, M.D., Donnelly, M.L.L., Lewis, A., Mehrotra, A.P., Wilkie, J. and Gani, D. 1999. A model for non-stoichiometric, co-translational protein scission in eukaryotic ribosomes. Bioorg. Chem. 27: 55–79.

Santa Cruz, S., Chapman, S., Roberts, A.G., Roberts, I.M., Prior, D.A.M. and Oparka, K.J. 1996. Assembly and movement of a plant-virus carrying a green fluorescent protein overcoat. Proc. Natl. Acad. Sci. USA 93: 6286–6290.

Sarni, F., Grand, C. and Boudet, A.M. 1984. Purification and properties of cinnamoyl-CoA reductase and cinnamyl alcohol-dehydrogenase from poplar stems (Populus × euramericana). Eur. J. Biochem. 139: 259–265.

Sewalt, V.J.H., Ni, W.T., Blount, J.W., Jung, H.G., Masoud, S.A., Howles, P.A., Lamb, C. and Dixon, R.A. 1977. Reduced lignin content and altered lignin composition in transgenic tobacco down-regulated in expression of L-phenylalanine ammonia-lyase or cinnamate 4-hydroxylase. Plant Physiol 115: 41–50.

Seymour, G.B., Fray, R.G., Hill, P. and Tucker, G.A. 1993. Down-regulation of two nonhomologous endogenous tomato genes with a single chimeric sense gene construct. Plant Mol. Biol. 23: 1–9.

Sijen, T. and Kooter, J.M. 2000. Post-transcriptional gene-silencing: RNAs on the attack or on the defense? Bioessays 22: 520–531.

Slater, S., Mitsky, T.A., Houmiel, K.L., Hao, M., Reiser, S.E., Taylor, N.B., Tran, M., Valentin, H.E., Rodriguez, D.J., Stone, D.A., Padgette, S.R., Kishore, G. and Gruys, K.J. 1999. Metabolic engineering of Arabidopsis and Brassica for poly(3-hydroxybutyrate-co-3-hydroxyvalerate) copolymer production. Nature Biotechnol. 17: 1011–1016.

Sugita, K., Matsunaga, E., Kasahara, T. and Ebinuma, H. 2000. Transgene stacking in plants in the absence of sexual crossing. Mol. Breed. 6: 529–536.

Suzuki, N., Geletka, L.M. and Nuss, D.L. 2000. Essential and dispensable virus-encoded replication elements revealed by efforts to develop hypoviruses as gene expression vectors. J. Virol. 74: 7568–7577.

Tang, K., Tinjuangjun, P., Xu, Y., Sun, X., Gatehouse, J.A., Ronald, P.C., Qi, H., Lu, X., Christou, P. and Kohli, A. 1999. Particle-bombardment-mediated co-transformation of elite Chinese rice cultivars with genes conferring resistance to bacterial blight and sap-sucking insect pests. Planta 208: 552–563.

Thierry, D. and Vaucheret, H. 1996. Sequence homology requirements for transcriptional silencing of 35S transgenes and post-transcriptional silencing of nitrite reductase (trans)genes by the tobacco 271 locus. Plant Mol. Biol. 32: 1075–1083.

Thomas, C.L. and Maule, A.J. 2000. Limitations on the use of fused green fluorescent protein to investigate structure-function relationships for the cauliflower mosaic virus movement protein. J. Gen. Virol. 81: 1851–1855.

Ulker, B., Allen, G.C., Thompson, W.F., Spiker, S. and Weissinger, A.K. 1999. A tobacco matrix attachment region reduces the loss of transgene expression in the progeny of transgenic tobacco plants. Plant J. 18: 235–263.

Urwin, P.E., McPherson, M.J. and Atkinson, H.J. 1998. Enhanced transgenic plant resistance to nematodes by dual proteinase inhibitor constructs. Planta 204: 472–479.

Vailhe, M.A.B., Besle, J.M., Maillot, M.P., Cornu, A., Halpin, C. and Knight, M. 1998. Effect of down-regulation of cinnamyl alcohol dehydrogenase on cell wall composition and on degradability of tobacco stems. J. Sci. Food Agric. 76: 505–514.

Vain, P., Worland, B., Kohli, A., Snape, J.W., Christou, P., Allen, G.C. and Thompson, W.F. 1999. Matrix attachment regions increase transgene expression levels and stability in transgenic rice plants and their progeny. Plant J. 18: 233–242.

van Engelen, F.A., Schouten, A., Molthoff, J.W., Roosien, J., Salinas, J., Dirkse, W.G., Schots, A., Bakker, J., Gommers, F.J., Jongsma, M.A., Bosch, D. and Stiekema, W.J. 1994. Coordinate expression of antibody subunit genes yields high-levels of functional antibodies in roots of transgenic tobacco. Plant Mol. Biol. 26: 1701–1710.

Varnavski, A.N., Young, P.R. and Khromykh, A.A. 2000. Stable high-level expression of heterologous genes in vitro and in vivo by noncytopathic DNA-based Kunjin virus replicon vectors. J. Virol. 74: 4394–4403.

Varshavsky, A. 1992. The N-end rule. Cell 69: 725–735.

Vaucheret, H., Nussaume, L., Palauqui, J-C., Quillere, J.I. and El-mayan, T. 1997. A transcriptionally active state is required for post-transcriptional silencing (co-suppression) of nitrate reductase host genes and transgenes. Plant Cell 9: 1495–1504.

Vaucheret, H., Elmayan, T., Thierry, D., van der Geest, A., Hall, T., Conner, A.J., Mylnarova, L. and Nap, J.P. 1998. Flank matrix attachment regions (MARs) from chicken, bean, yeast or tobacco do not prevent homology-dependent trans-silencing in transgenic tobacco plants. Mol. Gen. Genet. 259: 388–392.

Ye, Z.-H. 1997. Association of caffeoyl coenzyme A 3-O-methyltransferase expression with lignifying tissues in several dicot plants. Plant Physiol. 115: 1341–1350.

Ye, Z.-H. and Varner J.E. 1995. Differential expression of 2 O-methyltransferases in lignin biosynthesis in Zinnia elegans. Plant Physiol. 108: 459–467.

Ye, Z.-H., Kneusel, R.E., Matern, U. and Varner, J.E. 1994. An alternative methylation pathway in lignin biosynthesis. Plant Cell 6: 1427–1439.

Ye, X., Al-Babili, S., Klöti, A., Zhang, J., Lucca, P., Beyer, P. and Potrykus, I. 2000. Engineering the provitamin A (β-carotene) biosynthetic pathway into (carotenoid-free) rice endosperm. Science 287: 303–305.

Zhong, R., Morrison, W.H., Negrel, J. and Ye, Z.-H. 1998. Dual methylation pathways in lignin biosynthesis. Plant Cell 10: 2033–2045.

Plant Molecular Biology **47**: 311–340, 2001.
© 2001 *Kluwer Academic Publishers. Printed in the Netherlands.*

Cell wall metabolism in fruit softening and quality and its manipulation in transgenic plants

David A. Brummell* and Mark H. Harpster
*DNA Plant Technology, 6701 San Pablo Avenue, Oakland, CA 94608, USA (*author for correspondence at present address: Department of Pomology, University of California, Davis, CA 95616, USA)*

Key words: cell wall, fruit quality, fruit softening, *Lycopersicon esculentum*, matrix glycans, pectin, xyloglucan

Abstract

Excessive softening is the main factor limiting fruit shelf life and storage. Transgenic plants modified in the expression of cell wall modifying proteins have been used to investigate the role of particular activities in fruit softening during ripening, and in the manufacture of processed fruit products. Transgenic experiments show that polygalacturonase (PG) activity is largely responsible for pectin depolymerization and solubilization, but that PG-mediated pectin depolymerization requires pectin to be de-methyl-esterified by pectin methylesterase (PME), and that the PG β-subunit protein plays a role in limiting pectin solubilization. Suppression of PG activity only slightly reduces fruit softening (but extends fruit shelf life), suppression of PME activity does not affect firmness during normal ripening, and suppression of β-subunit protein accumulation increases softening. All these pectin-modifying proteins affect the integrity of the middle lamella, which controls cell-to-cell adhesion and thus influences fruit texture. Diminished accumulation of either PG or PME activity considerably increases the viscosity of tomato juice or paste, which is correlated with reduced polyuronide depolymerization during processing. In contrast, suppression of β-galactosidase activity early in ripening significantly reduces fruit softening, suggesting that the removal of pectic galactan side-chains is an important factor in the cell wall changes leading to ripening-related firmness loss. Suppression or overexpression of endo-$(1\rightarrow4)\beta$-D-glucanase activity has no detectable effect on fruit softening or the depolymerization of matrix glycans, and neither the substrate nor the function for this enzyme has been determined. The role of xyloglucan endotransglycosylase activity in softening is also obscure, and the activity responsible for xyloglucan depolymerization during ripening, a major contributor to softening, has not yet been identified. However, ripening-related expansin protein abundance is directly correlated with fruit softening and has additional indirect effects on pectin depolymerization, showing that this protein is intimately involved in the softening process. Transgenic work has shown that the cell wall changes leading to fruit softening and textural changes are complex, and involve the coordinated and interdependent activities of a range of cell wall-modifying proteins. It is suggested that the cell wall changes caused early in ripening by the activities of some enzymes, notably β-galactosidase and ripening-related expansin, may restrict or control the activities of other ripening-related enzymes necessary for the fruit softening process.

Abbreviations: CMC, carboxymethylcellulose; EGase, endo-$(1\rightarrow4)\beta$-D-glucanase; PG, endopolygalacturonase; PME, pectin methylesterase; RG, rhamnogalacturonan; XET, xyloglucan endotransglycosylase

Introduction

Fruit are not only an enjoyable component of a healthy diet but also a valuable source of vitamins, minerals, antioxidants and fiber. Improvements in fruit quality benefit both the consumer and, by reducing commercial losses, the producer. The main factor in determining the post-harvest deterioration of fruit crops is the rate of softening of the fruit, which influences shelf life, wastage, infection by post-harvest pathogens, frequency of harvest, and limits transportation and storage, all of which directly affect costs.

Controlling deterioration by the introduction of transgenes to suppress particular genes involved in fruit softening is an approach which could potentially add significant value by improving quality and reducing spoilage, and is of great commercial importance. Attempts to reduce fruit softening in transgenic plants are distinct from control of the entire ripening process by suppression of ethylene production (Hamilton *et al.*, 1990; Oeller *et al.*, 1991). Prevention of ripening can aid in harvesting and shipping, but when fruit are gassed with ethylene to induce ripening, shelf-life is often as short as in wild-type fruit. Reducing just the rate of fruit softening allows the desirable components of ripening (accumulation of pigments, sugars, volatiles, organic acids) to proceed as normal, but increases shelf life and decreases spoilage rate. Reducing fruit softening may also be useful in non-climacteric species where suppression of ethylene evolution is not feasible. In addition to softening, alterations in fruit texture can also be commercially desirable, particularly in processing where improved integrity is important in canning, and in chopped and diced fruit in products such as salsa and yoghurt. In processed tomatoes, both firmness and texture contribute to the viscosity of juice and paste, which is the most important factor determining commercial value.

Disassembly of the fruit cell wall is largely responsible for softening and textural changes during ripening, but the precise roles of particular cell wall alterations and of the cell wall-modifying enzymes bringing these about are not known. To investigate function, suppression or over-expression of cell wall enzyme levels *in vivo* by transgenic manipulation is preferred to *in vitro* treatment of cell walls or isolated polysaccharide preparations with particular enzymes. *In vitro* experiments can be limited by penetration of exogenous proteins to usual sites of action, or by other enzyme activities required for the action of the applied enzyme and, *in vitro*, cell walls or isolated wall polymers frequently are not degraded as much as *in vivo* (Hatfield and Nevins, 1986; McCollum *et al.*, 1989). However, a limitation of the transgenic analysis of the function of single genes is that most cell wall-modifying proteins are present as multi-gene families, and if another isoform is present that can complement the function of the suppressed isoform then no phenotype will be observed. Also, since the precise role of an enzyme is usually not known, effects other than those predicted frequently occur, and fruit must be carefully examined both biochemically and phenotypically in order to observe unexpected changes.

Tomato has become a model system for transgenic analysis of fruit softening and ripening, due to its ease of transformation, commercial importance and the availability of basic genetic and biochemical information. Since most of the transgenic plants that have been characterized for altered fruit softening are tomato, this review will emphasize the literature on tomato but with reference to other species where appropriate. It must be remembered, however, that although tomato is very well characterized, it may not be typical of fruit softening in other species. In sections describing particular cell wall enzymes where more than one species is discussed, the first letters of the binomial species name will be used in front of the gene name to distinguish between identical gene or gene product names. For example, tomato (*Lycopersicon esculentum*) *Cel*1 and *Exp*2 will be referred to as *Le-Cel*1 and *LeExp*2, to avoid confusion with strawberry (*Fragaria × ananassa*) *FaCel*1 and *FaExp*2, and so forth.

Cell wall changes during fruit ripening

The dicot cell wall consists of rigid, inextensible cellulose microfibrils held together by interpenetrating coextensive networks of matrix glycans, pectins and structural glycoproteins (McCann and Roberts, 1992; Carpita and Gibeaut, 1993). The primary cell wall is composed of numerous polymers which vary in structure somewhat between species, but eight polymeric components (cellulose, three matrix glycans composed of neutral sugars, three pectins rich in D-galacturonic acid, and structural proteins) are usually present.

1. Cellulose is composed of $(1\rightarrow4)\beta$-D-glucan chains assembled together by hydrogen bonding into very long crystalline microfibrils, each ca. 36 glucan chains in cross section but with many thousands of chains in total.
2. Xyloglucan possesses a $(1\rightarrow4)\beta$-D-glucan backbone like cellulose, but is substituted with α-D-xylose in a regular fashion on three consecutive glucose residues out of four, xylose occasionally being extended with β-D-galactosyl-α-L-fucose (or α-L-arabinose in some species). Xyloglucan is susceptible to cleavage by *Trichoderma* endo-$(1\rightarrow4)\beta$-D-glucanases (EGases) on the reducing-end side of unsubstituted glucose residues to produce approximately equal amounts of heptasaccharide (Glc$_4$.Xyl$_3$) and nonasac-

charide (Glc$_4$.Xyl$_3$.Gal.Fuc) xyloglucan subunit oligosacharides.

3. (Galacto)glucomannan has a backbone composed of regions of $(1{\rightarrow}4)\beta$-D-glucan and $(1{\rightarrow}4)\beta$-D-mannan in approximately equal amounts, with occasional side chains of single units of terminal α-D-galactose.

4. Glucuronoarabinoxylan has a backbone of $(1{\rightarrow}4)\beta$-D-xylan, with side chains of single units of non-reducing terminal α-L-arabinose and α-D-glucuronic acid.

5. Homogalacturonan is composed of long chains of $(1{\rightarrow}4)\alpha$-D-galacturonic acid, and is initially highly methyl-esterified.

6. Rhamnogalacturonan I (RG I) is made of alternating α-D-rhamnose and α-D-galacturonic acid residues, with long side-chains attached to the rhamnose residues of either unbranched $(1{\rightarrow}4)\beta$-D-galactan or branched α-L-arabinans or type I arabinogalactans.

7. Rhamnogalacturonan II (RG II) is made of a backbone of $(1{\rightarrow}4)\alpha$-D-galacturonic acid like homogalacturonan, but with complex side chains of several types of neutral sugar. It is a minor cell wall component but RG II monomers can dimerize together as boron di-esters and may affect the porosity of the wall.

8. Structural proteins, of four different types, some of which are heavily glycosylated.

In dicot species with Type I walls (Carpita and Gibeaut, 1993), cellulose microfibrils are coated with and cross-linked together by matrix glycans, of which the most abundant is xyloglucan. The glucan backbone of xyloglucan binds firmly to cellulose by hydrogen bonding, and xyloglucan molecules can span between adjacent microfibrils, linking them together. Glucomannan and glucuronoarabinoxylan are present in the wall in less abundant amounts, and also cross-link microfibrils by hydrogen bonding although more weakly than xyloglucan. The spaces in the cellulose/matrix glycan network are filled by the highly hydrated pectins which also form a network, held together by ester bonds between pectin molecules and by ionic calcium cross links between de-methyl-esterified homogalacturonans. The cellulose/matrix glycan and pectin networks may be locked together by covalent links between some xyloglucan molecules and pectins (Thompson and Fry, 2000). Structural proteins may form an additional network. Little information on the detailed composition or structure of particular fruit cell walls is available. The basic structure is proba-

bly as described above, although fruit cell walls are usually highly enriched in pectins (often more than 50% of the wall), and composition probably varies considerably between species. For example, tomato fruit xyloglucan is not fucosylated (Maclachlan and Brady, 1994), consistent with reports that xyloglucan from solanaceous species is lower in substitution with xylose and possesses side chains with terminal galactose or arabinose rather than galactosyl-fucose (York et al., 1996). Tomato matrix glycans contain xyloglucan (Sakurai and Nevins, 1993; Maclachlan and Brady, 1994) and probably glucomannan and xylans (Tong and Gross, 1988; Seymour et al., 1990), but in general insufficient is known about the structure of individual fruit cell wall polymers, particularly non-xyloglucan matrix glycans.

During ripening, cell wall architecture and the polymers of which it is composed are progressively modified, with the nature or extent of the changes varying between species. The cell wall structure becomes increasingly hydrated as the cohesion of the pectin gel changes, and this is the main factor influencing how easily cells can be split open or separated from one another, which determines fruit texture (Jarvis, 1984). In fruit such as strawberry and avocado which develop a soft melting texture during ripening, swelling and softening of the cell wall is evident, but in fruit such as apple, which ripen to a crisp, fracturable texture, cell wall swelling is not observed (Redgwell et al., 1997). A reduction in cell-to-cell adhesion is caused by a breakdown and dissolution of the pectin-rich middle lamella, and begins early in ripening in a soft fruit such as tomato (Crookes and Grierson, 1983) and late in softening in a crisp fruit such as apple (Ben-Arie et al., 1979). During ripening, the pH of the cell wall space declines and there are increases in the concentrations of some ions (Almeida and Huber, 1999).

Ripening is also usually accompanied by a reduction in cell turgor, due to increasing concentrations of solutes in the cell wall space and to wall loosening (Shackel et al., 1991).

In order to examine the ripening-related modifications in cell wall polymers which underly cell wall structural changes, and the effects of suppression of particular genes on wall polymers, cell walls must be isolated and sequentially extracted to produce fractions enriched in particular wall components. These extractions are usually with: (1) chelating agents, such as CDTA or EDTA, which remove calcium from the wall, solubilizing pectin held in the

314

wall by ionic bonds; (2) sodium carbonate, which by de-esterification releases pectin held in the wall by covalent bonds (this treatment also breaks ester bonds between solubilized pectin molecules, resulting in extracted pectin of relatively low molecular weight); (3) weak alkali, such as 1 M or 4% (0.7 M) potassium hydroxide, which solubilizes matrix glycans loosely bound in the wall (these fractions usually contain only small amounts of xyloglucan, and are mainly matrix glycans such as glucomannan and glucuronoarabinoxylan); (4) strong alkali, such as 4 M or 24% (4.3 M) potassium hydroxide (this releases matrix glycans tightly bound in the wall, both by breaking hydrogen bonds and by causing swelling, but not dissolution, of cellulose microfibrils). In tomato about two-thirds of this extract is xyloglucan. The residue of these treatments is mainly cellulose, but some matrix glycans and substantial amounts of pectin are always found associated with the insoluble residue. Note that these are arbitrary categories of polysaccharides based on solubilization; although extracts are enriched in particular polymers, each will contain a range of pectin and glycan polysaccharides with the precise composition depending on extraction conditions. Relatively large volumes of extractant, usually repeated, and long incubation times are necessary to completely extract components, and, if inadequate, the unsolubilized part of each fraction will become extracted by the next step.

Analysis of cell wall polysaccharides has revealed that large changes occur in both pectins and matrix glycans during ripening. Four of the largest changes are shown in Figure 1. During ripening pectins become increasingly depolymerized and soluble (Huber and O'Donoghue, 1993; Brummell and Labavitch, 1997; Chun and Huber, 1997). Strictly, only a portion of the polyuronides become water soluble, because, although there is hydrolysis of polyuronides from covalent attachment to the wall, most remain associated with the wall by ionic bonds to other insoluble pectic molecules. Thus, solubilization of polyuronides usually means an increase in the ease of extractability, measured as an increase in the amount of polyuronides that can be extracted from the wall by chelators. Pectin depolymerization and the loss of cell wall galactose and arabinose side chains to RG I (Gross, 1984; Gross and Sams, 1984), increases cell wall porosity, which initially may be quite low and limit access of cell wall hydrolases to glycan substrates (Baron-Epel et al., 1988). Methyl-esterified pectins become increasingly de-esterified by a process beginning in the

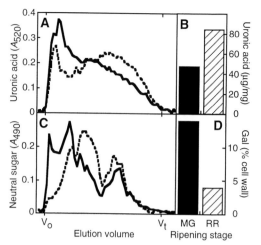

Figure 1. Cell wall changes during ripening of tomato fruit. Solid lines and black bars, mature green (MG); broken lines and hatched bars, red ripe (RR). A. Depolymerization of chelator-soluble pectin. Molecular weight profile of CDTA-soluble polyuronide after size exclusion chromatography on Sepharose CL-2B ($V_0 = 20$ MDa, $V_t = 100$ kDa). B. Amount of chelator-soluble pectin. Expressed as μg of CDTA-soluble uronic acid per mg ethanol-insoluble cell wall material. C. Depolymerization of matrix glycans tightly bound to cellulose. Molecular weight profile of 24% KOH-soluble matrix glycan after size exclusion chromatography on Sepharose CL-6B ($V_0 = 1$ MDa, $V_t = 10$ kDa). D. Galactose content of the cell wall. Non-cellulosic neutral sugar composition of crude cell walls was determined by gas chromatography and expressed as % of cell wall. Data redrawn from: A, Brummell and Labavitch (1997); B, Chun and Huber (1997); C, Brummell et al. (1999c); D, Gross (1984).

middle lamella at the mature green stage and spreading throughout the wall during ripening (Roy et al., 1992; Blumer et al., 2000). During cell development the degree of esterification of the wall declines as a whole, not all of which is due to loss of methyl groups (McCann and Roberts, 1994), suggesting that interpectate ester links are broken in addition to glycosidic bonds. As pectins are increasingly de-esterified to carboxylate ions, this causes the formation of charged surfaces in the wall, which may be important in modulating pH and ion balance, and may limit the movement of charged proteins (Carpita and Gibeaut, 1993). Changes in pH and ionic conditions in the apoplast may also affect the activity of cell wall-localized enzymes (Chun and Huber, 1998; Almeida and Huber, 1999).

Regions of de-esterified charged pectate molecules also associate together by calcium cross-links, which adds rigidity to the wall (Jarvis, 1984). The fraction of matrix glycans loosely bound to cellulose, which is probably mainly glucomannan and xylans, shows little change in molecular weight profile during ripening, and depolymerization of matrix glycans is limited to those tightly bound to cellulose (Tong and Gross, 1988; Brummell *et al.*, 1999c). The extract of tightly bound matrix glycans contains predominantly xyloglucan, with smaller amounts of probably glucomannan and xylans, but in this fraction both xyloglucan and non-xyloglucan polysaccharides are depolymerized during ripening (Tong and Gross, 1988; Sakurai and Nevins, 1993; Maclachlan and Brady, 1994; Brummell *et al.*, 1999c). In addition to cell wall disassembly, synthesis and incorporation of new components into the wall continues throughout ripening (Mitcham *et al.*, 1989; Greve and Labavitch, 1991).

Many different cell wall-modifying enzymes are present in fruit, and Figure 2 shows how the activities of some of these change during development in tomato in relation to other ripening parameters. Some activities are present throughout development and increase or decrease during ripening, whereas others are ripening specific and appear only during ripening. Almost all represent the combined activities of several isoforms, possibly with slightly different substrate specificities, some of which may be ripening specific. It is likely that fruit of all species have the same range of enzyme activities, but present at very different levels. For example, tomato is unusually high in PG activity and very low in endo-$(1{\rightarrow}4)\beta$-D-glucanase (EGase) activity, whereas avocado is unusually high in EGase activity. Many factors contribute to the texture of fruit, which varies between species from the gel-like softness of ripe avocado to the crispness of ripe apples, including cell morphology, size, shape, packing, contents and turgor (Harker *et al.*, 1997). However, one of the largest components of texture is provided by cell-to-cell adhesion and cell wall-thickness and structure, and their modification during ripening. Variations in cell wall composition and cell wall-modifying proteins thus contribute substantially to the variety of fruit textures in different species.

Ripening is associated with changes in sugar metabolism, degradation of chlorophyll and accumulation of pigments (Figure 2A). In climacteric fruit, such as tomato, banana, apple and kiwi fruit, the beginning of ripening and softening is marked by a transient burst of ethylene production and increased respiration (Figure 2B). Ethylene is the hormone controlling the expression of ripening-related genes, and coordinates and regulates the whole ripening process. Non-climacteric fruit, such as strawberry, bell pepper and pineapple, produce only basal levels of ethylene throughout ripening. Ripening in non-climacteric fruit

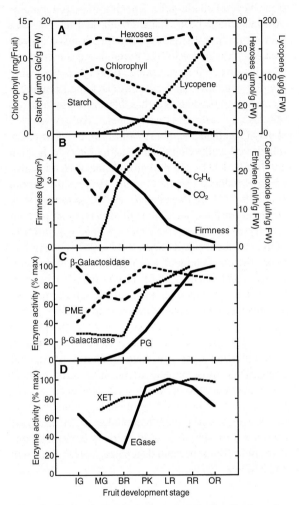

Figure 2. Changes in ripening parameters and cell wall polysaccharide-modifying enzyme activities during tomato fruit development. A. Pigments and sugars. Chlorophyll (*a+b*), McMahon *et al.* (1990); lycopene, Smith *et al.* (1988); hexoses (Glc+Fru) and starch, Yelle *et al.* (1988). B. Firmness and climacteric. Firmness, Maclachlan and Brady (1994); ethylene and carbon dioxide, Rothan and Nicolas (1989). C. Pectin-modifying enzymes. Polygalacturonase (PG), Smith *et al.* (1990); pectin methylesterase (PME), Harriman *et al.* (1991); β-galactanase and β-galactosidase, Carey *et al.* (1995). D. Matrix glycan-modifying enzymes. Endo-$(1{\rightarrow}4)\beta$-D-glucanase (EGase), Hall (1964); xyloglucan endotransglycosylase (XET), Maclachlan and Brady (1994). Data redrawn from sources indicated. Abbreviations of ripening stages: IG, immature green (expanding, not full size); MG, mature green (full size, locules developing); BR, breaker (first sign of red coloration on outside of fruit); PK, pink (30–70% light red); LR, light red (100% light red); RR, red ripe (100% dark red and edibly soft); OR, over ripe (senescing).

is little affected by treatments with exogenous ethylene, and how ripening is regulated in these fruit is unclear. Several single-gene, ripening-impaired mutants of tomato have been characterized, and have proven very useful in investigating the regulation and expression of cell wall-modifying proteins and the cell wall changes involved in softening. Fruit of *rin* (*ripening inhibitor*) and *nor* (*non-ripening*) lack the normal climacteric and produce only basal levels of ethylene, do not develop carotenes and lycopene, and soften very slowly (Tigchelaar *et al.*, 1978). In *rin*, depolymerization of cell wall xyloglucan is negligible (Maclachlan and Brady, 1994), and polyuronide metabolism is also severely reduced (Seymour *et al.*, 1987a; Giovannoni *et al.*, 1989; Della Penna *et al.*, 1990). Treatment with ethylene does not cause *rin* fruit to ripen or soften, but not all ethylene-inducible genes are impaired. mRNAs of the unknown ripening genes *E4* and *E8* accumulate upon ethylene treatment, although that of *E17* does not (Lincoln and Fischer, 1988). The *Nr* (*Never-ripe*) mutant is deficient in a wide range of ethylene responses, and this locus has been identified as an ethylene receptor (Yen *et al.*, 1995). Assessing mRNA abundances and enzyme activities in wild-type and mutant fruit is complicated since the mutants do not ripen, and comparisons must be made at the same chronological age rather than developmental stage. However, relative to wild-type, fruit of *rin* accumulate reduced levels of extractable carboxymethylcellulase (CMCase) and pectinesterase activity, and PG activity is virtually undetectable (Buescher and Tigchelaar, 1975; Poovaiah and Nukaya, 1979).

Polygalacturonase

Polygalacturonases (PGs, poly(1→4-α-D-galacturonide)glycanohydrolases) are enzymes that catalyze the hydrolytic cleavage of galacturonide linkages, and can be of the exo- or endo-acting types. The exo type (EC 3.2.1.67) removes single galacturonic acid units from the non-reducing end of polygalacturonic acid, whereas the endo type (EC 3.2.1.15) cleaves such polymers at random. The fruit ripening-specific enzyme usually referred to as PG is of the endo-acting type; however, both endo and exo types of these enzymes are found in fruit (Hadfield and Bennett, 1998). The substrate for PG in the cell wall is mainly homogalacturonans, which are secreted to the cell wall in a highly methyl-esterified form which must be de-

esterified before they can become a substrate for PG (Jarvis, 1984; Carpita and Gibeaut, 1993).

An increase in the activity of PG has long been associated with fruit ripening, although the amount detected varies widely with species (Hobson, 1962). Ripening avocado, tomato and peach possess relatively high levels of PG activity, although activity in peach is about 50 times less than in tomato (Huber and O'Donoghue, 1993; Pressey and Avants, 1978). PG activity has been reported to be absent in other species including strawberry, apple and melon, but PG activity and/or mRNA have subsequently been detected (Hadfield and Bennett, 1998) and it seems that in some species PG activity is present at low levels but is either very labile or needs to be assayed under particular conditions. In tomato, PG activity is not detectable in pre-ripe fruit, but PG mRNA appears at the onset of ripening or breaker stage (Della Penna *et al.*, 1986), and activity increases from early in ripening (Smith *et al.*, 1988; Biggs and Handa, 1989). PG activity and immunologically detectable PG protein accumulate rapidly with ripening (Brady *et al.*, 1982; Tucker and Grierson, 1982), and activity continues to increase as fruit become over ripe (Hobson, 1964; Tucker *et al.*, 1980). PGs are present in plants in very large gene families (Hadfield and Bennett, 1998), with over 50 genes in Arabidopsis (The Arabidopsis Genome Initiative, 2000).

However, in tomato the mRNA of only one gene family member accumulates in ripening fruit, in a manner paralleling the increases in PG protein and activity (Della Penna *et al.*, 1986). This PG mRNA achieves very high levels, exceeding 2% of the total poly(A)$^+$ RNA in ripe fruit (Della Penna *et al.*, 1987). Protein sequencing confirmed that the ripening-specific PG protein corresponds with this mRNA, which encodes a predicted polypeptide of 457 amino acids that is processed by removal of a 24 amino acid signal sequence (for targeting the polypeptide to the ER-Golgi endomembrane system for processing and secretion), an amino-terminal pro-sequence consisting of a further 47 amino acids, and 13 amino acids at the carboxyl terminus, to produce two glycosylated mature proteins, of 43 and 45 kDa respectively, differing only in the extent of glycosylation (Grierson *et al.*, 1986; Sheehy *et al.*, 1987; Della Penna and Bennett, 1988; Pogson *et al.*, 1991). Analysis of genomic clones suggests that the gene is probably present as a single gene per haploid genome (Bird *et al.*, 1988; Giovannoni *et al.*, 1989). The gene becomes transcriptionally activated as ripening begins and remains transcriptionally active through ripening (Della Penna et al., 1989). Accumulation of *PG* mRNA is regulated by ethylene, with low levels of ethylene being sufficient for induction and mRNA accumulation increasing with ethylene exposure (Sitrit and Bennett, 1998). However, the appearance of *PG* mRNA and PG protein are not tightly coupled, suggesting that increases in PG protein and enzyme activity may also be regulated by translational or posttranslational mechanisms (Theologis et al., 1993; Sitrit and Bennett, 1998).

Tomato PG was the first cell wall hydrolase to be examined using transgenic methods. Two groups independently downregulated *PG* mRNA accumulation by constitutive expression of an antisense *PG* transgene driven by the cauliflower mosaic virus 35S promoter (Sheehy et al., 1988; Smith et al., 1988). Suppression of *PG* mRNA accumulation was ca. 99% in homozygous progeny of these plants, with fruit retaining 0.5–1% of wild-type levels of PG enzyme activity (Smith et al., 1990; Kramer et al., 1992). Overall fruit ripening was not affected by suppression of *PG* mRNA, as determined by lycopene accumulation and ethylene production rates (Smith et al., 1990). During ripening in wild-type fruit, extensive depolymerization of polyuronide molecules occurs, and this was diminished in fruit with reduced levels of PG activity (Smith et al., 1990). Analysis of the physical properties of the polymers in the cell wall by ^{13}C nuclear magnetic resonance found that a proportion of the polyuronides showed increased mobility during ripening, and that in antisense-PG fruit this mobility loss was slightly reduced (Fenwick et al., 1996). Consistent with these observations, size exclusion chromatography showed that, relative to controls, in antisense-PG fruit depolymerization of CDTA-soluble polyuronides was slightly delayed at the red ripe and over ripe ripening stages, and reduced in a Na_2CO_3-soluble fraction at the over ripe stage, resulting in a slightly greater amount of high-molecular-weight molecules and lower amount of medium-sized and small molecules (Brummell and Labavitch, 1997). Solubilization of CDTA-extractable polyuronides during ripening was not reduced, and was actually a little greater than in wild type (Carrington et al., 1993; Brummell and Labavitch, 1997). However, fruit suppressed in PG activity showed reduced amounts of water-soluble polyuronide, matched by an equivalent increase in sodium carbonate-soluble polyuronide, suggesting that PG acts to depolymerize covalently bound pectin and solubilize it into a water-

soluble fraction (Carrington et al., 1993). Fruit suppressed in PG were slightly firmer than non-transgenic controls at all ripening stages, but the difference was small and less than that due to variability in genotype, growing conditions and fruit handling methods (Kramer et al., 1992; Langley et al., 1994).

A related study made use of the non-softening, ripening-impaired *rin* mutant to examine the role of PG in ripening-related cell wall metabolism. In *rin*, mRNA of the endogenous *PG* gene accumulates at much reduced levels (Della Penna et al., 1987, 1989; Biggs and Handa, 1989; Knapp et al., 1989), and both PG activity (Buescher and Tigchelaar, 1975) and polyuronide solubilization (Seymour et al., 1987a) are very low. Treatment of *rin* fruit with ethylene does not increase *PG* mRNA accumulation (Giovannoni et al., 1989; Knapp et al., 1989), suggesting that the endogenous *PG* promoter is not able to activate gene transcription in this mutant. To circumvent this problem, Giovannoni et al. (1989) fused a sense *PG* transgene to the promoter of the *E8* gene, a gene which is expressed and activated by ethylene or propylene in *rin*. Propylene treatment of transgenic *rin* fruit transformed with the chimeric *E8::PG* transgene resulted in accumulation of up to 60% of wild-type levels of PG enzyme activity and almost wild-type levels of chelator-soluble polyuronide but did not restore softening. The extent of polyuronide depolymerization in these non-softening transgenic fruit was also similar to wild-type (Della Penna et al., 1990).

The above experiments show that PG is responsible for a major component of polyuronide solubilization and depolymerization during ripening but demonstrate that PG is neither necessary nor sufficient for fruit softening. However, in the antisense experiments, suppression of PG activity was not complete and 0.5% of wild-type PG activity remained, which in tomato is significant due to the high levels of PG present in wild-type fruit. Almost wild-type levels of polyuronide depolymerization occurred in these fruit (Brummell and Labavitch, 1997). More recently tomato lines have been identified in which the *PG* gene has been disrupted, and thus functionally inactivated, by insertion of a *Ds* transposable element, almost completely eliminating PG activity (Cooley and Yoder, 1998). Preliminary observations indicated that fruit softening behavior was not obviously different from wild-type (Cooley and Yoder, 1998), but it would be very interesting to examine the extent of polyuronide depolymerization in these fruit.

Examination of other aspects of fruit quality in antisense-PG lines revealed unexpected phenotypes. Significant improvements were noted in several post-harvest factors, including shelf-life, resistance to cracking, handling characteristics and resistance to post-harvest pathogens (Schuch *et al.*, 1991; Kramer *et al.*, 1992; Langley *et al.*, 1994). These enhanced physical properties may have been due in large part to reduced cell separation in PG-suppressed fruit, resulting in improved tissue integrity (Langley *et al.*, 1994). Furthermore, large increases in the viscosity of juice or paste derived from these fruit were noted. When no attempt was made to inactivate PG activity during paste preparation (a cold break), or in fruit processed in a commercial pilot plant with a hot break of limited effectiveness, the viscosity of the total juice or paste and of the serum alone were significantly increased relative to controls (Schuch *et al.*, 1991; Kramer *et al.*, 1992; Brummell and Labavitch, 1997). The large increase in serum viscosity in antisense-PG material is due to the retention of high-molecular-weight polyuronides, which in controls had become highly depolymerized (Brummell and Labavitch, 1997). In experiments where a very effective hot break rapidly and completely inactivated PG activity in both control and antisense-PG fruit prior to homogenization, no difference in paste viscosity was observed between the genotypes (Schuch *et al.*, 1991). Thus, suppression of PG activity can help prevent the extensive depolymerization of polyuronides that occurs during processing if the hot break is not completely effective, and which is the major cause of paste viscosity loss.

Overall, the data show that PG activity is responsible for polyuronide depolymerization and solubilization during ripening but that this makes only a small contribution to fruit softening. However, the integrity of stored fruit and the textural properties of paste were improved by suppression of PG. The role of PG in fruit ripening may thus be mainly concerned with fruit textural changes and quality properties, and ultimately in fruit deterioration to allow seed dispersal.

Polygalacturonase β-subunit

PG protein extracted from the cell walls of tomato fruit is recovered in three forms. PG2A and PG2B are the 43 and 45 kDa differentially processed products of the *PG* gene mentioned above, whereas PG1 is a higher-molecular-weight form consisting of one, or possibly two, PG2A or PG2B proteins in combination with a 38 kDa glycoprotein known as the PG β-subunit or converter (Pressey and Avants, 1973; Ali and Brady, 1982; Moshrefi and Luh, 1983; Pogson *et al.*, 1991). All three forms are capable of polyuronide degradation *in vitro* (Ali and Brady, 1982; Moshrefi and Luh, 1983; Moore and Bennett, 1994), but β-subunit protein alone does not possess PG activity, and no enzyme activity has yet been found associated with it. The binding between the β-subunit and PG2 (PG2A or PG2B) is extremely strong, and comparisons of PG1 and PG2 activity have shown that *in vitro* the β-subunit substantially modifies PG2 activity in regard to heat stability, pH optimum, salt requirements and substrate requirements (Pressey and Avants, 1973; Tucker *et al.*, 1980, 1981; Pressey 1984; Knegt *et al.*, 1988; Pogson *et al.*, 1991; Moore and Bennett, 1994). The β-subunit either alone or in combination with PG2 (i.e., as PG1) also shows binding to polydextrans (Knegt *et al.*, 1991). Based on these observations, it has been proposed that the function of the β-subunit *in muro* is to modify PG2 activity, either by binding to it and altering its activity characteristics, or by binding to its substrate, or by localizing PG2 to particular regions in the wall. However, there is also evidence from the differential extractability of PG1 and PG2 in salt solutions and at different pH and temperatures to suggest that PG1 may not exist *in vivo*, and that the association between the β-subunit and PG2 is formed artifactually during extraction (Pressey, 1986a, b, 1988; Knegt *et al*, 1991; Moore and Bennett, 1994).

The β-subunit is synthesized as a 69 kDa precursor protein which possesses a signal sequence of 30 amino acids and undergoes extensive posttranslational processing (Zheng *et al.*, 1992). An amino-terminal propeptide of 78 amino acids is removed, as well as a large carboxyl-terminal domain of 25.6 kDa, to yield a mature domain of ca. 31.5 kDa. This is heavily glycosylated and many of the phenylalanine residues are post-translationally modified, to form the mature 37 to 39 kDa protein. Sequence analysis has revealed that almost the entire mature protein consists of an imperfectly repeated 14 amino acid motif (Zheng *et al.*, 1992). β-Subunit mRNA accumulates early in young green fruit, gradually increases during fruit development and peaks at around the beginning of ripening, then declines at the same time that *PG* mRNA appears and increases (Zheng *et al.*, 1992, 1994). Immunologically detectable β-subunit protein also gradually increases during fruit development, but remains relatively constant throughout fruit ripening after β-subunit mRNA abundance has declined (Pogson and

Brady, 1993; Zheng *et al.*, 1994). Thus, β-subunit mRNA and protein are present in the fruit well before ripening-related ethylene and *PG* mRNA, protein and activity appear (Tucker *et al.*, 1981; Pressey, 1986b, 1988; Pogson and Brady 1993; Zheng *et al.*, 1992, 1994). The presence of β-subunit mRNA and protein in young green fruit before ripening-related ethylene evolution begins implies that ethylene is not required for β-subunit expression. Also, β-subunit protein can be extracted from the ripening-impaired *rin* and *nor* mutants (Pressey, 1988; Della Penna *et al.*, 1990; Pogson and Brady, 1993), which produce only background levels of ethylene and trace amounts of PG. In fruit of *rin*, β-subunit mRNA and protein abundance are similar to wild-type (Zheng *et al.*, 1994). Treatment of immature green fruit of *rin* with ethylene only slightly affects β-subunit mRNA and protein levels relative to wild-type, suggesting that β-subunit expression is unchanged by ethylene, is unrelated to *PG* expression, and is presumably regulated by other developmental cues (Zheng *et al.*, 1994). mRNA, protein and activity related to the β-subunit have also been detected at low levels in root, leaf and flower, suggesting that the β-subunit is a member of a small gene family, although the other gene family members may be quite divergent (Pressey, 1984; Zheng *et al.*, 1992, 1994).

Introduction of a β-subunit antisense transgene under the control of the 35S promoter suppressed β-subunit mRNA and protein accumulation to almost undetectable levels during tomato fruit development (Watson *et al.*, 1994). Overall fruit ripening, as determined by lycopene accumulation, and *PG* mRNA and protein abundance were not affected by the presence of the β-subunit transgene. In controls, PG enzyme activity recovered from fruit was mainly in the high-molecular-weight thermostable PG1 form early in ripening, with increasing amounts of low-molecular-weight PG2 later in ripening as β-subunit protein levels became limiting. In transgenic lines expressing the β-subunit antisense transgene, almost all the PG activity was present as PG2 due to the lack of β-subunit protein. Relative to controls, fruit suppressed in β-subunit protein accumulation had slightly greater amounts of extractable PG enzyme activity at all ripening stages. Analysis of fruit cell walls found that in four independent transgenic lines the amount of chelator-extractable polyuronide was ca. 50% greater than in wild-type during fruit ripening (Watson *et al.*, 1994). Independent analysis additionally found a 90% increase in the amount of chelator-soluble polyuronides in green fruit (Chun and Huber,

1997). This last observation is significant because it shows that the β-subunit can cause a modification of pectin metabolism both in the presence and absence of PG activity.

Fruit suppressed in β-subunit protein accumulation were of similar firmness to controls at the mature green stage, but softened more extensively as ripening progressed and at the red ripe stage were 22% softer than control fruit (Chun and Huber, 2000). Scanning electron microscopy observations suggested that this increased softening was at least partly due to reduced tissue integrity brought about by weakening of the middle lamella, presumably due to increased PG-mediated polyuronide degradation. However, size exclusion chromatography of chelator-soluble polyuronides from antisense β-subunit fruit found either no difference to controls (Chun and Huber, 1997), or a slightly increased amount of low molecular weight species and a reduced amount of high-molecular-weight species (Watson *et al.*, 1994). In *in vitro* experiments to examine the role of the β-subunit in PG-mediated pectin metabolism, cell walls were prepared from wild-type and antisense β-subunit green fruit (Chun and Huber, 1997). These cell walls were exposed to PG2, and polyuronides solubilized both by the PG treatment itself, and by extraction with CDTA after the PG treatment, were recovered. Size exclusion chromatography showed that the molecular weight profiles of polyuronides solubilized by the PG2 treatment were the same in wild-type and antisense β-subunit cell walls, but polyuronides extractable by CDTA after the PG2 treatment were much more depolymerized in antisense β-subunit cell walls than in wild-type. In excised discs of mature green fruit pericarp, treatment with purified PG2 protein brought about pronounced intracellular weaking in antisense β-subunit fruit, but had only minor effects on control fruit (Chun and Huber, 2000). Thus, suppression of β-subunit protein accumulation in the cell wall allows increased polyuronide depolymerization and middle lamella degradation by PG2 added *in vitro*, suggesting that the β-subunit can impede PG2 activity in the cell wall. How this finding relates to the very small or no difference in molecular weight profile of chelator-soluble polyuronides extracted from whole ripening fruit of wild-type and antisense β-subunit lines is not clear. However, in both these size exclusion chromatography studies (Watson *et al.*, 1994; Chun and Huber, 1997) the polyuronides fractionated mainly in the void volume outside the separation range of the column, and it would be interesting to re-examine these samples on a column of increased size separation range.

Analysis of transgenic plants found that suppression of β-subunit protein accumulation increases fruit softening during ripening, extractable PG activity, polyuronide solubilization and possibly polyuronide depolymerization, and together show that the β-subunit plays a role in pectin metabolism *in vivo*. When β-subunit protein levels are substantially reduced in transgenic plants, the middle lamella loses cohesion during ripening and tissue integrity is reduced, suggesting that the β-subunit acts to limit the degradation of cell wall polyuronides by PG. Whether it does this by affecting PG activity, substrate accessibility or mobility is currently unclear, but in pericarp discs of mature green fruit the mobility of added PG2 is increased in antisense β-subunit tissue (Chun and Huber, 2000). Since β-subunit protein accumulates early in fruit development before the accumulation of PG protein begins, the data are consistent with the β-subunit being distributed throughout the developing wall to control the subsequent diffusion of PG in the wall matrix during ripening. This restriction of access of PG to polyuronide substrate sites would protect cell wall pectin from excessive PG-mediated degradation early in ripening, at least until free β-subunit is depleted and new PG is present in the wall as PG2.

Pectin methylesterase

Polygalacturonans are secreted to the cell wall in a highly methyl-esterified form, and are de-esterified during cell development. During ripening in tomato, the degree of methyl-esterification of cell wall pectin declines from 90% in mature green fruit to 35% in red ripe fruit (Koch and Nevins, 1989). This is accomplished by pectin methylesterase (PME; EC 3.1.1.11), which de-esterifies polyuronides by removing methyl groups from the C6 position of galacturonic acid residues of high-molecular-weight pectin. Demethylation of pectin to their free carboxyl groups changes the pH and charge in the cell wall, allows the aggregation of polyuronides into a calcium-linked gel structure, and makes the polyuronides susceptible to degradation by PG (Pressey and Avants, 1982; Jarvis 1984; Seymour *et al.*, 1987b; Koch and Nevins, 1989; Carpita and Gibeaut, 1993).

In tomato, PME is present as a small gene family consisting of at least four genes, some of which are highly homologous (Ray *et al.*, 1988; Harriman

et al., 1991; Hall *et al.*, 1994; Turner *et al.*, 1996; Gaffe *et al.*, 1997). Three of the genes are present in the genome as a tandem repeat (Hall *et al.*, 1994, Turner *et al.*, 1996). PME proteins are synthesized as preproteins of 540–580 amino acids possessing a signal sequence and a large amino-terminal extension of ca. 22 kDa, both of which are removed to yield a mature protein of 34–37 kDa (Gaffe *et al.*, 1997). Most lack consensus sequences for N-glycosylation in the mature protein, although at least one site is present in the amino-terminal extension of all the members described to date. PME protein is found in most tissues of the plant and exists in multiple isoforms (Pressey and Avants, 1972; Delincee, 1976; Tucker *et al.*, 1982; Warrilow *et al.*, 1994). Three immunologically related isoforms are specific to fruit, plus several other isoforms found in all tissues including fruit (Gaffe *et al.*, 1994). PME protein and activity are present throughout fruit development, increasing from the early stages of green fruit to the mature green stage and then increasing again during ripening by two- to three-fold with a peak early in ripening, and then declining slightly (Harriman *et al.*, 1991; Tieman *et al.*, 1992). The abundance of *PME* mRNA shows a different pattern of accumulation, increasing to a maximum in mature green fruit and declining rapidly as ripening progresses (Ray *et al.*, 1988; Harriman *et al.*, 1991). PME protein and activity thus reach a maximum after *PME* mRNA abundance has declined substantially. It is likely in both these last studies (Ray *et al.*, 1988; Harriman *et al.*, 1991) that the *PME* mRNA and immunologically detectable protein measured is a composite of the product of two or more highly homologous genes. When gene-specific probes prepared from 3′-untranslated regions were used, mRNA from one *PME* gene (*PME2/PEC2*/clone B16) was found at constant low abundance throughout fruit development and ripening, while mRNA from another gene (*PME1/PEC1*/clone B8) increased to very high levels in green fruit to peak at the breaker stage and then declined in ripe fruit (Hall *et al.*, 1994). It is not known what factors control PME gene expression in fruit development. In the *Nr* and *nor* ripening mutants, levels of PME activity, protein and mRNA are similar to wild type, but in *rin* all three decline rapidly at a time equivalent to when ripening begins in wild-type fruit (Harriman *et al.*, 1991). The *rin* mutation thus strongly affects PME expression.

Two groups independently suppressed PME activity in tomato by introduction of antisense *PME2/PEC2* transgenes under the control of the constitutive 35S

321

promoter, either a genomic DNA fragment with introns or a cDNA fragment (Tieman *et al.*, 1992; Hall *et al.*, 1993). In both cases, although *PME*2 mRNA and immunodetectable protein were reduced to undetectable or trace levels in fruit, PME activity was not eliminated and was present at almost 10% of wild type. PME activity in leaf and root was not reduced by introduction of this *PME*2 transgene, although no mRNA or immunodetectable protein homologous to the wild-type *PME*2 gene were detected (Hall *et al.*, 1993). The coding region of *PME*2 hybridizes with the more highly expressed *PME*1 (Hall *et al.*, 1994), but mRNA of neither gene was detected in fruit of antisense plants. This suggests that the antisense *PME*2 transgenes suppressed mRNA accumulation of *PME*2 and homologous genes including *PME*1, and that the residual activity in fruit and activity in other tissues was due to more divergent *PME* gene products.

The degree of pectin methylesterification in transgenic antisense PME fruit was higher than controls by 15–40% throughout ripening (Tieman *et al.*, 1992; Hall *et al.*, 1993), but the fruit otherwise ripened normally as judged by ethylene and lycopene production (Tieman *et al.*, 1992). Increased pectin methylesterification resulted in reduced polyuronide depolymerization in red ripe fruit, and decreased the amount of chelator-soluble pectin during ripening by 20–30% (Tieman *et al.*, 1992). Presumably the former is due to the resistance of methyl-esterified pectin to PG-mediated hydrolysis, and the latter to reduced amounts of pectin bound ionically to the wall which were instead attached by linkages not affected by removal of calcium. Suppression of PME activity and consequent changes in pectin metabolism, including reduced pectin depolymerization, did not affect fruit softening during normal ripening, but in over-ripe fruit caused an almost complete loss of tissue integrity (Tieman and Handa, 1994). This was correlated with an increase in soluble calcium and reductions in bound calcium, soluble sodium and bound and soluble magnesium, suggesting that a lowered ability of cell walls to bind divalent cations has deleterious effects on tissue integrity, probably in part due to reduced interpectate calcium cross-bridges. The changed ionic and physical conditions in the wall may also have affected the activity of other cell wall modifying enzymes, including PG (Buescher and Hobson, 1982; Brady *et al.*, 1985; Chun and Huber, 1998). Suppression of PME activity thus has a negative effect on fruit integrity during senesence in prolonged storage, resembling the effect of suppression of the β-subunit,

but opposite to the improved fruit integrity and shelf life observed after suppression of PG. However, large improvements in several fruit processing attributes were observed. Raw juice prepared from antisense PME fruit showed an almost 20% increase in soluble solids content (Tieman *et al.*, 1992). Processed juice showed significantly higher total and soluble solids, serum viscosity, paste viscosity and reduced serum separation (Thakur *et al.*, 1996a). This was associated with a large increase in polyuronide molecular weight relative to controls, larger than the difference in ripening fruit, presumably due to the the high degree of pectin methyl-esterification protecting pectin from PG-mediated hydrolysis during fruit homogenization (Thakur *et al.*, 1996b). PME activity thus plays little role in fruit softening during ripening, but substantially affects tissue integrity during senescence and fruit processing characteristics.

β-Galactosidase

One of the largest changes that occurs in the cell walls of ripening fruit is a loss of galactosyl residues from cell wall polymers (Gross and Sams, 1984). In tomato, the decline in polymeric galactose and rise in free galactose begins early and increases with ripening (Wallner and Bloom, 1977; Gross and Wallner, 1979; Gross, 1983), and is observed mainly in pectic fractions, although smaller changes are observed in matrix glycan and cellulose fractions (Gross, 1984; Seymour *et al.*, 1990). Much of the galactose in the wall is present as side chains attached to rhamnose residues in the backbone of RG I, where the galactose is present either in type I $(1{\rightarrow}4)\beta$-D-galactan chains or in type II branched $(1{\rightarrow}3),(1{\rightarrow}6)\beta$-D-galactan chains (Carpita and Gibeaut, 1993). The latter branched chains are also found on cell wall arabinogalactan proteins, but in tomato the $(1{\rightarrow}4)\beta$-D-galactan predominates and is degraded during ripening (Seymour *et al.*, 1990). Since endo-galactanases have not been detected in higher plants, the enzyme activity most likely to be responsible for degradation of cell wall β-galactan is exo-β-D-galactosidase (EC 3.2.1.23), an enzyme which removes terminal non-reducing β-D-galactosyl residues from β-D-galactosides. β-Galactosidase purified from tomato fruit can be separated into three forms, termed β-galactosidase I, II and III (Pressey, 1983). All three forms are active against the model substrate *p*-nitrophenyl-β-D-galactopyranoside, but only β-galactosidase II is active against a $(1{\rightarrow}4)\beta$-D-galactan-rich polymer prepared from tomato cell walls (Pressey, 1983). The characteristics of this activity are consistent with β-galactosidase II being an exo-$(1\rightarrow4)\beta$-D-galactanase (Pressey, 1983; Carey *et al.*, 1995), but since the enzyme is active against a variety of galactoside substrates it should strictly be termed a β-D-galactosidase/exo-galactanase (Smith and Gross, 2000). The polymeric galactan side chains of RG I, and possibly the terminal galactosyl residues of xyloglucan side-chains, are the most likely substrates of β-galactosidases in the wall.

Total β-galactosidase activity is high and does not change appreciably during tomato fruit development and ripening (Wallner and Walker, 1975; Pharr *et al.*, 1976; Carey *et al.*, 1995). However, when the activities of individual isoforms are examined, the activities of β-galactosidase I and III are high in green fruit and decline during ripening, whereas β-galactosidase II activity is low or undetectable in green fruit and increases up to 7-fold during ripening (Pressey, 1983; Carey *et al.*, 1995; Carrington and Pressey, 1996). In tomato, β-galactosidases are encoded by a gene family of at least seven members (Carey *et al.*, 1995; Smith *et al.*, 1998; Smith and Gross, 2000). The genes have been termed *TBG1–TBG7*, and encode predicted polypeptides of 90–97 kDa except for *TBG4*, which is ca. 100 amino acids shorter at the carboxyl-terminal end and encodes a predicted polypeptide of 78 kDa (Smith and Gross, 2000). The deduced amino acid sequence of *TBG4* corresponds to the protein termed β-galactosidase II (Smith *et al.*, 1998). All seven gene family members possess at least one consensus sequences for N-linked glycosylation with some having as many as six, and all are predicted to possess either signal sequences for ER entry or, in the case of *TBG7*, a transit peptide targeting the protein to the chloroplast (Smith and Gross, 2000). Transcripts derived from all seven genes are detected during fruit development, although different patterns of expression are observed (Carey *et al.*, 1995; Smith *et al.*, 1998; Smith and Gross, 2000). mRNA of *TBG6* reaches the highest abundance, but is present only in green fruit and disappears as ripening begins. The mRNA accumulation of four genes, *TBG*1, 3, 4 and 5, is ripening related. mRNAs of *TBG*1 and 3 are present at constant low levels from the breaker to over ripe stages, whereas mRNAs of *TBG*4 and 5 reach higher levels which peak at the turning stage and decline as ripening progresses. Transcripts of each of these four genes are also detected in other tissues of the plant.

In the non-softening *rin* and *nor* ripening mutants, the loss of cell wall galactose and rise in free galactose are much reduced (Gross, 1983; Gross, 1984), and the ripening-related rise in β-galactanase (β-galactosidase II) activity is not observed (Carey *et al.*, 1995). Consistent with this observation, the mRNA abundance of *TBG4* is markedly reduced in *rin* and *nor* relative to wild-type (Smith *et al.*, 1998; Smith and Gross, 2000). mRNA accumulation of *TBG* 1, 3 and 5 is similar to wild-type, whereas mRNA of *TBG6*, which in wild-type disappears as ripening begins, persists in *rin* and *nor* of the same chronological age as wild-type ripening fruit (Smith and Gross, 2000). Together, these observations suggest that *TBG4* is responsible for ripening-related exo-galactanase activity and a large proportion of the loss of cell wall galactose that occurs during ripening, and may contribute to fruit softening. Due to their unaltered expression in *rin* and *nor* relative to wild-type, it seems unlikely that mRNA accumulation of *TBG* 1, 3 and 5 is regulated solely by ethylene, but it is possible that *TBG4* expression is ethylene-regulated.

Heterologous expression in yeast showed that TBG1 is an exo-galactanase active against tomato cell wall polysaccharides (Carey *et al.*, 2001). Sense suppression by a short gene-specific region of *TBG1* cDNA reduced *TBG1* mRNA abundance to 10% of wild-type levels in ripe fruit, but did not reduce total exo-galactanase activity assayed against lupin galactan, or total β-galactosidase activity assayed against a model substrate, and did not affect cell wall galactose content or fruit softening (Carey *et al.*, 2001). Assuming that TBG1 protein accumulation was suppressed equivalently, this suggests that TBG1 contributes either only a small proportion of total β-galactosidase activity, or that its activity is directed specifically to a particular, possibly minor, cell wall component.

Antisense suppression of *tEG1A* (*TBG3*) resulted in a reduction in extractable exo-galactanase activity of up to 75% , as assayed against lupin galactan (de Silva and Verhoeyen, 1998). These transgenic plants were prepared using a construct containing a long region (1082 bp) of the *TBG3* coding sequence, and in addition to *TBG3*, mRNA abundances of *TBG4* and *TBG1* were also suppressed. All these genes contain regions of high DNA sequence homology, and the reduced activity of exo-galactanase in protein extracts was presumably due to suppression of multiple β-galactosidase gene products. Reduced exo-galactanase activity was correlated in both mature green and ripe fruit with increased cell wall galactose content and with increased cell wall labeling by an antibody specific for polymeric galactan, showing that cell wall galactose loss was strongly reduced. No obvious differences in fruit softening behavior during ripening were noted, but fruit deteriorated more slowly during long-term storage and juice prepared from ripe fruit contained an increased proportion of insoluble solids and slightly increased viscosity.

Heterologous expression of TBG4 in yeast confirmed that this gene product was both a galactosidase and exo-galactanase, and showed that the heterologously-expressed enzyme was more active against sodium carbonate-soluble pectins than against chelator-soluble pectins or matrix glycans (Smith and Gross, 2000). This is consistent with studies of purified galactosidase II, which concluded that the natural substrate is galactan-rich side chains of pectin found mainly in the sodium carbonate-soluble pectin fraction (Carrington and Pressey, 1996). Antisense suppression of TBG4 using a long region (1500 bp) of TBG4 cDNA strongly reduced TBG4 mRNA accumulation, and reduced extractable exo-galactanase activity by up to 90% (D. Smith and K. Gross, personal communication). It seems likely that homologous β-galactosidase genes, such as TBG1 and TBG3, would also have been suppressed. Strong suppression of β-galactosidase activity early in ripening was correlated with reduced galactose loss, and reduced softening by up to 40% later in ripening. In other transgenic lines, a suppression of exo-galactanase only late in ripening also reduced galactose loss, but was not correlated with increased firmness. Thus, suppression of β-galactosidase activity early in ripening can prevent cell wall changes required for softening, but suppression of activity later in ripening after softening has begun is without effect on firmness.

These transgenic experiments show that the loss of cell wall galactose during ripening is due to β-galactosidase/exo-galactanase activity, of which TBG4 may be the most important isoform. Cell wall galactose loss appears to play a substantial role in fruit softening, and also contributes to fruit textural changes as detected by altered juice properties. One possibility is that the increased firmness of exo-galactanase-suppressed fruit is directly due to the reduced loss and thus increased content of pectic galactan side-chains, which in pea cotyledons increases the mechanical strength of the wall (McCartney et al., 2000). However, it is interesting that reduced exo-galactanase activity early in ripening rather than late in ripening is correlated with reduced fruit softening. This

suggests that the effect of suppression of β-galactan degradation on fruit softening may be indirect, perhaps by preventing increases in porosity of the wall during ripening, which would impede the access of other hydrolases to pectic or glycan substrates and thus prevent or retard depolymerization of structural polysaccharides.

Endo-(1 →4)β-D-glucanase

Cell wall matrix glycans undergo considerable depolymerization during fruit ripening (Huber, 1983, 1984; Gross et al., 1986; Tong and Gross, 1988; McCollum et al., 1989; O'Donoghue and Huber, 1992; Maclachlan and Brady, 1994, Brummell et al., 1999c), which is believed to contribute substantially to fruit softening. The enzymes causing this depolymerization have not been unambiguously identified, but may include EGase (EC 3.2.1.4). These enzymes are often referred to as cellulases, but in higher plants most lack the cellulose binding domains found in microbial cellulases, and thus alone are probably not capable of degrading crystalline cellulose (Brummell et al., 1994). EGases hydrolyze internal linkages of (1→4)β-D-linked glucan chains adjacent to unsubstituted residues, and in vitro are active against xyloglucan, cello-oligosaccharides, non-crystalline cellulose and the model substrate CMC (Wong et al., 1977; Hayashi et al., 1984; Hatfield and Nevins, 1986; Nakamura and Hayashi, 1993; Ohmiya et al., 1995). In the cell wall, their substrates probably include xyloglucan, integral and peripheral regions of non-crystalline cellulose (particularly the outer layers of cellulose microfibrils where glucan chains are interwoven with xyloglucan chains), and possibly glucomannan where sufficient consecutive (1→4)β-D-linked glucan residues occur for substrate binding.

EGase cDNA and genomic clones have been isolated from many species and all share conserved amino acid motifs presumably involved in catalysis or secondary structure (Brummell et al., 1994), but the predicted proteins possess primary structures of three different types: (1) proteins of 50–55 kDa with signal sequences for ER targeting and secretion, and usually but not always possessing sequences for N-glycosylation (most of the EGases characterized to date are of this type); (2) proteins of 80–90 kDa lacking a signal sequence but with a long amino-terminal extension possessing a membrane-spanning domain anchoring the protein to internal or plasma mem-

324

branes, and with a high degree of *N*-glycosylation (Brummell *et al.*, 1997b); (3) proteins of ca. 65 kDa with a signal sequence and a long carboxyl-terminal extension possessing a putative carbohydrate-binding domain of unknown function showing similarity to microbial cellulose-binding domains (Trainotti *et al.*, 1999b). Multiple isoforms of EGase have been detected in ripening fruit (Kanellis and Kalaitzis, 1992; Maclachlan and Brady, 1992). EGases are encoded by large, very divergent multigene families, which in tomato consists of at least eight members (Lashbrook *et al.*, 1994; Milligan and Gasser, 1995; del Campillo and Bennett, 1996; Brummell *et al.*, 1997b; Catalá *et al.*, 1997; Catalá and Bennett, 1998).

EGase activity has been found in fruit of all species examined (Brummell *et al.*, 1994) but the amount varies considerably, avocado having 160 times more activity than peach and 770 times more activity than tomato on a fresh weight basis (Lewis *et al.*, 1974). EGase mRNAs accumulate to relatively low levels, and in ripening tomato are present below the level of detection by hybridization with total RNA gel blots and accumulate to abundances several hundred-fold less than tomato PG mRNA (Lashbrook *et al.*, 1994). Even in avocado, where EGase activity is high, detection of *PaCel*1 requires the use of poly(A)$^+$ RNA (Christoffersen *et al.*, 1984). In tomato, EGase activity is present throughout fruit development, with highest levels in young expanding green fruit and during ripening and, unlike PG, with activity declining in over ripe fruit (Hall, 1964; Hobson, 1968). The measured extractable EGase activity is presumably a composite of the products of several different genes. mRNAs of tomato *LeCel*1, *LeCel*2 and *LeCel*4 accumulate at the earliest stages of fruit cell expansion (Lashbrook *et al.*, 1994; Gonzalez-Bosch *et al.*, 1996; Brummell *et al.*, 1997a), and mRNA of *LeCel*7 accumulates during the later stages of cell expansion in young green fruit and in full-size mature green fruit, but declines to undetectable levels as ripening begins (Catalá *et al.*, 2000). Both *LeCel*1 and *LeCel*2 exhibit a ripening-related accumulation of mRNA in fruit pericarp with large increases at the onset of ripening (Lashbrook *et al.*, 1994; Gonzalez-Bosch *et al.*, 1996). *LeCel*1 mRNA abundance reaches a maximum early in ripening and then declines, whereas *LeCel*2 mRNA abundance continues to increase as ripening progresses and reaches higher levels than *LeCel*1. Expression of neither gene is specific for fruit pericarp, with mRNA of *LeCel*1 accumulating in anthers, *LeCel*2 in fruit locules and some flower parts, and both in abscising flower abscis-

sion zones (Lashbrook *et al.*, 1994; Gonzalez-Bosch *et al.*, 1996). *LeCel*5 also shows an overlapping expression in these tissues, its mRNA accumulating in abscission zones and pistils, and in fruit at the light red ripening stage (Kalaitzis *et al.*, 1999).

In hypocotyls, the mRNA abundance of *LeCel*7 is increased by treatment with auxin (Catalá *et al.*, 1997), suggesting that expression of this gene may also be positively regulated by auxin in green fruit. Accumulation of both *LeCel*1 and *LeCel*2 mRNAs in ripening fruit is promoted by exogenous ethylene and inhibited by an ethylene antagonist, suggesting that expression of both these genes is ethylene regulated (Lashbrook *et al.*, 1994; Gonzalez-Bosch *et al.*, 1996). However, studies using fruit of the *rin* mutant show that expression of the two genes is not coordinated (Gonzalez-Bosch *et al.*, 1996). In *rin*, *LeCel*1 mRNA is present at low but detectable levels, and after ethylene treatment is strongly increased to levels higher than those in wild-type fruit. In contrast, *LeCel*2 mRNA is barely detectable in *rin*, either with or without ethylene treatment. The *rin* mutation thus affects the expression of *LeCel*1 and *LeCel*2 mRNA differently, and suggests that they are subject to distinct regulatory control. The reduced expression relative to wild type but strong ethylene responsiveness of *LeCel*1 in *rin* is similar to that of the unknown ripening genes *E4* and *E8* (Lincoln and Fischer, 1988), whereas the lack of expression and ethylene response of *LeCel*2 in *rin* is similar to that of *PG* (Giovanonni *et al.*, 1989). The high expression of *LeCel*1 in ethylene-treated non-softening *rin* fruit shows that *LeCel*1 mRNA accumulation alone is not sufficient to cause fruit softening, but the lack of *LeCel*2 mRNA in *rin* is consistent with the product of this gene playing a role in softening.

In non-climacteric fruit, the regulation of ripening-related gene expression is different from climacteric fruit. In strawberry, expression of ripening-related *FaCel*1 is suppressed by the high levels of auxin present in young fruit, and its mRNA does not accumulate until levels of auxin decline as fruit ripen (Harpster *et al.*, 1998). *FaCel*2 (also called *FaEG3*), which possesses a carboxyl-terminal carbohydrate-binding domain, is also expressed during fruit development but before softening begins, its mRNA accumulating in green fruit and remaining at relatively constant levels in white and ripening fruit (Llop-Tous *et al.*, 1999; Trainotti *et al.*, 1999a). Like *FaCel*1, mRNA abundance of this gene is also down-regulated by auxin treatment (Trainotti *et al.*, 1999b). It is possible that in strawberry the activities of both these genes, perhaps acting synergistically, contribute to cell wall disassembly and fruit softening. In bell pepper, ripening and the accumulation of ripening-related *CaCel*1 mRNA and protein is increased by treatment with high concentrations of ethylene (Ferrarese *et al.*, 1995; Harpster *et al.*, 1997). Although pepper is generally regarded as non-climacteric, there is considerable variability between different cultivars in ethylene evolution and respiration rates and in responses to exogenous ethylene (Gross *et al.*, 1986; Villavicencio *et al.*, 1999).

In tomato, mRNA accumulation of the highly divergent EGases *LeCel*1 and *LeCel*2 was suppressed individually by constitutive expression of antisense transgenes (Lashbrook *et al.*, 1998; Brummell *et al.*, 1999b). In both cases, in the most suppressed lines mRNA accumulation of the gene was decreased in fruit pericarp by 99% relative to wild type, without affecting the expression of the other EGase. However, the softening of transgenic fruit suppressed in *LeCel*1 or *LeCel*2 mRNA accumulation was indistinguishable from that of azygous segregants, and no changes in normal ripening behavior were observed. Thus, neither of these genes are a primary determinant of softening in ripening tomato fruit. Since both these EGases are expressed in ripening fruit, together with at least one other EGase gene family member, *LeCel*5 (Kalaitzis *et al.*, 1999), it is possible that some redundancy of function is present between different EGase gene products, and that the absence of one gene product may be compensated for by another. The suppression of multiple EGases may be necessary in order to observe phenotypic changes related to EGase action in tomato fruit. Although suppression of *LeCel*1 and *LeCel*2 mRNA accumulation did not result in altered fruit softening, reduced mRNA abundance of either of these genes in abscission zones reduced cell separation during abscission, measured as the abscision incidence of excised flowers (*LeCel*1) or the breakstrength of fruit abscission zones (*LeCel*2).

In bell pepper, sense suppression of *CaCel*1 reduced immunodetectable CaCel1 protein and extractable CMCase activity to undetectable levels in ripe fruit (Harpster and Dunsmuir, unpublished). This suggests that in pepper, unlike in tomato, extractable CMCase activity in ripe fruit is probably due to the product of a single EGase gene. During ripening, depolymerization of matrix glycans including xyloglucan in controls and in suppressed fruit lacking detectable CMCase activity was indistinguishable (Figure 3A). This suggests that CaCel1 EGase plays a minor or no role in matrix glycan disas-

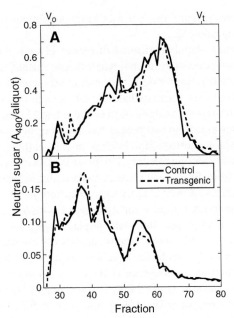

Figure 3. Effect of transgenic modification of EGase activity on cell wall matrix glycan depolymerization. A. Molecular weight profiles of matrix glycans from cell walls of red ripe bell pepper either wild-type (solid lines) or suppressed in activity of *CaCel*1 (dotted line). B. Molecular weight profiles of matrix glycans from cell walls of mature green tomato either control (transformed with empty vector, solid lines) or overexpressing activity of pepper *CaCel*1 (dotted line). Cell walls were sequentially extracted with CDTA and sodium carbonate to remove pectins, then with KOH to extract matrix glycans, which were neutralized and separated by size exclusion chromatography on Sepharose CL-6B as described (Brummell *et al.*, 1999c). (V_o = 1 MDa, V_t = 10 kDa).

sembly, and implies the presence of other hydrolytic enzymes active against cell wall polysaccharides but not against CMC, perhaps related to the xyloglucan-specific EGase found in growing stems (Matsumoto *et al.*, 1997). In a related series of experiments, the pepper *CaCel*1 gene was overexpressed in tomato under the control of the constitutive 35S promoter (Harpster, Brummell and Dunsmuir, unpublished). The deduced protein of pepper CaCel1 is highly homologous to that of tomato LeCel1, the mature proteins being 86% identical and 93% similar at the amino acid level, suggesting that the pepper gene product may be capable of performing the same function in the fruit as the tomato gene product. Extractable CMCase activity of over-expressing green and ripe tomato fruit was increased by at least 20-fold, and immunodetectable CaCel1 protein was associated with a cell wall fraction. However, the over-expression of high levels of CaCel1 in green fruit did not substantially affect the molecular weight profile of 24% KOH-soluble matrix glycans (Figure 3B), nor accelerate depolymerization during ripening.

Thus, depolymerization of matrix glycans during ripening occurs as normal when EGase mRNA accumulation is suppressed, and over-expression of high levels of EGase activity has little effect on this process. The transgenic analysis of ripening-related EGase activities in tomato and pepper has not identified a role, or even a substrate, for EGases in cell wall changes occurring during ripening. Purification of EGase activities from diverse plant tissues has revealed substantial variations in their activity against different cell wall polymers, but affinity for xyloglucan is usually low. EGases from pea stem and poplar suspension culture are highly active against acid-swollen cellulose or cello-oligosaccharides, but less so against xyloglucan although xyloglucan is degraded (Wong *et al.*, 1977; Hayashi *et al.*, 1984; Nakamura and Hayashi, 1993; Ohmiya *et al.*, 1995), and may even act predominantly against non-crystalline cellulose (Ohmiya *et al.*, 2000). EGase purified from avocado fruit is also active against cello-oligosaccharides but only weakly active against xyloglucan (Hatfield and Nevins, 1986). *In vitro*, an avocado protein extract enriched in CM-Case activity causes no observable depolymerization of avocado cell wall xyloglucan, but does bring about limited depolymerization of non-xyloglucan matrix glycans (O'Donoghue and Huber, 1992). Conversely, a CMCase-depleted enzyme extract causes depolymerization of xyloglucan, but has little effect on non-xyloglucan matrix glycans. This suggests that in avocado, as in pepper, xyloglucan may not be the major substrate for EGase action, and that the activity of an unknown xyloglucanase is largely responsible for xyloglucan depolymerization. During ripening in avocado, there is a disorganization of the wall as the cellulose microfibrils lose cohesiveness and become dissociated from matrix polymers. This was concluded to be due to the action of EGase in degrading non-crystalline regions of cellulose microfibrils, resulting in a large loss of microfibrillar material and disruption of the cell wall interpolymeric complex with only limited depolymerization of xyloglucan (O'Donoghue *et al.*, 1994). Avocado is not typical of other fruit, since the ripe avocado shows a marked loss of textural integrity with very low firmness and no elasticity. Nevertheless, this extreme example of fruit softening may give clues to the role of EGases in other species. In tomato, cellulose undergoes only very slight depolymerization during ripening (Maclachlan and Brady, 1994), but depolymerization could be more

extensive in particular glucan chains which are only a small proportion of the total, for example those on the surface of microfibrils. The role of EGases in cell wall disassembly is still unclear, but perhaps the activities of EGases with a carbohydrate binding domain (of Type 3, above) and secreted cell wall ripening-related EGases (of Type 1, above) are directed together against the glucan chains of cellulose microfibrils, and to some extent xyloglucan, at the microfibril/xyloglucan interface to cause cell wall loosening.

Xyloglucan endotransglycosylase

The endo-cleavage of xyloglucans can be achieved not only irreversibly by EGases, but also reversibly by the activity of xyloglucan endotransglycosylase (XET, EC 2.4.1.207; also called endo-xyloglucan transferase or EXGT) (Smith and Fry, 1991). These enzymes cleave internal linkages of the $(1 \rightarrow 4)\beta$-D-glucan backbones of xyloglucan and transfer the newly formed potentially reducing end to the C-4 position of the glucose unit at the non-reducing end of another xyloglucan polymer or oligosaccharide, with net retention of the anomeric configuration of the glycosidic bond. XETs are highly specific for xyloglucan both as donor substrate and acceptor substrate, although the degree of side-chain substitution of acceptor xyloglucan subunit oligosaccharides strongly affects reaction efficiency (Fry *et al.*, 1992; Lorences and Fry, 1993, Campbell and Braam, 1999a). During the reaction, XET forms a relatively stable glycosyl-enzyme intermediary complex which decomposes by transfer of the attached glycosyl preferentially to a xyloglucan acceptor (transglycosylation) or more slowly to water (hydrolysis) (Sulova *et al.*, 1998).

XETs are synthesized as polypeptides of about 300 amino acids, possessing a signal sequence and usually one consensus sequence for *N*-glycosylation, giving rise to mature proteins of 31–34 kDa (Nishitani and Tominaga, 1992; de Silva *et al.*, 1993; Okazawa *et al.*, 1993). An amino acid motif related to microbial β-glucanases and four cysteine residues in the carboxyl-terminal region of the protein are highly conserved (Okazawa *et al.*, 1993; Campbell and Braam, 1999b). However, the characteristics of XET enzymes vary considerably and they can be divided into two types based on biochemical properties, those which possess hydrolase activity and those which lack it. First, a specialized form of XET found in the cotyledons of germinating seeds which possess xyloglucan as a storage reserve, such as nasturtium, acts to hydrolyze xyloglucan to single subunit oligosaccharides (four glucosyl residues in the backbone) at low substrate concentrations, but transglycosylation predominates over hydrolysis when acceptor concentrations are high enough (Fanutti *et al.*, 1993). This activity can use xyloglucan as small as two subunit oligosaccharides as donor substrate. XETs that can depolymerize xyloglucan in the absence of xyloglucan oligosaccharides by hydrolysis, and in the presence of oligosaccharides by hydrolysis and endotransglycosylation, have also been purified from azuki bean epicotyl and kiwi fruit (Tabuchi *et al.*, 1997; Schroder *et al.*, 1998). Both the azuki bean and the kiwi enzymes are active mainly on high-molecular-weight xyloglucan, producing fragments with an average molecular mass of 40–60 kDa. Second, some activities from growing tissues exhibit no glycosidase or glycanase activity, and use only xyloglucans above 10 kDa (equivalent to eight xyloglucan subunit oligosaccharides) as donor substrate (Nishitani and Tominaga, 1992). Recombinant enzymes derived from an XET gene expressed in ripe tomato fruit and from four XET genes expressed in *Arabidopsis* vegetative tissues also were not active as hydrolases (Arrowsmith and de Silva, 1995; Campbell and Braam, 1999a). Such activities cannot cause a net depolymerization of xyloglucan in the absence of xyloglucan oligosaccharides. XETs with and without hydrolase activity have been purified from azuki bean epicotyl (Nishitani and Tominaga; 1992; Tabuchi *et al.*, 1997), and XET isozymes isolated from the same tissue can differ significantly in biochemical properties (Steele and Fry, 2000). XETs are encoded by large gene families, consisting of over 20 genes in *Arabidopsis* (Campbell and Braam, 1999b; The Arabidopsis Genome Initiative, 2000).

In fruit of tomato, apple and kiwi, XET activity is very high in young expanding fruit, declines during fruit maturation and rises again slightly during ripening (Maclachlan and Brady, 1992, 1994; de Silva *et al.*, 1994; Percy *et al.*, 1996). These activities are due to the products of different genes, since in tomato mRNA of *LeEXGT*1, which accumulates at high levels in growing hypocotyls, also accumulates in fruit but only at the young expanding green stage (Catalá *et al.*, 1997, 2000). mRNA of a different XET gene, *LeXET*B1 (*tXET*B1), accumulates to a maximum in ripening pink fruit and then declines slightly at the red ripe stage (Arrowsmith and de Silva, 1995). Presumably the product of this gene contributes to the

small rise in XET activity observed during ripening (Maclachlan and Brady, 1994). mRNA accumulation of tomato *LeEXGT*1 is positively regulated by auxin in hypocotyls, and by inference in green fruit (Catalá *et al.*, 1997, 2000). Ripening-related XET activity and mRNA abundance are positively regulated by ethylene in kiwi fruit (Redgwell and Fry, 1993; Schroder *et al.*, 1998). In tomato, fruit of *rin* possess reduced activity of XET relative to wild-type (Maclachlan and Brady, 1994). Although it has not yet been shown definitively, it seems likely that ripening-related XET gene expression is positively regulated by ethylene in climacteric fruit.

In tomato, modification of XET mRNA accumulation in transgenic plants has been carried out for two different XET genes, *LeEXGT*1 which is expressed in expanding green fruit and *LeXET*B1 which is expressed in ripening fruit. Expression of a constitutive antisense transgene of *LeEXGT*1 suppressed its mRNA accumulation in growing green fruit and decreased final fruit size, and overexpression of a constitutive sense *LeEXGT*1 transgene increased its mRNA accumulation in growing green fruit and increased final fruit size (Asada *et al.*, 1999). *LeEXGT*1 mRNA abundance and fruit size were inversely correlated with fruit solids content, showing that smaller fruit had increased concentrations of sugars. These findings show an important role for *LeEXGT*1 in cell expansion in young green fruit, which strongly affects the final size achieved by the fruit. mRNA accumulation of the ripening-related *LeXET*B1 has also been suppressed in ripening fruit (de Silva *et al.*, 1994). However, in a population of plants where the antisense transgene was under the control of the ripening-specific *PG* promoter, no correlation between *LeXET*B1 mRNA abundance and fruit softening behavior or paste viscosity was observed (J. de Silva, personal communication). Since XETs are encoded by large gene families, it is possible that more than one XET gene is expressed in ripening fruit, and that suppression of multiple XET genes may be necessary in order to establish the role of ripening-related XET activity in fruit softening or quality.

XET is believed to be involved in the incorporation of newly-synthesized xyloglucan into the wall, in cell wall loosening during growth, in the consolidation of cell wall structure after growth, and probably in other aspects of cell wall metabolism (Campbell and Braam, 1999b). High levels of XET activity present in very young fruit are involved in cell division, cell expansion and fruit growth, but the role of XET activity in fruit softening remains unclear. In kiwi, XET activity from ripening fruit possesses hydrolase activity (Schroder *et al.*, 1998), and in the presence or absence of xyloglucan oligosaccharides could contribute to xyloglucan depolymerization occurring during softening. In tomato, XET activity from ripe fruit lacks hydrolase activity (de Silva *et al.*, 1994; Arrowsmith and de Silva, 1995; Faik *et al.*, 1998). This activity could not bring about a net depolymerization of xyloglucan by transglycosylation unless xyloglucan oligosaccharides are present, but could still rearrange xyloglucan crosslinks in the wall and contribute to wall loosening. This XET activity could also play a maintenance role during ripening, by the incorporation of new xyloglucan into the wall.

Expansin

Not all cell wall-modifying enzymes act by cleavage of covalent glycosidic or ester bonds. Expansins are cell wall localized proteins that were originally identified by their ability to cause cell wall loosening in *in vitro* assays (McQueen-Mason *et al.*, 1992). Purified expansins lack detectable hydrolase or transglycosylase activity, and they do not bring about observable depolymerization of CMC, cell wall matrix glycans or pectins (McQueen-Mason *et al.*, 1992, 1993; McQueen-Mason and Cosgrove, 1994, 1995; see Darley *et al.*, 2001, this issue). However, without detectable hydrolysis they can bring about the mechanical weakening of paper, which is essentially pure cellulose held together by hydrogen bonding and physical entanglement between cellulose microfibrils (McQueen-Mason and Cosgrove, 1994). Expansins bind weakly to crystalline cellulose *in vitro* but more strongly to cellulose coated with matrix glycans. They probably act by causing a reversible disruption of hydrogen bonding between cellulose microfibrils and matrix polysaccharides, particularly xyloglucan, resulting in a loosening of the wall and allowing the turgor-driven slippage of microfibrils relative to one another (McQueen-Mason and Cosgrove, 1995; Cosgrove, 2000; Whitney *et al.*, 2000).

Expansins are small proteins, synthesized as primary translation products of about 240–270 amino acids which includes a signal sequence of 15–25 amino acids (Shcherban *et al.*, 1995). The mature proteins are 25–27 kDa and, in dicots, usually lack glycosylation. Expansin protein sequences are highly conserved, and possess amino acid residues with spac-

ings similar to those of other carbohydrate-binding proteins (Shcherban *et al.*, 1995). In the amino-terminal half of the protein are eight conserved cysteine residues similar to those of the chitin-binding domain of wheat germ agglutinin, which presumably are involved in disulfide bridges important for protein conformation. Near the carboxyl terminus of the protein are four conserved tryptophan residues similar to microbial cellulose binding domains, which presumably are involved in protein-polysaccharide binding. There is also an amino acid domain with some similarity to the active site of a microbial EGase, but although barely measurable EGase activity has been detected for some expansins, it seems unlikely that expansin action involves hydrolysis (Cosgrove, 1999, 2000). Two types of expansins have been described, both possessing wall-loosening activity. α-Expansins, as described above, are the major expansins in dicots but are also present in monocots (Cosgrove, 1999, 2000). β-Expansins are glycoproteins of ca. 30 kDa that are divergent from but structurally related to α-expansins and are found at high levels in the pollen of graminaceous monocots, but also have vegetative homologues in dicots (Cosgrove *et al.*, 1997). It seems likely that α- and β-expansins act on different components of the wall (Cosgrove *et al.*, 1997). Expansins are encoded by large gene families, consisting of at least 25 genes in *Arabidopsis* (Cosgrove, 2000; The Arabidopsis Genome Initiative, 2000).

During the growth and maturation of green tomato fruit, at least six expansin genes show staggered and overlapping periods of mRNA accumulation (Brummell *et al.*, 1999a; Catalá *et al.*, 2000). mRNAs of *LeExp*2 and *LeExp*4 are present only during the earliest stages of fruit development at the time at which cell expansion is most rapid. mRNAs of *LeExp*5 and *LeExp*6 also accumulate early in development but reach highest levels slightly later, at the immature and mature green stages when cell expansion is slowing in rate. mRNA of *LeExp*7 is detected at low levels transiently in young green fruit, whereas mRNA of *LeExp*3 is present throughout green fruit development and persists into ripening. At the onset of ripening mRNA and protein of *LeExp*1 accumulate to high levels, and are present at high abundance through to the over-ripe stage (Rose *et al.*, 1997, 2000; Brummell *et al.*, 1999a, c). In ripening fruit mRNAs of at least three expansin genes are present, mRNA of *LeExp*5 declining to undetectable levels very early in ripening, mRNA of *LeExp*3 declining slowly in abundance during ripening, and *LeExp*1 being present at high levels throughout ripening (Rose *et al.*, 1997; Brummell *et al.*, 1999a). The most abundant expansin mRNA in ripening tomato fruit is thus that of the ripening-related *LeExp*1. Expansins expressed in green fruit are presumably involved in cell expansion, whereas ripening tomato fruit do not show any significant increase in size, and ripening-related expansin protein is presumably involved in other aspects of cell wall modification. Expansion-related and ripening-related expansin proteins are immunologically distinct (Rose *et al.*, 2000), suggesting that these proteins are divergent in structure and possibly function. Expansin proteins immunologically related to tomato LeExp1 have been found in ripening fruit of a range of species, suggesting that accumulation of ripening-related expansin protein is a common feature of fruit ripening (Civello *et al.*, 1999; Rose *et al.*, 2000).

In addition to green fruit, tomato *LeExp*2 is also expressed in hypocotyl, where its mRNA abundance is positively regulated by auxin, suggesting that *LeExp*2 mRNA abundance in green fruit may also be regulated by auxin (Catalá *et al.*, 2000). In contrast, ripening-related *LeExp*1 mRNA abundance is positively regulated by ethylene in ripening tomato fruit, since it is strongly diminished by the ethylene antagonist 2,4-norbornadiene, is increased by ethylene treatment of mature green ACC synthase-antisense fruit lacking endogenous ethylene production, and is barely detectable in *rin* and *nor* (Rose *et al.*, 1997). In strawberry, which is non-climacteric, accumulation of *FaExp*2 mRNA begins at the white stage just as softening begins, and remains at high levels during ripening (Civello *et al.*, 1999). Although several ripening-related genes in strawberry are negatively regulated by auxin in young fruit, including *FaCel*1 (Harpster *et al.*, 1998), abundance of *FaExp*2 mRNA is not affected by auxin treatment (Civello *et al.*, 1999), suggesting that expression of this gene is controlled either developmentally or by other hormones.

Suppression and over-expression of *LeExp*1 mRNA and protein accumulation during ripening by constitutive expression of sense *LeExp*1 transgenes resulted in altered fruit softening during ripening, and in complex changes in cell wall polysaccharide metabolism (Brummell *et al.*, 1999c). Suppression of LeExp1 protein in ripening fruit to barely detectable levels, ca. 3% of wild-type, reduced fruit softening during ripening by 15–20% and prevented the final stages of polyuronide depolymerization, but did not affect matrix glycan depolymerization. Conversely, over-expression of LeExp1 protein to 3-fold of wild-type

abundance did not affect polyuronide depolymerization but substantially increased fruit softening and the depolymerization of matrix glycans, including xyloglucan, tightly bound to cellulose, particularly in green fruit. As control fruit ripened, their firmness and matrix glycan molecular weight profile became more similar to the overexpressing transgenic fruit, suggesting that in green fruit LeExp1 protein prematurely brings about changes that occur slowly during ripening in wild-type fruit.

The increased firmness of LeExp1-suppressed fruit was probably not due to large alterations in matrix glycan depolymerization, since suppression of LeExp1 protein accumulation did not detectably alter the molecular weight profile of matrix glycans during ripening. Thus, the observed reduction in fruit softening was probably due predominantly to some aspect of wall loosening caused directly by LeExp1, rather than indirectly through modification of hydrolase activity. Whether this wall relaxation is a separation of load-bearing polymers held together by non-covalent bonds, or involves some matrix glycan depolymerization, or both, is not known. Matrix glycan changes limited to a small fraction of the total molecules, or occurring in minor matrix glycan components, or in cellulose, cannot be eliminated. Over-expression in green fruit shows that LeExp1 protein can bring about matrix glycan disassembly, perhaps by acting at the microfibril/matrix glycan interface to expose matrix glycan substrate sites to hydrolases which otherwise are not able to attack them. LeExp1-mediated wall loosening also appears to be necessary for the final component of polyuronide depolymerization. Since there is no evidence that expansin itself causes pectin depolymerization, effects on pectin disassembly must be due to indirect factors. The results are consistent with the suggestion (Watson *et al.*, 1994; Chun and Huber, 1997) that there are two populations of pectins in the wall, the first being solubilized or loosely bound pectins that are degraded early in ripening, and the second being recalcitrant more tightly bound pectins that are normally degraded late in ripening. Perhaps LeExp1-mediated wall loosening is necessary to allow pectinase enzymes, such as PG, access to these latter polyuronides later in ripening. The reduction in polyuronide depolymerization caused indirectly by suppression of LeExp1 (Brummell *et al.*, 1999c) was greater than that caused directly by suppression of PG itself (Brummell and Labavitch, 1997). Since in PG-suppressed fruit reduced pectin breakdown is correlated with slightly reduced fruit softening (Langley

330

et al., 1994), the more attenuated polyuronide depolymerization seen in LeExp1-suppressed fruit probably also contributes to some extent to reduced softening, particularly late in ripening.

These data show that matrix glycan depolymerization is an important contributor to fruit softening, but the precise mechanism of expansin action cannot yet be elucidated. In green fruit, the presence of high levels of LeExp1 protein can potentiate matrix glycan disassembly similar to that which occurs during ripening, but in ripening fruit matrix glycan depolymerization was not detectably altered by suppression of LeExp1 protein to low levels, indicating that matrix glycan depolymerization during ripening requires either very small amounts or LeExp1 protein or could be independent of it. Whatever the nature of the changes, the results indicate key roles for LeExp1 in the cell wall changes leading to fruit softening, with a direct effect on cell wall loosening and indirect effects on pectin depolymerization and possibly on matrix glycan depolymerization. The enhanced firmness of LeExp1-suppressed fruit was detectable throughout ripening, even at the over ripe stage. Postharvest evaluation of LeExp1-suppressed fruit relative to azygous controls found no significant difference in fruit size, but fruit shelf life and the viscosity of paste derived from ripe fruit were significantly increased (Brummell, Howie, Ma and Dunsmuir, unpublished).

Suppression of *LeExp*1 mRNA accumulation by an *LeExp*1 transgene did not affect the mRNA accumulation of three related expansin genes expressed in green fruit (Figure 4), which are potentially involved in cell expansion and fruit growth. The coding sequences of *LeExp*3, *LeExp*4 and *LeExp*5 are 54%, 71% and 62% identical in DNA sequence to *LeExp*1, respectively, although regions of much higher identity exist. However, in these same fruit mRNA accumulation of *PG*, which is unrelated in sequence to *LeExp*1, was reduced throughout ripening and was almost 5-fold less in red ripe fruit (Figure 5). This suggests that that some sort of feedback regulation may be occurring, and that the reduced access of PG to its substrate in LeExp1-suppressed fruit is perhaps causing accumulation of PG protein at the wrong place in the wall, which brings about reductions in PG secretion, synthesis and ultimately mRNA abundance. A complex relationship between the final demand for a cell wall protein and its synthesis may thus exist, although further characterization of this interesting result is necessary. Such a small reduction in PG mRNA abundance could not explain reduced polyuronide depolymerization directly,

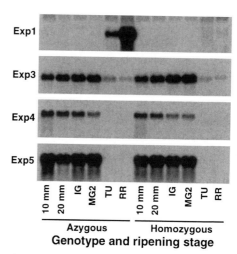

Figure 4. Accumulation of mRNA of three tomato expansin gene family members in fruit of azygous controls and in fruit suppressed in *LeExp*1 mRNA accumulation by constitutive expression of an *LeExp*1 transgene. Total RNA (10 μg per lane) was separated in a denaturing gel, blotted to nylon membranes and four identical blots probed with 3'-untranslated regions of each gene labeled by PCR. Final wash was at 16 °C below T_m. These probes used at this stringency were specific for the genes indicated (Brummell *et al.*, 1999a). Abbreviations of ripening stages: 10 mm, fruit 10 mm in diameter; 20 mm, fruit 20 mm in diameter; IG, immature green; MG2, mature green stage 2; TU, turning (10–30% light red); RR, red ripe.

since PG is present in large excess in tomato and even a reduction to only 0.5% of wild-type activity has little effect on pectin breakdown (Brummell and Labavitch, 1997).

Conclusions

Although there is obviously much that remains unknown, it has become clear that the disassembly of the cell wall during fruit ripening is not a haphazard breakdown of the wall structure, but specific goals are accomplished by an ordered series of modifications brought about by a range of enzyme activities acting in a controlled way, often limited by or requiring the activity of other enzymes. Analysis of transgenic plants has helped elucidate the role in fruit softening or quality of enzymes involved in cell wall ripening changes, but in virtually every case has also resulted in additional, unexpected findings that are equally informative. The function of proteins involved in the modification of cell wall pectins has become relatively clear, but the role of enzymes involved in the modification of matrix glycans, and possibly cellulose, remains obscure.

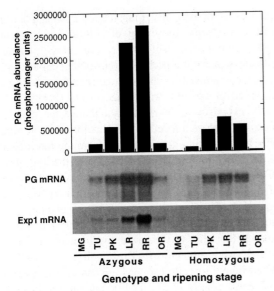

Figure 5. Accumulation of *PG* mRNA in ripening tomato fruit of azygous controls and in fruit suppressed in *LeExp*1 mRNA accumulation by constitutive expression of an *LeExp*1 transgene. Total RNA (5 μg per lane) was separated in a denaturing gel, blotted to nylon membranes and two identical blots probed with random-prime labeled cDNAs of *LeExp*1 or *PG*. Final wash was at 4 °C below T_m. The histogram shows relative *PG* mRNA abundance as quantified using a phosphorimager. Abbreviations of ripening stages: MG, mature green; TU, turning; PK, pink; LR, light red; RR, red ripe; OR, over-ripe.

Two of the earliest changes occurring in ripening are the loss of pectic galactan side-chains, and the solubilization of polyuronides. Treatment with purified β-galactosidases increases the solubility and decreases the degree of polymerization of avocado and papaya cell wall pectin (De Veau *et al.*, 1993; Ali *et al.*, 1998), and in potato tubers over-expression of a fungal endo-galactanase reduces galactan content to 30% of wild type and increases the accessibility of exogenous PME and PG to pectin substrates (Sorensen *et al.*, 2000). This suggests that during ripening the degradation of galactan side chains of RG I increases pectin solubilization, directly by affecting RG I solubility and indirectly by increasing the porosity of the wall and allowing increased access of PME and PG to homogalacturonan. In tomato, antisense suppression of β-galactanase activity has a substantial effect on fruit firmness and paste properties, implying that cell wall galactose loss early in ripening is important for subsequent firmness and textural changes in the fruit. The PG β-subunit protein may also be involved in limiting pectin solubilization, since its suppression results in increased extractability of polyuronides (Watson *et al.*,

1994; Chun and Huber, 1997). This may indirectly restrict PG action, whether the β-subunit binds to PG in the wall or not.

In peach and melon, PG activity increases relatively late in ripening, well after softening has begun (Pressey *et al.*, 1971; Orr and Brady, 1993; Hadfield *et al.*, 1998) and, correspondingly, depolymerization of pectins occurs late in ripening (Dawson *et al.*, 1992; Rose *et al.*, 1998). This is not the case, however, in avocado or tomato, where PG activity increases shortly after the onset of ripening (Awad and Young, 1979; Biggs and Handa, 1989), and although polyuronide depolymerization is most pronounced late in softening, it occurs throughout the ripening process (Huber and O'Donoghue, 1993). In tomato fruit, the extent of polyuronide breakdown is quite limited relative to avocado (Huber and O'Donoghue, 1993). In unripe avocado fruit, polyuronides are highly methylesterified and resistant to attack by PG but become highly depolymerized late in ripening, and the degree of pectin methylesterification appears to entirely explain the limitations to PG activity early in ripening (Wakabayashi *et al.*, 2000). During ripening in both species, the degree of pectin methlesterification is reduced by PME, and in tomato if this is prevented by suppression of PME activity, then polyuronide depolymerization is reduced (Tieman *et al.*, 1992). There may also be physical restrictions to PG activity. Transgenic fruit suppressed in β-subunit protein accumulation show biochemical and physical changes indicative of altered PG-mediated pectin degradation (Watson et al., 1994; Chun and Huber, 1997, 2000), and when tomato fruit PG is constitutively expressed in tobacco, PG protein is processed and secreted to the cell wall, and is active against tobacco pectin *in vitro*, but tobacco leaf cell wall polyuronide is not depolymerized (Osteryoung *et al.*, 1990). Homogenization of fruit (Brummell and Labavitch, 1997) or autolysis occurring in enzymically active cell wall preparations (Seymour *et al.*, 1987b; Huber and O'Donoghue, 1993) removes physical restrictions in tomato, resulting in the depolymerization of pectin to very small fragments. Furthermore, in whole fruit, the inhibition of pectin depolymerization caused indirectly by suppression of LeExp1 was greater than that caused directly by suppression of *PG* activity (Brummell and Labavitch, 1997; Brummell *et al.*, 1999c). Taken together, these observations suggest that access of PG to its substrate rather than the amount of PG activity is a major factor controlling pectin disassembly.

Much of the strength of the wall is due to the crosslinking of cellulose microfibrils by matrix glycans, particularly xyloglucan, and softening during ripening is associated with a progressive disassembly of this network. During cell growth, there is extensive depolymerization of xyloglucan and loss of xyloglucan fragments, combined with incorporation of new xyloglucan into the wall (Labavitch and Ray, 1974; Talbott and Ray, 1992). During fruit ripening, xyloglucan and other matrix glycans are depolymerized, but the total content of xyloglucan does not change appreciably, and xyloglucan depolymerization occurs to a limited extent with the average molecular weight range declining from 300–1000 to 100–300 kDa (as dextran equivalents) with little low molecular weight material detected even in over ripe fruit (Maclachlan and Brady, 1994; Brummell et al., 1999c). This suggests that ripening involves different changes to the cell wall than growth, consisting of limited xyloglucan depolymerization without large-scale solubilization of low-molecular-weight xyloglucan fragments. It is interesting that EGase purified from growing pea stem is quite highly active against xyloglucan (Hayashi et al., 1984), whereas EGase from avocado fruit is active at a much slower rate (Hatfield and Nevins, 1986; O'Donoghue and Huber, 1992). In pea stem, xyloglucan is present in three domains in the wall, the first susceptible to enzymatic attack, presumably the crosslinks between microfibrils, the second extractable by KOH, presumably bound to the surface of microfibrils, and the third releasable by microbial cellulase, presumably embedded in the microfibrils (Pauly et al., 1999). It is not known during ripening if the cuts in the backbone leading to xyloglucan depolymerization are predominantly in the spanning region, in the bound regions, or both.

Suppression of LeExp1 accumulation significantly reduces fruit softening, indicating a key role for this protein in the cell wall changes leading to softening. Expansion-related expansins are believed to function by loosening glucan-glucan hydrogen bonds between cellulose and xyloglucan, but act only when the wall is under tension (McQueen-Mason and Cosgrove, 1995). LeExp1 protein does not cause observable changes in the growth of vegetative tissues when overexpressed ectopically in transgenic plants (Brummell et al., 1999c), suggesting that ripening-related expansins have a slightly different mechanism of action than expansion-related expansins. However, ripening-related expansin may still act at the same site, the microfibril/matrix interface, with different consequences to the cell. Ripening-related expansin action may cause some loosening of cellulose-xyloglucan hydrogen bonds, causing wall relaxation without significant expansion, perhaps contributing along with increased apoplastic solutes to the reduction in turgor observed in ripening fruit (Shackel et al., 1991). In vitro, expansins synergistically enhance the depolymerization of cellulose by fungal cellulases, presumably by separating glucan chains from the microfibril surface and making them accessible to the hydrolase (Cosgrove et al., 1998). Similarly, overexpression of high levels of LeExp1 in green tomato fruit can potentiate depolymerization of matrix glycan, presumably mediated by EGases or other hydrolases, but whether this is the usual mechanism of action of LeExp1 is not known (Brummell et al., 1999c). When bound to cellulose, xyloglucan is inaccessible to EGases (Pauly et al., 1999), but the presence of ripening-related expansin may allow both loosening of glucan-xyloglucan hydrogen bonds and sufficient separation between the chains, perhaps coupled with a conformational change of the xyloglucan, to allow EGases to bind. In cells of ripening fruit, where turgor is declining, perhaps this action of ripening-related expansins is more important than the slippage of glucan-xyloglucan hydrogen bonds caused by expansion-related expansins in high-turgor expanding cells. Ripening-related expansin may also promote some hydrolysis of amorphous non-crystalline glucan chains by EGases, potentially destabilizing the microfibril/xyloglucan network if these are the chains to which xyloglucan are attached, but substantial depolymerization of cellulose during ripening of tomato has not been observed (Maclachlan and Brady, 1994). Cleavage of xyloglucan within the region bound to cellulose, resulting in shortened lengths of xyloglucan bound to cellulose, may result in a loosened wall but would weaken the wall less substantially than would extensive breakage of cross-linking regions. Perhaps the cleavage of xyloglucan cross links occurs slowly with ripening, limited by the low affinity of EGases for xyloglucan, which instead act preferentially, and thus earlier in ripening, on the xyloglucan bound to cellulose microfibrils. This may also be the site of action of the Type 3 EGases (described above) which possess a carbohydrate-binding domain (Trainotti et al., 1999b), and which in strawberry are expressed very early in ripening (Llop-Tous et al., 1999; Trainotti et al., 1999a). The role of XET activity in ripening is unknown, and it could contribute to xyloglucan depolymerization by transglycosylation using small xyloglucan molecules resulting from EGase action. XET could also modify the wall by rearrangements, or strengthen it by incorporation of new xyloglucan polymers since cell wall synthesis proceeds well into ripening (Mitcham et al., 1989; Greve and Labavitch, 1991).

The data described above show that the disassembly of tomato cell wall components during ripening is brought about by a range of enzyme activities, with the action of some enzymes being necessary for the action of others (Table 1). β-Galactosidase activity may be necessary to allow several other enzymes to reach their sites of action, PG activity is limited by PME and ripening-related expansin action, ripening related expansin activity may be required for EGase activity, EGase activity may be necessary to produce xyloglucan fragments for XET, XET may incorporate new xyloglucan into the wall if it is available, or rearrange xyloglucan crosslinks when it is not, and so forth. Cell wall pectins are thought to control the pore size within the wall and limit the movement of large molecules (Baron-Epel et al., 1988), but late in ripening after the actions of β-galactosidase, PME and PG have depolymerized and solubilized pectins, loosening of the cellulose/matrix glycan network mediated by ripening-related expansin appears to control access of pectinase to polyuronide. The picture that emerges is of a range of different enzymes, whose expression is regulated both in time and amount during ripening, acting together in an interdependant way to achieve controlled changes in softening and texture. In tomato, suppression of β-galactosidase and ripening-related expansin activity had substantial effects on fruit firmness, suggesting that the changes brought about by these cell wall-modifying proteins are key steps in the softening process, perhaps each controlling a series of other changes. Suppression of PG and PME activity had only small effects on softening during ripening, but caused substantial improvements in paste viscosity, which is believed to reflect fruit texture. Suppression of the β-subunit actually increased fruit softening, apparently largely though reducing cell-to-cell adhesion, which is a major contributor to fruit texture. In green tomato fruit, over-expression of ripening-related expansin brought about precocious matrix glycan depolymerization and fruit softening. These findings indicate that cell wall pectin metabolism may contribute mainly to textural changes, and modifications to the cellulose/matrix glycan network mainly to softening, although each obviously affects the other.

However, it must be remembered that observations on one species, such as tomato, may not necessarily extend to other species. Polyuronides are depolymerized to a very small size during ripening in avocado (Huber and O'Donoghue, 1993; Wakabayashi et al., 2000), but in strawberry and banana little if any polyuronide depolymerization is evident (Huber, 1984; Wade et al., 1992). Matrix glycans become highly depolymerized in strawberry, but not in avocado (Huber, 1984; Huber and O'Donoghue, 1993). Non-crystalline cellulose disappears during ripening in avocado (O'Donoghue et al., 1994), but in tomato little cellulose depolymerization is observed (Maclachlan and Brady, 1994). Galactose loss from cell walls was noted to varying extents in numerous species, but not in plum or cucumber (Gross and Sams, 1984). Differences in cell wall modification have even been noted within a species, between different tomato cultivars (Carrington and Pressey, 1996; Blumer et al., 2000). Although some processes may be fundamental to firmness decrease, the highly variable softening behavior and textural changes of fruit from different species are thus due in part to differences in the relative extent of the various ripening-related cell wall changes that occur, coupled with differences in initial cell wall composition. The wide diversity in resulting cell wall physicochemical properties has been correlated with final fruit texture (Redgwell et al., 1997). Consequently, it is likely that attempts to modify softening and texture in different species may require the manipulation of different enzyme activities.

Future prospects

Attempts to control fruit softening and quality by suppression of a single cell-wall-modifying protein may have been excessively optimistic. Cell-wall-modifying proteins are usually present in plants in large gene families, in excess of 20 genes, of which many are expressed in fruit development, and several during ripening. Fruit softening is a multi-gene trait with each enzyme activity having its own role to play in softening and textural changes, and suppression of multiple isoforms of particular activities or genes of several different types may be necessary in order to exert commercially significant effects. However, cell wall metabolism during ripening cannot be completely prevented, even if that were possible, because a fruit must retain sufficient softening and textural changes to make it desirable to eat. Analysis of transgenic plants

Table 1. Interdependence of cell wall-localized enzyme activities involved in cell wall modification during ripening of tomato fruit.

Enzyme	Requirements for activity
β-Galactosidase	none?
PME	increase in wall porosity by β-galactosidase?
PG	increase in wall porosity by β-galactosidase?
	de-methylesterification of pectin by PME
	wall loosening by ripening-related expansin
	depletion of PG β-subunit?
Expansin	none?
EGase	increase in wall porosity by β-galactosidase and PG?
	ripening-related expansin action?
	removal of xyloglucan side chains by glycosidases?
XET	increase in wall porosity by β-galactosidase and PG?
	availability of newly synthesized xyloglucan
	xyloglucan fragments from EGase action

with altered expression of cell wall-modifying proteins has provided much insight into cell wall changes occurring during ripening, but the field is still young. Some activities remain undiscovered, and the enzymes which depolymerize xyloglucan and non-xyloglucan matrix glycans during ripening are still not known for certain. In avocado, a cell wall extract depleted of CMCase (EGase) activity brought about xyloglucan depolymerization (O'Donoghue and Huber, 1992), but the activity responsible has not been identified. There are also many enzymes present in fruit whose activity or sequence is described but whose function in ripening has not yet been investigated, such as numerous glycosidases (Wallner and Walker, 1975; Pharr *et al.*, 1976), pectate lyase in strawberry and banana (Medina-Escobar *et al.*, 1997; Dominguez-Puigjaner *et al.*, 1997), EGases with a carbohydrate-binding domain (Trainotti *et al.*, 1999b), and endomannanase (Bewley *et al.*, 2000). Modified expression of enzymes of different types, either by the introduction of multiple transgenes or by suppression of regulatory proteins such as transcription factors, may prove to be the way to precisely control changes in fruit softening behavior.

Acknowledgements

We thank Kiyozo Asada, Jacqueline de Silva, Ken Gross, Graham Seymour and David Smith for communicating data in advance of publication, and Diane Burgess and Paul Oeller for thoughtful comments on the manuscript. We apologize to numerous colleagues whose work could not be cited due to space constraints, particularly those who work on species other than tomato. Preparation of this article and part of the original work described herein were supported by Seminis Vegetable Seeds.

References

Ali, Z.M. and Brady, C.J. 1982. Purification and characterization of the polygalacturonases of tomato fruits. Aust. J. Plant Physiol. 9: 155–169.

Ali, Z.M., Ng, S.-Y., Othman, R., Goh L.-Y. and Lazan, H. 1998. Isolation, characterization and significance of papaya β-galactanases to cell wall modification and fruit softening during ripening. Physiol. Plant. 104: 105–115.

Almeida, D.P.F. and Huber, D.J. 1999. Apoplastic pH and inorganic ion levels in tomato fruit: a potential means for regulation of cell wall metabolism during ripening. Physiol. Plant. 105: 506–512.

Arrowsmith, D.A. and de Silva, J. 1995. Characterisation of two tomato fruit-expressed cDNAs encoding xyloglucan *endo*-transglycosylase. Plant Mol. Biol. 28: 391–403.

Asada, K., Ohba, T., Takahashi, S. and Kato, I. 1999. Alteration of fruit characteristics in transgenic tomatoes with modified gene expression of endo-xyloglucan transferase. HortScience 34: 533.

Awad, M. and Young, R.E. 1979. Postharvest variation in cellulase, polygalacturonase, and pectinmethylesterase in avocado (*Persea americana* Mill cv. Fuerte) fruits in relation to respiration and ethylene production. Plant Physiol. 64: 306–308.

Baron-Epel, O., Gharyal, P.K. and Schindler, M. 1988. Pectins as mediators of wall porosity in soybean cells. Planta 175: 389–395.

Ben-Arie, R., Kislev, N. and Frenkel, C. 1979. Ultrastructural changes in the cell walls of ripening apple and pear fruit. Plant Physiol. 64: 197–202.

Bewley, J.D., Banik, M., Bourgault, R., Feurtado, J.A., Toorop, P. and Hilhorst, H.W.M. 2000. Endo-β-mannanase activity increases in the skin and outer pericarp of tomato fruits during ripening. J. Exp. Bot. 51: 529–538.

Biggs, M.S. and Handa, A.K. 1989. Temporal regulation of polygalacturonase gene expression in fruits of normal, mutant, and heterozygous tomato genotypes. Plant Physiol. 89: 117–125.

Bird, C.R., Smith, C.J.S., Ray, J.A., Moureau, P., Bevan, M.W., Bird, A.S., Hughes, S., Morris, P.C., Grierson, D. and Schuch, W. 1988. The tomato polygalacturonase gene and ripening-specific expression in transgenic plants. Plant Mol. Biol. 11: 651–662.

Blumer, J.M., Clay, R.P., Bergmann, C.W., Albersheim, P. and Darvill, A.G. 2000. Characterization of changes in pectin methylesterase expression and pectin esterification during tomato fruit ripening. Can. J. Bot. 78: 607–618.

Brady, C.J., MacAlpine, G., McGlasson, W.B. and Ueda, Y. 1982. Polygalacturonase in tomato fruits and the induction of ripening. Aust. J. Plant Physiol. 9: 171–178.

Brady, C.J., McGlasson, W.B., Pearson, J.A., Meldrum, S.K. and Kopeliovitch, E. 1985. Interactions between the amount and molecular forms of polygalacturonase, calcium, and firmness in tomato fruit. J. Am. Soc. Hort. Sci. 110: 254–258.

Brummell, D.A. and Labavitch, J.M. 1997. Effect of antisense suppression of endopolygalacturonase activity on polyuronide molecular weight in ripening tomato fruit and in fruit homogenates. Plant Physiol. 115: 717–725.

Brummell, D.A., Lashbrook, C.C. and Bennett, A.B. 1994. Plant endo-1,4-β-D-glucanases: structure, properties and physiological function. In: M.E. Himmel, J.O. Baker and R.P. Overend (Eds.) Enzymatic Conversion of Biomass for Fuels Production (American Chemical Society Symposium Series 566), American Chemical Society, pp. 100–129.

Brummell, D.A., Bird, C.R., Schuch, W. and Bennett, A.B. 1997a. An endo-1,4-β-glucanase expressed at high levels in rapidly expanding tissues. Plant Mol. Biol. 33: 87–95.

Brummell, D.A., Catalá, C., Lashbrook, C.C. and Bennett, A.B. 1997b. A membrane-anchored E-type endo-1,4-β-glucanase is localized on Golgi and plasma membranes of higher plants. Proc. Natl. Acad. Sci. USA 94: 4794–4799.

Brummell, D.A., Harpster, M.H. and Dunsmuir, P. 1999a. Differential expression of expansin gene family members during growth and ripening of tomato fruit. Plant Mol. Biol. 39: 161–169.

Brummell, D.A., Hall, B.D. and Bennett, A.B. 1999b. Antisense suppression of tomato endo-1,4-β-glucanase Cel2 mRNA accumulation increases the force required to break fruit abscission zones but does not affect fruit softening. Plant Mol. Biol. 40: 615–622.

Brummell, D.A., Harpster, M.H., Civello, P.M., Palys, J.M., Bennett, A.B. and Dunsmuir, P. 1999c. Modification of expansin protein abundance in tomato fruit alters softening and cell wall polymer metabolism during ripening. Plant Cell 11: 2203–2216.

Buescher, R.W. and Hobson, G.E. 1982. Role of calcium and chelating agents in regulating the degradation of tomato fruit tissue by polygalacturonase. J. Food Biochem. 6: 147–160.

Buescher, R.W. and Tigchelaar, E.C. 1975. Pectinesterase, polygalacturonase, Cx-cellulase activities and softening of the *rin* tomato mutant. HortScience 10: 624–625.

Campbell, P. and Braam, J. 1999a. *In vitro* activities of four xyloglucan endotransglycosylases from *Arabidopsis*. Plant J. 18: 371–382.

Campbell, P. and Braam, J. 1999b. Xyloglucan endotransglycosylases: diversity of genes, enzymes and potential wall-modifying functions. Trends Plant Sci. 4: 361–366.

Carey, A.T., Holt, K., Picard, S., Wilde, R., Tucker, G.A., Bird, C.R., Schuch, W. and Seymour, G.B. 1995. Tomato exo-

(1→4)β-D-galactanase. Isolation, changes during ripening in normal and mutant tomato fruit, and characterization of a related cDNA clone. Plant Physiol. 108: 1099–1107.

Carey, A.T., Smith, D.L., Harrison, E., Bird, C.R., Gross, K.C., Seymour, G.B. and Tucker, G.A. 2001. Down-regulation of a ripening-related β-galactosidase gene (TBG1) in transgenic tomato fruits. J. Exp. Bot. 52, 663–668.

Carpita, N.C. and Gibeaut, D.M. 1993. Structural models of primary cell walls in flowering plants: consistency of molecular structure with the physical properties of the walls during growth. Plant J. 3: 1–30.

Carrington, C.M.S. and Pressey, R. 1996. β-Galactosidase II activity in relation to changes in cell wall galactosyl composition during tomato ripening. J. Am. Soc. Hort. Sci. 121: 132–136.

Carrington, C.M.S., Greve, L.C. and Labavitch, J.M. 1993. Cell wall metabolism in ripening fruit. VI. Effect of the antisense polygalacturonase gene on cell wall changes accompanying ripening in transgenic tomatoes. Plant Physiol. 103: 429–434.

Catalá, C. and Bennett, A.B. 1998. Cloning and sequence analysis of Tomcel8, a new plant endo-β-1,4-D-glucanase gene encoding a protein with a putative carbohydrate-binding domain (accession no. AF098292). Plant Physiol. 118: 1535.

Catalá, C., Rose, J.K.C. and Bennett, A.B. 1997. Auxin regulation and spatial localization of an endo-1,4-β-D-glucanase and a xyloglucan endotransglycosylase in expanding tomato hypocotyls. Plant J. 12: 417–426.

Catalá, C., Rose, J.K.C. and Bennett, A.B. 2000. Auxin-regulated genes encoding cell wall-modifying proteins are expressed during early tomato fruit growth. Plant Physiol. 122: 527–534.

Christoffersen, R.E., Tucker, M.L. and Laties, G.G. 1984. Cellulase gene expression in ripening avocado fruit: the accumulation of cellulase mRNA and protein as demonstrated by cDNA hybridization and immunodetection. Plant Mol. Biol. 3: 385–391.

Chun, J.-P. and Huber, D.J. 1997. Polygalacturonase isozyme 2 binding and catalysis in cell walls from tomato fruit: pH and β-subunit effects. Physiol. Plant. 101: 283–290.

Chun, J.-P. and Huber, D.J. 1998. Polygalacturonase-mediated solubilization and depolymerization of pectic polymers in tomato fruit cell walls. Regulation by pH and ionic conditions. Plant Physiol. 117: 1293–1299.

Chun, J.-P. and Huber, D.J. 2000. Reduced levels of β-subunit protein influence tomato fruit firmness, cell-wall ultrastructure, and PG2-mediated pectin hydrolysis in excised pericarp tissue. J. Plant Physiol. 157: 153–160.

Civello, P.M., Powell, A.L.T., Sabehat, A. and Bennett, A.B. 1999. An expansin gene expressed in ripening strawberry fruit. Plant Physiol. 121: 1273–1279.

Cooley, M.B. and Yoder, J.I. 1998. Insertional inactivation of the tomato polygalacturonase gene. Plant Mol. Biol. 38: 521–530.

Cosgrove, D.J. 1999. Enzymes and other agents that enhance cell wall extensibility. Annu. Rev. Plant Physiol. Plant Mol. Biol. 50: 391–417.

Cosgrove, D.J. 2000. Loosening of plant cell walls by expansins. Nature 407: 321–326.

Cosgrove, D.J., Bedinger, P. and Durachko, D.M. 1997. Group I allergens of grass pollen as cell wall-loosening agents. Proc. Natl. Acad. Sci. USA 94: 6559–6564.

Cosgrove, D.J., Durachko, D.M. and Li, L.-C. 1998. Expansins have cryptic endoglucanase activity and can synergize the breakdown of cellulose by fungal cellulases. Annu. Meet. Am. Soc. Plant Physiol., Abstract 171.

Crookes, P.R. and Grierson, D. 1983. Ultrastructure of tomato fruit ripening and the role of polygalacturonase isozymes in cell wall degradation. Plant Physiol. 72: 1088–1093.

336

Dawson, D.M., Melton, L.D. and Watkins, C.B. 1992. Cell wall changes in nectarines (*Prunus persica*). Solubilization and depolymerization of pectic and neutral polymers during ripening and in mealy fruit. Plant Physiol. 100: 1203–1210 (and correction in Plant Pshysiol. 102: 1062–1063).

Darley, C.P., Forrester, A.M. and McQueen-Mason, S.J. 2001. The molecular basis of plant cell wall extension. Plant Mol. Biol., this issue.

de Silva, J. and Verhoeven, M.E. 1998. Production and characterisation of antisense-exogalactanase tomatoes. In: Report of the Demonstration Programme on Food Safety Evaluation of Genetically Modified Foods as a Basis for Market Introduction. Netherlands Ministry of Economic Affairs, The Hague, pp. 99–106.

de Silva, J., Jarman, C.D., Arrowsmith, D.A., Stronach, M.S., Chengappa, S., Sidebottom, C. and Reid, J.S.G. 1993. Molecular characterization of a xyloglucan-specific endo-(1→4)β-D-glucanase (xyloglucan endotransglycosylase) from nasturtium seeds. Plant J. 3: 701–711.

de Silva, J., Arrowsmith, D., Hellyer, A., Whiteman, S. and Robinson, S. 1994. Xyloglucan endotransglycosylase and plant growth. J. Exp. Bot. 45: 1693–1701.

De Veau, E.J.I., Gross, K.C., Huber, D.J. and Watada, A.E. 1993. Degradation and solubilization of pectin by β-galactosidases purified from avocado mesocarp. Physiol. Plant. 87: 279–285.

del Campillo, E. and Bennett, A.B. 1996. Pedicel breakstrength and cellulase gene expression during tomato flower abscission. Plant Physiol. 111: 813–820.

Delincee, H. 1976. Thin-layer isoelectric focusing of multiple forms of tomato pectinesterase. Phytochemistry 15: 903–906.

Della Penna, D. and Bennett, A., B. 1988. *In vitro* synthesis and processing of tomato fruit polygalacturonase. Plant Physiol. 86: 1057–1063.

Della Penna, D., Alexander, D.C. and Bennett, A.B. 1986. Molecular cloning of tomato fruit polygalacturonase: analysis of polygalacturonase mRNA levels during ripening. Proc. Natl. Acad. Sci. USA 83: 6420–6424.

Della Penna, D., Kates, D.S. and Bennett, A.B. 1987. Polygalacturonase gene expression in Rutgers, *rin*, *nor*, and *Nr* tomato fruits. Plant Physiol. 85: 502–507.

Della Penna, D., Lincoln, J.E., Fischer, R.L. and Bennett, A.B. 1989. Transcriptional analysis of polygalacturonase and other ripening associated genes in Rutgers, *rin*, *nor*, and *Nr* tomato fruit. Plant Physiol. 90: 1372–1377.

Della Penna, D., Lashbrook, C.C., Toenjes, K., Giovannoni, J.J., Fischer, R.L. and Bennett, A.B. 1990. Polygalacturonase isozymes and pectin depolymerization in transgenic *rin* tomato fruit. Plant Physiol. 94: 1882–1886.

Dominguez-Puigjaner, E., Llop I., Vendrell, M. and Prat, S. 1997. A cDNA clone highly expressed in ripe banana fruit shows homology to pectate lyases. Plant Physiol 114: 1071–1076.

Faik, A., Desveaux, D. and Maclachlan, G. 1998. Enzymic activities responsible for xyloglucan depolymerization in extracts of developing tomato fruit. Phytochemistry 49: 365–376.

Fanutti, C., Gidley, M.J. and Reid, J.S.G. 1993. Action of a pure xyloglucan endo-transglycosylase (formerly called xyloglucan-specific endo-(1→4)β-D-glucanase) from the cotyledons of germinated nasturtium seeds. Plant J. 3: 691–700.

Fenwick, K.M., Jarvis, M.C., Apperley, D.C., Seymour, G.B. and Bird, C.R. 1996. Polymer mobility in cell walls of transgenic tomatoes with reduced polygalacturonase activity. Phytochemistry 42: 301–307.

Ferrarese, L., Trainotti, L., Moretto, P., Polverino de Laureto, P., Rascio, N. and Casadoro, G. 1995. Differential ethylene-inducible expression of cellulase in pepper plants. Plant Mol. Biol. 29: 735–747.

Fry, S.C., Smith, R.C., Renwick, K.F., Martin, D.J., Hodge, S.K. and Matthews, K.J. 1992. Xyloglucan endotransglycosylase, a new wall-loosening enzyme activity from plants. Biochem. J. 282: 821–828.

Gaffe, J., Tieman, D.M. and Handa, A.K. 1994. Pectin methylesterase isoforms in tomato (*Lycopersicon esculentum*) tissues. Plant Physiol. 105: 199–203.

Gaffe, J., Tiznado, M.E. and Handa, A.K. 1997. Characterization and functional expression of a ubiquitously expressed tomato pectin methylesterase. Plant Physiol. 114: 1547–1556.

Giovannoni, J.J., Della Penna, D., Bennett, A.B. and Fischer, R.L. 1989. Expression of a chimeric polygalacturonase gene in transgenic *rin* (ripening inhibitor) tomato fruit results in polyuronide degradation but not fruit softening. Plant Cell 1: 53–63.

Gonzalez-Bosch, C., Brummell, D.A. and Bennett, A.B. 1996. Differential expression of two endo-1,4-β-glucanase genes in pericarp and locules of wild-type and mutant tomato fruit. Plant Physiol. 111: 1313–1319.

Greve, L.C. and Labavitch, J.M. 1991. Cell wall metabolism in ripening fruit. V. Analysis of cell wall synthesis in ripening tomato pericarp tissue using a D-[U-^{13}C]glucose tracer and gas chromatography-mass spectrometry. Plant Physiol. 97: 1456–1461.

Grierson, D., Tucker, G.A., Keen, J., Ray, J., Bird, C.R. and Schuch, W. 1986. Sequencing and identification of a cDNA clone for tomato polygalacturonase. Nucl. Acids Res. 14: 8595–8603.

Gross, K.C. 1983. Changes in free galactose, *myo*-inositol and other monosaccharides in normal and non-ripening mutant tomatoes. Phytochemistry 22: 1137–1139.

Gross, K.C. 1984. Fractionation and partial characterization of cell walls from normal and non-ripening mutant tomato fruit. Physiol. Plant. 62: 25–32.

Gross, K.C. and Sams, C.E. 1984. Changes in cell wall neutral sugar composition during fruit ripening: a species survey. Phytochemistry 23: 2457–2461.

Gross, K.C. and Wallner, S.J. 1979. Degradation of cell wall polysaccharides during tomato fruit ripening. Plant Physiol. 63: 117–120.

Gross, K.C., Watada, A.E., Kang, M.S., Kim, S.D., Kim, K.S. and Lee, S.W. 1986. Biochemical changes associated with the ripening of hot pepper fruit. Physiol. Plant. 66: 31–36.

Hadfield, K.A. and Bennett, A.B. 1998. Polygalacturonases: many genes in search of a function. Plant Physiol. 117: 337–343.

Hadfield, K.A., Rose, J.K.C., Yaver, D.S., Berka, R.M. and Bennett, A.B. 1998. Polygalacturonase gene expression in ripe melon fruit supports a role for polygalacturonase in ripening-associated pectin disassembly. Plant Physiol 117: 363–373.

Hall, C.B. 1964. Cellulase activity in tomato fruits according to portion and maturity. Bot. Gaz. 125: 156–157.

Hall, L.N., Tucker, G.A., Smith, C.J.S., Watson, C.F., Seymour, G.B., Bundick, Y., Boniwell, J.M., Fletcher, J.D., Ray, J.A., Schuch, W., Bird, C. and Grierson, D. 1993. Antisense inhibition of pectin esterase gene expression in transgenic tomatoes. Plant J. 3: 121–129.

Hall, L.N., Bird, C.R., Picton, S., Tucker, G.A., Seymour, G.B. and Grierson, D. 1994. Molecular characterisation of cDNA clones representing pectin esterase isozymes from tomato. Plant Mol. Biol. 25: 313–318.

Hamilton, A.J., Lycett, G.W. and Grierson, D. 1990. Antisense gene that inhibits synthesis of the hormone ethylene in transgenic plants. Nature 346: 284–287.

Harker, F.R., Redgwell, R.J., Hallett, I.C., Murray, S.H. and Carter, G. 1997. Texture of fresh fruit. Hort. Rev. 20: 121–224.

Harpster, M.H., Lee, K.Y. and Dunsmuir, P. 1997. Isolation and characterization of a gene encoding endo-β-1,4-glucanase from pepper (*Capsicum annuum* L.). Plant Mol. Biol. 33: 47–59.

Harpster, M.H., Brummell, D.A. and Dunsmuir, P. 1998. Expression analysis of a ripening-specific, auxin-repressed endo-1,4-β-glucanase gene in strawberry. Plant Physiol. 118: 1307–1316.

Harriman, R.W., Tieman, D.M. and Handa, A.K. 1991. Molecular cloning of tomato pectin methylesterase gene and its expression in Rutgers, ripening inhibitor, nonripening, and Never Ripe tomato fruits. Plant Physiol. 97: 80–87.

Hatfield, R. and Nevins, D.J. 1986. Characterization of the hydrolytic activity of avocado cellulase. Plant Cell Physiol. 27: 541–552.

Hayashi, T., Wong, Y.-S. and Maclachlan, G. 1984. Pea xyloglucan and cellulose. II. Hydrolysis by pea endo-1,4-β-glucanase. Plant Physiol. 75: 605–610.

Hobson, G.E. 1962. Determination of polygalacturonase in fruits. Nature 195: 804–805.

Hobson, G.E. 1964. Polygalacturonase in normal and abnormal tomato fruit. Biochem. J. 92: 324–332.

Hobson, G.E. 1968. Cellulase activity during the maturation and ripening of tomato fruit. J. Food Sci. 33: 588–592.

Huber, D.J. 1983. Polyuronide degradation and hemicellulose modifications in ripening tomato fruit. J. Am. Soc. Hort. Sci. 108: 405–409.

Huber, D.J. 1984. Strawberry fruit softening: the potential roles of polyuronides and hemicelluloses. J. Food Sci. 49: 1310–1315.

Huber, D.J. and O'Donoghue, E.M. 1993. Polyuronides in avocado (*Persea americana*) and tomato (*Lycopersicon esculentum*) fruits exhibit markedly different patterns of molecular weight downshifts during ripening. Plant Physiol. 102: 473–480.

Jarvis, M.C. 1984. Structure and properties of pectin gels in plant cell walls. Plant Cell Envir. 7: 153–164.

Kalaitzis, P., Hong, S.-B., Solomos, T. and Tucker, M.L. 1999. Molecular characterization of a tomato endo-β-1,4-glucanase gene expressed in mature pistils, abscission zones and fruit. Plant Cell Physiol. 40: 905–908.

Kanellis, A.K. and Kalaitzis, P. 1992. Cellulase occurs in multiple active forms in ripe avocado fruit mesocarp. Plant Physiol. 98: 530–534.

Knapp, J., Moureau, P., Schuch, W. and Grierson, D. 1989. Organization and expression of polygalacturonase and other ripening related genes in Ailsa Craig 'Neverripe' and 'Ripening inhibitor' tomato mutants. Plant Mol. Biol. 12: 105–116.

Knegt, E., Vermeer, E. and Bruinsma, J. 1988. Conversion of the polygalacturonase isoenzymes from ripening tomato fruits. Physiol. Plant. 72: 108–114.

Knegt, E., Vermeer, E., Pak, C. and Bruinsma, J. 1991. Function of the polygalacturonase convertor in ripening tomato fruit. Physiol. Plant. 82: 237–242.

Koch, J.L. and Nevins, D.J. 1989. Tomato fruit cell wall. 1. Use of purified tomato polygalacturonase and pectinmethylesterase to identify developmental changes in pectins. Plant Physiol. 91: 816–822.

Kramer, M., Sanders, R., Bolkan, H., Waters, C., Sheehy, R.E. and Hiatt, W.R. 1992. Postharvest evaluation of transgenic tomatoes with reduced levels of polygalacturonase: processing, firmness and disease resistance. Postharvest Biol. Technol. 1: 241–255.

Labavitch, J.M. and Ray, P.M. 1974. Relationship between promotion of xyloglucan metabolism and induction of elongation by indoleacetic acid. Plant Physiol. 54: 499–502.

Langley, K.R., Martin, A., Stenning, R., Murray, A.J., Hobson, G.E., Schuch, W.W. and Bird, C.R. 1994. Mechanical and optical assessment of the ripening of tomato fruit with reduced polygalacturonase activity. J. Sci. Food Agric. 66: 547–554.

Lashbrook, C.C., Gonzalez-Bosch, C. and Bennett, A.B. 1994. Two divergent endo-β-1,4-glucanase genes exhibit overlapping expression in ripening fruit and abscising flowers. Plant Cell 6: 1485–1493.

Lashbrook, C.C., Giovannoni, J.J., Hall, B.D., Fischer, R.L. and Bennett, A.B. 1998. Transgenic analysis of tomato endo-β-1,4-glucanase gene function. Role of *cel1* in floral abscission. Plant J. 13: 303–310.

Lewis, L.N., Linkins, A.E., O'Sullivan, S. and Reid, P.D. 1974. Two forms of cellulase in bean plants. In: Plant Growth Substances 1973, Hirokawa Publishing, Tokyo, pp. 708–718.

Lincoln, J.E. and Fischer, R.L. 1988. Regulation of gene expression by ethylene in wild-type and *rin* tomato (*Lycopersicon esculentum*) fruit. Plant Physiol. 88: 370–374.

Llop-Tous, I., Dominguez-Puigjaner, E., Palomer, X. and Vendrell, M. 1999. Characterization of two divergent endo-β-1,4-glucanase cDNA clones highly expressed in nonclimacteric strawberry fruit. Plant Physiol. 119: 1415–1421.

Lorences, E.P. and Fry, S.C. 1993. Xyloglucan oligosaccharides with at least two α-D-xylose residues act as acceptor substrates for xyloglucan endotransglycosylase and promote the depolymerisation of xyloglucan. Physiol. Plant. 88: 105–112.

Maclachlan, G. and Brady, C. 1992. Multiple forms of 1,4-β-glucanase in ripening tomato fruits include a xyloglucanase activatable by xyloglucan oligosaccharides. Aust. J. Plant Physiol. 19: 137–146.

Maclachlan, G. and Brady, C. 1994. Endo-1,4-β-glucanase, xyloglucanase and xyloglucan endo-transglycosylase activities versus potential substrates in ripening tomatoes. Plant Physiol. 105: 965–974.

Matsumoto, T., Sakai, F. and Hayashi, T. 1997. A xyloglucan-specific endo-1,4-β-glucanase isolated from auxin-treated pea stems. Plant Physiol. 114: 661–667.

McCann, M.C. and Roberts, K. 1992. Architecture of the primary cell wall. In: C.W. Lloyd (Ed.) The Cytoskeletal Basis of Plant Growth and Form, Academic Press, London, pp. 109–129.

McCann, M.C. and Roberts, K. 1994. Changes in cell wall architecture during cell elongation. J. Exp. Bot. 45: 1683–1691.

McCartney, L., Ormerod, A.P., Gidley, M.J. and Knox, J.P. 2000. Temporal and spatial regulation of pectic (1→4)β-D-galactan in cell walls of developing pea cotyledons: implications for mechanical properties. Plant J. 22: 105–113.

McCollum, T.G., Huber, D.J. and Cantliffe, D.J. 1989. Modification of polyuronides and hemicelluloses during muskmelon fruit softening. Physiol. Plant. 76: 303–308.

McMahon, R.W., Stewart, C.R. and Gladon, R.J. 1990. Relationship of porphobilinogen deaminase activity to chlorophyll content and fruit development in 'Heinz 1350' tomato. J. Am. Soc. Hort. Sci. 115: 298–301.

McQueen-Mason, S.J. and Cosgrove, D.J. 1994. Disruption of hydrogen bonding between plant cell wall polymers by proteins that induce wall extension. Proc. Natl. Acad. Sci. USA 91: 6574–6578.

McQueen-Mason, S.J. and Cosgrove, D.J. 1995. Expansin mode of action on cell walls. Analysis of wall hydrolysis, stress relaxation, and binding. Plant Physiol. 107: 87–100.

McQueen-Mason, S.J., Durachko, D.M. and Cosgrove, D.J. 1992. Two endogenous proteins that induce cell wall extension in plants. Plant Cell 4: 1425–1433.

338

McQueen-Mason, S.J., Fry, S.C., Durachko, D.M. and Cosgrove, D.J. 1993. The relationship between xyloglucan endotransglycosylase and in vitro cell wall extension in cucumber hypocotyls. Planta 190: 327–331.

Medina-Escobar, N., Cardenas, J., Moyano, E., Caballero, J.L. and Munoz-Blanco, J. 1997. Cloning, molecular characterization and expression pattern of a strawberry ripening-specific cDNA with sequence homology to pectate lyase from higher plants. Plant Mol. Biol. 34: 867–877.

Milligan, S.B. and Gasser, C.S. 1995. Nature and regulation of pistil-expressed genes in tomato. Plant Mol. Biol. 28: 691–711.

Mitcham E.J., Gross, K.C. and Ng, T.J. 1989. Tomato fruit cell wall synthesis during development and senescence. In vivo radiolabeling of wall fractions using [^{14}C]sucrose. Plant Physiol. 89: 477–481.

Moore, T. and Bennett, A.B. 1994. Tomato fruit polygalacturonase isozyme 1. Characterization of the β subunit and its state of assembly in vivo. Plant Physiol. 106: 1461–1469.

Moshrefi, M. and Luh, B.S. 1983. Carbohydrate composition and electrophoretic properties of tomato polygalacturonase isoenzymes. Eur. J. Biochem. 135: 511–514.

Nakamura, S. and Hayashi, T. 1993. Purification and properties of an extracellular endo-1,4-β-glucanase from suspension-cultured poplar cells. Plant Cell Physiol. 34: 1009–1013.

Nishitani, K. and Tominaga, R. 1992. Endo-xyloglucan transferase, a novel class of glycosyltransferase that catalyzes transfer of a segment of xyloglucan molecule to another xyloglucan molecule. J. Biol. Chem. 267: 21058–21064.

O'Donoghue, E.M. and Huber, D.J. 1992. Modification of matrix polysaccharides during avocado (Persea americana) fruit ripening: an assessment of the role of C_x-cellulase. Physiol. Plant. 86: 33–42.

O'Donoghue, E.M., Huber, D.J., Timpa, J.D., Erdos, G.W. and Brecht, J.K. 1994. Influence of avocado (Persea americana) C_x-cellulase on the structural features of avocado cellulose. Planta 194: 573–584.

Oeller, P.W., Wong, L.-M., Taylor, L.P., Pike, D.A. and Theologis, A. 1991. Reversible inhibition of tomato fruit senescence by antisense RNA. Science 254: 437–439.

Ohmiya, Y., Takeda, T., Nakamura, S., Sakai, F. and Hayashi, T. 1995. Purification and properties of a wall-bound endo-1,4-β-glucanase from suspension-cultured poplar cells. Plant Cell Physiol. 36: 607–614.

Ohmiya, Y., Samejima, M., Shiroishi, M., Amano, Y., Kanda, T., Sakai, F. and Hayashi, T. 2000. Evidence that endo-1,4-β-glucanases act on cellulose in suspension-cultured poplar cells. Plant J. 24: 147–158.

Okazawa, K., Sato, Y., Nakagawa, T., Asada, K., Kato, I., Tomita, E. and Nishitani, K. 1993. Molecular cloning and cDNA sequencing of endoxyloglucan transferase, a novel class of glycosyltransferase that mediates molecular grafting between matrix polysaccharides in plant cell walls. J. Biol. Chem. 268: 25364–25368.

Orr, G. and Brady, C.J. 1993. Relationship of endopolygalacturonase activity to fruit softening in a freestone peach. Postharvest Biol. Technol. 3: 121–130.

Osteryoung, K.W., Toenjes, K., Hall, B.D., Winkler, V. and Bennett, A.B. 1990. Analysis of tomato polygalacturonase expression in transgenic tobacco. Plant Cell 2: 1239–1248.

Pauly M., Albersheim P., Darvill A. and York W.S. 1999. Molecular domains of the cellulose/xyloglucan network in the cell walls of higher plants. Plant J. 20: 629–639.

Percy, A.E., O'Brien, I.E.W., Jameson, P.E., Melton, L.D., MacRae, E.A. and Redgwell, R.J. 1996. Xyloglucan endotransglycosy-

lase activity during fruit development and ripening of apple and kiwifruit. Physiol. Plant. 96: 43–50.

Pharr, D.M., Sox, H.N. and Nesbitt, W.B. 1976. Cell wall-bound nitrophenylglycosides of tomato fruits. J. Am. Soc. Hort. Sci. 101: 397–400.

Pogson, B.J. and Brady, C.J. 1993. Accumulation of the β-subunit of polygalacturonase 1 in normal and mutant tomato fruit. Planta 191: 71–78.

Pogson, B.J., Brady, C.J. and Orr, G.R. 1991. On the occurrence and structure of subunits of endopolygalacturonase isoforms in mature-green and ripening tomato fruits. Aust. J. Plant Physiol. 18: 65–79.

Poovaiah, B.W. and Nukaya, A. 1979. Polygalacturonase and cellulase enzymes in the normal Rutgers and mutant rin tomato fruits and their relationship to the respiratory climacteric. Plant Physiol. 64: 534–537.

Pressey, R. 1983. β-Galactosidases in ripening tomatoes. Plant Physiol. 71: 132–135.

Pressey, R. 1984. Purification and characterization of tomato polygalacturonase converter. Eur. J. Biochem. 144: 217–221.

Pressey, R. 1986a. Extraction and assay of tomato polygalacturonases. HortScience 21: 490–492.

Pressey, R. 1986b. Changes in polygalacturonase isoenzymes and converter in tomatoes during ripening. HortScience 21: 1183–1185.

Pressey, R. 1988. Reevaluation of the changes in polygalacturonases in tomatoes during ripening. Planta 174: 39–43.

Pressey, R. and Avants, J.K. 1972. Multiple forms of pectinesterase in tomatoes. Phytochemistry 11: 3139–3142.

Pressey, R. and Avants, J.K. 1973. Two forms of polygalacturonase in tomato. Biochim. Biophys. Acta 309: 363–369.

Pressey, R. and Avants, J.K. 1978. Difference in polygalacturonase composition of clingstone and freestone peaches. J. Food Sci. 43: 1415–1423.

Pressey, R. and Avants, J.K. 1982. Solubilization of cell walls by tomato polygalacturonases: effects of pectinesterases. J. Food Biochem. 6: 57–74.

Pressey, R., Hinton, D.M. and Avants, J.K. 1971. Development of polygalacturonase activity and solubilization of pectin in peaches during ripening. J. Food Sci. 36: 1070–1073.

Ray, J., Knapp, J., Grierson, D., Bird, C. and Schuch, W. 1988. Identification and sequence determination of a cDNA clone for tomato pectinesterase. Eur. J. Biochem. 174: 119–124.

Redgwell, R.J. and Fry, S.C. 1993. Xyloglucan endotransglycosylase activity increases during kiwifruit (Actinidia deliciosa) ripening: Implications for fruit softening. Plant Physiol. 103: 1399–1406.

Redgwell, R.J., MacRae, E.A., Hallett, I., Fischer, M., Perry, J. and Harker, R. 1997. In vivo and in vitro swelling of cell walls during fruit ripening. Planta 203: 162–173.

Rose, J.K.C., Lee, H.H. and Bennett, A.B. 1997. Expression of a divergent expansin gene is fruit-specific and ripening-regulated. Proc. Natl. Acad. Sci. USA 94: 5955–5960.

Rose, J.K.C., Hadfield, K.A., Labavitch, J.M. and Bennett, A.B. 1998. Temporal sequence of cell wall disassembly in rapidly ripening melon fruit. Plant Physiol. 117: 345–361.

Rose, J.K.C., Cosgrove, D.J., Albersheim, P., Darvill, A.G. and Bennett, A.B. 2000. Detection of expansin proteins and activity during tomato fruit ontogeny. Plant Physiol. 123: 1583–1592.

Rothan, C. and Nicolas, J. 1989. Changes in acidic and basic peroxidase activities during tomato fruit ripening. HortScience 24: 340–342.

Roy, S., Vian, B. and Roland, J.-C. 1992. Immunocytochemical study of the deesterification patterns during cell wall autolysis in the ripening of cherry tomato. Plant Physiol. Biochem. 30: 139–146.

Sakurai, N. and Nevins, D.J. 1993. Changes in physical properties and cell wall polysaccharides of tomato (*Lycopersicon esculentum*) pericarp tissue. Physiol. Plant. 89: 681-686.

Schroder, R., Atkinson, R.G., Langenkamper, G. and Redgwell, R.J. 1997. Biochemical and molecular characterisation of xyloglucan endotransglycosylase from ripe kiwi fruit. Planta 204: 242–251.

Schuch, W., Kanczler, J., Robertson, D., Hobson, G., Tucker, G., Grierson, D., Bright, S. and Bird, C. 1991. Fruit quality characteristics of transgenic tomato fruit with altered polygalacturonase activity. HortScience 26: 1517–1520.

Seymour, G.B., Harding, S.E., Taylor, A.J., Hobson, G.E. and Tucker, G.A. 1987a. Polyuronide solubilization during ripening of normal and mutant tomato fruit. Phytochemistry 26: 1871–1875.

Seymour, G.B., Lasslett, Y. and Tucker, G.A. 1987b. Differential effects of pectolytic enzymes on tomato polyuronides *in vivo* and *in vitro*. Phytochemistry 26: 3137–3139.

Seymour, G.B., Colquhoun, I.J., DuPont, M.S., Parsley, K.R. and Selvendran, R.R. 1990. Composition and structural features of cell wall polysaccharides from tomato fruits. Phytochemistry 29: 725–731.

Shackel, K.A., Greve, C., Labavitch, J.M. and Ahmadi, H. 1991. Cell turgor changes associated with ripening in tomato pericarp tissue. Plant Physiol. 97: 814–816.

Shcherban, T.Y., Shi, J., Durachko, D.M., Guiltinan, M.J., McQueen-Mason, S.J., Shieh, M. and Cosgrove, D.J. 1995. Molecular cloning and sequence analysis of expansins, a highly conserved, multigene family of proteins that mediate cell wall extension in plants. Proc. Natl. Acad. Sci. USA 92: 9245–9249.

Sheehy, R.E., Pearson, J., Brady, C.J. and Hiatt, W.R. 1987. Molecular characterization of tomato fruit polygalacturonase. Mol. Gen. Genet. 208: 30–36.

Sheehy, R.E., Kramer, M. and Hiatt, W.R. 1988. Reduction of polygalacturonase activity in tomato fruit by antisense RNA. Proc. Natl. Acad. Sci. USA 85: 8805–8809.

Sitrit, Y. and Bennett, A.B. 1998. Regulation of tomato fruit polygalacturonase mRNA accumulation by ethylene: a re-examination. Plant Physiol. 116: 1145–1150.

Smith, R.C. and Fry, S.C. 1991. Endotransglycosylation of xyloglucans in plant cell suspension cultures. Biochem. J. 279: 529–535.

Smith, D.L. and Gross, K.C. 2000. A family of at least seven β-galactosidase genes is expressed during tomato fruit development. Plant Physiol. 123: 1173–1183.

Smith, C.J.S., Watson, C.F., Ray, J., Bird, C.R., Morris, P.C., Schuch, W. and Grierson, D. 1988. Antisense RNA inhibition of polygalacturonase gene expression in transgenic tomatoes. Nature 334: 724–726.

Smith, C.J.S., Watson, C.F., Morris, P.C., Bird, C.R., Seymour, G.B., Gray, J.E., Arnold, C., Tucker, G.A., Schuch, W., Harding, S. and Grierson, D. 1990. Inheritance and effect on ripening of antisense polygalacturonase genes in transgenic tomatoes. Plant Mol. Biol. 14: 369–379.

Smith, D.L., Starrett, D.A. and Gross, K.C. 1998. A gene coding for tomato fruit β-galactosidase II is expressed during fruit ripening. Plant Physiol. 117: 417–423.

Sorensen, S.O., Pauly, M., Bush, M., Skjot, M., McCann, M.C., Borkhardt, B. and Ulvskov, P. 2000. Pectin engineering: modification of potato pectin by *in vivo* expression of an endo-1,4-β-D-galactanase. Proc. Natl. Acad. Sci. USA 97: 7639–7644.

Steele, N.M. and Fry, S.C. 2000. Differences in catalytic properties between native isoenzymes of xyloglucan endotransglycosylase (XET). Phytochemistry 54: 667–680.

Sulova, Z., Takacova, M., Steele, N.M., Fry, S.C. and Farkas, V. 1998. Xyloglucan endotransglycosylase: evidence for the existence of a relatively stable glycosyl-enzyme intermediate. Biochem J. 330: 1475–1480.

Tabuchi, A., Kamisaka, S. and Hoson, T. 1997. Purification of xyloglucan hydrolase/endotransferase from cell walls of azuki bean epicotyls. Plant Cell Physiol 38: 653–658.

Talbott, L.D. and Ray, P.M. 1992. Changes in molecular size of previously deposited and newly synthesized pea cell wall matrix polysaccharides. Effects of auxin and turgor. Plant Physiol. 98: 369–379.

Thakur, B.R., Singh, R.K., Tieman, D.M. and Handa, A.K. 1996a. Tomato product quality from transgenic fruits with reduced pectin methylesterase. J. Food Sci. 61: 85–108.

Thakur, B.R., Singh, R.K. and Handa, A.K. 1996b. Effect of an antisense pectin methylesterase gene on the chemistry of pectin in tomato (*Lycopersicon esculentum*) juice. J. Agric. Food Chem. 44: 628–630.

The Arabidopsis Genome Initiative. 2000. Analysis of the genome sequence of the flowering plant *Arabidopsis thaliana*. Nature 408: 796–815.

Theologis, A., Oeller, P.W., Wong, L.-M., Rottmann, W.H. and Gantz, D.M. 1993. Use of a tomato mutant constructed with reverse genetics to study fruit ripening, a complex developmental process. Devel. Genet. 14: 282–295.

Thompson, J.E. and Fry, S.C. 2000. Evidence for covalent linkage between xyloglucan and acidic pectins in suspension-cultured rose cells. Planta 211: 275–286.

Tieman, D.M. and Handa, A.K. 1994. Reduction in pectin methylesterase activity modifies tissue integrity and cation levels in ripening tomato (*Lycopersicon esculentum* Mill.) fruits. Plant Physiol. 106: 429–436.

Tieman, D.M., Harriman, R.W., Ramamohan, G. and Handa, A.K. 1992. An antisense pectin methylesterase gene alters pectin chemistry and soluble solids in tomato fruit. Plant Cell 4: 667–679.

Tigchelaar, E.C., McGlasson, W.B. and Buescher, R.W. 1978. Genetic regulation of tomato fruit ripening. HortScience 13: 508–513.

Tong, C.B.S. and Gross, K.C. 1988. Glycosyl-linkage composition of tomato fruit cell wall hemicellulosic fractions during ripening. Physiol. Plant. 74: 365–370.

Trainotti, L., Ferrarese, L., Dalla Vecchia, F., Rascio, N. and Casadoro, G. 1999a. Two different endo-β-1,4-glucanases contribute to the softening of strawberry fruit. J. Plant Physiol. 154: 355–362.

Trainotti, L., Spolaore, S., Pavanello, A., Baldan, B. and Casadoro, G. 1999b. A novel E-type endo-β-1,4-glucanase with a putative cellulose-binding domain is highly expressed in ripening strawberry fruits. Plant Mol. Biol. 40: 323–332.

Tucker, G.A. and Grierson, D. 1982. Synthesis of polygalacturonase during tomato fruit ripening. Planta 155: 64–67.

Tucker, G.A., Robertson, N.G. and Grierson, D. 1980. Changes in polygalacturonase isoenzymes during the 'ripening' of normal and mutant tomato fruit. Eur. J. Biochem. 112: 119–124.

Tucker, G.A., Robertson, N.G. and Grierson, D. 1981. The conversion of tomato-fruit polygalacturonase isoenzyme 2 into isoenzyme 1 *in vitro*. Eur. J. Biochem. 115: 87–90.

Tucker, G.A., Robertson, N.G. and Grierson, D. 1982. Purification and changes in activities of tomato pectinesterase isoenzymes. J. Sci. Food Agric. 33: 396–400.

340

Turner, L.A., Harriman, R.W. and Handa, A.K. 1996. Isolation and nucleotide sequence of three tandemly arranged pectin methylesterase genes (Accession Nos. U70675, U70676 and U70677) from tomato. Plant Physiol. 112: 1398.

Villavicencio, L., Blankenship, S.M., Sanders, D.C. and Swallow, W.H. 1999. Ethylene and carbon dioxide production in detached fruit of selected pepper cultivars. J. Am. Soc. Hort. Sci. 124: 402–406.

Wade, N.L., Kavanagh, E.E., Hockley, D.G. and Brady, C.J. 1992. Relationship beteen softening and the polyuronides in ripening banana fruit. J. Sci. Food Agric. 60: 61–68.

Wakabayashi, K., Chun, J.-P. and Huber, D.J. 2000. Extensive solubilization and depolymerization of cell wall polysaccharides during avocado (Persea americana) ripening involves concerted action of polygalacturonase and pectinmethylesterase. Physiol. Plant. 108: 345–352.

Wallner, S.J. and Bloom, H.L. 1977. Characteristics of tomato cell wall degradation in vitro. Implications for the study of fruit-softening enzymes. Plant Physiol 60: 207–210.

Wallner, S.J. and Walker, J.E. 1975. Glycosidases in cell wall-degrading extracts of ripening tomato fruits. Plant Physiol. 55: 94–98.

Warrilow, A.G.S., Turner, R.J. and Jones, M.G. 1994. A novel form of pectinesterase in tomato. Phytochemistry 35: 863–868.

Watson, C.F., Zheng, L. and Della Penna, D. 1994. Reduction of tomato polygalacturonase β subunit expression affects pectin solubilization and degradation during fruit ripening. Plant Cell 6: 1623–1634.

Whitney, S.E.C., Gidley, M.J., and McQueen-Mason S.J. 2000. Probing expansin action using cellulose/hemicellulose composites. Plant J. 22: 327–334.

Wong, Y.-S., Fincher, G.B. and Maclachlan, G.A. 1977. Kinetic properties and substrate specificities of two cellulases from auxin-treated pea epicotyls. J. Biol. Chem. 252: 1402–1407.

Yelle, S., Hewitt, J.D., Robinson, N.L., Damon, S. and Bennett, A.B. 1988. Sink metabolism in tomato fruit. III. Analysis of carbohydrate assimilation in a wild species. Plant Physiol. 87: 737–740.

Yen, H.-C., Lee, S., Tanksley, S.D., Lanahan, M.B., Klee, H.J. and Giovannoni, J.J. 1995. The tomato Never-ripe locus regulates ethylene-inducible gene expression and is linked to a homolog of the Arabidopsis ETR1 gene. Plant Physiol. 107: 1343–1353.

York, W.S., Kolli, V.S.K., Orlando, R., Albersheim, P. and Darvill, A.G. 1996. The structures of arabinoxyloglucans produced by solanaceous plants. Carbohydrate Res. 285: 99–128.

Zheng, L., Heupel, R.C. and Della Penna, D. 1992. The β subunit of tomato fruit polygalacturonase isoenzyme 1: isolation, characterization, and identification of unique structural features. Plant Cell 4: 1147–1156.

Zheng, L., Watson, C.F. and Della Penna, D. 1994. Differential expression of the two subunits of tomato polygalacturonase isoenzyme 1 in wild-type and rin tomato fruit. Plant Physiol. 105: 1189–1195.

Plant Molecular Biology **47**: 2001.

Index, Vol. 47 Nos. 1 & 2 (2001)